Textbook of Geomorphology and Geodynamics

Textbook of Geomorphology and Geodynamics

Edited by **Ken Shaw**

SYRAWOOD
PUBLISHING HOUSE

New York

Published by Syrawood Publishing House,
750 Third Avenue, 9th Floor,
New York, NY 10017, USA
www.syrawoodpublishinghouse.com

Textbook of Geomorphology and Geodynamics
Edited by Ken Shaw

International Standard Book Number: 978-1-68286-116-5 (Hardback)

Contents

Preface   VII

Chapter 1   **Scale effect on runoff and soil loss control using rice straw mulch under laboratory conditions**   1
S. H. R. Sadeghi, L. Gholami, E. Sharifi, A. Khaledi Darvishan and M. Homaee

Chapter 2   **Geothermal investigations in western Anatolia using equilibrium temperatures from shallow boreholes**   9
K. Erkan

Chapter 3   **Changes in soil quality after converting *Pinus* to *Eucalyptus* plantations in southern China**   20
K. Zhang, H. Zheng, F. L. Chen, Z. Y. Ouyang, Y. Wang, Y. F. Wu, J. Lan, M. Fu and X. W. Xiang

Chapter 4   **Soil organic carbon along an altitudinal gradient in the Despeñaperros Natural Park, southern Spain**   29
L. Parras-Alcántara, B. Lozano-García and A. Galán-Espejo

Chapter 5   **Impact of the addition of different plant residues on nitrogen mineralization–immobilization turnover and carbon content of a soil incubated under laboratory conditions**   39
M. Kaleeem Abbasi, M. Mahmood Tahir, N. Sabir and M. Khurshid

Chapter 6   **Kinetics of potassium release in sweet potato cropped soils: a case study in the highlands of Papua New Guinea**   48
B. K. Rajashekhar Rao

Chapter 7   **Soil microbiological properties and enzymatic activities of long-term post-fire recovery in dry and semiarid Aleppo pine (*Pinus halepensis* M.) forest stands**   57
J. Hedo, M. E. Lucas-Borja, C. Wic, M. Andrés-Abellán and J. de Las Heras

Chapter 8   **Effects of rodent-induced land degradation on ecosystem carbon fluxes in an alpine meadow in the Qinghai–Tibet Plateau, China**   67
F. Peng, Y. Quangang, X. Xue, J. Guo and T. Wang

Chapter 9   **Aggregate breakdown and surface seal development influenced by rain intensity, slope gradient and soil particle size**   75
S. Arjmand Sajjadi and M. Mahmoodabadi

Chapter 10   **Cobalt, chromium and nickel contents in soils and plants from a serpentinite quarry**   86
M. Lago-Vila, D. Arenas-Lago, A. Rodríguez-Seijo, M. L. Andrade Couce and F. A. Vega

Chapter 11   **Adsorption, desorption and fractionation of As(V) on untreated and mussel shell-treated granitic material**   **99**
N. Seco-Reigosa, L. Cutillas-Barreiro, J. C. Nóvoa-Muñoz, M. Arias-Estévez, E. Álvarez-Rodríguez, M. J. Fernández-Sanjurjo and A. Núñez-Delgado

Chapter 12   **Identifying areas susceptible to desertification in the Brazilian northeast**   **109**
R. M. S. P. Vieira, J. Tomasella, R. C. S. Alvalá, M. F. Sestini, A. G. Affonso, D. A. Rodriguez, A. A. Barbosa, A. P. M. A. Cunha, G. F. Valles, E. Crepani, S. B. P. de Oliveira, M. S. B. de Souza, P. M. Calil, M. A. de Carvalho, D. M. Valeriano, F. C. B. Campello and M. O. Santana

Chapter 13   **Use of phytoremediation and biochar to remediate heavy metal polluted soils: a review**   **123**
J. Paz-Ferreiro, H. Lu, S. Fu, A. Méndez and G. Gascó

Chapter 14   **Thermal shock and splash effects on burned gypseous soils from the Ebro Basin (NE Spain)**   **134**
J. León, M. Seeger, D. Badía, P. Peters and M. T. Echeverría

Chapter 15   **Short-term changes in soil Munsell colour value, organic matter content and soil water repellency after a spring grassland fire in Lithuania**   **144**
P. Pereira, X. Úbeda, J. Mataix-Solera, M. Oliva and A. Novara

Chapter 16   **Characterization of hydrochars produced by hydrothermal carbonization of rice husk**   **161**
D. Kalderis, M. S. Kotti, A. Méndez and G. Gascó

Chapter 17   **Variations of soil profile characteristics due to varying time spans since ice retreat in the inner Nordfjord, western Norway**   **168**
A. Navas, K. Laute, A. A. Beylich and L. Gaspar

Chapter 18   **Crop residue decomposition in Minnesota biochar-amended plots**   **182**
S. L.Weyers and K. A. Spokas

Chapter 19   **Factors driving the carbon mineralization priming effect in a sandy loam soil amended with different types of biochar**   **191**
P. Cely, A. M. Tarquis, J. Paz-Ferreiro, A. Méndez and G. Gascó

Permissions

List of Contributors

Preface

Every book is initially just a concept; it takes months of research and hard work to give it the final shape in which the readers receive it. In its early stages, this book also went through rigorous reviewing. The notable contributions made by experts from across the globe were first molded into patterned chapters and then arranged in a sensibly sequential manner to bring out the best results.

The various researches and developments taking place in the disciplines of geomorphology and geodynamics have enabled us to understand the origin and evolution of earth's surface. This book consists of contributions made by international experts in the form of detailed discussions of various theories and concepts revolving around geochemistry, geophysics, sedimentology, seismology, tectonophysics, etc. Different approaches, evaluations, methodologies and advanced studies on geomorphology and geodynamics have been included in this book. Those in search of detailed information to further their knowledge will be greatly assisted by this book. This book is highly recommended for students, academicians and researchers pursuing geomorphology and associated sciences.

It has been my immense pleasure to be a part of this project and to contribute my years of learning in such a meaningful form. I would like to take this opportunity to thank all the people who have been associated with the completion of this book at any step.

Editor

Scale effect on runoff and soil loss control using rice straw mulch under laboratory conditions

S. H. R. Sadeghi[1], L. Gholami[1,*], E. Sharifi[1], A. Khaledi Darvishan[1], and M. Homaee[2]

[1]Department of Watershed Management Engineering, Faculty of Natural Resources, Tarbiat Modares University, P.O. Box 46417-76489, Noor, Iran

[2]Department of Soil Science, Faculty of Agriculture, Tarbiat Modares University, Tehran, Iran

[*]now at: Department of Rangeland and Watershed Management, Faculty of Natural Resources, Sari Agricultural Sciences and Natural Resources University, Sari, Iran

Correspondence to: S. H. R. Sadeghi (sadeghi@modares.ac.ir)

Abstract. Amendments can control the runoff and soil loss by protecting the soil surface. However, scale effects on runoff and soil loss control have not been considered yet. The present study has been formulated to determine the efficiency of two plot sizes of 6 and $0.25\,\mathrm{m}^2$ covered by $0.5\,\mathrm{kg\,m}^{-2}$ of straw mulch with regard to changing the time to runoff, runoff coefficient, sediment concentration and soil loss under laboratory conditions. The study used a sandy-loam soil taken from summer rangeland, Alborz Mountains, northern Iran, and was conducted under simulated rainfall intensities of 50 and $90\,\mathrm{mm\,h}^{-1}$ and in three replicates. The results of the study showed that the straw mulch had a more significant effect on reducing the runoff coefficient, sediment concentration and soil loss on a $0.25\,\mathrm{m}^2$ plot scale. The maximum effectiveness in time to runoff for both the scales was observed at a rainfall intensity of $90\,\mathrm{mm\,h}^{-1}$. The maximum increasing and decreasing rates in time to runoff and runoff coefficient were observed at a rainfall intensity of $90\,\mathrm{mm\,h}^{-1}$, with 367.92 and 96.71 % for the $0.25\,\mathrm{m}^2$ plot and 110.10 and 15.08 % for the $6\,\mathrm{m}^2$ plot. The maximum reduction in the runoff coefficient was in the $0.25\,\mathrm{m}^2$ plot for the two rainfall intensities of 50 and $90\,\mathrm{mm\,h}^{-1}$, with rates of -89.34 and -96.71 %. The maximum change in soil loss at the intensities of both 50 and $90\,\mathrm{mm\,h}^{-1}$ occurred in the $0.25\,\mathrm{m}^2$ plot, with 100 %, whereas in the $6\,\mathrm{m}^2$ plot, decreasing rates of soil loss for the intensities of both 50 and $90\,\mathrm{mm\,h}^{-1}$ were 46.74 and 63.24 %, respectively.

1 Introduction

The soil erosion rates are accelerated by tillage and low vegetation cover (Cerdà et al., 2009 and 2010). Population increase and a growing demand for agricultural products (Prokop and Poręba, 2012; Zhao et al., 2013) and intensive dry land (Biro et al., 2013) has generated changes in land use and resulted in erosion and land degradation. There are various methods for soil conservation, but biological methods in bare and degraded slopes need a long time to become established (Adekalu et al., 2007; Smets et al., 2008a). In this context, various natural and organic mulches viz. crop residues, leaf litter, woodchips, bark chips, biological geo-textiles, gravel and crushed stones (Ruy, 2006; Smets et al., 2008a; Ruiz-Sinoga et al., 2010) have been applied for soil conservation. Mulches have extraordinary potential in soil erosion control (Morgan, 1986) and runoff reduction (Poesen and Lavee, 1991). However, the establishment of degraded areas and bare slopes by vegetation cover takes a long time (Adekalu et al., 2007; Smets et al., 2008a). The effect of mulches depends on many factors, including raindrop erosivity, soil condition, steepness and length of the slope, and the mulch rate and type (Amimoto, 1981; Cogo et al., 1984; Poesen and Lavee, 1991; Morgan, 1995; Auerswald et al., 2003; Adekalu et al., 2007; Kukal and Sarkar, 2010; Jordán et al., 2010; Choi et al., 2012; Gholami et al., 2013). Straw mulch as an organic amendment reduces soil erosion but also recovers the main soil properties lost due to agriculture

(García-Orenes et al., 2009, 2010); this is also done by other materials (Giménez Morera et al., 2010).

Although there are a lot of studies about soil amendments as a means of soil conservation, e.g., Fernández et al. (2012), Jiménez et al. (2012), García-Moreno et al. (2013), Robichaud et al. (2013), Lieskovský and Kenderessy (2014) and Martins et al. (2014), the effects of the study scale on the effectiveness of various mulch covers have rarely been considered. There are a few studies about the spatial-scale variations in mulches on runoff and soil loss. Poesen et al. (1994) reviewed the effects of rock fragments on soil erosion and stated that the spatial scale has an important impact on the soil erosion. They showed that, on the microplot scale ($4\,mm^2$ to $1\,m^2$), sediment yield reached a maximum value with 0 % rock fragment cover and reached a minimum value with 100 % rock fragment cover. On the mesoplot scale (i.e., interrill areas), negative, positive as well as convex upward relationships with cover percentages have been observed, depending on the fine-earth structure, on the vertical position in the topsoil, on the size of rock fragments and on the surface slope. Finally, on the macroplot scale (i.e., interrill and rill areas; $10–10\,000\,m^2$), sediment yield decreased exponentially with rock fragment cover. Cerdan et al. (2002) investigated the scale effect (plot to catchment) on runoff in agricultural areas of Normandy, France. Three databases – $450\,m^2$ plots, a 90 ha catchment and an 1100 ha catchment – were selected to collect runoff data. Between the three scales, a significant decrease in the runoff coefficient was observed as the area increased. Mingguo et al. (2007) also studied the effect of vegetation on the runoff–sediment-yield relationship on different spatial scales (plot to watershed) in hilly areas of the Loess Plateau, northern China, and found that vegetation could reduce runoff and soil loss on both scales but that, on a plot scale, the reduction rate of sediment was more than the runoff. Smets et al. (2008a) reviewed the impact of plot length on the effectiveness of different soil surface covers in reducing runoff and soil loss. The results indicated that, for plot lengths < 11 m, there was a large variation in the runoff and erosion-reducing effectiveness of each soil cover, depending on various factors. Smets et al. (2008b) also examined the spatial-scale effects on the effectiveness of organic mulches in reducing soil erosion in field and laboratory experiments (plot length ranges between 0.1 and 30.5 m). Results verified the effectiveness of mulches in reducing soil erosion by water on various scales. In addition, they showed a positive linear relation between the erosion-reducing effectiveness of an organic mulch cover and plot length. In short plots, the response of a soil surface cover to runoff and soil loss was immediately observed. Nevertheless, on longer plots, the runoff and soil loss response could be compensated for due to the longer plot length. Fernández et al. (2012) studied the seeding and the mulching and seeding effects on post-fire runoff and soil erosion in Galicia (NW Spain), with a rainfall rate of $67\,mm\,h^{-1}$ at plot scale. They showed that the conserved treatments did not significantly increase soil cover or affect runoff, but soil losses were low in all cases. García-Orenes et al. (2012) demonstrated that the use of a cover (plants or straw) contributes to increases in soil quality and reduces the risk of erosion. Liu et al. (2012) evaluated the effects of rice straw mulch and plastic film mulching at plot scale over 2 years in the Xiaofuling watershed in the Danjiangkou Reservoir area, China. The straw mulch treatment significantly decreased the sediment yield from 18 to 22 %. The results showed that the straw mulch was beneficial for controlling runoff and sediment.

Scrutinizing the available literature showed that, although there are lots of references to using straw as mulch for runoff and soil erosion control, there is no literature that reports the effectiveness of straw mulch on various plot scales. The present study was therefore planned to determine the efficiency of two plot sizes covered by straw mulch in changing the important runoff and soil loss components under laboratory conditions.

2 Materials and methods

The laboratory experiments were conducted by using two sets of 6×1 m and 0.5×0.5 m plots installed in the rainfall simulator laboratory, Faculty of Natural Resources of Tarbiat Modares University (TMU), located in Noor, Mazandaran Province, northern Iran. The experiments were carried out to study the effect of rice straw mulch on runoff and soil loss processes by using simulated rainfall at an intensity of 50 and $90\,mm\,h^{-1}$ and in three replicates (there were 24 experiments in total: 12 experiments for a rainfall intensity of $50\,mm\,h^{-1}$, of which 6 experiments were control treatments and 6 experiments were conservation treatments, and another 12 treatments for a rainfall intensity of $90\,mm\,h^{-1}$, of which 6 experiments were control treatments and 6 experiments were conservation treatments). The entire number of eight treatments in three replicates was formulated as a factorial design as shown in the following: 2 plot sizes (0.25 and $6\,m^2$) $\times$ 2 rainfall intensities (50 and $90\,mm\,h^{-1}$) $\times$ 2 treatments (control and straw mulch) = 8 treatments; 3 replicates = total of 24 rainfall simulations

The rainfall simulator consists of a 4000 L water tank and 27 precalibrated nozzles in three parallel lines designed to simulate raindrops of an average size of 1.3 mm. The drops fall from a height of between 4 and 6 m at the upper and lower parts of the plot, respectively, reaching a speed of $7\,m\,s^{-1}$ (Duiker et al., 2001).

A sandy-loam (14 % clay, 24 % silt and 62 % sand) topsoil was collected from a depth of 0–20 cm (Kukal and Sarkar, 2010) in the Alborz Mountains, northern Iran. The soil was transported to the lab and air-dried to the optimum moisture content to maintain the relative stability of soil aggregates and decrease the breakdown of the aggregates in the sieving process (Khaledi Darvishan et al., 2013). The coarse rock fragments and plant residues were removed from the soil

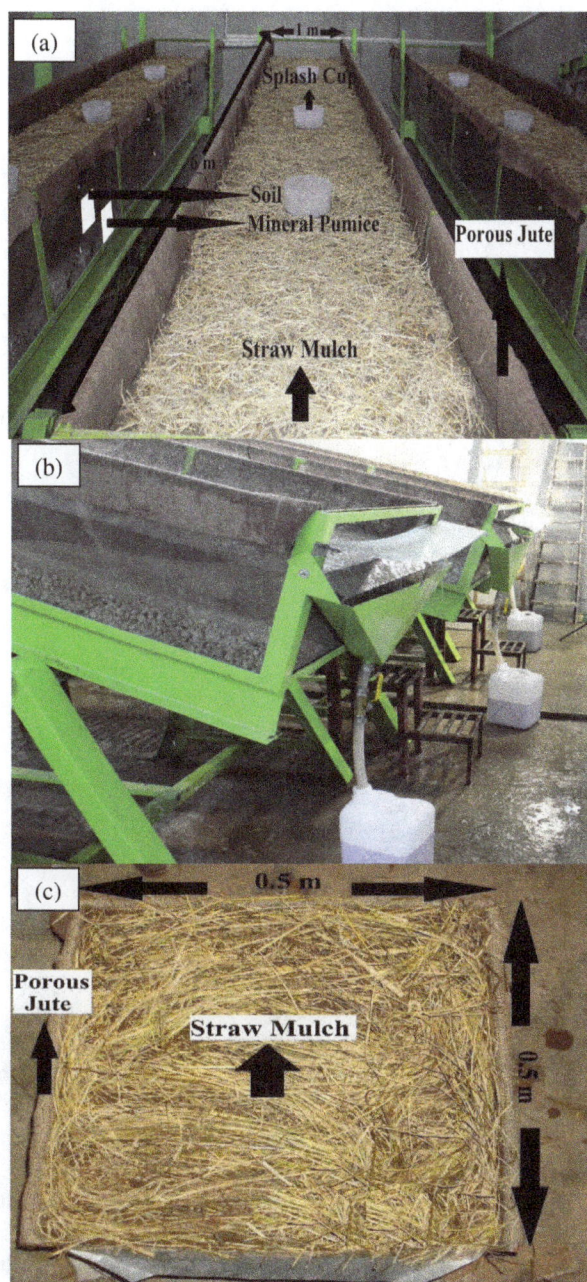

Figure 1. A general view of the treated plots of $6\,\mathrm{m}^2$ (**a**), the runoff collection system at a $6\,\mathrm{m}^2$ plot outlet (**b**) and $0.25\,\mathrm{m}^2$ (**c**) with rice straw mulch under lab conditions.

through passing it through an 8 mm sieve to obtain maximum homogeneity in the soil profile (Hawke et al., 2006). The pH, electrical conductivity (EC) and organic matter of the experimental soil were $7.95, 75.5\,\mu\mathrm{moh\,cm^{-1}}$ and $2.167\,\%$, respectively.

Three layers of mineral pumice grains with different sizes and a total thickness of 15 cm were used as a filter layer and placed at the bottom of the plots in order to simulate natural drainage conditions and decreasing plot weight (Defersha et al., 2011). A 15 cm thick soil layer was then placed on the top and separated from the mineral pumice by a sheet of porous jute (Defersha et al., 2011). The soil was ultimately compacted by a small PVC roller (a handmade roller filled with cement and sand) to achieve the bulk density of $1.38\,\mathrm{g\,cm^{-3}}$, almost equal to that measured for the soil under natural conditions (Romkens et al., 2001; Saedghi et al., 2010; Gholami et al., 2013). Each experiment was also covered using new soil and straw mulch (Adekalu et al., 2007). The rainfall intensities of 50 and $90\,\mathrm{mm\,h^{-1}}$ with a duration of 15 min were considered to correspond to climatological conditions at the location of origin of the soil; these conditions were obtained through intensity-duration-frequency (IDF) curve analysis for data collected in the nearest synoptic station (Kojour; longitude $51°44'$, latitude $36°23'$; height 1550 m) with a return period of less than 20 years.

Finally, the air-dried rice straw mulch was spread over the soil surface 5 days before treatments, with a cover, thickness and dry weight of about $90\,\%$ (Das and Agrawal, 2002; Adekalu et al., 2007; Kukal and Sarkar, 2010), $\sim 8\,\mathrm{cm}$ and $0.5\,\mathrm{kg\,m^{-2}}$, respectively. A general view of the experimental plots and setups is shown in Fig. 1. The control plots subjected to the study rain storms were monitored under identical lab conditions on bare soils and just before applying the straw mulch.

The time to runoff, runoff coefficient and soil loss were measured at the outlet of each plot for control (before mulching) and treated plots (after mulching) at intervals of 2 min (Ruiz-Sinoga et al., 2010). To establish the runoff and sediment fluxes in all experiments, the 2 min interval was considered appropriate because of the short whole duration of the experiments (15 min). The amount of soil loss was then measured using a decantation procedure; the soil was oven-dried at $105\,°\mathrm{C}$ for 24 h and weighed by means of high-precision scales (Kukal and Sarkar, 2011; Gholami et al., 2013).

The general linear model (GLM) using the SPSS 17 software (SPSS Inc. Released 2009) was applied to statistically analyze the main (individual) and interaction effects of spatial scale (plot size), conservation treatments and rainfall intensity on the quantitative characteristics of runoff, sediment concentration and soil loss. The necessary prerequisites were also fulfilled before applying the GLM.

3 Results and discussion

The amount of time to runoff and the runoff coefficient before and after the conservation treatment in each plot output and each scale are shown in Table 1. The percentage of

Table 1. Time to runoff and coefficient before and after conservation treatment on study scales.

Plot area (m^2)	Rainfall intensity (mm h^{-1})	Kinetic energy of rainfall (j m^{-2})	Time to runoff (s) (s)		Runoff coefficient (%) (%)	
			Control	Treated	Control	Treated
0.25	50	23.41	420.00	480.00	24.56	2.03
			609.6	368.4	19.60	2.94
			432.00	372.00	23.86	2.07
		SD	106.17	63.42	2.68	0.51
	90	24.10	69.00	480.00	34.18	1.30
			120.00	564.00	49.56	1.18
			126.00	300.00	37.91	1.39
		SD	31.32	134.88	8.02	0.11
6	50	23.22	38.51	72.52	69.35	60.20
			30.27	68.11	68.45	62.95
			34.34	70.44	69.48	62.48
		SD	4.12	2.21	0.56	1.47
	90	21.15	23.15	56.11	79.42	66.85
			30.32	52.27	78.32	72.18
			26.70	57.28	77.65	60.90
		SD	3.59	2.45	0.90	5.64

Table 2. Changes (%) in time to runoff and coefficient in plots treated with rice straw mulch.

Plot area (m^2)	Variable	Rainfall intensity (mm h^{-1}) 50	90
0.25	Time to runoff	−13.06	+367.92
	Coefficient	−89.34	−96.71
6	Time to runoff	+106.15	+110.10
	Coefficient	−10.43	−15.08

changes in study variables in treated plots and in comparison with control plots has been summarized in Table 2.

Tables 1 and 2 showed that the straw mulch increased time to runoff compared to untreated plots except at a rainfall intensity of 50 mm h^{-1} for the 0.25 m^2 plot; it also decreased the runoff coefficient on both the scales. This might be due to the water-storing effects of straw and also an increasing ponding time on the plot surface. This finding is in line with that reported by Poesen and Lavee (1991), Mingguo et al. (2007) and Smets et al. (2008a, b). However, the maximum change in effectiveness in time to runoff, for the two scales, could be found at a rainfall intensity of 90 mm h^{-1}.

These effects were more serious in the 0.25 m^2 plot, with a rate of + 367.92 %, while the 6 m^2 plot, compared to the 0.25 m^2 plot, could reduce the time to runoff at a rainfall intensity of 50 mm h^{-1}, with a rate of +106.15 %. Figures 2 and 3 also show the average rates of time to runoff and the coefficient on both scales.

Scrutinizing Table 2 and Figs. 2 and 3 also verified the varying effect of straw mulch on the runoff coefficient, which ranged from −10.43 to −96.71 % on the two scales. The minimum and the maximum effects also occurred at rainfall intensities of 50 mm h^{-1} in the 6 m^2 plot, with a rate of −10.43 %, and of 90 mm h^{-1} in the 0.25 m^2 plot, with a rate of 96.71 % mm h^{-1}. The 0.25 m^2 plot had the maximum reduction in runoff coefficient when rainfall intensities were 50 and 90 mm h^{-1}. These results showed that the 0.25 m^2 plot had the maximum impact on decreasing the runoff coefficient and increasing time to runoff, except in the case of a rainfall intensity of 50 mm h^{-1}. It verified that the straw mulch pieces could store more runoff, leading to more infiltration, as already reported by Poesen and Lavee (1991), Choi et al. (2012) and Liu et al. (2012). The results showed that there was large variation in the runoff coefficient (Smets et al., 2008a) and time to runoff on the 0.25 m^2 plots compared to those recorded for the 6 m^2 plots at variousrainfallintensities.

Table 3. Sediment concentration and soil loss measured at the outlet of the study plots before and after applying conservation treatment.

Plot area (m²)	Rainfall intensity (mm h⁻¹)	Sediment concentration (g L⁻¹)		Soil loss (g)	
		Control	Treated	Control	Treated
0.25	50	2.04	0.00	1.61	0.00
		1.13	0.00	0.98	0.00
		1.88	0.00	1.54	0.00
	SD	0.49	0.00	0.35	0.00
	90	2.69	0.00	3.78	0.00
		1.56	0.00	3.42	0.00
		2.00	0.00	3.27	0.00
	SD	0.57	0.00	0.26	0.00
6	50	6.13	3.87	226.27	131.38
		7.43	3.69	266.64	128.94
		8.27	4.70	302.82	161.62
	SD	1.08	0.54	38.29	18.20
	90	10.28	4.39	756.69	286.37
		10.71	4.47	787.94	315.10
		10.15	4.01	738.20	239.42
	SD	0.29	0.25	25.14	38.20

Table 4. Reduction rates (%) in average sediment concentration and soil loss in plots treated with rice straw mulch.

Plot area (m²)	Variable	Rainfall intensity (mm h⁻¹) 50	90
0.25	Sediment concentration	−100	−100
	Soil loss	−100	−100
6	Sediment concentration	−43.47	−58.69
	Soil loss	−46.74	−63.24

In this study the effectiveness of mulch in reducing runoff was influenced by the plot size. Thus, the runoff amount increased with increasing plot size, while Poesen et al. (1994), Cerdan et al. (2002) and Smets et al. (2008a, b) showed that the runoff amount decreased with increasing plot size. The differences between mulch type, application manner and density as well as soil type and rainfall intensity could be supposed as potential reasons behind the disagreement. However, according to McGregor et al. (1988), plot border effects on runoff rates were much more important in small plots compared to large ones.

Tables 3 and 4 showed that the conservation treatment essentially reduced soil loss, which is consistent with Poesen and Lavee (1991), Fernández et al. (2012), García-Orenes et al. (2014) and Fernández and Vega (2014). Sediment concentration also decreased in treated plots, as also reported by Poesen and Lavee (1991) and Smets et al. (2008a and b). This indicated that the flow was not powerful enough to detach particles. A similar finding has been reported by Poesen and Lavee (1991).

The sediment concentration and soil loss amounts before and after the conservation treatment on each scale are shown in Table 3. The relative effectiveness of straw mulch on sediment concentration and soil loss for the two scales is also summarized in Table 4. Figures 4 and 5, respectively, show the average rates of sediment concentration and soil loss in the two study plots.

Table 4 and Figs. 4 and 5 also show that the amounts of sediment concentration on the two study scales changed from −43.47 to −100 %. The maximum change occurred in the 0.25 m² plot at the intensities of both 50 and 90 mm h⁻¹. Thus, soil loss was found to be negligible after mulching in a small plot of 0.25 m² (Poesen et al., 1994). The results also showed that soil loss was reduced in the 0.25 and 6 m² plots; moreover, the variation ranged from −58.69 to −100 %. Poesen et al. (1994) found that soil loss was reduced by 100 % in small plots of 1 m² with a cover of 100 %. It was also

Table 5. Results of GLM test for plot size and conservation treatment effects on the quantitative characteristics of runoff and soil loss.

Source	Dependent variable	Type III sum of squares	d f	Mean square	F	Significant level
Plot	Time to runoff (s)	595 564.22	1	595 564.22	40.92	0.00
	Runoff coefficient (%)	16 413.83	1	16 413.83	381.42	0.00
	Sediment concentration (g L^{-1})	185.59	1	185.59	194.67	0.00
	Soil loss (g)	780 024.69	1	780 024.69	38.46	0.00
Treatment	Time to runoff (s)	40 142.53	1	40 142.53	2.76	0.11
	Runoff coefficient (%)	2317.91	1	2317.91	53.86	0.00
	Sediment concentration (g L^{-1})	63.64	1	63.64	66.75	0.00
	Soil loss (g)	139 578.68	1	139 578.68	6.88	0.02
Plot treatment	Time to runoff (s)	14 704.47	1	14 704.47	1.01	0.33
	Runoff coefficient (%)	616.72	1	616.72	14.33	0.001
	Sediment concentration (g L^{-1})	11.48	1	11.48	12.04	0.002
	Soil loss (g)	135 178.56	1	135 178.56	6.67	0.02

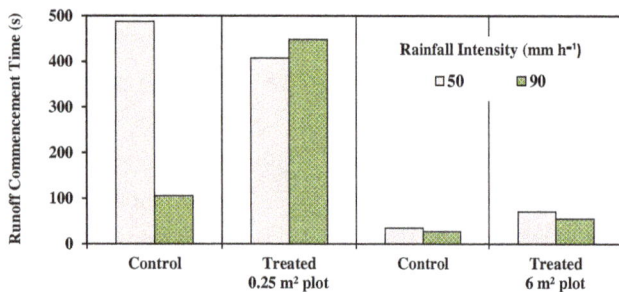

Figure 2. Average time to runoff for the two study scales and the two rainfall intensities.

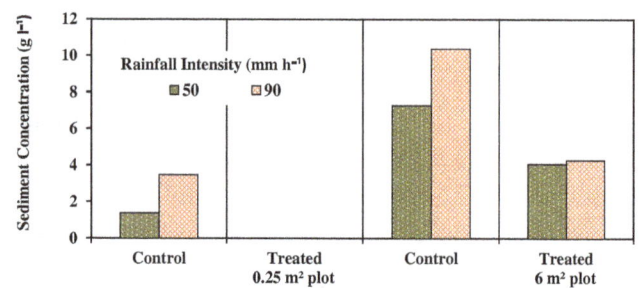

Figure 4. Average sediment concentration for the two study scales and the two rainfall intensities.

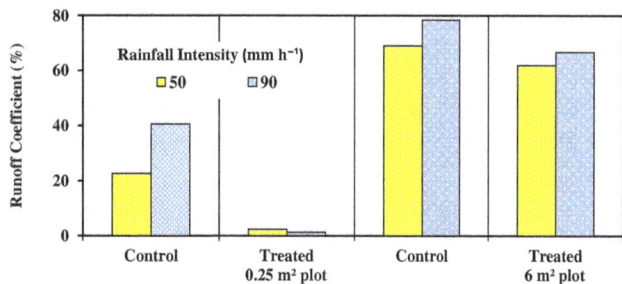

Figure 3. Average runoff coefficient for the two study scales and the two rainfall intensities.

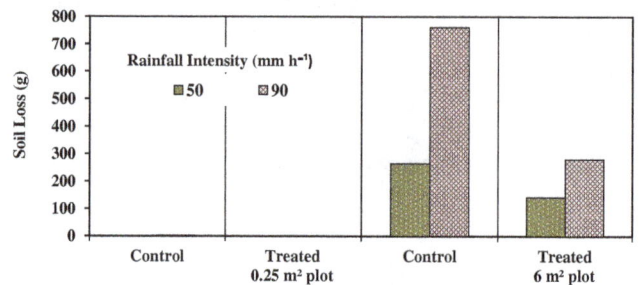

Figure 5. Average soil loss for the two study scales and the two rainfall intensities.

observed that both the study variables achieved the maximum effect in a small plot of 0.25 m^2 with regard to decreasing sediment concentration and soil loss. It has also been verified by Mingguo et al. (2007) that soil loss by water erosion in laboratory conditions reduced as plot size decreased. Poesen and Lavee (1994) and Smets et al. (2008a, b) also stated that soil loss by water erosion was influenced by the plot length. They showed that the small plots with mulch cover were significantly less effective in reducing relative soil loss compared to longer plots. By contrast, this study states that the small plot with straw mulch was more effective in reducing runoff and soil loss amounts (Mingguo et al., 2007). Therefore, the effectiveness of mulch cover in reducing runoff and soil loss by water erosion decreased with increasing plot size. These results were not consistent with Poesen et al. (1994) and Smets et al. (2008a, b), whereas they agreed with Mingguo et al. (2007).

Poesen et al. (1994), Cerdan et al. (2002), Boix-Fayos et al. (2006) and Smets et al. (2008a, b) showed that plot length (or spatial scale) can be important in variations in runoff or soil loss rates and in the effectiveness of surface covers.

These results were found to be important in designing runoff production and erosion plots and modeling runoff and soil loss rates (Smets et al., 2008a).

The results of the statistical analysis based on the GLM is summarized in Table 5. According to Table 5, changing plot size could have a significant effect ($P > 0.01$) on time to runoff and the coefficient, sediment concentration and soil loss. The runoff coefficient ($p = 0.00$), sediment concentration ($p = 0.00$) and soil loss ($p = 0.02$) were significantly influenced by plot size as well as by conservation treatment with rice straw mulch. The interaction effect of plot size and conservation treatment on the runoff coefficient, sediment concentration and soil loss were also significant, with respective p-values of 0.001, 0.002 and 0.02. However, time to runoff was only influenced by plot size.

4 Conclusions

The present study was conducted to study the effects of plot size on runoff and soil loss control. In order to so, two plot scales (0.25 and $6\,\mathrm{m}^2$) were treated with $0.5\,\mathrm{kg\,m}^{-2}$ of rice straw mulch under two rainfall intensities (50 and $90\,\mathrm{mm\,h}^{-1}$). The straw mulch increased the time to runoff compared to untreated plots, except at a rainfall intensity of $50\,\mathrm{mm\,h}^{-1}$ for the $0.25\,\mathrm{m}^2$ plot, and it also decreased the runoff coefficient on both the scales. The maximum change in effectiveness in the time to runoff, for the two scales, could be found at a rainfall intensity of $90\,\mathrm{mm\,h}^{-1}$. The maximum change in soil loss occurred in the $0.25\,\mathrm{m}^2$ plot at the intensities of both 50 and $90\,\mathrm{mm\,h}^{-1}$. The results showed that the $0.25\,\mathrm{m}^2$ plot had a better effectiveness in reducing the runoff coefficient, sediment concentration and soil loss. The results of the study clearly proved the different responses of the plots with regard to runoff soil loss components; these results can be practically applied when setting up experimental studies. The results further showed that the plots should mainly be used for comparative studies rather than for those aimed at obtaining accurate data on larger-scale outcomes.

Acknowledgements. The authors would like to thank Professor K. Banasik for his valuable scientific and technical assistance to second and fourth authors at time of their sabbatical programs in Poland led to some new approaches and outcomes. They also appreciate the journal editor, technical assistants and anonymous reviewers for their persistent and accurate endeavors in processing the manuscript.

Edited by: A. Cerdà

References

Adekalu, K. O., Olorunfemi, I. A., and Osunbitan, J. A.: Grass mulching effect on infiltration, surface runoff and soil loss of three agricultural soils in Nigeria, Bioresour. Technol., 98, 912–917, 2007.

Amimoto, P. Y.: Erosion and sediment control handbook, California Department of Conservation Report No. EPA 4 40/3-78-003, 197 pp., 1981.

Auerswald, K., Kainz, M., and Fiener, P.: Soil erosion potential of organic versus conventional farming evaluated by USLE modeling of cropping statistics for agricultural districts in Bavaria, Soil Use Manage., 19, 305–311, 2003.

Biro, K., Pradhan, B., Buchroithner, M., and Makeschin, F.: Land use/land cover change analysis an its impact on soil properties in the Northern part of Gadarif region, Sudan, Land Degradat. Develop., 24, 90–102, 2013.

Boix-Fayos, C., Martínez-Mena, M., Arnau-Rosalén, E., Calvo-Cases, A., Castillo, V., and Albaladejo, J.: Measuring soil erosion by field plots: understanding the sources of variation, Earth-Sci. Rev., 78, 267–85, 2006.

Cerdà, A., Giménez-Morera, A., and Bodí, M. B.: Soil and water losses from new citrus orchards growing on sloped soils in the western Mediterranean basin, Earth Surf. Proc. Land., 34, 1822–1830, 2009.

Cerdà, A., Hooke, J., Romero-Diaz, A., Montanarella, L., and Lavee, H.: Soil erosion on Mediterranean Type-Ecosystems, Land Degrad. Develop., 21, 71–74, doi:10.1002/ldr.968, 2010.

Cerdan, O. Y., Le Bissonnais, V., Souchere, P. M., and Lecomte, V.: Sediment concentration in interrill flow: interactions between soil surface conditions, vegetation and rainfall, Earth Surf. Proc. Land., 27, 193–205, 2002.

Choi, J., Shin, M. H., Yoon, J. S., and Jang, J. R.: Effect of rice straw mulch on runoff and NPS pollution discharges from a vegetable field, International Conference of Agriculture Engineering, July 8–12, Spain, 4 pp., 2012.

Cogo, N. P., Moldenhauer, W. C., and Foster, G. R.: Soil loss reductions from conservation tillage practices, Soil Sci. Soc. Am. J., 48, 368–373, 1984.

Das, D. K. and Agrawal, R. P.: Physical properties of soils, in: Fundamentals of Soil Science, New Delhi, J. Indian Soc. Soil Sci., 283–295, 2002.

Defersha, M. B., Quraishi, S., and Mellese, A. M.: The effect of slope steepness and antecedent moisture content on interrill erosion, runoff and sediment size distribution in the highlands of Ethiopia, Hydrol. Earth Syst. Sci., 15, 2367–2375, 2011.

Duiker, S. W., Flanagan, D. C., and Lal, R.: Erodibility and infiltration characteristics of five major soils of southwest Spain, Catena, 45, 103–121, 2001.

Fernández, C. and Vega, J. A.: Efficacy of bark strands and straw mulching after wildfire in NW Spain: Effects on erosion control and vegetation recovery, Ecol. Engin., 63, 50–57, 2014.

Fernández, C., Vega, J. A., Jiménez, E., Vieira, D. C. S., Merino, A., Ferreiro, A., and Fonturbel, T.: Seeding and mulching+seeding effects on post-fire runoff, soil erosion and species diversity in Galicia (NW Spain), Land Degrad. Develop., 23, 150–156, 2012.

García-Moreno, J., Gordillo-Rivero, A., Zavala, L. M., Jordán, A., and Pereira, P.: Mulch application in fruit orchards increases the persistence of soil water repellency during a 15-years period, Soil Till. Res., 130, 62–68, 2013.

García-Orenes, F., Cerdà, A., Mataix-Solera, J., Guerrero, C., Bodí, M. B., Arcenegui, V., Zornoza, R., and Sempere, J. G.: Effects of agricultural management on surface soil properties and soil-water losses in eastern Spain, Soil Till. Res., 106, 117–123, 2009.

García-Orenes, F., Guerrero, C., Roldán, A., Mataix-Solera, J., Cerdà, A., Campoy, M., Zornoza, R., Bárcenas, G., and Caravaca, F.: Soil microbial biomass and activity under different agricultural management systems in a semiarid Mediterranean agroecosystem, Soil Till. Res., 109, 110–115, 2010.

García-Orenes, F. Roldán, A., Mataix-Solera, J., Cerdà, A., Campoy, M., Arcenegui, V., and Caravaca, F.: Soil structural stability and erosion rates influenced by agricultural management practices in a semi-arid Mediterranean agro-ecosystem, Soil Use Manag., 28, 571–579, 2012.

Gholami, L., Sadeghi, S. H. R., and Homaee, M.: Straw mulching effect on splash erosion, runoff and sediment yield from eroded plots, Soil Sci. Soc. Am. J., 77, 268–278, 2013.

Giménez Morera, A., Ruiz Sinoga, J. D., and Cerdà, A.: The impact of cotton geotextiles on soil and water losses in Mediterranean rainfed agricultural land, Land Degrad. Develop., 210–217, 2010.

Hawke, R. M., Price, A. G., and Bryan, R. B.: The effect of initial soil water content and rainfall intensity on near-surface soil hydrologic conductivity: A laboratory investigation, Catena, 65, 237–246, 2006.

Jordán, A., Zavala, L. M., and Gil, J.: Effects of mulching on soil physical properties and runoff under semi-arid conditions in southern Spain, Catena , 81, 77–85, 2010.

Jordán, A., Zavala, L. M., and Muñoz-Rojas, M.: Mulching, effects on soil physical properties, in: Encyclopedia of Agrophysics, edited by: Gliński, J., Horabik, J., and Lipiec, J., Berlin, Springer, 492–496, 2011.

Khaledi Darvishan, A. V., Sadeghi, S. H. R. Homaee, M., and Arabkhedri, M.: Measuring sheet erosion using synthetic color-contrast aggregates, Hydrol.Proc., 9 pp., 2013.

Kukal, S. S. and Sarkar, M.: Splash erosion and infiltration in relation to mulching and polyviny alcohol application in semi-arid tropics, Archiv. Agron. Soil Sci., 56, 697–705, 2010.

Kukal, S. S. and Sarkar, M.: Laboratory simulation studies on splash erosion and crusting in relation to surface roughness and raindrop size, J. Ind. Soc. Soil Sci., 59, 87–93, 2011.

Lieskovský, J. and Kenderessy, P.: Modelling the effect of vegetation cover and different tillage practices on soil erosion in vineyards: a case study in Vráble (Slovakia) using watem/sedem, Land Degrad. Develop. , 25, 288–296, 2014.

Liu, Y., Taoa, Y., Wana, K. Y., Zhanga, G. S., Liub, D. B., Xiongb, G. Y., and Chena, F.: Runoff and nutrient losses in citrus orchards on sloping land subjected to different surface mulching practices in the danjiangkou reservoir area of China, Agr. Water Manag., 110, 34–40, 2012.

Mandal, D. and Sharda, V. N.: Appraisal of soil erosion risk in the Eastern Himalayan region of India for soil conservation planning, Land Degrad. Develop., 24, 430–437, 2013.

McGregor, K. C., Bengtson, R. L., and Mutchler, C. K.: Effects of surface straw on interrill runoff and erosion of grenada silt loam, Transact. ASAE, 31, 111–116, 1988.

Mingguo, Z., Qiangguo, C., and Hao, C.: Effect of vegetation on runoff-sediment yield relationship at different spatial scales in hilly areas of the Loess Plateau, North China, Ac. Ecol. Sin., 27, 3572–3581, 2007.

Morgan, R. P. C.: Soil erosion and conservation. Longman Scientific and Technical, Burnt Mile, Harlow, UK, 298 pp., 1986.

Morgan, R. P. C.: Soil erosion and conservation. Longman, Essex, England, 198 pp., 1995.

Prats, S. A., MacDonald, L. H., Monteiro, M., Coelho, C. O. A., and Keizer, J. J.: Effectiveness of forest residue mulching in reducing post-fire runoff and erosion in a pine and a eucalypt plantation in north-central Portugal, Geoderma , 191, 115–124, 2012.

Prats, S. A., Martins, M. A. D. S., Malvar, M. C., Ben-Hur, M., and Keizer, J. J.: Polyacrylamide application versus forest residue mulching for reducing post-fire runoff and soil erosion, Sci. Total Environ., 468/469, 464–474, 2014.

Poesen, J. W. A. and Lavee, H.: Effects of size and incorporation of synthetic mulch on runoff and sediment yield from interrills in a laboratory study with simulated rainfall, Soil Till. Res., 21, 209–223, 1991.

Poesen, J. W., Torri, D., and Bunte, K.: Effects of rock fragments on soil erosion by water at different spatial scales: a review, Catena, 23, 141–66, 1994.

Prokop, P. and Poręba, G. J.: Soil erosion associated with an upland farming system under population pressure in Northeast India, Land Degrad. Develop., 23, 310–321, 2012.

Robichaud, P. R., Lewis, S. A., Wagenbrenner, J. W., Ashmun, L. E., and Brown, R. E.: Post-fire mulching for runoff and erosion mitigation – Part I: Effectiveness at reducing hillslope erosion rates, Catena , 105, 75–92, 2013.

Romkens, M. J. M., Helming, K., and Prasad, S. N.: Soil erosion under different rainfall intensities, surface roughnessand soil water regimes, Catena , 46, 103–123, 2001.

Ruiz-Sinoga, J. D., Romero-Diaz, A., Ferre-Bueno, E., and Martínez-Murillo, J. F.: The role of soil surface conditions in regulating runoff and erosion processes on a metamorphic hillslope (southern Spain) soil surface conditions, runoff and erosion in southern Spain, Catena, 80, 131–139, 2010.

Ruy, S., Findeling, A., and Chadoeuf, J.: Effect of mulching techniques on plot scale runoff: FDTF modeling and sensitivity analysis, J. Hydrol., 326, 277–294, 2006.

Smets, T., Poesen, J., and Bochet, E.: Impact of plot length on the effectiveness of different soil-surface covers in reducing runoff and soil loss by water, Prog. Phys. Geogr., 32, 654–677, 2008a.

Smets, T., Poesen, J., and Knapen, A.: Spatial scale effects on the effectiveness of organic mulches in reducing soil erosion by water, Earth-Sci. Rev., 89, 1–12, 2008b.

Zhao, G., Mu, X., Wen, Z., Wang, F., and Gao, P.: Soil erosion, conservation and Eco-environment changes in the Loess Plateau of China, Land Degrad. Develop., 24, 499–510, 2013.

2

Geothermal investigations in western Anatolia using equilibrium temperatures from shallow boreholes

K. Erkan

Department of Civil Engineering, Marmara University, 34722, Göztepe, Istanbul, Turkey

Correspondence to: K. Erkan (kamil.erkan@marmara.edu.tr)

Abstract. Determination of the conductive heat flow in western Anatolia has broad implications in many areas, including studies on the present-day extensional tectonic activity and assessments of the geothermal resources in the region. In this study, high-resolution equilibrium temperatures from 113 boreholes with depths of ~ 100 m were analyzed for determination of the conductive heat flow. Thermal conductivities were either determined by measurements on outcrops or estimated using lithologic records. By a detailed analysis of temperature–depth curves, a total of 55 sites were selected as being useful for further conductive gradient/heat flow calculations, while the remaining 58 sites were abandoned due to hydrological effects on temperatures. Heat flow values with formal errors were calculated for 24 sites where rock thermal conductivity information is available. Due to the shallow depths of the investigated boreholes and uncertainties in thermal conductivity information, the results include a large accumulated error. A preliminary heat flow map is generated using the results of this study and a previous study in the southern Marmara region. Elevated heat flow values of $85–95$ mW m^{-2} are observed in the coastal areas, including peninsular parts of Çanakkale and Izmir. The central part of the Menderes Massif also shows elevated heat flow values, the highest values (> 100 mW m^{-2}) being in the northeastern part of the Gediz Graben near the Kula volcanic center. Moderate heat flow values of $55–70$ mW m^{-2} are observed in the eastern part of Çanakkale, central part of Balıkesir, northwest of Manisa, and northeast end of Bursa including Yalova. Some of the observed moderate values may be related to unconstrained near surface phenomena due to shallow depth of measurements. Towards the south of the study region, moderate heat flow values are also observed in Muğla. Previously reported regional heat flow values exceeding ~ 120 mW m^{-2} is not observed in the region. The heat flow values reported in this study are comparable to the previously reported values in the Aegean Sea, as the two regions form the back-arc section of the Hellenic subduction zone.

1   Introduction

Determination of the conductive heat flow near the surface of the earth has many important applications, such as understanding the recent history of plate tectonic activity (Erkan and Blackwell, 2008, 2009), determining the depth of brittle/ductile transition in the crust (Bonner et al., 2003), and estimating the geothermal energy potential of a region (Tester et al., 2006; Serpen et al., 2009).

Western Anatolia is a unique region with intense present-day plate tectonic activity (Dilek and Altunkaynak, 2009). With the Aegean Sea, the two regions form the back-arc area of the Hellenic subduction zone (Jolivet et al., 2013). The region has been characterized by crustal extension and subduction-related andesitic volcanism since the Oligocene (Fytikas et al., 1984). Volcanism and extension have migrated southward with the southward retreat of the Hellenic subduction zone. The crustal extension in the region is considered to have two major phases: an early phase of nearly E–W directed extension from the Miocene to the early Pliocene, and a second phase of N–S extension during the Pliocene and the Quaternary. The latter resulted in the formation of modern horst/graben structures observed in the Menderes Massif (Koçyiğit et al., 1999).

The region has also been the locus of geothermal energy development, as it includes the highest enthalpy geothermal systems found in Turkey (Serpen et al., 2009). The highest temperature (120–240 °C) geothermal systems have formed along the margins of the deep grabens of the Menderes Massif. These high temperatures were linked to the circulation of surface waters through deeply incised faults and the high heat flow from the basement of the Menderes horst–graben system.

The conventional method of heat flow determination requires high-resolution temperatures-versus-depth ($T–D$) measurements in thermally stable boreholes, and thermal conductivity determinations on representative rocks (Beardsmore and Cull, 2001). In western Anatolia, heat flow studies based on the conventional techniques have been very limited. A heat flow map of the region is available as part of the heat flow map of Turkey (Tezcan and Turgay, 1991), which uses non-equilibrium bottom-hole-temperature (BHT) data and a constant thermal conductivity. In the southern Marmara region, Pfister et al. (1998) reported results of equilibrium $T–D$ data from shallow (~ 100 m) boreholes, and thermal conductivity measurements from surface outcrops. Their results were included in making the preliminary heat flow map in this study (Fig. 4).

In this work, $T–D$ measurements from 113 shallow (depths of ~ 100 m) boreholes in western Anatolia are studied. Thermal conductivities were either measured on surface outcrops, or estimated from borehole lithologic information. As a result of processing the data, gradients and/or heat flow values were calculated for 55 points. A preliminary heat flow map of the region is generated and compared with results of previous studies.

2 Data collection

From 1995 to 1999, a regional campaign of collection of temperatures in boreholes and rock thermal conductivities was conducted in western and central Anatolia, for the determination of conductive heat flow (İlkışık et al., 1996a, b). Data collection was performed by a group at the General Directorate of Mineral Research and Exploration (MTA) of Turkey. The boreholes were provided by the State Hydrological Works (DSI) and Rural Services (presently, out of service) regional offices. These boreholes were either drilled as water supply wells (but not producing), or as groundwater monitoring wells. Among the available boreholes, ones far from the known geothermal areas and located on bedrock were especially chosen (Öztürk et al., 2006). For each borehole, location, depth, lithologic records, static levels, etc., were obtained from the personnel of the state agencies. The holes were generally 6–8 inch (15–20 cm) in radius at the collar, and cased the entirety of their depths, with perforations at certain levels. Behind the casing, pebbles were used as the filling material to allow permeability between the borehole and the formation (H. M. Yenigün, personal communication, 2012).

$T–D$ data were collected for each meter of depth, using an Amerada surface read-out portable logging tool. Static water levels were determined by a sinker before each $T–D$ measurement. All $T–D$ measurements were done below the water table. For thermal conductivity analysis, rock samples were collected from surface outcrops in the vicinity of each borehole. The measurements were run by a QTM-500 thermal conductivity device on dry samples. Generally, more than one type of lithology was sampled for each site, resulting in a larger data set for thermal conductivity measurements. A statistical analysis of these thermal conductivity measurements versus lithology is given by Balkan et al. (2015).

3 Data analysis

3.1 Data quality classification

Not all boreholes are suitable for conductive heat flow determinations. Quality of determination depends on the physical conditions of the borehole site. In this study, data were divided into various quality classes by analyzing the general characteristics of the $T–D$ curves. The criteria for each class and the associated numerical error in gradient calculations are summarized in Table 1.

Class A or B data are the ones that strictly satisfy the solution of 1-D heat transfer along a borehole (Jaeger, 1965). These include linearly increasing temperatures with depth, and projected surface temperature matching the mean annual surface temperature (MAST) of the measurement site. The projected surface temperature (the extrapolated value of the linear gradient at the surface) primarily depends on the latitude and the elevation of the borehole site, if no secondary effects exist due to microclimatic conditions (Roy et al., 1972). Another indication of a conductive section is that an increase/decrease in rock thermal conductivity results in a decrease/increase in the gradient, giving a constant heat flow along the borehole.

In some boreholes, vertical motion of the borehole fluid in some sections results in disturbed $T–D$ profiles, even though sections outside of the disturbed zone indicate conductive heat transfer (Roy et al., 1972; Erkan et al., 2008). Intra-borehole fluid flow (IBF) occurs in open (not grouted) boreholes and causes sharp changes in $T–D$ curves where the fluids enter and exit the borehole. These types of data were rated class C, with a larger (25 %) relative error in gradient calculations. If IBF dominates most of the $T–D$ profile, gradient calculations have uncertain reliability (class D, no error bound). In these boreholes, gradients are either constrained from a few control points, or calculated at very shallow (< 50 m) depths.

Borehole sites not suitable for conductive heat flow analysis are rated class X. These sites show the effect of local hydrologic activity and conductive thermal regime is overprinted by the groundwater motion. $T–D$ curves for these holes were generally observed to show isothermal behavior, indicating fast vertical flow. Other types of hydrologically active sites are found near geothermal systems. These sites show the effect of local geothermal activity, which shows distinctly higher temperatures. These types of data are rated class G, and are also not suitable for conductive heat flow determinations.

The regional distribution of all the data according to the quality classes is shown in Fig. 1. Out of 113 sites, 58 of them fall into class X or G, and are not suitable for conductive heat flow analysis. 24 sites fall into class D, and the remaining 31 sites fall into classes A/B/C.

3.2 Temperature–depth curves

The administrative provinces in Turkey have moderate sizes, and are suitable for comparative analysis of $T-D$ curves (Fig. 1). For example, due to the proximity of the sites, projected surface temperatures can be compared directly with their elevations. $T-D$ curves for class A, B, and C data are shown in various panels of Fig. 2, based on the provinces in which they are located.

In Çanakkale (Fig. 2a), $T-D$ curves show generally conductive behavior for entire lengths, and the effect of IBF is minimal. A weak downflow (33–50 m) in Pazarkoy, a strong upflow (95–125 m) in Cavuskoy, and a strong upflow (90–130 m) in Ortuluce are inferred. Projected surface temperatures correlate well with elevations. (Fig. 1) For example, Intepe and Cavuskoy are located near the sea shore, and have the highest projected surface temperatures. On the other hand, Pazarkoy and Terzialan are located farther inland, at higher elevations; they show relatively lower projected surface temperatures (an adiabatic lapse rate of $5\,^{\circ}\mathrm{C\,km^{-1}}$ may be used for correlating elevations with surface temperatures). In Yapildak, a downflow from the surface to 25 m must have been occurring for a long time, so that the $z = 25$ m level acts as the apparent surface of the borehole. The $T-D$ curve below this level shows the conductive thermal regime. Another interesting feature is the sharp break in the gradients for Intepe at 65 m. Lithologic records show a change from claystone to diabase lithology around this depth, which should result in this abrupt change in the gradient.

$T-D$ curves for Bursa and Balıkesir are shown in Fig. 2b. In Bursa, when compared to the other two sites, Eyerce is located about 300 m higher, which may explain the lower projected temperature for this hole.

For Izmir (Fig. 2c), holes are relatively shallower, but they show a conductive thermal regime for their entire depths. For Yusufdere, the first 50 m of the hole seem to be affected by hydrologic disturbances; below 50 m, a conductive regime is apparent. Bademli penetrates a very highly conductive sandstone lithology ($4.1\,\mathrm{W\,m^{-1}\,K^{-1}}$), which may be responsible for the low gradients in this hole. For three holes in Fig. 2c, near-surface systematic changes toward higher temperatures at their first ~ 50 m in depth are interesting (see deviations

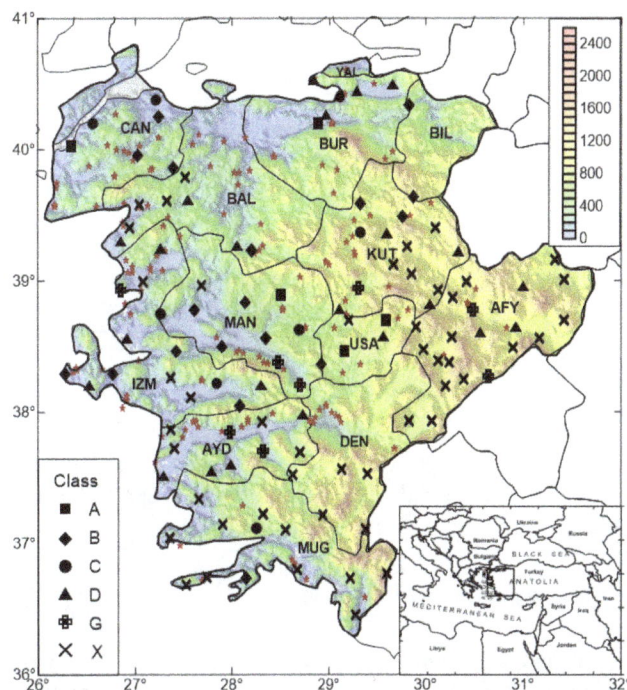

Figure 1. Data locations with the corresponding quality classes. See the text for details of the class definitions. Red star symbols show locations of hot springs. Elevations are in meters. Acronyms for administrative provinces are as follows: CAN: Çanakkale; BAL: Balıkesir; BUR: Bursa; BIL: Bilecik; KUT: Kütahya; MAN: Manisa; USA: Uşak; AFY: Afyon; IZM: Izmir; DEN: Denizli; AYD: Aydın; MUG: Muğla.

from the dashed lines in Yenmis, Ovaciki, and Ciftlikkoy). These changes may be due to recent changes in the MAST values. It is known that some holes (ones belonging to Rural Services) were protected in small rooms, which may cause a transient microclimatic effect (i.e., greenhouse heating) on the surface. The depths of these deviations indicate the time periods these rooms were built.

$T-D$ curves for Manisa (Fig. 2d) show some local fluctuations, but their general character shows conductive behavior. If these fluctuations are not instrumental, they may be due to cellular convections inside the borehole caused by the application of the sinker before the measurement. Cellular convections are formed when the borehole diameter is larger than a certain size, which may be the case for Manisa (see Pfister et al., 1998, for a more detailed discussion about cellular convections). The projected surface temperatures are consistent with each other, but they seem to have systematically higher values compared to the values in Izmir. If MAST values are not really higher in Manisa, this shift may indicate a calibration problem in the $T-D$ measurement tool.

$T-D$ curves for Kütahya and Uşak provinces show generally linear behavior (Fig. 2e). An elevation difference of ~ 500 m between the two provinces results in a 3–4 $^{\circ}\mathrm{C}$ difference for projected surface temperatures. In Koprucek, a

Table 1. Definitions of the data quality classes used in this study.

Class	Description	Estimated relative error in gradient
A	> 100 m conductive (linear) $T-D$ section	5 %
B	> 50 m conductive (linear) $T-D$ section	10 %
C	Disturbed $T-D$ curve due to intra-borehole fluid activity	
	Intermittent conductive sections	25 %
D	Intense intra-borehole fluid activity; conductive section too shallow	–
G	$T-D$ curve overprinted by geothermal activity	–
X	$T-D$ curve overprinted by groundwater activity	–

downflow or lateral flow seems to be occurring in the first 75 m of the borehole; below this depth, it is conductive. In Muğla (Fig. 2f), conductive behavior was observed in two holes among many visited holes (see Fig. 1).

4   Results

Calculated temperature gradients and their interval depths are given in Table 2, along with other useful information. Errors for gradient calculations were calculated based on the criteria in Table 1. For some sites, mostly for class D data, gradients were calculated by drawing a hypothetical line for the entire depth, and their intervals are shown to start at the surface in Table 2. Terrain correction was applied to some boreholes using Lee's topographic correction model (Beardsmore and Cull, 2001). In this model, the topography is fitted to a 2-D mountain range or a hill characterized by a certain height and width. The correction resulted in changes of up to $10\,°C\,km^{-1}$ in gradient. The error from the 2-D assumption of the topography is considered to be negligible compared to the error bounds for the gradients in Table 1. The regional distribution of the (corrected) gradient calculations and their errors is shown in Fig. 3a.

Thermal conductivities were determined based on the lithologic data for the interval where gradients were calculated. Measurements were made on surface outcrops under dry conditions, so they had to be corrected for wet conditions. The geometric mixing model was applied for the porosity correction (Fuchs et al., 2013). For porosities of sedimentary rocks, the values reported by Fuchs et al. (2013) were used. For volcanic and metamorphic rocks, porosity values of 5 and 4 % were used, respectively (JICA, 1987). In order to account for uncertainties in porosity assumptions, an additional error of 25 % was assumed for all porosities, and this error was propagated to the error in bulk thermal conductivity estimations. When no thermal conductivity measurements were available, literature values and their respective error bounds were used (Balkan et al., 2015; Blackwell and Steele, 1989; Clark, 1966). For data located in Quaternary alluvium, a generic value of $1.5 \pm 0.3\,W\,m^{-1}\,K^{-1}$ was used based on a statistical analysis (see the interactive comment, SED, 6, C78–C81).

The calculated heat flow values are given in Table 2 (25 points for classes A/B/C with their propagated error, and 11 points for class D data with no error bound). Regional distribution of the heat flow values is shown in Fig. 3b. With the exception of two holes in areas of active sedimentation (see discussion below), the average heat flow is calculated to be $73 \pm 22\,mW\,m^{-2}$, based on class A/B/C-type data.

Depending on the geographic location, recent climatic changes can have a significant effect on measured heat flow values, and require further correction (Majorowicz and Wybraniec, 2010). In particular, at high latitudes, the Pleistocene ice age and the subsequent warming require corrections of up to $20\,mW\,m^{-2}$ in heat flow determinations. For Turkey, effects of Holocene climatic change on heat flow measurements were calculated to be $2–4\,mW\,m^{-2}$ (Majorowicz and Wybraniec, 2010). The correction would be smaller for gradients measured at shallow depths, as the warming would affect the entirety of the section (see Majorowicz and Wybraniec, 2010, Fig. 2). On the other hand, more recent changes (last 100 years) in the climate require more attention for gradients measured at shallow boreholes. Tayanç et al. (2009) report no significant change in the mean annual temperatures in Turkey until 1993, and an accumulated warming of $\sim 0.5\,°C$ since then. The effect of this warming trend on gradients may be estimated using the chart given by Pollack and Huang (2000, Fig. 3). According to this, a unit change in the surface temperature in the last ~ 20 years penetrates into the subsurface down to $\sim 50\,m$, with exponentially decreasing magnitude. Gradient measurements below 50 m of the borehole would not be affected by the recent warming trend at all. Above this depth, and below the zone of annual temperature effects (below 20 m), the recent warming is expected to have a disturbance of $\sim 0.1\,°C$. For a 30 m long linear section and thermal conductivity of $1.5\,W\,m^{-1}\,K^{-1}$, this would result in an error of up to $5\,mW\,m^{-2}$ in heat flow calculations. This value is well within the error bounds reported for the present heat flow determinations.

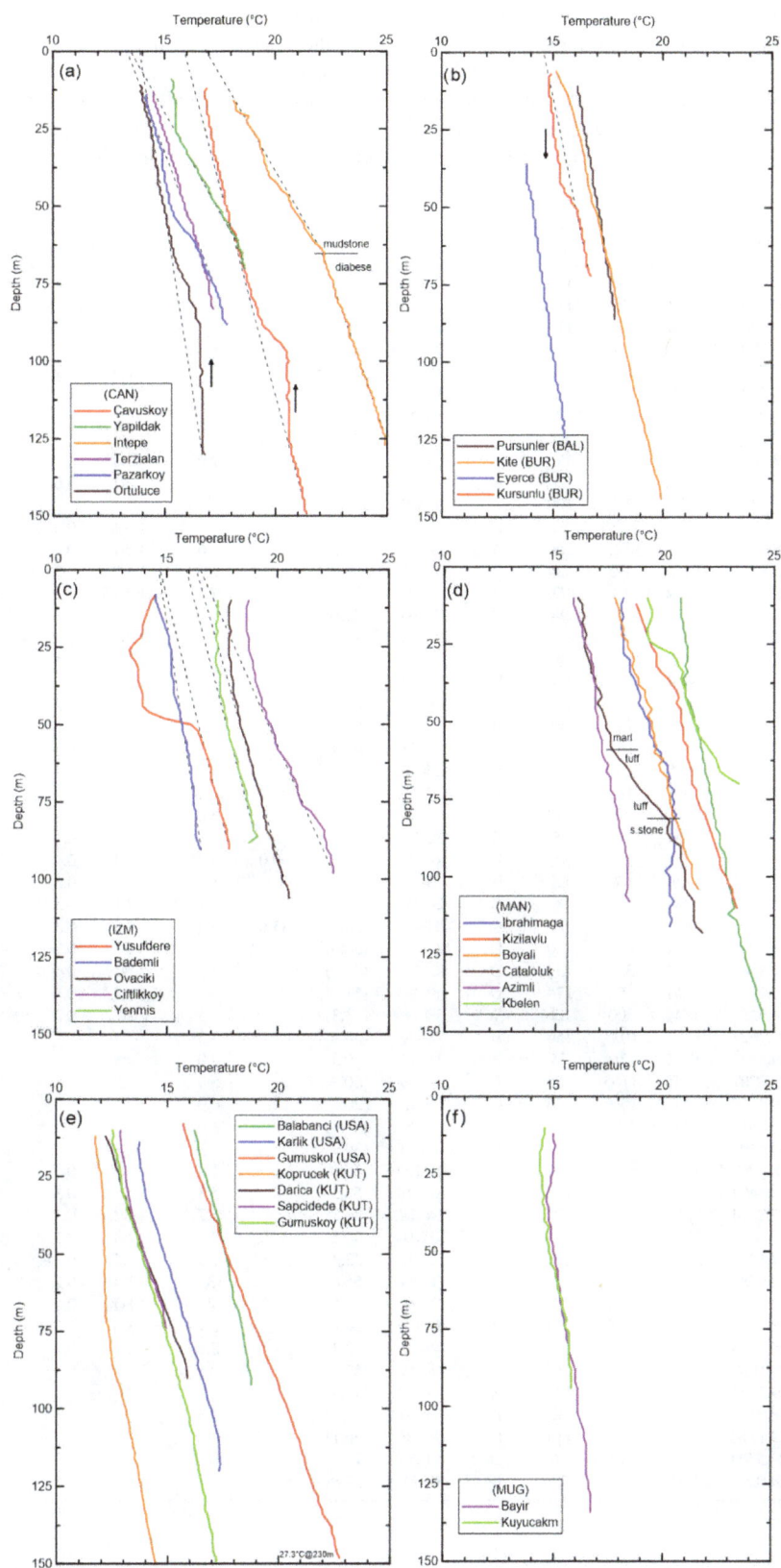

Figure 2. Temperature–depth (T–D) curves for classes A/B/C data for the provinces of (**a**) Çanakkale, (**b**) Balıkesir/Bursa, (**c**) Izmir, (**d**) Manisa, (**e**) Uşak/Kütahya, and (**f**) Muğla. Vertical arrows show the inferred direction of intra-borehole fluid flow.

Table 2. Class (A/B/C/D)-type data used in this study, along with gradients, thermal conductivities, heat flow values, and their respective errors. For sites where thermal conductivities cannot be quantified, only gradients are listed. Gradient/heat flow values in parentheses are estimated values without formal error (for class D-type data). Literature thermal conductivities are indicated by (L) next to the value, and are obtained from Balkan et al. (2015), Blackwell and Steele (1989), and Clark (1966). Depth intervals starting with $z = 0$ indicate that the gradient is calculated based on a hypothetic line using the projected mean annual surface temperature. See the caption of Fig. 1 for administrative province names; G: geothermal gradient; K: thermal conductivity; Q: (corrected) heat flow.

Site name	Latitude ($°$ E)	Longitude ($°$ N)	Prov.	Elev. (m)	Depth (m)	Class	Interval (m)	G ($°C\,km^{-1}$)	Corr. G	σG	K ($W\,m\,K^{-1}$)	σK	Q ($mW\,m^{-2}$)	σQ	Lithology
Kadikoy	38.6365	30.9175	AFY	979	106	D	0–106	(49.1)							
Agzikara	38.5900	30.5600	AFY	1284	110	D	0–110	(36.4)							
Calislar	38.8100	30.0400	AFY	1228	114	D	0–114	(30.4)	(36.4)						
Derbent	38.9400	31.0000	AFY	1238	176	D	120–156	(31.9)							
Tekeler	37.5406	27.7799	AYD	546	94	D	0–94	(21.3)			1.9	0.2	(41)		Schist
Ortakci	37.9700	28.7200	AYD	211	112	D	66–112	(28.3)							
Kargili	37.5877	27.9921	AYD	81	100	D	0–98	(26.5)							
Balat	37.4978	27.2848	AYD	20	96	D	80–95	(40.0)			1.6(L)	0.4	(64)		Marl
Pursunler	39.2270	28.2017	BAL	294	86	B	13–82	24.6	28.5	2.9	2.0	0.2	57	11	Andesite
Alacaatli	39.2534	28.0488	BAL	262	71	D	0–71	(19.7)	(24.5)		1.8(L)	0.6	(44)		Andesite
Akcal	39.6038	27.5416	BAL	250	100	D	0–100	(23.0)	(37.1)						
Bulutlucesme	39.2851	26.8492	BAL	328	92	D	0–40	(29.3)	(42.0)		1.8(L)	0.6	(76)		Andesite
Kite	40.1972	28.8763	BUR	74	156	A	20–148	32.5		1.6	1.5(L)	0.3	49	12	Q. aluvium
Eyerce	40.3375	29.8281	BUR	372	124	B	38–124	36.6		2.0	3.7	0.2	73	11	Marble
Kursunlu	40.4014	29.1105	BUR	15	72	C	50–70	30.0		7.5	1.5(L)	0.3	45	20	Q. aluvium
Linyit	40.2512	28.9616	BUR	91	94	D	72–94	(22.6)							
Cakirca	40.4762	29.6630	BUR	94	124	D	0–124	(29.0)			1.5(L)	0.3	(44)		Q. aluvium
As.Vet.	40.3980	29.0986	BUR	11	40	D	24–38	(47.5)			1.5(L)	0.3	(71)		Q. aluvium
Gurle	40.4313	29.2987	BUR	102	118	D	54–118	(87.5)							
Intepe	40.0279	26.3434	CAN	151	127	A	69–124	42.6	43.6	2.2	2.1(L)	0.3	92	18	Diabase
Pazarkoy	39.8647	27.3855	CAN	162	88	B	15–82	50.7		5.1	1.5(L)	0.3	76	23	Q. aluvium
Terzialan	39.9565	27.0234	CAN	152	83	B	17–73	41.4		4.1	1.0(L)	0.2	41	12	Claystone
Cavuskoy	40.2480	27.2407	CAN	21	162	B	125–162	32.4		3.2	1.5(L)	0.3	49	15	Q. aluvium
Yapildak	40.2005	26.5561	CAN	140	70	C	27–65	76.3	85.3	21.3	1.0(L)	0.2	85	38	Claystone
Ortuluce	40.3780	27.2111	CAN	58	130	C	0–130	23.1		5.8	2.5(L)	0.5	58	26	Conglomerate
Ciftlikkoy	38.2879	26.2796	IZM	51	98	B	32–74	50.0		5.0	1.7	0.1	85	14	Marl
Ovaciki	38.2898	26.7599	IZM	137	106	B	46–106	38.3	49.0	4.9	1.7(L)	0.4	83	28	Marl
Yenmis	38.4597	27.4172	IZM	189	88	B	48–82	35.3		3.5	1.5(L)	0.3	53	16	Q. aluvium fan
Bademli	38.0500	28.0792	IZM	364	90	B	20–90	21.4		2.1	4.1	0.5	88	19	Sandstone
Yusufdere	38.2172	27.8396	IZM	128	90	C	52–88	38.9	33.6	8.4	1.5	0.3	50	23	Q. aluvium fan
Haliller	38.1883	28.2960	IZM	328	100	D	86–100	(28.6)			1.5	0.3	(43)		Q. aluvium fan
Y. Kiriklar	39.2315	27.2549	IZM	357	154	D	46–134	(48.9)			1.6(L)	0.4	(78)		Marl
Seyrek	38.5500	26.9173	IZM	5	174	D	40–96	(51.8)			1.6(L)	0.4	(83)		Marl
Zeytineli	38.1917	26.5250	IZM	300	82	D	38–68	(33.3)			2.7	0.3	(90)		Limestone
Gumuskoy	39.4882	29.7627	KUT	1037	156	B	28–89	34.5		3.5					
Sapcidede	39.5884	29.3348	KUT	1014	74	B	36–74	40.3		4.0					
Darica	39.6380	29.8707	KUT	1165	90	B	40–78	36.6		5.0					
Koprucek	39.3660	29.3349	KUT	1046	158	C	100–150	26.8	27.7	6.9					
Esatlar	39.3439	29.6016	KUT	938	88	D	0–88	(47.0)							
Tepekoy	39.2100	30.3300	KUT	1100	182	D	0–182	(30.9)							
Cataloluk	38.8943	28.4907	MAN	676	122	A	90–122	25.0		1.3	3.5	0.5	88	17	Sandstone
Kizilavlu	38.5649	28.3404	MAN	289	110	B	70–110	52.5		5.3	1.5(L)	0.3	79	24	Q. aluvium
Alahidir	38.5000	27.8974	MAN	145	182	B	114–182	36.8		3.7	1.5(L)	0.3	55	17	Q. aluvium fan
Boyali	38.8338	28.1418	MAN	502	104	B	20–104	40.5		4.1	1.3	0.1	53	9	Alluvium
Azimli	38.7774	27.6073	MAN	101	108	B	52–94	33.3		3.3	1.5	0.3	50	15	Q. aluvium
Ibrahimaga	38.6284	28.6784	MAN	509	152	C	28–64	55.6		13.9	2.4	0.2	133	45	Schist
K. Belen	38.7500	27.2583	MAN	370	74	C	0–66	57.6		14.4	1.8(L)	0.6	104	60	Andesite/tuff
Bayir	36.7347	28.1509	MUG	185	134	B	40–126	20.9		2.1	3.5	0.3	73	14	Limestone
Kuyucakm	37.1119	28.2496	MUG	760	94	C	46–76	32.3		8.1	2.0	0.2	65	23	Limestone
Gumuskol	38.4627	29.1657	USA	895	230	A	19–108	52.1		2.6					
Karlik	38.7001	29.5954	USA	1066	120	A	34–104	42.3		2.1					
Balabanci	38.3618	28.9149	USA	716	92	B	20–50	38.0		3.8					
Karakuyu	38.7680	29.1116	USA	789	114	D	0–108	(56.1)							
Salmanlar	38.5600	29.5700	USA	925	56	D	44–52	(52.0)							
Armutlu	40.5158	28.8264	YAL	9	79	D	0–79	(27.8)			1.5(L)	0.3	(42)		Q. aluvium

5   Discussion

A preliminary contour map of the heat flow values was generated by combining data in Table 2 (using class A/B/C data) and the previous results of Pfister et al. (1998) in the southern Marmara region. For the Pfister et al. (1998) data, values outside the range of 40–140 mW m^{-2} were eliminated due to possible hydrologic disturbances. Furthermore, for three

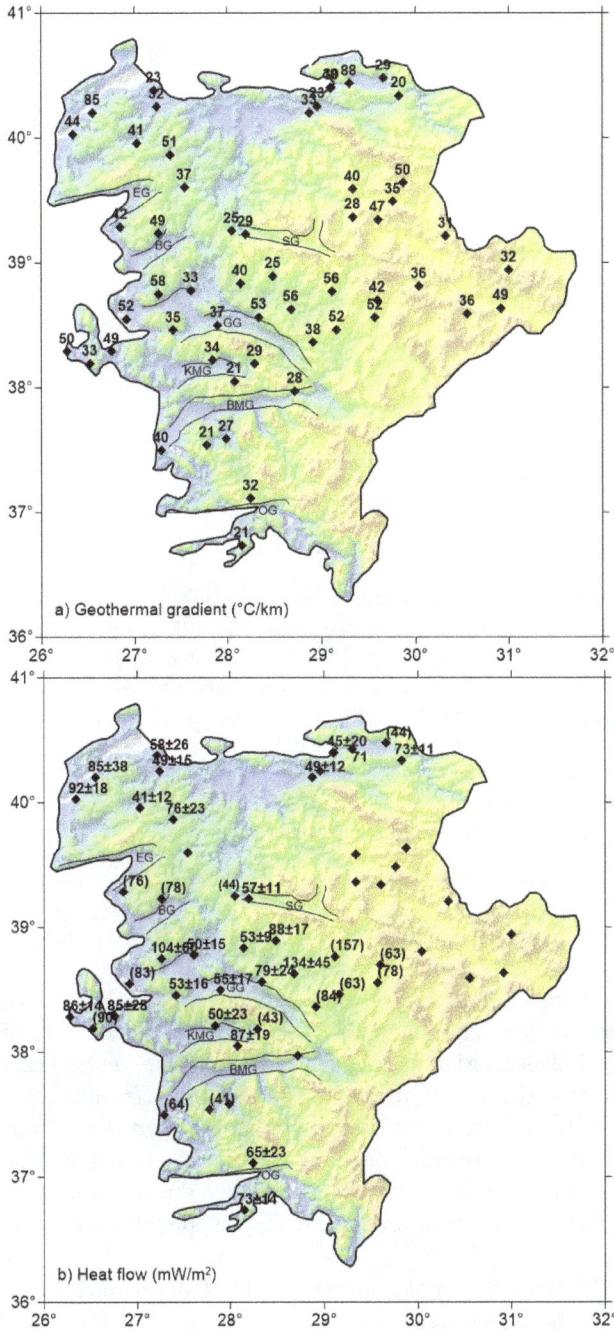

a) Geothermal gradient (°C/km)

b) Heat flow (mW/m²)

Figure 3. Regional distribution of the (**a**) geothermal gradients and (**b**) heat flow values in Table 2. Note that some sites only have gradient values due to the unavailability of thermal conductivity information. Black lines indicate boundaries of horst–graben structures, which are the dominant structural features in the region. GG: Gediz Graben; BMG: Büyük Menderes Graben; KMG: Küçük Menderes Graben; EG: Edremit Graben; BG: Bakırçay Graben; SG: Simav Graben; OG: Gökova Graben.

points located on alluvial fans within the Menderes Massif (Table 2), corrections for sedimentation and thermal refraction were applied before mapping (discussed below in de-

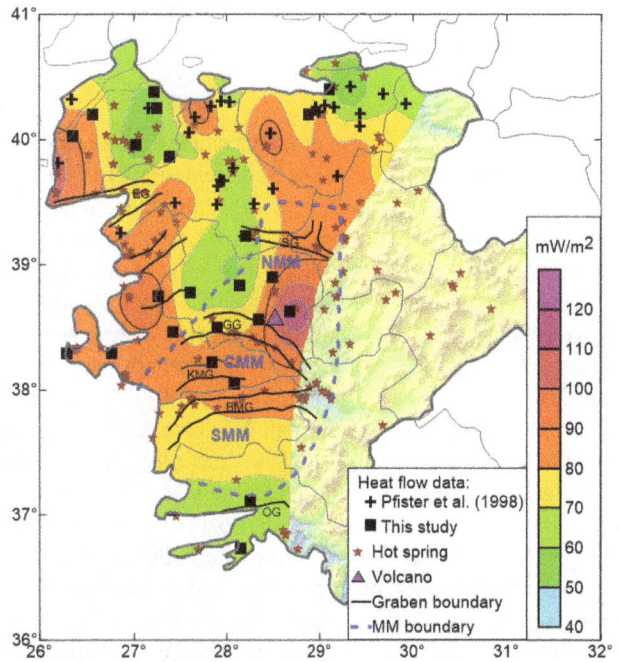

Figure 4. The preliminary heat flow map of the region using the results of this study (class A/B/C data in Table 2), and of Pfister et al. (1998). Blue lines outline the boundary of the Menderes Massif. The blue triangle indicates the location of the Kula volcanic center.

tail). The gridding was done using the minimum curvature technique with a grid spacing of 0.02° in both directions. Then, a 2-D isotropic Gaussian filter of radius 30 km was applied. The resulting heat flow map is shown in Fig. 4. The boundary of the Menderes Massif is also shown in Fig. 4 (blue dashed lines). In the discussions below, the Menderes Massif is divided into three units: the northern Menderes Massif (NMM), the central Menderes Massif (CMM), and the southern Menderes Massif (SMM), separated by two major graben units, the Gediz Graben (GG), which are and the Büyük Menderes Graben (BMG).

The preliminary heat flow map outlines regions with moderate (55–70 mW m^{-2}) and elevated (85–95 mW m^{-2}) heat flow values. Moderate heat flow values are observed in the interior parts of the southern Marmara region (east of Çanakkale, center of Balıkesir, and northeast of Bursa including Yalova), and northwest of Manisa. Also, at the southern end of the mapped area (in Muğla), heat flow values get moderate values. On the other hand, elevated heat flow values are observed in the western part of Çanakkale, the peninsular part of Izmir, and the central part of the Menderes Massif. The area of highest (> 100 mW m^{-2}) heat flow is in the northeastern part of the Gediz Graben, which is near with the observed area of the most recent volcanic activity (Kula volcanic field, see the triangle in Fig. 4); however, data control is low there.

5.1   Effects of sedimentation/erosion and thermal refraction

Heat flow measured in areas of active extension shows near-surface variations due to active sedimentation/erosion and thermal refraction (Blackwell, 1983). Within the studied region, the horst–graben system of Menderes Massif is expected to show the significant thermal effects of sedimentation/erosion and thermal refraction.

Sedimentation causes a downward motion of low temperatures at the surface, resulting in lower-than-normal surface heat flow values (Beardsmore and Cull, 2001). Sedimentation rates within the grabens can be calculated using the results of sedimentological studies. According to this, the modern Gediz Graben is Plio-Quaternary (~ 5 Myr) in age (Koçyiğit, et al., 1999; Bozkurt and Sözbilir, 2004), and the thicknesses of the sediments accumulated during this time interval are 200–1000 m, depending on the location (Seyitoğlu and Scott, 1996; Bozkurt and Sözbilir, 2004; Çiftci and Bozkurt, 2010). A model of the surface heat flow versus sedimentation rate for the Menderes Massif is shown in Fig. 5 (the blue curve). The calculated sedimentation rates of 40–200 m Myr^{-1} correspond to surface heat flow values 5–10 mW m^{-2} below the background values. However, the actual difference depends on the location where the heat flow is measured.

Erosion and denudation (collectively termed "erosion" in this text) cause an upward motion of high temperatures at depth, resulting in higher than normal surface heat flow values (Beardsmore and Cull, 2001). Erosion histories of different horst units of the Menderes Massif were estimated by radiometric dating and fission-track techniques (Gessner et al., 2001; Ring et al., 2003; see Seyitoğlu et al., 2004, for a review). These studies show an early phase of significant cooling during the late Oligocene and the early Miocene in NMM and SMM, but minimal present-day erosion rates. On the other hand, they report rapid cooling rates for the last 5 Myr at the northern and southern edges (called detachment zones) of the CMM. The reported cooling rates of $\sim 50\,°C$ km^{-1} (Gessner et al., 2001) correspond to erosion rates of ~ 1000 m Myr^{-1} for the last 5 Myr in these areas. For a background heat flow of 85 mW m^{-2}, surface surface heat flow in these detachment zones can have values of up to 130 mW m^{-2}. However, no data points are available on these zones in the present data set.

Thermal refraction occurs near the boundaries of the horst and graben units, as a result of the thermal conductivity contrasts between the two structural units. By their low conductivity values, grabens act as thermal lenses, refracting the heat to their surroundings. Thakur et al. (2012) show the effect of thermal refraction in Dixie Valley in Nevada (North America), which has a comparable size and depth to Gediz Graben in Turkey. Compared to the background heat flow, their model shows up to 15 mW m^{-2} lower values in the

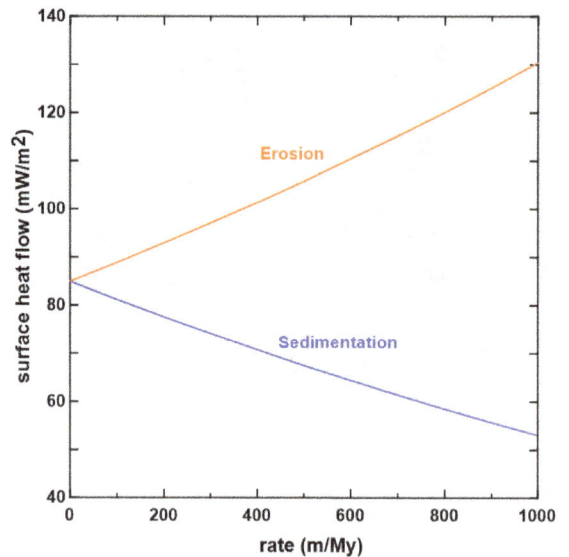

Figure 5. Changes in the surface heat flow in the Menderes Massif for increasing rates of sedimentation (blue line) and erosion (orange line). The graphs were prepared using the modules of G. Beardsmore (http://monash.edu/science/about/schools/geosciences/heatflow/). See Beardmore and Cull (2001) for the formulation of the problem. The model assumes extensional activity for the last 5 Myr, a background heat flow of 85 mW m^{-2}, and a thermal diffusivity of 1×10^{-6} m^2 s^{-1}. Note that these models use a 1-D assumption for sedimentation/erosion.

graben, and up to 30 mW m^{-2} higher values in the ranges, as a result of thermal refraction.

In the present data set, three sites on the alluvial fans (see the lithology information in Table 2) within graben units of the Menderes Massif are expected to experience a significant thermal effect of sedimentation and thermal refraction. As a result, a constant cumulative correction of 30 mW m^{-2} was applied to these points before generating the heat flow map. No correction for erosion was necessary, due to the unavailability of data on the areas of significant present-day erosion.

5.2   Heat flow versus maximum depth of seismicity in western Anatolia

The heat flow map in Fig. 4 can be compared independently by the observed maximum depth of seismicity in the region. For this purpose, high-quality (depth errors less than 2 km) hypocenter data from seismic studies are needed. Akyol et al. (2006) report hypocenter locations across a N–S profile on the Menderes Massif. Their class-A events show a maximum depth of seismicity of ~ 15 km in the central part of the CMM. Also, Aktar et al. (2007) report the results of a high-resolution survey in the southern part of the peninsular part of Izmir. The event depths go down to ~ 13 km there. Bonner et al. (2003) report a statistical correlation between heat flow and seismic depths by comparing two separate high-resolution data sets in California. According to this, seismic depths of ~ 15 km imply heat flow of ~ 100 mW m^{-2} or less, and are in general agreement with the results of this study. One of the limitations of these comparisons is that the depth of the brittle zone cannot be constrained if seismicity does not cover the entire brittle zone, so the inferred heat flow values by seismic event depths only put a maxiumum limit on the heat flow value.

5.3 Tectonic implications

The region in this study is part of the back-arc area of the Hellenic subduction zone (Jolivet et al., 2013). With the Aegean Sea, the general area is characterized by extensional tectonic activity and elevated heat flow values. The general area is also characterized by a single low P-wave velocity anomaly (Piromallo and Morelli, 2003), indicating high temperatures at the lithospheric levels. In the Aegean Sea, Erickson et al. (1977) report an average heat flow of 80 ± 22 mW m^{-2}. The average heat flow 73 ± 22 mW m^{-2} for western Anatolia reported in this study is comparable with the values measured in the Aegean Sea. Relatively lower average heat flow for western Anatolia can generally be attributed to moderate values observed in the southern Marmara region (see Fig. 5). On the other hand, parts of western Anatolia under active extension show heat flow values of 85–90 mW m^{-2}.

Heat flow values reported in this study for western Anatolia is somewhat lower than the values given by Tezcan and Turgay (1991, > 120 mW m^{-2} for the majority of the region). Based on a comparison of Cenozoic volcanism and heat flow distribution in North America, Blackwell (1978) suggests that heat flow values in excess of ~ 105 mW m^{-2} in continental regions imply partial melting in the upper crust and related silisic magmatic/volcanic activity. The absence of present-day silisic magmatic/volcanic activity in western Anatolia suggests that regional heat flow values of more than 120 mW m^{-2} are unlikely. On the other hand, some of the moderate heat flow values observed in this study (e.g., the eastern part of Çanakkale) are not observed by Tezcan and Turgay (1991). It is not clear whether these relatively low heat flow areas are due to some near-surface effects, or are representative of a crustal thermal regime. Indeed, equilibrium temperatures from deeper boreholes are needed to make more conclusive statements.

6 Conclusions

A total of 113 borehole sites were investigated in western Anatolia, and 55 of them were found to be useful for conductive gradients and/or heat flow analysis. 24 data points fall into class A/B/C quality, and can be used for heat flow mapping. The average heat flow is calculated to be 73 ± 22 mW m^{-2} in the region. These values are in agreement with the average heat flow of 80 ± 22 mW m^{-2} measured in

the Aegean Sea, as both regions form the back-arc section of the Hellenic subduction zone. The preliminary heat flow map of the region indicates elevated heat flow values (85–95 mW m^{-2}) in the coastal areas of the study region, including the western part of Çanakkale and the peninsular part of Izmir. The central part of the Menderes Massif is also characterized by elevated heat flow values, the highest values (> 100 mW m^{-2}) being near the Kula volcanic center. Moderate heat flow (55–70 mW m^{-2}) values are observed in the eastern part of Çanakkale, the central part of Balıkesir, and north of Manisa. Towards the south, moderate heat flow values are also observed in Muğla. With the present data set, it is not clear whether these moderate values represent crustal thermal conditions, or are caused by some near-surface effects.

Due to the shallow depths of the gradient measurements, and uncertainties in the thermal conductivity determinations, results of this study are preliminary. However, heat flow values with their formal errors are reported for the first time in western Anatolia using standard measurement and processing techniques.

Acknowledgements. The author is indebted to M. İlkışık for providing the field data, which enabled the realization of this study. The author would like to thank S. Ergintav and S. İnan for their support in this research while he was a visiting scholar at Marmara Research Center. This study was supported by the TÜBİTAK BIDEB 2232 program and Marmara University, Scientific Research Commission (FEN-A-100413-0127). Travel support for this project was provided by Marmara University, Scientific Research Commission (FEN-D-130313-0093). The author would also like to thank V. Ediger, B. Erkan, and C. Tapırdamaz (Marmara Research Center) for providing some useful data and software tools. The database of hot springs was obtained from the ATAG Earth Sciences Catalog (coordinator: M. C. Tapırdamaz). The manuscript was greatly improved by constructive reviews of C. Pascal, N. Balling, K. Gessner, M. Ilkışık, M. Richards, Z. Frone, R. Dingwall, and C. Mauroner. Some of the heat flow data processing was performed using the modules of G. R. Beardsmore (http://monash.edu/science/about/schools/geosciences/heatflow/).

Edited by: C. Juhlin

References

Aktar, M., Karabulut, H., Özalaybey, S., and Childs, D.: A conjugate strike-slip fault system within the extensional tectonics of western Turkey, Geophys. J. Int., 171, 1363–1375, 2007.

Akyol, N., Zhu, L., Mitchell, B. J., Sözbilir, H., and Kekovalı, K.: Crustal structure and local seismicity in western Anatolia, Geophys. J. Int., 166, 1259–1269, 2006.

Balkan, E., Erkan, K., and Şalk, M.: A statistical analysis of thermal conductivities of common rock types in western Turkey, Geothermics, submitted, 2015.

Beardsmore, G. R. and Cull, C. P.: Crustal heat flow: a guide to measurement and modeling, Cambridge Univ. Press, Cambridge, UK, 2001.

Blackwell, D. D.: Heat flow and energy loss in the Western United States, GSA Memoirs, 152, 175–208, 1978.

Blackwell, D. D.: Heat flow in the northern Basin and Range province, The Role of Heat in the Development of Energy and Mineral Resources in the Northern Basin and Range Province, Geothermal Resources Council, Special Report, 13, 81–93, 1983.

Blackwell, D. D. and Steele, J. L.: Thermal conductivity of sedimentary rocks: measurement and significance, in: Thermal history of sedimentary basins, Springer, New York, 13–36, 1989.

Bonner, J. L., Blackwell, D. D., and Herrin, E. T.: Thermal constraints on earthquake depths in California, B. Seismol. Soc. Am., 93, 2333–2354, 2003.

Bozkurt, E. and Sozbilir, H.: Tectonic evolution of the Gediz Graben: field evidence for an episodic, two-stage extension in western Turkey, Geol. Mag., 141, 63–79, 2004.

Clark, S. P.: Thermal conductivity, in: Handbook of physical constants, edited by: Clark, S., Geol. Soc. Am. Bull. Mem., 90, 587 pp., 1966.

Çiftçi, N. B. and Bozkurt, E.: Structural evolution of the Gediz Graben, SW Turkey: temporal and spatial variation of the graben basin, Basin Res., 22, 846–873, 2010.

Dilek, Y. and Altunkaynak, Ş.: Geochemical and temporal evolution of Cenozoic magmatism in western Turkey: mantle response to collision, slab break-off, and lithospheric tearing in an orogenic belt, Geological Society, London, Special Publications, 311, 213–233, 2009.

Erickson, A., Simmons, J. G., and Ryan, W. B. F.: Review of heat flow data from the Mediterranean and Aegean seas, in: International Symposium on Structural history of the Mediterranean Basins, edited by: Biju-Duval, B. and Montadert, L., Editions Technip, Paris, 263–280, 1977.

Erkan, K. and Blackwell, D. D.: A thermal test of the post-subduction tectonic evolution along the California transform margin, Geophys. Res. Lett., 35, L07309, doi:10.1029/2008GL033479, 2008.

Erkan, K. and Blackwell, D. D.: Transient thermal regimes in the Sierra Nevada and Baja California outer arcs following the cessation of Farallon subduction, J. Geophys. Res., 114, B02107, doi:10.1029/2007JB005498, 2009.

Erkan, K., Holdmann, G., Benoit, W., and Blackwell, D.: Understanding the Chena Hot Springs, Alaska, geothermal system using temperature and pressure data from exploration boreholes, Geothermics, 37, 565–585, 2008.

Fuchs, S., Schütz, F., Förster, H. J., and Förster, A.: Evaluation of common mixing models for calculating bulk thermal conductivity of sedimentary rocks: correction charts and new conversion equations, Geothermics, 47, 40–52, 2013.

Fytikas, M., Innocenti, F., Manetti, P., Peccerillo, A., Mazzuoli, R., and Villari, L.: Tertiary to Quaternary evolution of volcanism in the Aegean region, Geological Society, London, Special Publications, 17, 687–699, 1984.

Gessner, K., Ring, U., Johnson, C., Hetzel, R., Passchier, C. W., and Güngör, T.: An active bivergent rolling-hinge detachment system: Central Menderes metamorphic core complex in western Turkey, Geology, 29, 611–614, 2001.

İlkışık, O. M., Yalçýn, M. N., Sarý, C., Okay, N., Bayrak, M., Öztürk, S., Sener, Ç., Yenigün, H. M., Yemen, H., Sözen, I., and Karamanderesi, I. H.: Ege Bölgesi'nde IsıAkısıAraştırmaları, TÜBİTAK Proje No: YDABÇAG-233/G, Ankara, 1996a (in Turkish).

İlkışık, O. M., Sarý, C., Bayrak, M., Öztürk, S., Sener, Ç., Yenigün, H. M., and Karamanderesi, I. H.: Ege Bölgesinde Jeotermik Araştırmalar, TÜBİTAK, Proje No: YDABÇAG-430/G, Ankara, 1996b (in Turkish).

Jaeger, J. C.: Application of the theory of heat conduction to geothermal measurements, in: Terrestrial heat flow, edited by: Lee, H. K., American Geophysical Union, Geophysical Monograph Series, 8, 7–23, 1965.

JICA: The pre-feasibility study on the Dikili-Bergama geothermal development project, Final Report, Japan International Cooperation Agency, Tokyo, MPN 87-160, 1987.

Jolivet, L., Faccenna, C., Huet, B., Labrousse, L., Le Pourhiet, L., Lacombe, O., Lecomte, E., Burov, E., Danele, Y., Brun, J.-P., Philippon, M., Paul, A., Salaün, G., Karabulut, H., Piromallo, C., Monie, P., Gueydan, F., Okay, A. I., Oberhandsli, H., Pourteau, A., Augier, R., Gadenne, L., and Driussi, O.: Aegean tectonics: Strain localisation, slab tearing and trench retreat, Tectonophysics, 597, 1–33, 2013.

Koçyiğit, A., Yusufoğlu, H., and Bozkurt, E.: Evidence from the Gediz Graben for episodic two-stage extension in western Turkey, J. Geol. Soc. London, 156, 605–616, 1999.

Majorowicz, J. and Wybraniec, S.: New terrestrial heat flow map of Europe after regional paleoclimatic correction application, International Journal of Earth Sciences, 100, 881–887, 2011.

Öztürk, S., Karlı, R., and Destur, M.: Türkiye ısı haritası projesi raporu: MTA Derleme No 10937, 2006 (in Turkish).

Piromallo, C. and Morelli, A.: P wave tomography of the mantle under the Alpine-Mediterranean area, J. Geophys. Res., 108, 2065, doi:10.1029/2002JB001757, 2003.

Pfister, M., Ryback, L., and Şimşek, Ş.: Geothermal reconnaissance of the Marmara Sea region (NW Turkey): surface heat flow density in an area of active continental extension, Tectonophysics, 291, 77–89, 1998.

Pollack, H. N. and Huang, S.: Climate reconstruction from subsurface temperatures, Annu. Rev. Earth Pl. Sc., 28, 339–365, 2000.

Ring, U., Johnson, C., Hetzel, R., and Gessner, K.: Tectonic denudation of a Late Cretaceous–Tertiary collisional belt: regionally symmetric cooling patterns and their relation to extensional faults in the Anatolide belt of western Turkey, Geol. Mag., 140, 421–441, 2003.

Roy, R. F., Blackwell, D. D., and Decker, E. R.: Continental heat flow, in: The Nature of the Solid Earth, McGraw Hill, New York, 506–543, 1972.

Serpen, Ü., Aksoy, N., Öngür, T., and Korkmaz, E. D.: Geothermal Energy in Turkey: 2008 update, Geothermics, 38, 227–237, 2009.

Seyitoğlu, G. and Scott, B. C.: Age of the Alaşehir graben (west Turkey) and its tectonic implications, Geol. J., 31, 1–11, 1996.

Seyitoglu, G., Işık, V., and Cemen, I.: Complete Tertiary exhumation history of the Menderes Massif, western Turkey: an alternative working hypothesis, Terra Nova, 16, 358–364, 2004.

Tayanç, M., İm, U., Doğruel, M., and Karaca, M.: Climate change in Turkey for the last half century, Climatic Change, 94, 483–502, 2009.

Tester, J. W., Anderson, B., Batchelor, A., Blackwell, D., DiPippo, R., Drake, E., Garnish, J., Livesay, B., Moore, M. J., Nichols, K., Petty, S., Toksöz, M. N., Veatch Jr., R. W., Baria, R., Augustine C., Enda, M., Negraru, P., and Richards, M.: The Future of Geothermal Energy: Impact of Enhanced Geothermal Systems (EGS) on the United States in the 21st Century, Massachusetts Institute of Technology, 358 pp., 2006.

Tezcan, A. K. and Turgay, M. I.: Heat flow and temperature distribution in Turkey, edited by: Cermak, V., Haenal, R., and Zui, V., Geothermal atlas of Europe, Herman Haack Verlag, Gotha, Germany, 84–85, 1991.

Thakur, M., Blackwell, D. D., and Erkan, K.: The Regional Thermal Regime in Dixie Valley, Nevada, USA, Geothermal Resources Council Trans., 36, 59–67, 2012.

3

Changes in soil quality after converting *Pinus* to *Eucalyptus* plantations in southern China

K. Zhang[1], **H. Zheng**[1], **F. L. Chen**[1], **Z. Y. Ouyang**[1], **Y. Wang**[1], **Y. F. Wu**[2], **J. Lan**[2], **M. Fu**[2], and **X. W. Xiang**[2]

[1]State Key Laboratory of Urban and Regional Ecology, Research Center for Eco-Environmental Sciences, Chinese Academy of Sciences, Beijing 100085, China
[2]Guangxi Dongmen Forest Farm, Fusui 532108, Guangxi, China

Correspondence to: H. Zheng (zhenghua@rcees.ac.cn)

Abstract. Vegetation plays a key role in maintaining soil quality, but long-term changes in soil quality due to plant species change and successive planting are rarely reported. Using the space-for-time substitution method, adjacent plantations of *Pinus* and first, second, third and fourth generations of *Eucalyptus* in Guangxi, China were used to study changes in soil quality caused by converting *Pinus* to *Eucalyptus* and successive *Eucalyptus* planting. Soil chemical and biological properties were measured and a soil quality index was calculated using principal component analysis. Soil organic carbon, total nitrogen, alkaline hydrolytic nitrogen, microbial biomass carbon, microbial biomass nitrogen, cellobiosidase, phenol oxidase, peroxidase and acid phosphatase activities were significantly lower in the first and second generations of *Eucalyptus* plantations compared with *Pinus* plantation, but they were significantly higher in the third and fourth generations than in the first and second generations and significantly lower than in *Pinus* plantation. Soil total and available potassium were significantly lower in *Eucalyptus* plantations (1.8–$2.5\,\mathrm{g\,kg^{-1}}$ and 26–$66\,\mathrm{mg\,kg^{-1}}$) compared to the *Pinus* plantation ($14.3\,\mathrm{g\,kg^{-1}}$ and $92\,\mathrm{mg\,kg^{-1}}$), but total phosphorus was significantly higher in *Eucalyptus* plantations (0.9–$1.1\,\mathrm{g\,kg^{-1}}$) compared to the *Pinus* plantation ($0.4\,\mathrm{g\,kg^{-1}}$). As an integrated indicator, soil quality index was highest in the *Pinus* plantation (0.92) and lowest in the first and second generations of *Eucalyptus* plantations (0.24 and 0.13). Soil quality index in the third and fourth generations (0.36 and 0.38) was between that in *Pinus* plantation and in first and second generations of Eucalyptus plantations. Changing tree species, reclamation and fertilization may have contributed to the change observed in soil quality during conversion of *Pinus* to *Eucalyptus* and successive *Eucalyptus* planting. Litter retention, keeping understorey coverage, and reducing soil disturbance during logging and subsequent establishment of the next rotation should be considered to help improving soil quality.

1 Introduction

Vegetation plays a key role in soil development due to its influence on nutrient cycling, hydrological processes and soil erosion (de la Paix et al., 2013; Zhao et al., 2013). Degradation of soil quality is a serious problem (Miao et al., 2012; Zhao et al., 2013). *Eucalyptus* is an important tree species for afforestation in tropical and subtropical regions and has been introduced to many countries around the world. In southern China, millions of hectares of degraded land, cropland and natural secondary forest have been converted into *Eucalyptus* plantations and successive planting has been undertaken (Wen et al., 2009). However, due to nutrient limitations in many areas (LeBauer and Treseder, 2008) and a high demand for nutrients by *Eucalyptus* (Laclau et al., 2010), this kind of land use change may exhaust soil nutrients and decrease soil quality (Yu et al., 2000b). Inappropriate plantation management also accelerates the decline in soil quality (Yu et al., 2009). There is an urgent need to assess the effects that *Eucalyptus* planting has on soil quality since it plays a key role in sustaining forest productivity.

Soil quality includes soil physical, chemical and biological properties, as well as soil processes and their interactions (Andrews and Carroll, 2001). Many studies have focused on

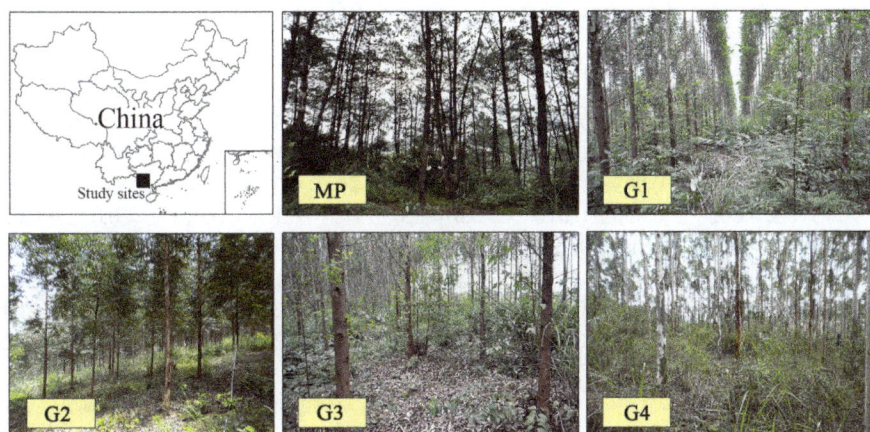

Figure 1. Study site and pictures of the *Pinus* and the successive *Eucalyptus* plantations. MP, G1, G2, G3 and G4 refer to the *Pinus* plantation and the first, second, third and fourth generation *Eucalyptus* plantations, respectively.

soil physic-chemical properties (Garay et al., 2004; Muñoz-Rojas et al., 2012; Parras-Alcántara et al., 2013), microbial communities (Wu et al., 2012) or enzyme activities (Wang et al., 2008), which only reflect some aspects of soil quality. A soil quality index (SQI) was proposed for quantifying the combined biological, chemical and physical response of soil to land use and soil/crop management practices (Andrews and Carroll, 2001; Andrews et al., 2002). It provides an intelligible and more holistic measurement of soil quality and, in recent years, the SQI has been used to assess the impacts of land use change, forest and cropland management and ecological restoration (Navas et al., 2011; Morugán-Coronado et al., 2013; Tesfahunegn, 2013). Methods used to calculate SQI include expert opinion and principal component analysis (PCA) (Andrews et al., 2002), with the latter more widely used in recent studies (Navas et al., 2011).

The ecological consequences of *Eucalyptus* planting are important and have been studied in depth. For example, it was reported that soil organic carbon, nitrogen, microbial biomass and the metabolic quotient were significantly lower in *Eucalyptus* plantations compared to natural and regenerated forests or pastures (Behera and Sahani, 2003; Sicardi et al., 2004; Araújo et al., 2010; Chen et al., 2013). It was also found that the conversion of native savanna or sugarcane fields to *Eucalyptus* plantations did not cause impacts on soil organic carbon, microbial biomass carbon or nitrogen contents (Binkley et al., 2004). Fialho and Zinn (2014) compiled paired-plot studies on how soil organic carbon stocks under native vegetation change after planting fast-growth *Eucalyptus* species in Brazil and found that *Eucalyptus* plantations on average had no net effect on soil organic carbon stocks. The results of these different studies were not consistent and the effect of successive *Eucalyptus* planting has rarely been reported.

Here we accessed the effects of converting *Pinus* to *Eucalyptus* and subsequent successive *Eucalyptus* planting on

soil quality by examining adjacent plantations of local *Pinus massoniana* Lamb. (*Pinus*) and first, second, third and fourth generations of *Eucalyptus urophylla x grandis* (*Eucalyptus*). The changes in soil quality were investigated by measuring the soil chemical and biological properties and by calculating a SQI using the PCA method for each plantation. Exhaustion of soil nutrients was the main problem considered in the studied *Eucalyptus* plantation; soil physical attributes were not considered in this study. The study aimed to test the effect on soil bio-chemical quality after (1) converting *Pinus* to *Eucalyptus* plantations and (2) successive *Eucalyptus* planting with rotation time of 5 years in southern China.

2 Materials and methods

2.1 Study area

This study was conducted at Fusui, Guangxi, China (22°14′–22°21′ N, 107°47′–107°56′ E; Fig. 1). Elevation of study sites was between 140 and 250 m above sea level. The dominant aspect of slopes was southeast, with slope below 15°. The region has a typical subtropical monsoon climate with mean annual temperature of 21.2–22.3 °C. Annual rainfall is 1100–1300 mm, concentrated during June–August. Soils in the region are mainly lateritic red earth with a profile depth of more than 80 cm. Soil pH ranged from 4 to 5 (Chen et al., 2013).

Before the 1980s, this area was dominated by *Pinus massoniana* Lamb. (*Pinus*), with 30-year rotation (clearing and new planting), used for fire wood, timber and oil production. In the 1980s, fast-growing *Eucalyptus urophylla x grandis* (*Eucalyptus*) with a 5-year rotation period began to replace *Pinus*. The first generation of *Eucalyptus* was planted with a density of about 1400 trees ha^{-1} after clear-cutting, fire clearance and full reclamation (plowed to 50 cm depth). The second generation of *Eucalyptus* was regenerated by

Table 1. Soil quality indicator scores (mean ± standard error) for soil samples taken from the *Pinus* and *Eucalyptus* plantations.

	MP	G1	G2	G3	G4
SOC	0.962 ± 0.022^a	0.206 ± 0.048^{bc}	0.064 ± 0.038^c	0.328 ± 0.011^b	0.419 ± 0.103^b
TN	0.918 ± 0.053^a	0.104 ± 0.022^d	0.022 ± 0.011^d	0.494 ± 0.022^c	0.649 ± 0.037^b
TK	0.929 ± 0.056^a	0.003 ± 0.002^b	0.020 ± 0.004^b	0.018 ± 0.002^b	0.052 ± 0.006^b
AK	0.996 ± 0.002^a	0.450 ± 0.010^c	0.614 ± 0.017^b	0.121 ± 0.013^d	0.032 ± 0.020^e
CBH	0.972 ± 0.014^a	0.816 ± 0.025^{ab}	0.252 ± 0.142^c	0.491 ± 0.016^c	0.528 ± 0.044^{bc}
PO	0.861 ± 0.080^a	0.032 ± 0.017^d	0.062 ± 0.009^{cd}	0.235 ± 0.021^{bc}	0.369 ± 0.026^b
POD	0.831 ± 0.085^a	0.039 ± 0.015^c	0.022 ± 0.020^c	0.260 ± 0.004^b	0.317 ± 0.033^b
ACP	0.938 ± 0.035^a	0.038 ± 0.022^c	0.084 ± 0.018^c	0.227 ± 0.031^b	0.239 ± 0.024^b

MP, G1, G2, G3 and G4 refer to the *Pinus* plantation and the first, second, third and fourth generation *Eucalyptus* plantations, respectively. SOC, soil organic carbon; TN, total nitrogen; TK, total potassium; AK, available potassium; CBH, cellobiosidase; PO, phenoloxidase; POD, peroxidase and ACP, acid phosphatase. Means followed by the same letter (a–d) within a row are not significantly different at $p < 0.05$.

sprouts. The third generation of *Eucalyptus* was planted with seedlings after strip reclamation (50 cm depth). Finally, the fourth generation was regenerated by sprouts again. Leaves, branch and bark litter were kept in the plantation during the plant growth period. However, at harvest time, most branch litter was removed and burned before the next rotation, and soil erosion happened easily due to the lack of soil protection during the crop transition periods.

Before *Eucalyptus* planting, base fertilizer (500 g seedling^{-1}, N : P : K = 10 : 15 : 5) was added into a 20 cm depth soil hole under each new *Eucalyptus* tree and covered with soil. Then, 6, 12 and 24 months after planting, 250, 500 and 500 g, respectively, of fertilizer (N : P : K = 15 : 10 : 8) per tree was separately added in soil holes that were 30 cm away from each tree. Herbicide (glyphosate) was applied once a year during the first 3 years after *Eucalyptus* planting and consequently the coverage of understory plants was less than 50 %. However, the understory coverage increased gradually during the fourth and fifth years after *Eucalyptus* planting. During sampling time, tree, shrub and grass cover in the *Pinus* plantation was about 60, 25 and 70 %, respectively. Tree, shrub and grass cover in the *Eucalyptus* plantations was similar, about 40, 10 and 45 %, respectively. The litter layer was about 3 cm in the *Pinus* plantation and 1 cm deep in the *Eucalyptus* plantations.

2.2 Experimental design and sampling

The adjacent plantations of local *Pinus* and first, second, third and fourth generations of *Eucalyptus* at the Dongmen Forestry Farm were selected (Fig. 1) to represent *Pinus* planting and first, second, third and fourth generations of successive *Eucalyptus* planting. Three 20 m × 20 m plots were randomly selected and marked out at each plantation site, with an average distance of 20 m between plots. During soil sampling, the ages of *Pinus* and *Eucalyptus* was 20 and 3 years, respectively.

In October 2010, soil from the top 0–10 cm layer was collected from the study sites. Ten soil cores were collected at randomly selected points at each plot using a 3.6 cm diameter soil auger and mixed together as a composite sample. Three composite samples were collected at each point. Stones and roots were removed from the soil samples by hand and soil samples were sieved through 2 mm sieves. Soil samples were stored at 4 °C for soil microbial biomass and enzyme activity analyses, or air dried for chemical analyses.

2.3 Soil chemical and biological analyses

Soil water content was determined gravimetrically after oven-drying at 105 °C for 24 h in order to correct sample weights in biochemical property measurements. Soil pH was measured in deionized water (1 : 2.5 w/v) using a Delta 320 pH-meter (Mettler-Toledo Instruments (Shanghai) Co., Ltd.). Soil organic carbon (SOC) was measured using the Walkley and Black wet oxidation method as outlined in Bao (2000). Soil total nitrogen (TN) was determined by combustion in a Vario EL III Elemental Analyser (Elementar Analysensysteme GmbH, Germany). Soil alkaline hydrolytic nitrogen (AN) was measured according to Bao (2000). For the assessment of AN, 1 g of air-dried soil was incubated in 5 mL sodium hydroxide solution (1.2 M) in the outside ring of an airtight Conway diffusion cell; 3 mL boric acid (0.3 M) were added to the inner well at 40 °C for 24 h. With methyl red-bromocresol green as indicator, 0.01 M hydrochloric acid was used to titrate ammonia absorbed in the boric acid. For total phosphorus (TP) and total potassium (TK), air-dried soil samples were digested using 18.4 M sulfuric acid (1 : 10 w/v) and 12.7 M perchloric acid at 275 °C for 6 h, TP and TK were measured using a Prodigy High Dispersion ICP-OES (Teledyne Technologies Incorporated, USA). Available phosphorus (AP) was measured after extraction with 0.05 N sulfuric and 0.025 N hydrochloric acids (1 : 5 w/v), with shaking during 5 min. Available potassium (AK) was measured after extraction with 1 N ammonium acetate solution ($1:10\,w/v$), shaken for 30 min (Bao, 2000). AP and AK were measured using the Prodigy High Dispersion ICP-OES.

Soil microbial biomass carbon and nitrogen (MBC and MBN, respectively) were estimated using the chloroform fumigation extraction method (Vance et al., 1987). Soil β-glucosidase (BG), phenol oxidase (PO), peroxidase (POD) and acid phosphatase (ACP) activities were measured according to Waldrop et al. (2000). Soil protease (PRO) activity was estimated according to Ladd and Butler (1972). Soil urease (URE) activity was assessed according to Kandeler and Gerber (1988). Soil cellobiosidase (CBH) activity was measured using the fluorimetric method according to Saiya-Cork et al. (2002).

2.4 Calculation of the soil quality index

The SQI was calculated according to Andrews and Carroll (2001). Three steps were involved in the elaboration of this quality index: (i) definition of a minimum data set (MDS), (ii) assignment of a score to each indicator by linear scoring functions and (iii) data integration into an index.

Three steps were used to identify the MDS in our study. (1) Data screening: one-way analysis of variance was performed for soil chemical and biological properties; only variables with significant differences between treatments ($p < 0.05$) were chosen for the next step. (2) Selection of representative variables: PCA was performed on the variables chosen from step (1); only principal components (PCs) that explained at least 5 % of the variation in the data up to 85 % of the cumulative variation were examined, within each PC, only weighted factors with absolute values within 10 % of the highest weight were retained for the MDS. (3) Redundancy reduction: multivariate correlation coefficients were used to determine the strength of the relationships among variables. Highly correlated variables (correlation coefficient > 0.70) were considered redundant and candidates for elimination from the data set. In order to choose variables within well-correlated groups, we summed the absolute values of the correlation coefficients for these variables and assumed that the variable with the highest correlation sum best represented the group. The choice among well-correlated variables was also based on the published references and expert opinion about the soils and sites (Xu, 2000; Wang et al., 2008; Yu et al., 2009). Any uncorrelated, highly weighted variables were considered important and retained in the MDS.

Linear scoring was used in our study. Indicators were ranked in ascending or descending order depending on whether a higher value was considered "good" or "bad" in terms of soil function. In our study, all variables were "more is better". The linear scoring function used for converting measured values to scored values is as follows (Zheng et al., 2005):

$$S_{ij} = \frac{V_{ij} - V_{i\min}}{V_{i\max} - V_{i\min}}$$

where S_{ij} is the score of soil variable i of sample j, V_{ij} is the observed variable value of sample j, $V_{i\max}$ is the highest value of variable i, and $V_{i\min}$ is the lowest value of variable i.

The scores of the indicators (Table 1) were integrated into a SQI according to Andrews et al. (2002) as follows:

$$SQI = \sum_{i=1}^{n} \frac{S_i}{n},$$

where S_i is the score assigned to indicator i, and n is the number of indicators included in the MDS.

2.5 Statistical analyses

One-way variance analyses were used to test the significant differences in soil chemical and biological properties and SQIs among these treatments. Normality of the data was tested using the Kolmogorov–Smirnov test and the soil indicators obeyed standard normal distribution. Tukey's test was used for multiple comparison analysis. All statistical analyses were performed using SPSS 10.0 (SPSS Inc., Chicago, IL, USA) and SigmaPlot for Windows version 11.0 (Systa Software Inc.). The three plots established per site do not constitute true replicates, because they are located within the same area of the five plantations (Pinus, first, second, third and fourth generations Eucalyptus plantations). However, these plantation stands occur in similar topographic conditions and soil parent material, and were similar in planting history (Pinus plantation before the 1980s), which allowed us to consider them as different treatments.

3 Results

3.1 Soil chemical properties

SOC, TN and AN ranges at our study sites were 10.8–26.9 g kg^{-1}, 0.9–1.6 g kg^{-1} and 41.4–60.3 mg kg^{-1}, respectively. They were high in the Pinus plantation, low in the first and second generations of Eucalyptus plantations, and moderate in the third and fourth generations (Fig. 2a–c). SOC and TN in the fourth generation of Eucalyptus plantations were significantly lower than in the Pinus plantation, but significantly higher than in the second generation of Eucalyptus plantations. Soil AN in the fourth generation of Eucalyptus was significantly higher than in the second generation, but not different from that in the Pinus plantation.

Soil TP content was significantly lower in the Pinus plantation (0.41 g kg^{-1}) than in the Eucalyptus plantations (0.90–1.07 g kg^{-1}). In the Eucalyptus plantations, soil TP content was significantly lower in the second generation than in the third and fourth generations (Fig. 2d). Soil AP content was 2.98 mg kg^{-1} in the Pinus plantation, which is lower than 5.03 and 5.14 mg kg^{-1} in the first and second generations of Eucalyptus plantations, respectively, a little higher than 2.56

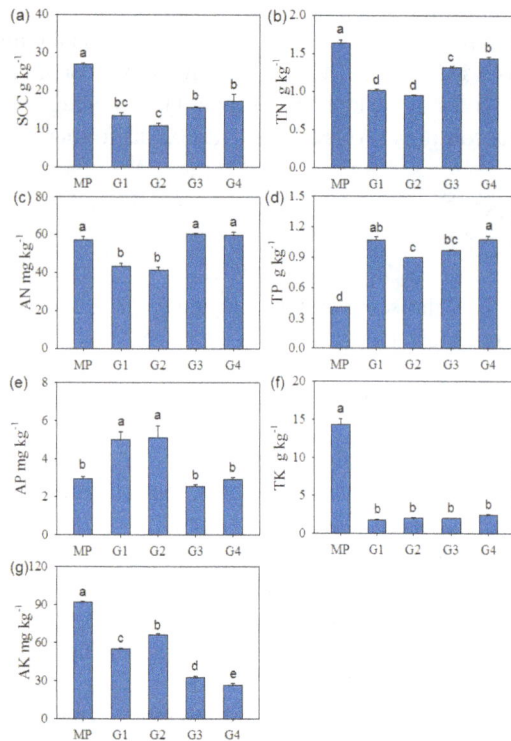

Figure 2. Soil chemical properties in the *Pinus* and the successive *Eucalyptus* plantations. MP, G1, G2, G3 and G4 refer to the *Pinus* plantation and the first, second, third and fourth generation *Eucalyptus* plantations, respectively. SOC, soil organic carbon; TN, total nitrogen; TP, total phosphorus; TK, total potassium; AN, alkaline hydrolytic nitrogen; AP, available phosphorus and AK, available potassium. Soil chemical properties with the same letter are not significantly different at $p < 0.05$.

and 2.93 mg kg^{-1} in the third and fourth generations, respectively (Fig. 2e).

Soil TK and AK ranged from 1.8 to 6.3 g kg^{-1} and from 24 to 92 mg kg^{-1}, respectively. TK content was significantly lower in the *Eucalyptus* plantations than in the *Pinus* plantation, without significant differences between *Eucalyptus* generations (Fig. 2f). AK content was significantly higher in the *Pinus* plantation compared with in *Eucalyptus* plantations (Fig. 2g).

3.2   Soil microbial biomass carbon and nitrogen

Soil MBC and MBN contents ranged from 278 to 673 mg kg^{-1} and from 7 to 35 mg kg^{-1}, respectively. Soil MBC content was significantly lower in the first and second generation of *Eucalyptus* plantations than in the third and fourth generations (Fig. 3a). Soil MBN content was significantly lower in the second generation (Fig. 3b). The MBC / MBN ratio was significantly higher in the second generation *Eucalyptus* plantation than in other plantations (Fig. 3c).

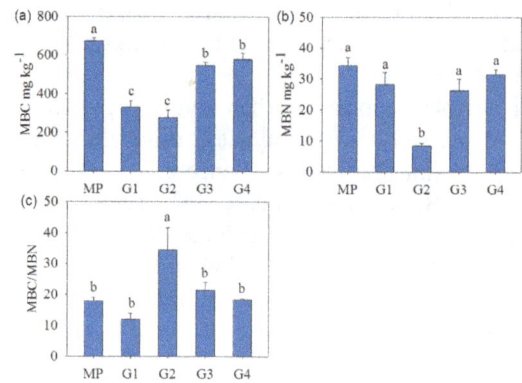

Figure 3. Soil microbial biomass carbon and nitrogen in the *Pinus* and the successive *Eucalyptus* plantations. MP, G1, G2, G3 and G4 refer to the *Pinus* plantation and the first, second, third and fourth generation *Eucalyptus* plantations, respectively. MBC, microbial biomass carbon and MBN, microbial biomass nitrogen. Microbial indicators with the same letter are not significantly different at $p < 0.05$.

3.3   Soil enzyme activities

BG activity was significantly higher in the *Pinus* plantation compared with in *Eucalyptus* plantations (Fig. 4a). Soil CBH, PO, POD, PRO and ACP activities were highest in the *Pinus* plantation, lowest in the first or second generations of *Eucalyptus* plantations and between the middle in the third or fourth generations of *Eucalyptus* plantations (Fig. 4b–f). Soil URE activity was significantly higher in the *Pinus* plantation than in the *Eucalyptus* plantations, with no significant differences among *Eucalyptus* plantations (Fig. 4g).

3.4   Changes in soil quality index

SQI was highest in the *Pinus* plantation (0.92) and lowest in the second generation of the *Eucalyptus* plantations (0.13). In the third and fourth *Eucalyptus* generations, SQI was significantly higher than in the second *Eucalyptus* generation (0.36 and 0.38, respectively, compared to 0.13), but still lower than in the *Pinus* plantation (0.92, Fig. 5).

4   Discussion

4.1   Decrease in soil quality in the first and second generations of Eucalyptus planting

Our study found that SOC, TN, AN, TK, AK, MBC, MBN and enzyme activities significantly decreased in the first and second generation *Eucalyptus* plantations after converting *Pinus* to *Eucalyptus* (Figs. 2–4). This is consistent with many studies which reported that SOC, TN, MBC, carbon metabolic activity and metabolic quotient were significantly lower in *Eucalyptus* plantations than in natural and regenerated forest or pastures (Behera and Sahani, 2003; Sicardi

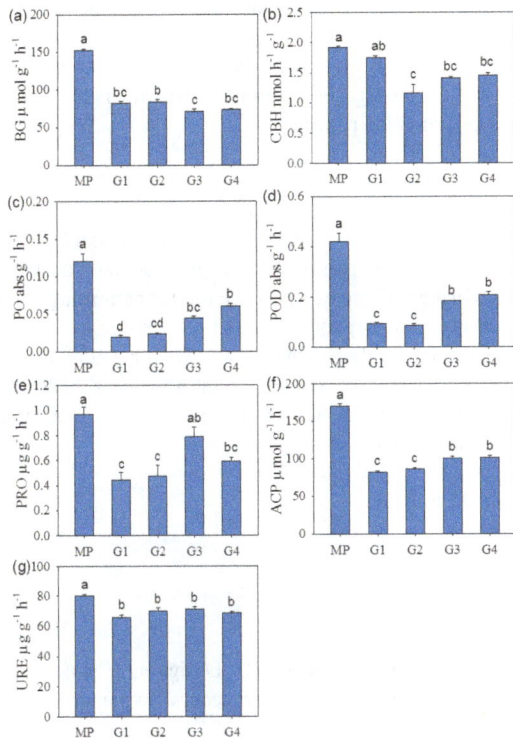

Figure 4. Soil enzyme activities in the *Pinus* and the successive *Eucalyptus* plantations. MP, G1, G2, G3 and G4 refer to the *Pinus* plantation and the first, second, third and fourth generation *Eucalyptus* plantations, respectively. BG, β-glucosidase; CBH, cellobiosidase; PO, phenoloxidase; POD, peroxidase; PRO, protease; URE, urease and ACP, acid phosphatase. Enzyme activities with the same letter are not significantly different at $p < 0.05$.

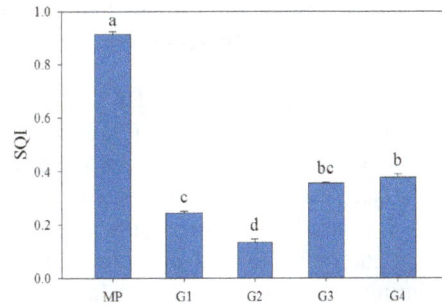

Figure 5. Soil quality index in the *Pinus* and the successive *Eucalyptus* plantations. MP, G1, G2, G3 and G4 refer to the *Pinus* plantation and the first, second, third and fourth generation *Eucalyptus* plantations, respectively. SQI, soil quality index. SQIs with the same letter are not significantly different at $p < 0.05$.

et al., 2004; Chen et al., 2013). However, in our study, soil TP was significantly higher in *Eucalyptus* plantation, which might be caused by fertilization in *Eucalyptus* plantations and the low mobility of phosphorus. The decreased soil quality in the first and second generations of *Eucalyptus* plantations after converting *Pinus* to *Eucalyptus* plantation may have been caused by tree species change, full reclamation, herbicide application, clear-cutting and short rotation.

Eucalyptus has a fast growth rate and a strong nutrient absorption capacity, which means that they absorb soil nutrients and store them in the biomass efficiently (Laclau et al., 2005), resulting in lower nutrient contents in the soil (Fig. 2). Tesfaye et al. (2014) evaluated seven tree species for fuelwood and soil restoration and found that *Eucalyptus* presented the highest growth rates and biomass, but depleted soil nitrogen. The depletion of nitrogen in soils under *Eucalyptus* plantations was also found by Wen et al. (2009). The lower soil nutrient contents in the second generation *Eucalyptus* plantation may lead to lower soil microbial biomass and enzyme activities and a higher MBC / MBN ratio (Figs. 3, 4).

Full reclamation normally impacts the soil structure, exposes more aggregates and accelerates organic matter de-

composition (Zinn et al., 2002), which would lead to lower soil organic matter contents in the *Eucalyptus* plantations. Understorey vegetation may provide a better microcosm for microorganisms and alleviate rainfall-induced erosion and nutrient leaching (Yu et al., 2000a). In our study, herbaceous vegetation was treated with herbicide during the first 3 years of the *Eucalyptus* planting exposing the uncovered soils to erosion.

At harvest, clear-cutting totally destroyed plant coverage and the subsequent fire clearance burned all the residues and the litter layer, which resulted in bare ground prior to *Eucalyptus* planting. Soil erosion and nutrient leaching can occur during heavy rainfall if there is no protection from plant and litter layers (Yu et al., 2000a). Fire also removes significant amounts of organic matter (Certini, 2005) and nutrients are lost through volatilization (Fisher and Binkley, 2000). Soil microbial biomass decreases during fire because of increased decay and death of heat-sensitive microbes or through alterations of soil physic-chemical properties (De Marco et al., 2005). These could have been the causes for the bio-chemical depletion observed in soils that underwent *Pinus* to *Eucalyptus* conversion.

Eucalyptus cultivation rotations are very short (only 5 years) compared to *Pinus* (30 years), which means that there is frequent biomass loss in *Eucalyptus* plantations. Furthermore, short rotations have an impact on the nutrient absorption patterns and the soil nutrient returns. During the early stages of plant life, nutrient absorption is high and litter production is small, but as the plant gets older nutrient absorption decreases and litter production increases (Xu, 2000). The 5-year rotation of *Eucalyptus* plantations and frequent cutting led to severe soil nutrient loss.

4.2 Increase in soil quality in the third and fourth generations of Eucalyptus planting

Our results showed that after the sudden decline following converting *Pinus* to *Eucalyptus*, SOC, TN, MBC, CBH, PO, POD and ACP activities recovered in the third and fourth generations, albeit lower than that in the native *Pinus* plantation (Figs. 2–4). This is inconsistent with Yu et al. (2000b), who recorded that soil physical and chemical properties decreased during successive *Eucalyptus* planting. However, Lima et al. (2006) recorded an increasing soil organic matter trend as the plantation aged after *Eucalyptus* afforestation of degraded pastures. The inconsistency between these studies might be caused by different climate, soil properties, *Eucalyptus* species and management practices amongst the study sites. The recovery of soil quality in the third and fourth generation *Eucalyptus* plantations may be attributed to reduced soil disturbance and fertilizer application.

Soil disturbance caused by strip reclamation in the third generation *Eucalyptus* plantation was much smaller than that due to full reclamation during the conversion from *Pinus* to *Eucalyptus*, which may have led to a reduction in soil organic matter decomposition, soil erosion and nutrient leaching. The decreased soil erosion risk, nutrient leaching and organic matter decomposition rate helped in improving soil organic matter accumulation and nutrient levels (Yu et al., 2000a).

Fertilization could increase plant growth and litter input (Madeira et al., 1995), so improving soil quality. If soil nutrient inputs through fertilization and litter fall equal the output caused by erosion, plant absorption and leaching, then soil quality would reach a steady status (Xu, 2000). Our results suggest a recovery in soil quality after the third and fourth generations of *Eucalyptus* planting (Fig. 5). To better understand the impacts of successive *Eucalyptus* planting on soil quality, evaluation of more generations of *Eucalyptus* plantations is needed.

5 Conclusions

Findings from our study suggest that soil quality decreased significantly in the first and second generation *Eucalyptus* plantations after converting *Pinus* to *Eucalyptus* plantations, partially recovering in the following third and fourth generations. Changes in tree species, reclamation, herbicide application and long-term fertilization might have contributed to the changes observed in the soil quality during successive *Eucalyptus* planting. Our results emphasize the importance of long-term soil quality monitoring. Improving management practices, such as maintenance of litter and herbaceous cover and reduction of soil disturbance during logging and subsequent establishment of the next planting rotation, should be considered to maintain soil quality.

Author contributions. H. Zheng and Z. Y. Ouyang designed the experiment and K. Zhang, F. L. Chen, and Y. Wang carried it out. Y. F. Wu, J. Lan, M. Fu and X. W. Xiang helped collecting soil samples and gave suggestion about the experiment. K. Zhang and H. Zheng prepared the paper.

Acknowledgements. We thank Y. F. Li, X. G. Pan and Q. B. Wu for helping us with the experimental investigation and soil sampling. We thank Guangxi Dongmen Forest Farm for permitting us to conduct the study in their plantations and for helping us collect data. This research was funded by the Knowledge Innovation Program of the Chinese Academy of Science (grant no. KZCX2-EW-QN406) and the National Natural Science Foundation of China (grant no. 31170425, 40871130).

Edited by: A. Jordán

References

Andrews, S. S. and Carroll, C. R.: Designing a soil quality assessment tool for sustainable agroecosystem management, Ecol. Appl., 11, 1573–1585, 2001.

Andrews, S. S., Karlen, D. L., and Mitchell, J. P.: A comparison of soil quality indexing methods for vegetable production systems in Northern California, Agr. Ecosyst. Environ., 90, 25–45, 2002.

Araújo, A. S. F., Silva, E. F. L., Nunes, L. A. P. L., and Carneiro, R.F.V.: The effect of converting tropical native savanna to Eucalyptus grandis forest on soil microbial biomass, Land Degrad. Dev., 21, 540–545, 2010.

Bao, S. D.: Soil Agro-chemical Analysis, China Agriculture Press, Beijing, 2000 (in Chinese).

Behera, N. and Sahani, U.: Soil microbial biomass and activity in response to Eucalyptus plantation and natural regeneration on tropical soil, Forest Ecol. Manag., 174, 1–11, 2003.

Binkley, D., Kaye, J., Barry, M., and Ryan, M. G.: First-rotation changes in soil carbon and nitrogen in a Eucalyptus plantation in Hawaii, Soil Sci. Soc. Am. J., 68, 1713–1719, 2004.

Certini, G.: Effects of fire on properties of forest soils: a review, Oecologia, 143, 1–10, 2005.

Chen, F. L., Zheng, H., Zhang, K., Ouyang, Z. Y., Lan, J., Li, H. L., and Shi, Q.: Changes in soil microbial community structure and metabolic activity following conversion from native Pinus massoniana plantations to exotic Eucalyptus plantations, Forest Ecol. Manag., 291, 65–72, 2013.

de la Paix, M. J., Lanhai, L., Xi, C., Ahmed, S., and Varenyam, A.: Soil degradation and altered flood risk as a consequence of deforestation, Land Degrad. Dev., 24, 478–485, 2013.

De Marco, A., Gentile, A. E., Arena, C., and De Santo, A. V.: Organic matter, nutrient content and biological activity in burned and unburned soils of a Mediterranean maquis area of southern Italy, Int. J. Wildland Fire, 14, 365–377, 2005.

Fialho, R. C. and Zinn, Y. L.: Changes in soil organic carbon under Eucalyptus plantations in Brazil: a comparative analysis, Land Degrad. Dev., 25, 428–437, 2014.

Fisher, R. F. and Binkley, D.: Fire effects, 241–261, edited by: Fisher, R. F. and Binkley, D., in : Ecology and management of forest soils, 3rd edition, John Wiley & Sons Inc, New York, 2000.

Garay, I., Pellens, R., Kindel, A., Barros, E., and Franco, A. A.: Evaluation of soil conditions in fast-growing plantations of Eucalyptus grandis and Acacia mangium in Brazil: a contribution to the study of sustainable land use, Appl. Soil Ecol., 27, 177–187, 2004.

Kandeler, E. and Gerber, H.: Short-term assay of soil urease activity using colorimetric determination of ammonium, Biol. Fert. Soils, 6, 68–72, 1988.

Laclau, J. P., Ranger, J., Deleporte, P., Nouvellon, Y., Saint-Andre, L., Marlet, S., and Bouillet, J. P.: Nutrient cycling in a clonal stand of Eucalyptus and an adjacent savanna ecosystem in Congo 3, Input-output budgets and consequences for the sustainability of the plantations, Forest Ecol. Manag., 210, 375–391, 2005.

Laclau, J. P., Ranger, J., Goncalves, J. L. D., Maquere, V., Krusche, A. V., M'Bou, A. T., Nouvellon, Y., Saint-Andre, L., Bouillet, J. P., Piccolo, M. D., and Deleporte, P.: Biogeochemical cycles of nutrients in tropical Eucalyptus plantations: Main features shown by intensive monitoring in Congo and Brazil, Forest Ecol. Manag., 259, 1771–1785, 2010.

Ladd, J. N. and Butler, J. H. A.: Short-term assays of soil proteolytic enzyme activities using proteins and dipeptide derivatives as substrates, Soil Biol. Biochem., 4, 19–30, 1972.

LeBauer, D. S. and Treseder, K. K.: Nitrogen limitation of net primary productivity in terrestrial ecosystems is globally distributed, Ecology, 89, 371–379, 2008.

Lima, A. M. N., Silva, I. R., Neves, J. C. L., Novais, R. F., Barros, N. F., Mendonca, E. S., Smyth, T. J., Moreira, M. S., and Leite, F. P.: Soil organic carbon dynamics following afforestation of degraded pastures with Eucalyptus in southeastern Brazil, Forest Ecol. Manag., 235, 219–231, 2006.

Madeira, M., Araújo, M. C., and Pereira, J. S.: Effects of water and nutrient supply on amount and on nutrient concentration of litterfall and forest floor litter in Eucalyptus globulus plantations, Plant Soil, 168, 287–295, 1995.

Miao, C., Yang, L., Chen, X., and Gao, Y.: The vegetation cover dynamics (1982–2006) in different erosion regions of the Yellow River basin, China, Land Degrad. Dev., 23, 62–71, 2012.

Morugán-Coronado, A., Arcenegui, V., García-Orenes, F., Mataix-Solera, J., and Mataix-Beneyto, J.: Application of soil quality indices to assess the status of agricultural soils irrigated with treated wastewaters, Solid Earth, 4, 119–127, doi:10.5194/se-4-119-2013, 2013.

Muñoz-Rojas, M., Jordán, A., Zavala, L. M., De la Rosa, D., Abd-Elmabod, S. K., and Anaya-Romero, M.: Organic carbon stocks in Mediterranean soil types under different land uses (Southern Spain), Solid Earth, 3, 375–386, doi:10.5194/se-3-375-2012, 2012.

Navas, M., Benito, M., Rodríguez, I., and Masaguer, A.: Effect of five forage legume covers on soil quality at the Eastern plains of Venezuela, Appl. Soil Ecol., 49, 242–249, 2011.

Parras-Alcántara, L., Martín-Carrillo, M., and Lozano-García, B.: Impacts of land use change in soil carbon and nitrogen in a Mediterranean agricultural area (Southern Spain), Solid Earth, 4, 167–177, doi:10.5194/se-4-167-2013, 2013.

Saiya-Cork, K. R., Sinsabaugh, R. L., and Zak, D. R.: The effects of long term nitrogen deposition on extracellular enzyme activity in an Acer saccharum forest soil, Soil Biol. Biochem., 34, 1309–1315, 2002.

Sicardi, M., García-Préchacb, F., and Frionic, L.: Soil microbial indicators sensitive to land use conversion from pastures to commercial Eucalyptus grandis (Hill ex Maiden) plantations in Uruguay, Appl. Soil Ecol., 27, 125–133, 2004.

Tesfahunegn, G. B.: Soil quality indicators response to land use and soil management systems in Northern Ethiopia's catchment, Land Degrad. Dev., doi:10.1002/ldr.2245, 2013

Tesfaye, M. A., Bravo-Oviedo, A., Bravo, F., Kidane, B., Bekele, K., and Sertse, D.: Selection of tree species and soil management for simultaneous fuelwood production and soil rehabilitation in the Ethiopian central highlands, Land Degrad. Dev., doi:10.1002/ldr.2268, 2014.

Vance, E. D., Brookes, P. C., and Jenkinson, D. S.: An extraction method for measuring soil microbial biomass C, Soil Biol. Biochem., 19, 703–707, 1987.

Waldrop, M. P., Balser, T. C., and Firestone, M. K.: Linking microbial community composition to function in a tropical soil, Soil Biol. Biochem., 32, 1837–1846, 2000.

Wang, Q. K., Wang, S. L., and Liu, Y.: Responses to N and P fertilization in a young Eucalyptus dunnii plantation: Microbial properties, enzyme activities and dissolved organic matter, Appl. Soil Ecol., 40, 484–490, 2008.

Wen, Y. G., Zheng, X., Li, M. C., Xu, H. G., Liang, H. W., Huang, C. B., Zhu, H. G., and He, B.: Effects of eucalypt plantation replacing Masson pine forest on soil physiochemical properties in Guangxi, southern China, Journal of Beijing Forestry University, 31, 145–148, 2009 (in Chinese).

Wu, J. P., Liu, Z. F., Sun, Y. X., Zhou, L. X., Lin, Y. B., and Fu, S. L.: Introduced Eucalyptus urophylla plantations change the composition of the soil microbial community in subtropical China, Land Degrad. Dev., 24, 400–406, 2012.

Xu, D. P.: An approach to nutrient balance in tropical and south subtropical short rotation plantations, 27–35, edited by: Yu, X. B., in: Studies on long-term productivity management of Eucalyptus plantation, China Forestry Publishing House, Beijing, 2000 (in Chinese).

Yu, F. K., Huang, X. H., Wang, K. Q., and Duang, C. Q.: An overview of ecological degradation and restoration of Eucalyptus plantations, Chinese Journal of Eco-Agriculture, 17, 393–398, 2009 (in Chinese).

Yu, X. B., Chen, Q. B., Wang, S. M., and Mo, X. Y.: Research on land degradation in forest plantation and preventive strategies, 1–7, edited by: Yu, X. B., in: Studies on long-term productivity management of Eucalyptus plantation, China Forestry Publishing House, Beijing, 2000a (in Chinese).

Yu, X. B., Yang, G. Q., and Li, S. K.: Study on the changes of soil properties in Eucalyptus plantation under continuous-planting practices, 94–103, edited by: Yu, X. B., in: Studies on long-term productivity management of Eucalyptus plantation, China Forestry Publishing House, Beijing, 2000b (in Chinese).

Zhao, G., Mu, X., Wen, Z., Wang, F., and Gao, P.: Soil erosion, conservation, and eco-environment changes in the Loess Plateau of China, Land Degrad. Dev., 24, 499–510, 2013.

Zheng, H., Ouyang, Z. Y., Wang, X. K., Miao, H., Zhao, T. Q., and Peng, T. B.: How different reforestation approaches affect red soil properties in Southern China, Land Degrad. Dev., 16, 387–396, 2005.

Zinn, Y. L., Resck, D. V. S., and Silva, J.E. da: Soil organic carbon as affected by afforestation with Eucalyptus and Pinus in the Cerrado region of Brazil, Forest Ecol. Manag., 166, 285–294, 2002.

Soil organic carbon along an altitudinal gradient in the Despeñaperros Natural Park, southern Spain

L. Parras-Alcántara, B. Lozano-García, and A. Galán-Espejo

Department of Agricultural Chemistry and Soil Science, Faculty of Science, Agrifood Campus of International Excellence – ceiA3, University of Córdoba, 14071 Córdoba, Spain

Correspondence to: L. Parras-Alcántara (qe1paall@uco.es)

Abstract. Soil organic carbon (SOC) is extremely important in the global carbon (C) cycle as C sequestration in non-disturbed soil ecosystems can be a C sink and mitigate greenhouse-gas-driven climate change. Soil organic carbon changes in space and time are relevant to understand the soil system and its role in the C cycle. This is why the influence of topographic position on SOC should be studied. Seven topographic positions from a toposequence between 607 and 1168 m were analyzed in the Despeñaperros Natural Park (Jaén, SW Spain). Depending on soil depth, one to three control sections (0–25, 25–50 and 75 cm) were sampled at each site. The SOC content in studied soils was below $30\,\mathrm{g\,kg^{-1}}$ and strongly decreases with depth. These results were related to the gravel content and to the bulk density. The SOC content from the topsoil (0–25 cm) varied largely through the altitudinal gradient ranging between 27.3 and $39.9\,\mathrm{g\,kg^{-1}}$. The SOC stock (SOCS) varied between 53.8 and $158.0\,\mathrm{Mg\,ha^{-1}}$ in the studied area, which had been clearly conditioned by the topographic position. Therefore, results suggest that elevation should be included in SOCS models and estimations at local and regional scales.

1 Introduction

Soils are an important carbon reservoir (Barua and Haque, 2013; Yan-Gui et al., 2013). In fact, the primary terrestrial pool of organic carbon (OC) is soil, which accounts for more than 71 % of the Earth's terrestrial OC pool (Lal, 2010). In addition, soils have the ability to store C for a long time (over the last 5000 years) (Brevik and Homburg, 2004). Soils play a crucial role in the overall C cycle, and small changes in the soil organic carbon stock (SOCS) could significantly affect atmospheric carbon dioxide (CO_2) concentrations, and through that global climate change. Within the C cycle, soils can be a source of greenhouse gases through CO_2 and methane (CH_4) emissions, or can be a sink for atmospheric CO_2 through C sequestration in soil organic matter (OM) (Breuning-Madsen et al., 2009; Brevik, 2012).

Climate, soil use and soil management affect OC variability, particularly in soils under Mediterranean type of climate, characterized by low OC content, weak structure and readily degradable soils (Hernanz et al., 2002). In temperate climates, recent studies show differences in C sequestration rates in soils depending on use and management (Muñoz-Rojas et al., 2012a, b), climate and mineralogical composition (Wang et al., 2010), texture, slope and elevation (Hontoria et al., 2004), and tillage intensity and no-till duration (Umakant et al., 2010). Soil conservation strategies are being seen as a strategy to increase soil OM content (Barbera et al., 2012; Batjes et al., 2014; Jaiarree et al., 2014; Srinivasarao et al., 2014; Fialho and Zinn, 2014).

Several studies have been carried out to estimate differences in soil organic carbon (SOC) dynamics in relation to soil properties, land uses and climate (Eshetu et al., 2004; Lemenih and Itanna, 2004; Muñoz-Rojas et al., 2013). Although the impact of topographic position on soil properties on SOC content is widely recognized (Venterea et al., 2003; Fu et al., 2004; Brevik, 2013), relatively few studies have been conducted to examine the role of topographic position (Fernández-Romero et al., 2014; Lozano-García et al., 2014).

The spatial variation of soil properties may also be significantly influenced by aspect (which may induce microclimate variations), physiography, parent material, and vegetation (López-Vicente et al., 2009; Brevik, 2013; Ashley et al., 2014; Bakhshandeh et al., 2014; Dingil et al., 2014; Gebrelibanos et al., 2014; Kirkpatrick et al., 2014). Ovales and Collins (1986) evaluated soil variability due to pedogenic processes across landscapes in contrasting climatic environments and concluded that topographic position and variations in soil properties were significantly related. McKenzie and Austin (1993) and Gessler et al. (2000) found that variations of some soil properties could be related to the slope steepness, length, curvature and the relative location within a toposequence. Both studies suggest that the assessment of the hillslope sequence helps to understand variations of soil properties in order to establish relationships among specific topographic positions and soil properties. Asadi et al. (2012) found that the integrated effect of topography and land use determined soil properties. Topography is a relevant factor controlling soil erosion processes through the redistribution of soil particles and soil OM (Cerdà and García Fayos, 1997; Ziadat and Taimeh, 2013).

The topographic factor has been traditionally included in the study of the spatial distribution of soil properties (Fernández-Calviño et al., 2013; Haregeweyn et al., 2013; Ozgoz et al., 2013; Wang and Shao, 2013). Over time, many researchers have quantified the relationships between topographic parameters and soil properties such as soil OM and physical properties such as particle size distribution, bulk density and depth to specific horizon boundaries (McKenzie and Austin, 1993; Gessler et al., 1995, 2000; Pachepsky et al., 2001; Ziadat, 2005). Soil OM content has been negatively correlated with the topographic gradient (Ruhe and Walker, 1968) and slope gradient (Nizeyimana and Bicki, 1992). However, quantitative relationships between soil topography and soil physical–chemical properties are not well established for a wide range of environments (Hattar et al., 2010).

Research along altitudinal gradients has shed light on the effects of climate on soil properties. Ruiz-Sinoga et al. (2012) found a strong relationship between soil OM and elevation, which was due to reduced decomposition rates with lower temperatures. High erosion rates have been found under dry climates and low altitudes in Israel (Cerdà, 1998a, b), which support the idea of high OM losses due to soil erosion in dry areas.

In line with this, in Mediterranean natural areas there is no information about the soil variability; also little data is available related to the control topography exerts on soil properties (Lozano-García and Parras-Alcántara, 2014). Therefore, the aims of this study are (i) to quantify SOC contents and their vertical distribution in a natural forest area, (ii) to assess the SOCS differences in soils along an altitudinal gradient and (iii) to assess their relationship with soil depth in a Mediterranean natural area.

2 Material and methods

2.1 Study site

The Despeñaperros Natural Park ($76.8 \, km^2$) is one of the best-preserved landscapes in southern Europe. It is located within the eastern Sierra Morena (province of Jaén, southeastern Spain), at coordinates 38°20′–38°27′ N, 3°27′–3°37′ W. The study area is characterized by warm dry summers and cool humid winters, and climate is temperate semiarid with continental features due to elevation. Average extreme temperatures range between $-10\,°C$ (winter) and $42\,°C$ (summer), with mean temperature $15\,°C$. The moisture regime is dry Mediterranean, with average annual rainfall of 800 mm. High temperatures and long drought periods cause water deficits up to 350 mm annually.

It is a mountainous area, with an altitudinal range of 540 m a.s.l. in the Despeñaperros River valley to 1250 m a.s.l at Malabrigo Mountain. The relief is steep with slopes ranging from 3 to 45 %, and the parent materials are primarily slate and quartzite. Most abundant soils in the area are Phaeozems (PH), Cambisols (CM), Regosols (RG) and Leptosols (LP), according to the classification by IUSS Working Group WRB (2006) (Table 1). Well-preserved Mediterranean woodlands and scrublands occupy the study area and deer hunting habitat is the main land use.

2.2 Soil sampling and analytical methods

Seven sites were selected along a topographic gradient in a south-facing slope in the Despeñaperros Natural Park (Table 1). Soil samples were collected at each site following a random sampling design according to FAO (2006). Each selected point was sampled using soil control sections (SCSs) at different depths (S1: 0–25, S2: 25–50 and S3: 50–75 cm). Soil control sections were used for a uniform comparison between studied soils. Four replicates of each soil sample were analyzed in laboratory (17 sampling points × 1, 2 or 3 SCS × 4 replicates).

Soil samples were air-dried at constant room temperature (25 °C) and sieved (2 mm) to discard coarse particles. The analytical methods used in this study are described in Table 2.

Statistical analysis was performed using SPSS Inc. (2004). The physical and chemical soil properties were analyzed statistically for each SCS of different soil groups (PH, CM, RG and LP), including the average and standard deviation (SD). The statistical significance of the differences in each variable between each sampling point and soil type was tested using the Anderson–Darling test at each control section for each soil type. Differences with $p < 0.05$ were considered statistically significant.

Table 1. Soil groups of the study area at each of the seven topographic positions with properties. The key refers to the reference soil groups of the IUSS Working Group WRB (2006) with lists of qualifiers.

Topographic position	m a.s.l.[a]	Slope %	Parent material		Vegetation series	Soil groups	Qualifiers	n[b]
A	1168	15.3	Quartzite sandstone	–	Maritime pine (*Pinus pinaster*) Holm oak (*Quercus ilex*) Gum rockrose (*Cistus ladanifer*)	Leptosols – LP	Mollic – mo	2
B	1009	16.5	Quartzite sandstone	–	Holm oak (*Quercus ilex*) Cork oak (*Quercus suber*) Strawberry tree (*Arbutus unedo*) Gum rockrose (*Cistus ladanifer*)	Regosols – RG Leptosols – LP Cambisols – CM	Eutric – eu Mollic – mo Humic – hu	3
C	945	20.8	Quartzite sandstone	–	Stone pine (*Pinus pinea*) Mastic (*Pistacia lentiscus*)	Cambisols – CM Regosols – RG Phaeozems – PH	Humic – hu Dystric – dy Luvic – lv	3
D	865		Quartzite		Portuguese oak (*Quercus faginea*) Strawberry tree (*Arbutus unedo*) Gum rockrose (*Cistus ladanifer*)	Regosols – RG	Umbric – um	2
E	778		Quartzite slates	–	Holm oak (*Quercus ilex*) Strawberry tree (*Arbutus unedo*) Gum rockrose (*Cistus ladanifer*)	Leptosols – LP	Umbric – um	3
F	695	12.0	Quartzite		Cork oak (*Quercus suber*) Holm oak (*Quercus ilex*) Strawberry tree (*Arbutus unedo*) Gum rockrose (*Cistus ladanifer*)	Leptosols – LP	Lithic – li	2
G	607	18.5	Slates		Holm oak (*Quercus ilex*) Mastic (*Pistacia lentiscus*)	Leptosols – LP	Mollic – mo	2

[a] Meters above sea level; [b] sample size.

Table 2. Methods used in field measurements, laboratory analysis and to make calculations from study data.

Parameters	Method
Field measurements Bulk density ($Mg\,m^{-3}$)	Cylindrical core sampler[a]; Blake and Hartge (1986)
Laboratory analysis Particle size distribution pH – H_2O Organic C (%)	Robinson pipette method; USDA (2004)[b] Volumetric with Bernard calcimeter; Duchaufour (1975) Walkley and Black method; Nelson and Sommers (1982)
Parameters calculated from study data SOC stock ($Mg\,ha^{-1}$) Total SOC stock ($Mg\,ha^{-1}$)	(SOC concentration $\times$ BD $\times\,d\,\times\,(1 - \delta_{2\,mm}\%) \times 0.1$)[c]; IPCC (2003) $\Sigma_{horizon}$ SOC Stock$_{horizon}$; IPCC (2003)

[a] 3 cm diameter, 10 cm length and 70.65 cm^3 volume. [b] Prior to determining the particle size distribution, samples were treated with H_2O_2 (6 %) to remove organic matter (OM). Particles larger than 2 mm were determined by wet sieving and smaller particles were classified according to USDA standards (2004). [c] Where SOC is the organic carbon content (g Kg^{-1}), d the thickness of the soil layer (cm), $\delta_{2\,mm}$ is the fractional percentage (%) of soil mineral particles > 2 mm in size in the soil, and BD the soil bulk density (Mg m^{-3}).

3 Results and discussion

3.1 Soil properties

The soils are stony soils, acidic, with low base concentrations, oligotrophic and with slightly unsaturated complex change and located in areas of variable slopes ranging between 5 and 38 %. Phaeozems are the most developed soils in the study area. They are deep, dark, and well humidified with high biological activity and high vegetation density on gentle slopes and shady side foothills. Cambisols are developed and deep soils; however, Leptosols are the least developed and shallowest soils.

Phaeozems are the most pedogenically developed soils in the study area. They are found on gentle slopes (< 3 %),

Table 3. Properties of the soils evaluated (average $\pm$SD*) in the Despeñaperros Natural Park.

Topographic position	m a.s.l.	SCS	Depth cm	Gravel %	Sand %	Silt %	Clay %	BD Mg m^{-3}	OM g kg^{-1}	pH H$_2$O
A	1168	S1	0–25	33.1 ± 13.8 aA	56.5 ± 1.1 aA	22.3 ± 3.0 aA	21.2 ± 4.1 aA	1.1 ± 0.19 aA	64.5 ± 8.9 aA	6.3 ± 0.7 aA
		S2	25–50	7.0 ± 3.1 bA	39.3 ± 0.81 bA	30.7 ± 4.2 aA	30.0 ± 6.1 aA	1.5 ± 0.21 bA	0.99 ± 0.21 bA	5.3 ± 0.5 bA
B	1009	S1	0–25	17.0 ± 10.0 aB	52.9 ± 29.8 aA	29.9 ± 30.6 aA	17.2 ± 5.3 aA	1.1 ± 0.10 aA	68.6 ± 5.2 aA	5.9 ± 0.4 aA
		S2	25–50	27.1 ± 6.4 bB	58.7 ± 20.1 aB	19.1 ± 12.2 bB	22.1 ± 8.0 aB	1.3 ± 0.12 aB	35.3 ± 3.4 bB	5.6 ± 0.7 aA
		S3	50–75	14.3 ± 16.9 aA	41.6 ± 18.1 bA	25.7 ± 15.2 aA	32.6 ± 2.9 bA	1.5 ± 0.12 bA	10.5 ± 2.8 cA	5.7 ± 0.5 aA
C	945	S1	0–25	34.0 ± 5.5 aA	59.2 ± 7.2 aA	24.7 ± 3.1 aA	16.1 ± 6.2 aA	1.2 ± 0.10 aA	58.0 ± 9.5 aA	5.9 ± 0.4 aA
		S2	25–50	14.4 ± 7.2 bC	36.1 ± 12.2 bA	28.2 ± 2.5 aA	35.7 ± 14.1 bA	1.3 ± 0.06 aB	30.9 ± 6.3 bB	5.5 ± 0.4 aA
		S3	50–75	14.9 ± 11.9 bA	24.4 ± 15.9 cB	30.4 ± 9.8 aA	45.2 ± 16.2 cB	1.5 ± 0.05 aA	0.99 ± 0.12 cB	5.2 ± 0.6 aA
D	865	S1	0–25	39.9 ± 6.2 aA	47.6 ± 19.3 aB	38.1 ± 7.5 aB	14.3 ± 2.1 aA	1.1 ± 0.09 aA	62.9 ± 10.4 aA	5.6 ± 1.0 aA
		S2	25–50	24.0 ± 4.5 bB	46.6 ± 18.2 aC	36.2 ± 7.9 aA	17.2 ± 5.4 aB	1.3 ± 0.10 aB	35.9 ± 7.6 bB	5.7 ± 0.8 aA
		S3	50–75	11.9 ± 10.2 cA	30.9 ± 11.1 bB	47.1 ± 5.4 bB	22.0 ± 6.8 aC	1.5 ± 0.13 bA	1.0 ± 0.30 cB	4.5 ± 0.4 bB
E	778	S1	0–25	25.5 ± 6.8 aC	52.2 ± 7.2 aA	30.2 ± 5.1 aA	17.6 ± 2.4 aA	1.2 ± 0.13 aA	56.3 ± 8.9 aA	5.7 ± 0.7 aA
F	695	S1	0–25	28.2 ± 7.4 aC	34.2 ± 5.3 aC	41.0 ± 9.8 aB	24.8 ± 2.8 aA	1.2 ± 0.14 aA	46.9 ± 7.4 aB	6.3 ± 0.5 aA
G	607	S1	0–25	42.9 ± 19.3 aD	54.9 ± 4.1 aA	27.7 ± 2.5 aA	17.3 ± 6.6 aA	1.3 ± 0.13 aB	54.9 ± 9.2 aB	6.2 ± 0.7 aA

m a.s.l.: meters above sea level; SCS: soil control section; BD: bulk density; OM: organic matter. * Standard deviation. Numbers followed by different lower-case letters within the same column have significant differences ($P < 0.05$) at different depths, considering the same topographic position. Numbers followed by different capital letters within the same column have significant differences ($P < 0.05$) considering the same SCS at different topographic positions.

usually in shaded areas on Ordovician sandstones. The gravel content is variable, ranging between 7 and 31 % (total weight). Texturally they are sandy soils at the surface and silty-clay-loam or silty-clay soils at depth, with a horizon sequence A0/A1/AB/Bt/C1. These soils show Luvic (lv) characteristics (Luvic Phaeozems (lv-PH)) and are > 1 m in depth with pH along the profile ranging from 6.3 to 5.6 at depth and about 4.3 % OM content (Tables 1 and 3).

Cambisols are less developed soils than Luvic Phaeozems; however, these soils are more developed and deeper than Regosols and Leptosols. They appear in areas of variable slope (3–38 %) and are > 1 m in depth characterized by a cambic horizon (Bw) on Ordovician quartzite (Table 1) with approximately 20 % gravel content. At the surface they are sandy soils (< 60 % sand content) with high clay content in the Bw horizon and increasing clay content with depth (Table 3). The horizon sequences were A0/A1/AB/BW/BC/C1 or A0/A1/AB/BW. These soils are characterized by low OM content at depth. Gallardo et al. (2000) showed that the low OM content could be explained by the semiarid Mediterranean conditions. In addition, Parras-Alcántara et al. (2013a) found there is less OM and fewer mineral aggregates in sandy soils, thus favoring high levels of OM transformation. Because of this, Hontoria et al. (2004) suggested that physical variables determine soil development in the driest areas of Spain to a greater degree than management or climatic variables. The Cambisols topsoil has humic (hu) characteristics, with > 5 % OM content (Table 3) due to plant debris accumulation in the A0 horizon. This OM is poorly structured and partially decomposed, thereby reducing the amount and increasing the OM evolution degree with depth. In line with this, Bech et al. (1983) reported that the free OM concentration in the surface horizon was higher than 90 %, while humic and fulvic acid concentrations were less than 2 % in soils with *Quercus ilex* spp. *ballota* vegetation. Free

OM was reduced and humidification increased up to 30 % in deeper layers.

Regosols can be found in steeply sloping areas (> 8 %) characterized by high water erosion and subject to rejuvenation processes. We found Eutric (eu), Dystric (dy) and Umbric (um) Regosols (Table 1) on sandstone and quartzite parent materials with > 25 % gravel content in surface layers that in some cases disappeared in depth. These soils are sandy loam in surface layers and silty clay in deep layers, with different horizon sequences (A0/A1/AB/BC/C1, A0/A1/AC/C1 and A1/AC/C1). Eutric Regosols are deeper soils (> 80 cm) that are loamy with high gravel content (25.1–32.2 %) at the surface decreasing with deep, acidic pH (5.9) and high OM content (6.7 %) at the surface. The Dystric Regosols are stony soils that are shallow (< 40 cm), loamy at the surface and sandy at depth with high gravel content (> 40 %) at the surface, acidic pH (6.2) and high OM content (7.3 %) in the surface horizon (Table 3). The Umbric Regosols are also stony soils that are deep soils (> 70 cm), loamy with high gravel content (40 %) in the surface decreasing to 11 % at depth, have acidic pH (5.6) and high OM content (6.5 %) (Table 3).

Leptosols are the least developed soils of the study area. Lithic (li), Mollic (mo) and Eutric (eu) Leptosols were identified (Table 1) formed in sandstone, quartzite and slate on variable slopes (1.5–46 %). Horizon sequences A1/AC/C1, A1/AC, and AC/C1 and A1 were found. The gravel content was variable (> 40 % in the topographically elevated areas and decreasing with depth) with high sand content (> 50 %) in the surface layers. One characteristic of these soils is that the clay content increased with depth, reaching up to 30 %. According to Recio et al. (1986), the physical–chemical properties of the soils in the study area are due to lithology, while their low edaphic development is conditioned by age. According to Nerger et al. (2007), the alteration and pedogenesis processes taking place in these soils

Table 4. Soil organic carbon (SOC) content and soil organic carbon stock (SOCS) (average ±SD*) in the Despeñaperros Natural Park.

Topographic position	m a.s.l.	SCS	SOC g kg^{-1}	T-SOC g kg^{-1}	SOCS Mg ha^{-1}	T-SOCS Mg ha^{-1}
A	1168	S1	37.5 ± 16.8 aA	38.1 ± 8.4 A	70.8 ± 33.5 aA	72.9 ± 17.0 A
		S2	0.58 ± 0.09 bA		2.1 ± 0.57 bA	
B	1009	S1	39.9 ± 10.3 aA		91.1 ± 13.2 aB	
		S2	20.5 ± 6.4 bB	66.6 ± 8.2 B	49.8 ± 14.9 bB	158.0 ± 15.8 B
		S3	6.1 ± 7.8 cA		19.1 ± 19.2 cA	
C	945	S1	33.7 ± 8.6 aA		67.4 ± 9.7 aA	
		S2	18.0 ± 9.1 bB	52.3 ± 5.9 C	50.1 ± 22.4 bB	119.3 ± 10.9 C
		S3	0.58 ± 0.09 cB		1.8 ± 0.26 cB	
D	865	S1	36.6 ± 7.9 aA		62.1 ± 8.9 aA	
		S2	20.9 ± 9.0 bB	58.1 ± 5.7 C	52.1 ± 16.7 bB	116.1 ± 8.6 C
		S3	0.57 ± 0.09 cB		1.9 ± 0.30 cB	
E	778	S1	32.7 ± 13.2 aA	32.7 ± 13.2 A	72.6 ± 25.0 aA	72.6 ± 0.65 A
F	695	S1	27.3 ± 15.1 aB	27.3 ± 15.1 A	59.3 ± 27.3 aC	59.3 ± 27.3 A
G	607	S1	31.9 ± 13.1 aB	31.9 ± 13.1 A	53.8 ± 18.3 aC	53.8 ± 18.3 A

m a.s.l.: meters above sea level; SCS: soil control section; SOC: soil organic carbon; T-SOC: total SOC; SOCS: soil organic carbon stock; T-SOCS: total SOCS. * Standard deviation. Numbers followed by different lower-case letters within the same column have significant differences ($P < 0.05$) at different depths, considering the same topographic position. Numbers followed by different capital letters within the same column have significant differences ($P < 0.05$) considering the same SCS at different topographic positions.

usually occur on low slopes. The Lithic Leptosols are the least developed soils at this study site, with thicknesses ranging between 10 and 15 cm in areas of steep slope. In flat areas, their low development is due to their extreme youth. These soils are loamy with a high gravel content (> 28 %), acidic pH and > 4 % OM content. Mollic Leptosols are characterized by mollic surface horizons (thick, well-structured, dark, high base saturation and high OM content), on variable slopes (18.5–38.5 %). According to Corral-Fernández et al. (2013) these soils are characterized by organic residue accumulation in the surface horizons; this OM is poorly structured and partially decomposed at the surface with increasing decomposition rate with depth. Umbric Leptosols are characterized by high OM content, are shallow, and either loamy with high stony content (> 20 % gravel content) or sandy (> 55 % sand content), have low bulk density conditioned by the OM content, high porosity and acidic pH (Table 3).

3.2   Distribution of soil organic carbon

Generally, soils in the study area are characterized by > 3 % OC content, making them part of the 45 % of the mineral soils of Europe that have between 2 and 6 % OC content (Rusco et al., 2001). Soil OM content decreased with depth at all topographic positions (A, B, C and D positions) (Table 4). However, this property cannot be observed in the lowest topographic positions (E, F and G positions) due to the low edaphic development (Umbric Leptosols, Lithic Leptosols

and Mollic Leptosols) as only one SCS exists (S1: 0–25 cm) (Tables 1 and 4).

The soils in this study are characterized by high sand content at the surface (S1) varying between 59.2 and 34.2 % for C and F positions respectively, and reduced sand content with depth in all studied soils (Table 3), affecting OM development. Clay content reaches 45 % in C: S3. In addition, the mineral medium may play an important role in soil humidification processes, so we can explain low soil OM concentrations with depth due in part to soil texture, because soil OM tends to decrease with depth in virtually all soils, regardless of textural changes. Clays over sands induce a decrease in soil OM with depth; also, the aggregate formation between OM and the mineral fraction is reduced favoring high OM levels in sandy soils at depth (González and Candás, 2004). Furthermore, Gallardo et al. (2000) argued that the relatively low concentrations of OM in depth could be explained by the climate (Mediterranean semiarid). Similar results have been found by Corral-Fernández et al. (2013), Parras-Alcántara et al. (2014) and Lozano-García and Parras-Alcántara (2013a) in the Pedroches Valley, near the study area.

Another key issue is that the clay fraction increased with depth in the B and C positions (reaching a clay content of as high as 45 % (C: S3)) and its relation with soil OM at depth (S2: 25–50 cm), which was characterized by high OM contents as compared to S3 (B: 2.0/0.6 %; C: 1.8/0.06 %) (Tables 3 and 4). Burke et al. (1989) and Leifeld et al. (2005) have identified high OM levels in soils with high clay content in depth indicating clay stabilization mechanisms in the soil. This effect can be observed in the B and C topographic positions, where an increase in clay content was observed at depth as compared to the upper horizons (B: S1–17.2 %/S2–22.1 %; C: S1–16.1 %/S2–35.7 %). This OM increase may be due to carbon translocation mechanisms (dissolved organic carbon), soil biological activity and/or the root depth effect (Sherstha et al., 2004).

Soil OM appears to be concentrated in the first 25 cm (S1), where the mineralization and immobilization C processes should be slightly active. In the surface layer (S1), OM was variable along the toposequence studied ranging between 39.9 and 27.3 g kg^{-1} at the B and F positions, respectively (Table 4). In this regard, it is important to point out that the S1 layer can reach over 60 % of the total soil organic carbon (T-SOC) values documented, corresponding to 60, 64.4 and 63 % for the B, C and D positions respectively as compared to the rest of the soil profile (S2 or S2+S3). Batjes (1996) states that for the 0 to 100 cm depth approximately 50 % of soil organic carbon (SOC) appears in the first 30 cm of the soil. Jobbágy and Jackson (2000) showed that 50 % of SOC is concentrated in the first 20 cm in forest soils to 1 m depth. Civeira et al. (2012) showed that SOC in the upper 30 cm of soils in Argentina is much higher than in the 30–100 cm interval. Data provided by these authors and the results obtained in this study may be comparable because in this study we used a 75 cm depth and the mentioned authors used a 1 m depth. Furthermore, Jobbágy and Jackson (2000) indicated that changes in SOC were conditioned by vegetation type (which determines the vertical distribution of roots) and to a lesser extent the effect of climate and clay content. Despite this, climatic conditions can be a determining factor in the SOC concentrations for surface horizons, whereas clay content may be the most important element in deeper horizons; also, clay contributes to stabilize OM by protecting physically microbial activity and reducing C outputs. This effect is important under homogeneous climate conditions (as those in the study area). At the regional–global scale, the precipitation contributes to maximize SOC and temperature accelerates mineralization process decreasing the SOC (Post et al., 1982).

Results of T-SOC analysis in the studied area did not show up great along the toposequence. T-SOC depended on the degree of development of the soil that appeared at each topographical position. The T-SOC was highest at the B (66.5 g kg^{-1}), D (58.1 g kg^{-1}) and C (52.3 g kg^{-1}) positions, corresponding to Cambisols–Regosols–Leptosols, Regosols, and Phaeozems–Cambisols–Regosols respectively. Leptosols showed the lowest T-SOC content with 27.3 g kg^{-1}, 31.9 g kg^{-1}, 32.7 g kg^{-1} and 38.1 g kg^{-1} at the F, G, E and A topographic positions, respectively. Similarly, > 60 % of SOC was concentrated in the S1 layer of deeper soils (B, C and D).

Precipitation and temperature varied through the studied toposequence, where precipitation increases and temperature decreases with increasing elevation (Parras-Alcántara et al.,

2004). T-SOC content was not affected by climatic variations, but depended on the soil development in each landscape position. Reduced T-SOC contents were observed at the lowest topographic positions, where soils were shallower. This is in agreement with Power and Schlesinger (2002), who concluded that topographic position affects T-SOC, due to low OM decomposition rates under low temperatures.

3.3   Soil organic carbon stocks

Soil organic carbon stocks in the study area showed a reduction with depth in all topographic positions (Table 4). This SOCS reduction along the profile is linked to OM reduction with depth, which also depends on the gravel content and the bulk density (Table 3).

Higher values of SOCS (up to 91.1 Mg ha^{-1}) were found in the upper SCS at elevated topographic positions (highest value at the B position). The lowest SOCS values were found at the G position (53.8 Mg ha^{-1}), the lowest site in the toposequence. This trend of decreasing SOCS with decreasing elevation is constant except at the A and E positions. Both are poorly developed soils with high OM content in the surface horizon.

In the S1 SCS, between 53.8 and 58.0 % of the SOCS were found at the D and B topographic positions. This constituted 63.0 and 60.0 % of T-SOC in these topographic positions. This shows that the gravel content and bulk density affects the SOCS in the surface horizons of the toposequence studied, and, therefore, SOCS decreases when SOC increases. In the most developed soil, similar SOC and SOCS concentrations (B: 60 % – SOC; 58 % – SOCS) were observed in the S1 layer, conditioned by bulk density and gravel content. In addition, SOCS decreased in depth conditioned by reduction of gravel content and increasing bulk density. This is not in agreement with Tsui et al. (2013) and Minasny et al. (2006), who suggested a negative relation between bulk density and depth as a consequence of high OM content at the surface, linked to low clay concentrations (Li et al., 2010). In this sense, we observed that high SOCS depended on the SOC concentration and the clay content. However, the SOC concentration affected the SOCS to a lesser degree so that in S2 (25–50 cm) we found > 10 % of SOCS related to SOC (C position).

In contrast, low SOCS can be found in S3 except at the B topographic position (19.1 Mg ha^{-1}). This situation could be due to the fact that pedological horizons were generally different than the SCS divisions (S1: 0–25 cm; S2: 25–50 and S3: 50–75 cm) (Hiederer, 2009); in other words, the SCS divisions often led to the mixing of two or more soil horizons (depending on thickness horizon) in any given SCS division.

In all studied soils, the clay content increased with depth. This clay content increase is associated with higher values of SOC (B: S2 and C: S2). In line with this, we can explain high SOCS concentrations in clayey soils caused by clay stabilization mechanisms on SOC; this effect can observed at the A

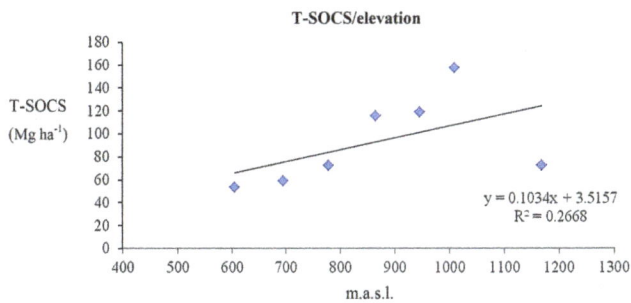

Figure 1. Exponential regression model for T-SOCS versus altitudinal gradient. T-SOCS: total soil organic carbon stock

topographic position which has higher clay content with respect to the B and D positions. However, a SOCS increase can be observed. This is the case at the D and C topographical positions with SOCS values of 52.1 and 50.1 Mg ha^{-1}, respectively in the S2 sampling layer (Table 4), showing a correlation between S1 and S2, due to carbon translocation processes as dissolved organic carbon, bioturbation and/or deep rooting (Sherstha et al., 2004).

3.4 Soil organic carbon stocks along the altitudinal gradient

The SOCS results along the toposequence were also studied. It is important to point out that total SOCS (T-SOCS) was influenced by topographical position in the toposequence analyzed. T-SOCS increased exponentially with elevation from G (607 m a.s.l.) to B site (1009 m a.s.l.), with the exception of the highest topographic position, A (1168 m a.s.l.), with an exponential regression relationship (Fig. 1). Similar results were found by Ganuza and Almendros (2003), Leifeld et al. (2005) and Fernández-Romero et al. (2014). These studies showed that the T-SOCS increased with elevation. However, Avilés-Hernández et al. (2009) found that T-SOCS from forest soils decreased with elevation in a toposequence in Mexico due to variations in the OM decomposition rate, and Lozano-García and Parras-Alcántara (2014) found that T-SOCS decreased with elevation in a traditional Mediterranean olive grove due to erosion. With respect to the A position in this study, the lower T-SOCS (72.9 Mg ha^{-1}) values with respect to the rest of the studied toposequence may be due to soil loss caused by erosion processes in soils with a low level of development. Similar results have been found by Parras-Alcántara et al. (2004) and Durán-Zuazo et al. (2013). Parras-Alcántara et al. (2004) explained their findings as a consequence of high soil erosion rates, caused by high erosivity of rainfall, high erosionability, steep slopes, low vegetation cover and the lack of conservation practices in the studied area. Durán-Zuazo et al. (2013) explained this effect by low vegetation densities in the upper parts of mountain areas that can cause high erosion with strong water runoff. Martínez-Mena et al. (2008) have emphasized the effects of

erosion on soil OM loss, especially under semiarid conditions. In this context, a low vegetation cover can accelerate OM decomposition, weakening soil aggregates (Balesdent et al., 2000; Paustian et al., 2000). Cerdà (2000) indicated that this effect could occur regardless of climatic conditions.

As can be seen in Table 4, T-SOCS decrease was not homogeneous. In some cases, rapid changes were found, while in other situations gradual changes were noted. Abrupt changes in T-SOCS occurred between the B/C and D/E topographic positions, showing T-SOCS differences of 38 Mg ha^{-1} and 44 Mg ha^{-1} respectively. Gradual changes in T-SOCS occurred between the C/D, E/F and F/G topographic positions with variations of 3 Mg ha^{-1}, 13 Mg ha^{-1} and 6 Mg ha^{-1} respectively. Some authors have concluded that the SOCS reduction can be explained by soil physical properties – mainly texture (Corral-Fernández et al., 2013; Parras-Alcántara et al., 2013b). The studied soils are sandy at the surface, with clay increasing with depth, except in E, F and G sites (soils that have S2 and/or S3 SCS); therefore, OM stabilizing mechanisms are produced, reducing the aggregate formation between SOC and mineral fraction at depth. As a result, the SOCS content is lower with sandy soils (Nieto et al., 2013). González and Candás (2004) and Parras-Alcántara et al. (2013a) obtained similar results: the first in sandy-loamy soils and the second in Mediterranean clayey soils. In addition, low SOC levels are conditioned by the climatic characteristics of southern Europe (Gallardo et al., 2000).

4 Conclusions

Soils found in the Despeñaperros Natural Park include Phaeozems, Cambisols, Regosols and Leptosols. Phaeozems are the deepest and most developed soils, and Leptosols are the least developed and shallowest soils. These soils are characterized by low OM content with depth due to the semiarid Mediterranean conditions and the high sand content. The studied soils are characterized by organic residue accumulation in the surface horizons.

The SOC content decreased with depth at all topographic positions and the clay fraction increased with depth. The mineral medium played an important role in soil humidification processes. In addition, the SOC in the S2 layers is characterized by high SOC values with respect to the S3 layers indicating clay stabilization mechanisms in the soil. We can explain this increase due to carbon translocation mechanisms (dissolved organic carbon), soil biological activity and/or the root depth effect.

With respect to T-SOC content, there is not a large difference between T-SOC along the toposequence. The T-SOC of these soils depends on the degree of development of the soils found at each topographic position. We can observe a T-SOC reduction at the lowest topographic positions for less developed soils and a T-SOC increase at the highest topographic positions in the more developed soils. SOCS in the study zone shows a reduction with depth in all topographic positions. This SOCS reduction along the profile is linked to OM and gravel content reduction and an increase in bulk density with depth. The T-SOCS increased with altitude, due to the higher turnover of organic material (plants) and the lower decomposition rate due to lower temperatures.

Acknowledgements. The authors thank to Eric C. Brevik for his contribution to improve this paper and Paulo Pereira (Handling Topical Editor) and the anonymous referees for the supportive and constructive comments that helped to improve the manuscript.

Edited by: P. Pereira

References

Asadi, H., Raeisvandi, A., Rabiei, B., and Ghadiri, H.: Effect of land use and topography on soil properties and agronomic productivity on calcareous soils of a semiarid region, Iran, Land Degrad. Develop., 23, 496–504, 2012.

Ashley, G. M., Beverly, E. J., Sikes, N. E., and Driese, S. G.: Paleosol diversity in the Olduvai Basin, Tanzania: Effects of geomorphology, parent material, depositional environment, and groundwater on soil development, Quat. Internat., 322/323, 66–77, 2014.

Avilés-Hernández, V., Velázquez-Martínez, A., Ángeles-Pérez, G., Etchevers-Barra, J., De los Santos-Posadas, H., and Llandera, T.: Variación en almacenes de carbono en suelos de una toposecuencia, Agrociencia, 43, 457–464, 2009.

Bakhshandeh, S., Norouzi, M., Heidari, S., and Bakhshandeh, S.: The role of parent material on soil properties in sloping areas under tea plantation in Lahijan, Iran, Carpathian, J. Earth Environ. Sci., 9, 159–170, 2014.

Balesdent, J., Chenu, C., and Balabane, M.: Relationship of soil organic matter dynamics to physical protection and tillage, Soil Till. Res., 53, 215–230, 2000.

Barbera, V., Poma, I., Gristina, L., Novara, A., and Egli, M.: Long-term cropping systems and tillage management effects on soil organic carbon stock and steady state level of C sequestration rates in a semiarid environment, Land Degrad. Develop., 23, 82–91, 2012.

Barua, A. K. and Haque, S. M. S.: Soil characteristics and carbon sequestration potentials of vegetation in degraded hills of Chittagong, Bangladesh, Land Degrad. Develop., 24, 63–71, 2013.

Batjes, N. H.: Total carbon and nitrogen in the soils of the world, Eur. J. Soil Sci., 47, 151–163, 1996.

Batjes, N. H.: Projected changes in soil organic carbon stocks upon adoption of recommended soil and water conservation practices in the Upper Tana River Catchment, Kenia, Land Degrad. Develop., 25, 278–287, 2014.

Bech, J., Hereter, A., and Vallejo, R.: Las tierras pardo ácidas sobre granodioritas de la zona nor-oriental del macizo del Montseny, An. Edaf. Agrob., 42, 371–393, 1983.

Blake, G. R. and Hartge, K. H.: Bulk density, in: Methods of soil analysis – Part I: Physical and mineralogical methods, edited by:

Klute, A., Agronomy Monography no. 9. ASA, SSSA, Madison WI, USA, 363–375, 1986.

Breuning-Madsen, H., Elberling, B., Balstroem, T., Holst, M., and Freudenberg, M.: A comparison of soil organic carbon stock in ancient and modern land use systems in Denmark, Eur. J. Soil Sci. 60, 55–63, 2009.

Brevik, E. C.: Soils and climate change: Gas fluxes and soil processes, Soil Horiz., 53, 12–23, 2012.

Brevik, E. C.: Forty years of soil formation in a South Georgia, USA borrow pit, Soil Horiz., 54, 20–29, 2013.

Brevik, E. C. and Homburg, J.: A 5000 year record of carbon sequestration from a coastal lagoon and wetland complex, Southern California, USA, Catena, 57, 221–232, 2004.

Burke, I., Yonker, C., Parton, W., Cole, C., Flach, K., and Schimel, D.: Texture, climate, and cultivation effects on soil organic matter content in U.S. grassland soils, Soil Sci. Soc. Am. J., 53, 800–805, 1989.

Cerdà, A.: Effect of climate on surface flow along a climatological gradient in Israel. A field rainfall simulation approach, J. Arid Environ. 38, 145–159, 1998a.

Cerdà, A.: Relationship between climate and soil hydrological and erosional characteristics along climatic gradients in Mediterranean limestone areas, Geomorphology, 25, 123–134, 1998b.

Cerdà, A.: Aggregate stability against water forces under different climates on agriculture land and scrubland in southern Bolivia, Soil Till. Res., 57, 159–166, 2000.

Cerdà, A. and García-Fayos, P.: The influence of slope angle on sediment, water and seed losses on badland landscapes, Geomorphology, 18, 77–90, 1997.

Civeira, G., Irigoin, J., and Paladino, I. R.: Soil organic carbon in Pampean agroecosystems: Horizontal and vertical distribution determined by soil great group, Soil Horiz., 53, 43–49, 2012.

Corral-Fernández, R., Parras-Alcántara, L., and Lozano-García, B.: Stratification ratio of soil organic C, N and C : N in Mediterranean evergreen oak woodland with conventional and organic tillage, Agric. Ecosyst. Environ., 164, 252–259, 2013.

Dingil, M., Öztekin, M. E., and Şenol, S.: Definition of the physiographic units and land use capability classes of soils in mountainous areas via satellite imaging, Fresen. Environ. Bull., 23, 952–955, 2014.

Duchaufour, P. H.: Manual de Edafología, Editorial Toray-Masson, Barcelona, 1975.

Durán-Zuazo, V. H., Francia-Martínez, J. R., García-Tejero, I., and Cuadros-Tavira, S.: Implications of land-cover types for soil erosion on semiarid mountain slopes: Towards sustainable land use in problematic landscapes, Acta Ecol. Sinica, 33, 272–281, 2013.

Eshetu, Z., Giesler, R., and Högberg, P.: Historical land use affects the chemistry of forest soils in the Ethiopian highlands, Geoderma, 118, 149–165, 2004.

FAO: Guidelines for soil description. Food and Agriculture Organization of the United Nations, Rome, Italy, 2006.

Fernández-Calviño, D., Garrido-Rodríguez, B., López-Periago, J. E., Paradelo, M., and Arias-Estévez, M.: Spatial distribution of copper fractions in a vineyard soil, Land Degrad. Develop., 24, 556–563, 2013.

Fernández-Romero, M. L., Parras-Alcántara, L., and Lozano-García, B.: Land use change from forest to olive grove soils in a toposequence in Mediterranean areas (South of Spain), Agric. Ecosyst. Environ., 195, 1–9, 2014.

Fialho, R. C. and Zinn, Y. L.: Changes in soil organic carbon under Eucaliptus plantations in Brazil: a comparative analysis, Land Degrad. Develop., 25, 428–437, 2014.

Fu, B. J., Liu, S. L., Ma, K. M., and Zhu, Y. G.: Relationships between soil characteristics, topography and plant diversity in a heterogeneous deciduous broad-leaved forest near Beijing, China, Plant Soil, 261, 47–54, 2004.

Gallardo, A., Rodríguez-Saucedo, J., Covelo, F., and Fernández-Ales, R.: Soil nitrogen heterogeneity in dehesa ecosystem, Plant Soil, 222, 71–82, 2000.

Ganuza, A. and Almendros, G.: Organic carbon storage of the Basques Country (Spain): the effect of climate, vegetation type and edaphic variables, Biol. Fert. Soils, 37, 154–162, 2003.

Gebrelibanos, T. and Assen, M.: Effects of slope aspect and vegetation types on selected soil properties in a dryland Hirmi watershed and adjacent agro-ecosystem, northern highlands of Ethiopia, Afr. J. Ecol., 52, 292–299, 2014.

Gessler, P. E., Moore, I. D., McKenzie, N. J., and Ryan, P. J.: Soil-landscape modeling and spatial prediction of soil attributes. Special issue: integrating GIS and environmental modeling, Int. J. GIS, 9, 421–432, 1995.

Gessler, P. E., Chadwick, O. A., Chamran, F., Althouse, and L., Holmes, K.: Modeling soil-landscape and ecosystem properties using terrain attributes, Soil Sci. Soc. Am. J., 64, 2046–2056, 2000.

González, J. and Candás, M.: Materia orgánica de suelos bajo encinas: mineralización de carbono y nitrógeno, Invest. Agrar., 75–83, 2004.

Haregeweyn, N., Poesen, J., Verstraeten, G., Govers, G., De Vente, J., Nyssen, J., Deckers, J., and Moeyersons, J.: Assessing the performance of a spatially distributed soil erosion and sediment delivery model (WATEM/SEDEM in Northern Ethiopia, Land Degrad. Develop. 24, 188–204, 2013.

Hattar, B. I., Taimeh, A. Y., and Ziadat, F. M.: Variation in soil chemical properties along toposequences in an arid region of the Levant, Catena, 83, 34–45, 2010.

Hernanz, J. T., López, R., Navarrete, T., and Sánchez-Girón, V.: Long-term effects of tillage systems and rotations on soil structural stability and organic carbon stratification in semiarid central Spain, Soil Till. Res., 66, 129–141, 2002.

Hiederer, R.: Distribution of Organic Carbon in Soil Profile Data, EUR 23980 EN, Luxembourg: Office for Official Publications of the European Communities, 126 pp., 2009.

Hontoria, C., Rodríguez-Murillo, J. C., and Saa, A.: Contenido de carbono orgánico en el suelo y factores de control en la España peninsular, Edafología, 11, 149–157, 2004.

IPCC, Intergovernmental Panel on Climate Change: Good practice guidance for land use, land use change and forestry, edited by: Penman, J., Gytarsky, M., Hiraishi, T., Krug, T., Kruger, D., Pipatti, R., Buendia, L., Miwa, K., Ngara, T., Tanabe, K., and Wagner, F., IPCC/OECD/IEA/IGES, Hayama, Japan, 2003.

IUSS Working Group WRB: World reference base for soil resources 2006, World Soil Resources Reports2nd edition, No. 103. FAO, Rome, Italy, 2006.

Jaiarree, S., Chidthaisong, A., Tangtham, N., Polprasert, C., Sarobol, E., and Tyler S. C.: Carbon Budget and sequestration potential in a sandy soil treated with compost, Land Degrad. Develop., 25, 120–129, 2014.

Jobbágy, E. G. and Jackson, R. B.: The Vertical Distribution of Soil Organic Carbon and Its Relation to Climate and Vegetation, Ecol. Appl., 10, 423–436, 2000.

Kirkpatrick, J. B., Green, K., Bridle, K. L., and Venn, S. E.: Patterns of variation in Australian alpine soils and their relationships to parent material, vegetation formation, climate and topography, Catena, 121, 186–194, 2014.

Lal, R.: Managing soils and ecosystems for mitigating anthropogenic carbon emissions and advancing global food security, Bioscience, 60, 708–721, 2010.

Leifeld, J., Bassin, S., and Fuhrer, J.: Carbon stocks in Swiss agricultural soils predicted by land use: soil characteristics and altitude, Agric. Ecosyst. Environ., 105, 255–266, 2005.

Lemenih, M. and Itanna, F.: Soil carbon stock and turnovers in various vegetation types and arable lands along an elevation gradient in southern Ethiopia, Geoderma 123, 177–188, 2004.

Li, P., Wang, Q., Endo, T., Chao, X., and Kakubari, Y.: Soil organic carbon stock is closely related to aboveground vegetation properties in cold-temperature mountainous forests, Geoderma, 154, 407–415, 2010.

López-Vicente, M., Navas, A., Machín, J.: Effect of physiographic conditions on the spatial variation of seasonal topsoil moisture in Mediterranean soils, Austr. J. Soil Res., 47, 498–507, 2009.

Lozano-García, B. and Parras-Alcántara, L.: Land use and management effects on carbon and nitrogen in Mediterranean Cambisols, Agric. Ecosyst. Environ., 179, 208–214, 2013.

Lozano-García, B. and Parras-Alcántara, L.: Variation in soil organic carbon and nitrogen stocks along a toposequence in a traditional Mediterranean olive grove, Land Degrad. Develop., 25, 297–304, 2014.

Martínez-Mena, M., López, J., Almagro, M., Boix-Fayos, C., and Albadalejo, K.: Effects of water erosion and cultivation on the soil carbon stock in a semiarid area of South-East Spain, Soil Till. Res., 99, 119–129, 2008.

McKenzie, N. J. and Austin, M. P.: A quantitative Australian approach to medium and small scale surveys based on soil stratigraphy and environmental correlation, Geoderma, 57, 329–355, 1993.

Minasny, B., McBratney, A. B., Mendonça-Santos, M. L., Odeh, I. O. A., and Guyon, B.: Prediction and digital mapping of soil carbon storage in the Lower Namoi Valley, Aust. J. Soil Res., 44, 233–244, 2006.

Muñoz-Rojas, M., Jordán, A., Zavala, L. M., De la Rosa, D., Abd-Elmabod, S. K., and Anaya-Romero, M.: Impact of land use and land cover changes on organic C stocks in Mediterranean soils (1956–2007), Land Degrad. Develop., 2012a.

Muñoz-Rojas, M., Jordán, A., Zavala, L. M., De la Rosa, D., Abd-Elmabod, S. K., and Anaya-Romero, M.: Organic carbon stocks in Mediterranean soil types under different land uses (Southern Spain), Solid Earth, 3, 375–386, doi:10.5194/se-3-375-2012, 2012b.

Muñoz-Rojas, M., Jordán, A., Zavala, L. M., González-Peñaloza, F. A., De la Rosa, D., Pino-Mejias, R., and Anaya-Romero, M.: Modelling soil organic carbon stocks in global change scenarios: a CarboSOIL application, Biogeosciences, 10, 8253–8268, doi:10.5194/bg-10-8253-2013, 2013.

Nelson, D. W. and Sommers, L. E.: Total carbon, organic carbon and organic matter, in: Methods of soil analysis, Part 2: Chemicaland microbiological properties, edietd by: Page, A. L., Miller, R. H., and Keeney, D., Agronomy monograph, vol. 9. ASA and SSSA, Madison WI, 539–579, 1982.

Nerger, R., Núñez, M. A., and Recio, J. M.: Presencia de carbonatos en suelos desarrollados sobre material granítico del Batolito de los Pedroches (Córdoba), in: Tendencias Actuales de la Ciencia del Suelo, edietd by: Jordán, A. and Bellifante, N., Universidad de Sevilla, 768–774, 2007.

Nieto, O. M., Castro, J., and Fernández-Ondoño, E.: Conventional tillage versus cover crops in relation to carbón fixation in Mediterranean olive cultivation, Plant Soil, 365, 321–335, 2013.

Nizeyimana, E. and Bicki, T. J.: Soil and soil-landscape relationships in the north central region of Rwanda, East-central Africa. Soil Sci., 153, 224–236, 1992.

Ovales, F. A. and Collins, M. E.: Soil-landscape relationships and soil variability in North Central Florida, Soil Sci. Soc. Am. J., 50, 401–408, 1986.

Ozgoz, E., Gunal, H., Acir, N., Gokmen, F., Birol, M., and Budak, M.: Soil quality and spatial variability assessment of effects in a typic Haplustall, Land Degrad. Develop., 24, 277–286, 2013.

Pachepsky, Y. A., Timlin, D. J., and Rawls, W. J.: Soil water retention as related to topographic variables, Soil Sci. Soc. Am. J., 65, 1787–1795, 2001.

Parras-Alcántara, L., Corral, L., and Gil, J.: Ordenación territorial del Parque Natural de Despeñaperros (Jaén): Criterios metodológicos, Ed. Instituto de Estudios Giennenses, Jaén, 2004.

Parras-Alcántara, L., Martín-Carrillo, M., and Lozano-García, B.: Impacts of land use change in soil carbon and nitrogen in a Mediterranean agricultural area (Southern Spain), Solid Earth, 4, 167–177, doi:10.5194/se-4-167-2013, 2013a.

Parras-Alcántara, L., Díaz-Jaimes, L., and Lozano-García, B.: Organic farming affects c and n in soils under olive groves in mediterranean areas, Land Degrad. Develop., available online in Wiley Online Library (wileyonlinelibrary.com) doi:10.1002/ldr.2231, 2013b.

Parras-Alcántara, L., Díaz-Jaimes, L., Lozano-García, B., Fernández, P., Moreno, F., and Carbonero, M.: Organic farming has little effect on carbon stock in a Mediterranean dehesa (southern Spain), Catena 113, 9–17, 2014.

Paustian, K., Six, J., Elliot, E. T., and Hunt, H. Q.: Management options for reducing CO_2 emissions from agricultural soils, Biogeochemistry, 48, 147–163, 2000.

Post, W. M., Emanuel W. R., Zinke P. J., and Stangenberger, A. J.: Soil carbon pools and world life zones, Nature, 298, 156–159, 1982.

Power, J. and Schlesinger, W. H.: Relationships among soil carbon distribution and biophysical factors at nested spatial scales in rain forest of northeastern Costa Rica, Geoderma, 109, 165–190, 2002.

Recio, J. M., Corral, L., and Paneque, G.: Estudio de suelos en la Comarca de los Pedroches (Córdoba.), An. Edaf. Agrob., 45, 989–1012, 1986.

Ruhe, R. V. and Walker, P. H.: Hillslope models and soil formation: I. open systems, in: Trans. Int. Congr. Soil Sci. 9th Adelaide, edited by: Holmes, J. W., 4. Elsevier, NY, 551–560, 1968.

Ruiz-Sinoga, J. D. and Diaz, A. R.: Soil degradation factors along a Mediterranean pluviometric gradient in Southern Spain, Geomorphology, 118, 359–368, 2010.

Ruiz-Sinoga, J. D., Pariente, S., Diaz, A. R., and Martínez-Murillo, F. J.: Variability of relationships between soil organic carbon and some soil properties in Mediterranean rangelands under different climatic conditions (South of Spain), Catena, 94, 17–25, 2012.

Rusco, E., Jones, R. J., and Bidoglio, G.: Organic Matter in the soils of Europe: Present status and future trends, EUR 20556 EN. JRC, Official Publications of the European Communities, Luxembourg, 2001.

Sherstha, B. M., Sitaula, B. K., Singh, B. R., and Bajracharya, R. M.: Soil organic carbon stocks in soil aggregates under different land use systems in Nepal, Nutr. Cycl. Agroecosys., 70, 201–213, 2004.

SPSS Inc.: SPSS for windows, Version 13.0. Chicago, SPSS Inc., 2004.

Srinivasarao, C. H., Venkateswarlu, B., Lal, R., Singh, A. K., Kundu, S., Vittal, K. P. R., Patel, J., and Patel, M. M.: Long-term manuring and fertilizer effects on depletion of soil organic stocks under Pearl millet-cluster vean-castor rotation in Western India, Land Degrad. Develop., 25, 173–183, 2014.

Tsui, C. C., Tsai, C. C., and Chen, Z. S.: Soil organic carbon stocks in relation to elevation gradients in volcanic ash soils of Taiwan, Geoderma, 209/210, 119–127, 2013.

Umakant, M., Ussiri, D., and Lal, R.: Tillage effects on soil organic carbon storage and dynamics in Corn Belt of Ohio USA, Soil Till. Res., 107, 88–96, 2010.

USDA: Soil survey laboratory methods manual, Soil survey investigation report no. 42, Version 4.0. USDA-NCRS, Lincoln, NE, 2004.

Venterea, R. T., Lovett, G. M., Groffman, P. M., and Schwarz, P. A.: Landscape patterns of net nitrification in a northern hardwood conifer forest, Soil Sci. Soc. Am. J., 67, 527–539, 2003.

Wang, Q., Wang, S., Xu, G., and Fan, B.: Conversion of secondary broadleaved forest into Chinese fir plantation alters litter production and potential nutrient returns, Plant Ecol., 209, 269–278, 2010.

Wang, Y. Q. and Shao, M. A.: Spatial variability of soil physical properties in a region of the loess plateau of PR China subjet to wind and water erosion, Land Degrad. Develop., 24, 296–304, 2013.

Yan-Gui, S., Xin-Rong, L., Ying-Wu, C., Zhi-Shan, Z., and Yan, L.: Carbon fixation of cyanobacterial-algal crusts after desert fixation and its implication to soil organic matter accumulation in Desert, Land Degrad. Develop., 24, 342–349, 2013.

Ziadat, F. M.: Analyzing digital terrain attributes to predict soil attributes for a relatively large area, Soil Sci. Soc. Am. J., 69, 1590–1599, 2005.

Ziadat, F. M. and Taimeh, A. Y.: Effect of rainfall intensity, slope and land use and antecedent soil moisture on soil erosion in an arid environment, Land Degrad. Develop., 24, 582–590, 2013.

5

Impact of the addition of different plant residues on nitrogen mineralization–immobilization turnover and carbon content of a soil incubated under laboratory conditions

M. Kaleeem Abbasi, M. Mahmood Tahir, N. Sabir, and M. Khurshid

Department of Soil and Environmental Sciences, University of Poonch, Rawalakot Azad Jammu and Kashmir, Pakistan

Correspondence to: M. Kaleem Abbasi (mkaleemabbasi@gmail.com)

Abstract. Application of plant residues as soil amendment may represent a valuable recycling strategy that affects carbon (C) and nitrogen (N) cycling in soil–plant systems. The amount and rate of nutrient release from plant residues depend on their quality characteristics and biochemical composition. A laboratory incubation experiment was conducted for 120 days under controlled conditions (25 °C and 58 % water-filled pore space) to quantify initial biochemical composition and N mineralization of leguminous and non-leguminous plant residues, i.e., the roots, shoots and leaves of *Glycine max*, *Trifolium repens*, *Zea mays*, *Populus euramericana*, *Robinia pseudoacacia* and *Elaeagnus umbellata*, incorporated into the soil at the rate of 200 mg residue N kg^{-1} soil. The diverse plant residues showed a wide variation in total N, C, lignin, polyphenols and C / N ratio with higher polyphenol content in the leaves and higher lignin content in the roots. The shoot of *Glycine max* and the shoot and root of *Trifolium repens* displayed continuous mineralization by releasing a maximum of 109.8, 74.8 and 72.5 mg N kg^{-1} and representing a 55, 37 and 36 % recovery of N that had been released from these added resources. The roots of *Glycine max* and *Zea mays* and the shoot of *Zea mays* showed continuous negative values throughout the incubation. After an initial immobilization, leaves of *Populus euramericana*, *Robinia pseudoacacia* and *Elaeagnus umbellata* exhibited net mineralization by releasing a maximum of 31.8, 63.1 and 65.1 mg N kg^{-1}, respectively, and representing a 16, 32 and 33 % N recovery, respectively. Nitrogen mineralization from all the treatments was positively correlated with the initial residue N contents ($r = 0.89$; $p \leq 0.01$) and negatively correlated with lignin content ($r = -0.84$; $p \leq 0.01$), C / N ratio ($r = -0.69$; $p \leq 0.05$), lignin / N ratio ($r = -0.68$; $p \leq 0.05$), polyphenol / N ratio ($r = -0.73$; $p \leq 0.05$) and (lignin + polyphenol) : N ratio ($r = -0.70$; $p \leq 0.05$) indicating a significant role of residue chemical composition and quality in regulating N transformations and cycling in soil. The present study indicates that incorporation of plant residues strongly modifies the mineralization–immobilization turnover (MIT) of soil that can be taken into account to develop synchronization between net N mineralization and crop demand in order to maximize N delivery and minimize N losses.

1 Introduction

Application of organic materials as soil amendments is an important management strategy that can improve and uplift soil-quality characteristics and alter the nutrient cycling through mineralization or immobilization turnover of added materials (Khalil et al., 2005; Campos et al., 2013; Baldi and Toselli, 2014; Novara et al., 2013; Hueso-González et al., 2014; Oliveira et al., 2014). Use of local organic materials derived either from livestock or plants have been attaining worldwide support for improving the fertility and productivity potential of degraded and nutrient-poor soils (Huang et al., 2004; Tejada and Benítez, 2014). Indeed, plant residues and animal manures are potentially important sources of nutrients for crop production in smallholder agriculture. However, the Hindu Kush Himalayan regions, including the state of Azad Jammu and Kashmir, have a wide diversity of leguminous species and non-leguminous plants compared to the livestock production. Hence, use of plant residues as organic nutrient source is relatively simple for the farmers compared to the application of manure. Incorporating plant residues into agricultural soils can sustain organic carbon content, improve soil physical properties, enhance biological activities and increase nutrient availability (Hadas et al., 2004; Cayuela et al., 2009). In the short-term, incorporation of plant residues provides the energy and nutrients for microbial growth and activity, acts as a driving force for the mineralization–immobilization processes in the soil and is a source of nitrogen (N) for plants (Jansson and Persson, 1982). In the long-term, incorporation of crop residues is important for the maintenance of organic carbon (C) and N stocks in the nutrient pool of arable soils (Rasmussen and Parton, 1994).

Incorporation of crop residues provides readily available C and N to soils depending upon the decomposition rates and synchrony of nutrient mineralization (Murungu et al., 2011). The N availability from these residues depends on the amount of N mineralized or immobilized during decomposition. However, previous studies demonstrated that the decomposition and nutrient release rates of residues are often regulated by environmental factors, such as temperature and soil moisture, and biochemical composition of plant materials and their interaction (Abiven et al., 2005; Khalil et al., 2005). The biochemical composition or quality parameters such as total N concentration, lignin (LG), polyphenols (PP), carbon : nitrogen (C / N) ratio, LG / N, PP / N and (LG + PP) / N ratios are considered useful indicators that control decomposition and N release of added residues (Nakhone and Tabatabai, 2008; Vahdat et al., 2011; Abera et al., 2012). However, it has not been clearly established which of these variables correlate best with N mineralization of plant residues (Nakhone and Tabatabai, 2008), as contrasting results have been reported in the literature (Nourbakhsh and Dick, 2005). On the one hand, it has been reported that N released from leguminous tree leaves indicated that the (lignin + polyphenol) : N ratio was the most important factor in predicting N mineralization (Mafongoya et al., 1998). On the other hand, Frankenberger and Abdelmagid (1985) suggested that lignin content of the legumes is not a good predictor of the N mineralization. Handayanto et al. (1994) suggested that the N concentration or lignin : N ratio of the leaves were not good indicators of N release for agroforestry materials. Palm and Sanchez (1991) attributed the differences in N mineralization rates of various tropical legumes to polyphenols. Handayanto et al. (1994) found, however, that the total N content of plant residues was not correlated with rates of N released under non-limiting N conditions.

Earlier studies clearly demonstrated the beneficial effects of plant residues on soil–plant systems (Huang et al., 2004; Cayuela et al., 2009; Khalil et al., 2005; Baldi and Toselli, 2014). However, there is still a scope to explore the possibilities for achieving maximum benefits in term of rate, time and amount of N released. For example, the synchronization of net N mineralization with plant/crop growth is desirable to maximize N delivery for the crop and minimize N losses.

Abiven et al. (2005) reported that one of the tools to achieve synchronization is the use of plant residues with different natures and qualities. Application of residues with a high C / N ratio results in immediate net N immobilization while residues with a low C / N ratio result in net N mineralization, showing that mineralization–immobilization turnover (MIT) can be influenced differently by chemical components of added plant materials. To achieve this target, the combination of legumes and non-legumes plant materials or different plant components of the same plant species, i.e., root, shoot and leaves, can be tested.

Keeping in mind the beneficial effects of plant residues on soil–plant systems, especially in the mountainous upland soils vulnerable to soil (water) erosion, the present work aims to (i) examine the initial biochemical composition and quality characteristics of on-farm available plant residues and to (ii) quantify the N-release potential (mineralization) of these residues added to a soil incubated under controlled laboratory conditions (25 °C) in Rawalakot, Azad Jammu and Kashmir, Pakistan.

2   Materials and methods

2.1   Soil sampling

The soil used in this study was collected from an arable field located at the research farm of the Faculty of Agriculture of the University of Poonch, Rawalakot, Azad Jammu and Kashmir, Pakistan. The study site is located at latitude $33°51'32.18''$ N, longitude $73°45'34.93''$ E and an elevation of 1638 m above sea level. The climate of the region is subtemperate. Mean daily maximum and minimum air temperatures ranged from 27 to 29 °C (June–July) and 1.0 to −3.5 °C (January–February). The mean annual rainfall ranged between 1100 and 1500 mm with more than 50 % of the total precipitation during monsoon each year. The soil in the study site was clay loam in texture, classified as Humic Lithic Eutrudepts (Inceptisols; Ali et al., 2006). The field was bare at the time of sampling but previously maize (*Zea mays* L.) and wheat (*Triticum aestivum* L.) were cultivated. The selected field was divided into 10 subplots to ensure proper and representative soil sampling. Soil samples were collected from a depth of 0–15 cm at random from three points in each plot using a soil auger of 5 cm in diameter. The soil samples from all the selected plots were thoroughly mixed to get a composite sample. The field-fresh soil was passed through a 4 mm sieve to eliminate coarse rock and plant material, thoroughly mixed to ensure uniformity and stored at 4 °C before use (not more than 2 weeks). A subsample of about 0.5 kg was taken, air dried, passed through a 2 mm sieve and used for the determination of physical and chemical characteristics. The original soil analysis is presented in Table 1.

Table 1. Selected physicochemical properties of the soil used in the study.

Soil properties	Values
Bulk density ($Mg\,m^{-3}$)	1.20
Particle density ($Mg\,m^{-3}$)	2.48
Porosity (%)	48.3
Sand ($g\,kg^{-1}$)	241
Silt ($g\,kg^{-1}$)	394
Clay ($g\,kg^{-1}$)	365
Texture class	clay loam
pH	7.2
CEC ($cmol\,kg^{-1}$)	7.3
Organic matter ($g\,kg^{-1}$)	10.4
Organic C ($g\,kg^{-1}$)	6.03
Total N ($g\,kg^{-1}$)	0.58
C : N ratio	10 : 1
Total mineral N ($mg\,kg^{-1}$)	8.7
Total organic N ($mg\,kg^{-1}$)	591.0
P ($mg\,kg^{-1}$)	3.4
K ($mg\,kg^{-1}$)	88.0
Fe ($mg\,kg^{-1}$)	15.7
Mn ($mg\,kg^{-1}$)	17.0
Cu ($mg\,kg^{-1}$)	1.02
Zn ($mg\,kg^{-1}$)	1.16

2.2 Collection of plant residues

Six predominant on-farm available plant species were selected. These included *Glycine max* shoot, *Glycine max* root, *Trifolium repens* shoot, *Trifolium repens* root, *Zea mays* shoot, *Zea mays* root, and leaves of *Populus euramericana, Robinia pseudoacacia* and *Elaeagnus umbellata*. Plant samples/residues were collected at different times during the year 2012. *Glycine max* and *Trifolium repens* samples were collected from the field before flowering (summer) while *Zea mays* samples were taken 1 week before crop harvest. The tree leaves were sampled in late fall. Plant residues were washed with running tap water, rinsed three times with distilled water, dried at 65 °C for 48 h, milled and passed through a 1 mm sieve. Triplicate samples of plant residue were taken and analyzed for their C, N, lignin and polyphenol concentrations. Total N contents of the residues were determined by Kjeldhal digestion, distillation and the titration method (Bremner and Mulvaney, 1982). Wet digestion method was used for organic C analysis (Nelson and Sommers, 1982). The lignin content was determined using Van Soest methods (Van Soest et al., 1991). Soluble polyphenols were extracted in hot water (100 °C, 1 h) and determined by colorimetry using a Folin–Denis reagent (Folin and Denis, 1915).

2.3 Laboratory incubation

The incubation methods used in this study were followed by the methods used in our previous studies (Abbasi et al., 2011; Abbasi and Khizar, 2012). Briefly stated, about 100 g of soil already stored in the refrigerator at 4 °C was weighed and transferred into 200 mL glass jars. The initial moisture content of the soil was 28 % (w/w), which was increased by adding distilled water to achieve a final water-filled pore space of 58 %. The treatments were comprised of a control (no N) and nine plant residues sources, i.e., *Glycine max* shoot, *Glycine max* root, *Trifolium repens* shoot, *Trifolium repens* root, *Zea mays* shoot, *Zea mays* root, and leaves of *Populus euramericana, Robinia pseudoacacia* and *Elaeagnus umbellata*; 10 incubation timings, i.e., 0, 7, 14, 21, 28, 42, 60, 80, 100 and 120 days; and three replications. Altogether, a total of 300 jars (10 treatments × 10 incubation timings × 3 replications) were arranged in a completely randomized design. Plant residues were weighed and added into the jars at a rate equivalent to $200\,mg\,N\,kg^{-1}$. After adding residues, all the jars were weighed and their weights were recorded. The soil was then incubated under controlled conditions at 25 °C. Soil moisture was checked/adjusted after every 2 days by weighing the glass jars and adding the required amount of distilled water when the loss was greater than 0.05 g.

2.4 Soil extraction and analysis

Samples of all 10 treatments were analyzed for total mineral nitrogen (TMN) as described previously (Abbasi and Khizar, 2012). Initial concentration of TMN ($NH_4^+-N + NO_3^--N$) on day 0 was determined by extracting soil samples with 200 mL of 1 M KCl added directly to the flask immediately after incorporation of each N source. Thereafter, triplicate samples from each treatment were removed randomly from the incubator at different incubation timings and extracted by shaking for 1 h with 200 mL of 1 M KCl followed by filtration. The total mineral N of the extract was determined by using the steam distillation and titration method (Keeney and Nelson, 1982). Net cumulative N mineralized (NCNM) from different plant-residue treatments was calculated following the method described previously (Sistani et al., 2008).

2.5 Statistical analysis

All data were statistically analyzed by multifactorial analysis of variance using the software package MSTATC Version 3.1 (1990). Least-significant differences (LSD) were used as a post hoc test to indicate significant variations within the values of either treatments or time intervals. Correlation (r) between initial quality characteristics of the plant residues (total nitrogen, LG, PP and their ratios) and net N mineralization and the correlation among quality traits were also conducted using SPSS Statistics version 20.0 for Mac (IBM

Table 2. Mean biochemical composition of the plant residues used in the experiment ($n = 3$).

Plant residues (treatments)	Plant organs	Total N	Total C (LG)	Lignin (PP)	Polyphenols	C/N	LG/N	PP/N	LG+PP/N
					g kg^{-1}				
Glycine max	shoot	35.2a	447c	11f	13.1f	12.7	0.3	0.4	0.7
Glycine max	root	12.8e	466b	29d	26.9d	36.4	2.3	2.1	4.4
Zea mays	shoot	9.6f	472ab	41b	29.5cd	49.2	4.3	3.1	7.3
Zea mays	root	4.0g	486a	48a	31.4c	121.5	12.0	7.9	19.9
Trifolium repens	shoot	27.4b	397g	13f	18.0e	14.4	0.4	0.6	1.1
Trifolium repens	root	16.0d	423de	21e	20.2e	26.4	1.3	1.2	2.5
Populus euramericana	leaves	20.8c	435cd	34c	53.8a	20.9	1.6	2.6	4.2
Robinia pseudoacacia	leaves	33.3a	404fg	28d	32.3c	12.1	0.8	1.0	1.8
Elaeagnus umbellata	leaves	34.7a	418ef	32cd	38.7b	12.1	0.9	1.1	2.0
LSD ($p \leq 0.05$)	–		3.14	14.16	4.53	3.77	–	–	–	–

Note: different letters in each column show significant differences among treatments with $p \leq 0.05$

Corp., 2011). A probability level of $p \leq 0.05$ was considered significant (Steel and Torrie, 1980).

3 Results and discussion

3.1 Chemical composition of the residues – residue quality

A significant difference ($p \leq 0.05$) among different residue treatments was observed for different components of the plant residues presented in Table 2. The total N ranged from a minimum of 4.0 to a maximum of 35.2 g kg^{-1}. Shoots of *Glycine max* and leaves of *Robinia pseudoacacia* and *Elaeagnus umbellata* displayed the highest N compared to the remaining treatments (Table 2). The total C contents varied between 397 g kg^{-1} in the *Trifolium repens* shoot and a maximum of 486 g kg^{-1} in the *Zea mays* root. *Zea mays* (both shoot and root) displayed the highest C contents compared to the remaining plant-residue treatments. The C:N showed a similar trend recorded for residue C content. The LG content varied between a minimum of 11 g kg^{-1} in the *Glycine max* shoot and a maximum of 48 g kg^{-1} in the *Zea mays* roots. Similarly, a minimum PP content (13.1 g kg^{-1}) was recorded in the *Glycine max* shoot while a maximum PP (52.8 g kg^{-1}) was found in the *Populus euramericana* leaves. The LG/N, PP/N and LG+PP/N ratios were highest in the *Zea mays* root while the lowest values were recorded in the *Glycine max* shoot. Generally, total N contents of the legume residues were higher compared to the non-legumes. Similarities could be observed between the same organs of the different species, i.e., all the roots were characterized by high C, LG and PP contents and lower N concentration. Leaves were particularly rich in PP and total N. The differences in the concentration of quality characteristics of residues according to plant components, i.e., shoot, root and leaves, have been reported previously (Abiven et al., 2005;

Nourbakhsh and Dick, 2005). It has been reported that high lignin content in root was due to the presence of suberin in the roots and its ability to form complex barriers when associated with lignin (Abiven et al., 2005). Plant residues used in this study provided a wide range of contrasted chemical composition and significant variation in quality characteristics because of the difference in (i) type of species, i.e., leguminous and non-leguminous, trees and crops, and (ii) plant components/organs, i.e., shoot, root and leaves.

3.2 Nitrogen mineralization

Analysis of variance showed that N mineralization was significantly ($p \leq 0.05$) affected by the treatments and the incubation timings, while the interaction between the treatments and the timings was also significant. Results indicated that the control soil without any amendment released a maximum of 77.7 mg N kg^{-1} on day 100 compared to 13.7 mg kg^{-1} at the start, showing a substantial release of N into mineral N pool (Table 3). Expressed as the total N initially present, the net N mineralized during the incubation was 14 %. The mineralization of native soil N observed here was in accordance with our previous study where a maximum of 90 mg kg^{-1} mineral N was released from the control soil, representing 16 % of the initial N of the soil (Abbasi and Khizar, 2012). Among different plant materials added, the legumes, i.e., the shoot of *Glycine max* and shoot and root of *Trifolium repens*, exhibited significantly higher TMN compared to the non-legumes. The maximum TMN released from these amendments varied between 150 and 189 mg kg^{-1}. The mean values indicated that these legumes were collectively able to release 85 mg N kg^{-1} compared to 20 mg kg^{-1} by maize and 58 mg N kg^{-1} by leaves of the non-legumes trees. As expected, the plant organs also affected N mineralization and, in general, roots displayed significantly lower TMN compared to the shoot and leaves. Incorporation of *Glycine max*

Table 3. Mean changes in the concentration of total mineral N of a soil amended with different plant residues and incubated at 25 °C under controlled laboratory conditions during a 120-day period ($n = 3$).

Treatments	Days after plant-residue addition										LSD ($p \leq 0.05$)
	0	7	14	21	28	42	60	80	100	120	
	mg N kg^{-1} soil										
Control	13.7	13.9	12.9	17.1	30.9	65.9	63.1	75.6	77.7	51.7	2.88
T_1	14.8	39.2	49.2	76.8	96.7	158.1	165.2	174.1	188.7	160.9	7.90
T_2	13.7	8.1	5.2	8.3	11.8	13.8	28.4	50.4	49.4	27.7	8.15
T_3	13.7	7.4	6.2	6.9	10.5	23.1	21.2	36.1	46.7	21.0	5.34
T_4	14.3	7.4	9.4	7.7	8.8	15.3	22.2	21.4	32.4	26.4	4.30
T_5	14.1	19.0	21.6	55.5	62.5	86.8	127.6	150.8	145.8	93.3	7.31
T_6	15.5	8.2	5.2	23.9	34.0	85.3	98.0	149.9	130.2	85.8	9.46
T_7	13.0	5.7	4.1	8.6	22.6	55.5	73.1	106.8	87.3	66.9	8.39
T_8	13.9	7.4	9.2	23.6	46.6	91.3	111.0	138.9	127.8	93.7	7.83
T_9	12.9	9.4	14.5	25.3	51.1	80.1	92.7	140.0	116.4	93.5	6.88
LSD ($p \leq 0.05$)	2.43	4.77	3.12	5.11	7.63	8.23	6.87	9.23	8.27	7.34	

T_0 is the control; T_1 is *Glycine max* shoot, T_2 is *Glycine max* root; T_3 is *Zea mays* shoot, T_4 is *Z. mays* root; T_5 is *Trifolium repens* shoot; T_6 is *Trifolium repens* root; T_7 are *Populus euramericana* leaves; T_8 are *Robinia pseudoacacia* leaves; T_9 are *Elaeagnus umbellata* leaves. LSD represents the least significant difference ($p \leq 0.05$) among incubation periods (within rows) and among the treatments (within column).

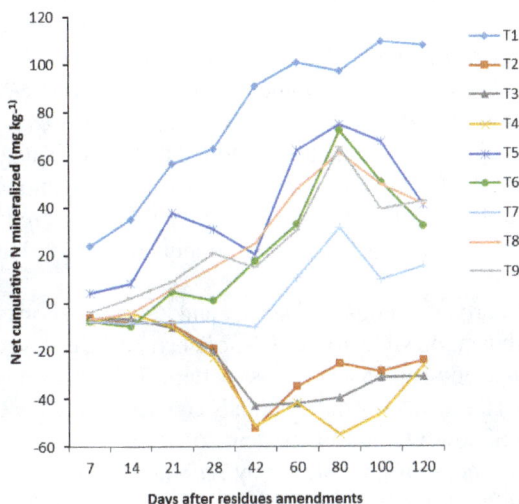

Figure 1. Net cumulative N mineralized from the added plant residues at different incubation periods. Legend: T_1 is *Glycine max* shoot, T_2 is *Glycine max* root; T_3 is *Zea mays* shoot, T_4 is *Zea mays* root; T_5 is *Trifolium repens* shoot; T_6 is *Trifolium repens* root; T_7 are *Populus euramericana* leaves; T_8 are *Robinia pseudoacacia* leaves; T_9 are *Elaeagnus umbellata* leaves.

root and *Zea mays* shoot and root resulted in a constant decrease in TMN, and the maximum values ranged between 32 and 49 mg kg^{-1} compared to 78 mg kg^{-1} in the control treatment. However, after initial negative values until day 14 and 21, leaves of *Populus euramericana*, *Robinia pseudoacacia* and *Elaeagnus umbellata* continuously increased TMN until reaching between 107 and 140 mg kg^{-1} (highest values).

3.3 Net cumulative N mineralization

Nitrogen mineralization of added plant residues was determined on the basis of net cumulative N mineralized. The N mineralization from *Glycine max* and *Trifolium repens* shoot showed positive values throughout the incubation, ranging from 24 to 110 mg kg^{-1} for *Glycine max* and 5 to 75 mg kg^{-1} for *Trifolium repens* (Fig. 1). Considering the NCNM at the end day 120, the net N mineralized as percentage of total N applied from *Glycine max* and *Trifolium repens* shoot was 54 and 21 %, respectively. The percent of N mineralized from *Glycine max* shoot had been reported previously and ranged from 39 to 43 % of applied N residues (Nakhone and Tabatabai, 2008). However, the NCNM from *Glycine max* roots, *Zea mays* shoot and *Zea mays* roots exhibited negative values throughout the incubation, indicating net immobilization. Among the three residues, *Zea mays* roots displayed higher negative values leading to higher immobilization. Roots of *Glycine max* and leaves of *Populus euramericana*, *Robinia pseudoacacia* and *Elaeagnus umbellata* showed four phases of mineralization–immobilization turnover: initial negative values from days 7 to 21, slow mineralization from days 21 to 60, a rapid mineralization between days 60 and 80 and a decline in net between days 100 and 120. The net N mineralized as percentage of total N applied from roots of *Glycine max* and leaves of *Populus euramericana*, *Robinia pseudoacacia* and *Elaeagnus umbellata* was 16, 8, 21 and 21 %, respectively. Net nitrogen mineralization (% of added N) from different organic materials during 110 days of incubation was in the range of −35 % in *Triticum aestivum* (wheat) residues to 81 % in *Trifolium repens* (white clover) residues (Kumar and Goh, 2003). Similarly, a 44, 38 and 35 % of N added had been released from

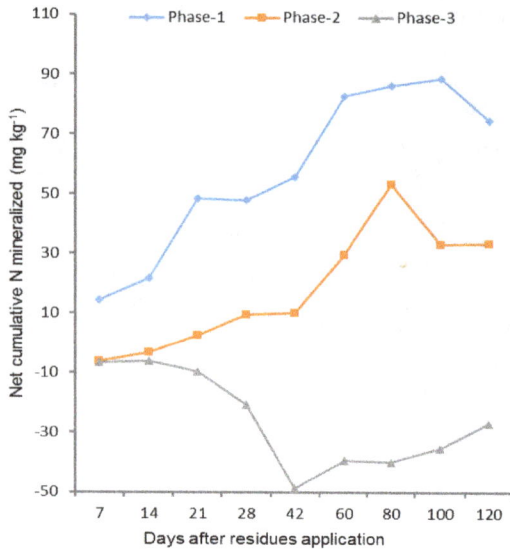

Figure 2. The mineralization–immobilization turnover of added plant residues representing three phases during 120 days incubation.

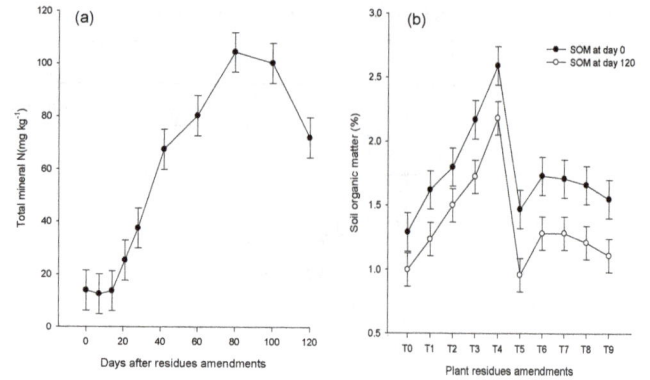

Figure 3. Mineralization trend of added plant residues across timings (**a**) and soil organic matter (SOM) turnover of different plant residues recorded at the start of the experiment on day 0 and at the end of incubation on day 120 (**b**). The hanging bar on each major line represents the LSD ($p \leq 0.05$) between incubation periods and between each treatment.

the leaves of peanut, pigeon pea and hairy indigo, respectively (Thippayarugs et al., 2008).

All legumes (except *Glycine max* root) exhibited the highest NCNM (average 30 % of added plant N residues) compared to non-legumes (17 %). Similarly, the cereal crop *Zea mays* shoot and root exhibited net immobilization compared to net mineralization observed in the legumes and tree leaves. The plant components also showed variation in NCNM. For example, shoots of *Glycine max* and *Trifolium repens* mineralized an average of 74 mg N kg^{-1} compared to 4 mg N kg^{-1} from the roots. Likewise, leaves of forest trees showed higher NCNM compared to the roots of legumes and non-legumes crop.

The shoots of *Glycine max* and *Trifolium repens* exhibited the highest NCNM without any negative value during incubation because of high N concentration and a low C / N ratio. However, it is interesting to note that the total N concentration of the leaves of *Robinia pseudoacacia* and *Elaeagnus umbellata* was higher and C / N ratio was lower compared to the *Trifolium repens* shoot, but the net mineralization (averaged) of *Trifolium repens* shoot was higher (47 and 58 %) compared to the leaves of *Robinia pseudoacacia* and *Elaeagnus umbellata*, respectively. The low mineralization in leaves in spite of high N content and low C / N ratio was attributed to higher concentration of LG, PP, LG / N, PP / N and LG+PP / N. These results demonstrated the effect of other factors in addition to total N and C / N ratio on plant-residue decomposition and N mineralization kinetics. As indicated in a previous study (Trinsoutrot et al., 2000), the net accumulation (whether positive or negative) of mineral N in soil during decomposition of organic residues is directly related to the residue N content. However, our results clearly

indicated that N was not the only factor affecting the mineralization of added residues; some additional quality characteristics also influenced MIT of plant residues. Likewise, the total N content and C / N ratio of the leaves of *Robinia pseudoacacia* and *Elaeagnus umbellata* were on par with *Glycine max* shoot but the net mineralization of *Glycine max* shoot was 3-fold higher. It had been reported that organic materials with similar C / N ratios may mineralize different amounts of N because of differences in composition that are not reflected by the C / N ratio (e.g., different lignin contents) (Mohanty et al., 2011).

Similarly, roots of *Glycine max* and *Zea mays* showed net immobilization while roots of *Trifolium repens* displayed fast decomposition and net N-release pattern. This discrepancy in root MIT was mainly due to high N concentration, low C / N ratio and low LG and PP contents of the roots of *Trifolium repens*. The N turnover shown by *Trifolium repens* roots confirmed the strong below-ground N dynamics and residual effect of *Trifolium repens* when grown in the soil.

Among the leaves of different trees tested, leaves of *Robinia pseudoacacia* and *Elaeagnus umbellata* released a substantial amount of N into the mineral N pool. Leaf residues have been described as high-quality litter materials in terms of high N and low lignin contents (Thippayarugs et al., 2008) and have been found to decompose easily and release mineral N substantially (Mtambanengwe and Kirchmann, 1995) as observed in our study. However, *Populus euramericana* leaves exhibited higher net immobilization (for a longer period) and lower net mineralization. The variation was again due to disparity in the biochemical composition. The low N content, high C / N ratio and high PP content may have been largely responsible for the slow decomposition and low net mineralization of *Populus euramericana* leaves. These results inferred that the same plant components may not necessarily show similar decomposition and

Table 4. Pearson linear correlation coefficients between initial quality characteristics of the plant residues and net N mineralization and correlation within plant-quality characteristics.

	N_{min}	TN	LG	PP	C:N	LG:N	PP:N
TN	0.89**						
LG	−0.84**	−0.66*					
PP	−0.42ns	−0.10ns	0.62*				
C:N	−0.69*	−0.80**	0.73*	0.07ns			
LG:N	−0.68*	−0.76**	0.77**	0.14ns	0.99**		
PP:N	−0.73*	−0.77**	0.82**	0.29ns	0.99**	0.98**	
LG+PP:N	−0.70*	−0.76**	0.79**	0.19ns	0.99**	1.00**	0.99**

** and * represent significant levels at $p \leq 0.01$ and $p \leq 0.05$, respectively; the correlation significance and non-significance level was calculated at $p \leq 0.05$. The abbreviations represent N mineralization (N_{min},), total nitrogen (TN), lignin (LG) and polyphenols (PP).

mineralization turnover because of the variation in biochemical composition.

In general, the added plant residues increased organic matter stock in soil and thereby increased N mineralization and N transformation processes in soil. Plant or crop residues, when added or incorporated into the soil, increase the organic matter (avoid the climate change), reduce the soil and water losses and increase the biological activity in the soils. Such changes bring a substantial improvement in the physical, chemical and microbial properties of soil and eventually in the soil quality (Giménez Morera et al., 2010; Jiménez et al., 2013; Zhao et al., 2013; Singh et al., 2014; Prats et al., 2014)

3.4   Pattern and trend of N mineralization

The patterns of N mineralization varied among plant residues and plant components. After incorporation into soil and during incubation, the added residues exhibited three main patterns of cumulative net mineralization (Fig. 2): (i) a pattern of the continuous and rapid release of net N throughout the incubation without showing any negative value indicating net mineralization, shown by the *Glycine max* shoot and *Trifolium repens* shoot; (ii) a pattern shown by the *Trifolium repens* roots and *Populus euramericana*, *Robinia pseudoacacia* and *Elaeagnus umbellata* leaves indicated initial negative values of net cumulative immobilization for variable periods followed by slow and then a rapid release of N, indicating immobilization–mineralization turnover; (iii) a pattern of continuous negative values throughout the incubation, indicating net N immobilization as seen in the case of the *Glycine max* root and the *Zea mays* shoot and root. The MIT and N-release patterns by plant residues observed here were in accordance with those reported previously in both leguminous and non-leguminous plant residues (Kumar and Goh, 2003).

The N mineralization trend over time showed wide variation (Fig. 3a). These results highlighted the time taken for releasing N into the mineral N pool by the added plant residues. Results showed an initial lag phase where most of the applied residues endured immobilization with little mineralization;

only the *Glycine max* and *Trifolium repens* shoots showed mineralization during 0 to 21 days of incubation. The rapid mineralization phase occurred from day 28 to day 80. Thereafter a declining phase of mineralization started toward the later part of the incubation from day 100 to day 120.

3.5   Changes in soil organic matter

In order to examine the changes in soil organic matter (SOM) in response to added plant residues, a comparison between the SOM at the start of day 0 and the end of incubation on day 120 has been shown (Fig. 3b). Soil organic matter contents of all the treatments recorded on day 120 were lower than those recorded on day 0. The unaccounted SOM ranged between 32 and 67 % compared to that recorded on day 0. The decreasing trend of SOM was substantially higher for the treatments showing mineralization (54–67 %) compared to those showing immobilization (32–38 %). By the end of day 120, the loss of SOM was in the following order: *Trifolium repens* shoot > *Elaeagnus umbellata* leave > *Trifolium repens* root = *Robinia pseudoacacia* leaves > *Populus euramericana* > *Glycine max* shoot > *Zea mays* shoot > *Zea mays* root = *Glycine max* root. The SOM turnover observed here coincided with net mineralization. In the initial lag phase when mineralization was either very low or displayed negative values, on average only 8 % of the initial SOM had been utilized (7–21 days). The SOM utilization during days 28–80 when mineralization was rapid was 31 % of the initial amount, while 43 % of initial SOM was utilized in the later part of incubation (between days 100 and 120) when mineralization start showing a declining trend.

3.6   Relationship between cumulative N mineralization and residue-quality characteristics

Results of the study showed highly significant positive correlation between N mineralization and plant-residue N concentrations ($r = 0.89$; $p \leq 0.01$) (Table 4). In contrast, a negative significant correlations existed between net cumulative N mineralized and LG ($r = −0.84$; $p \leq 0.01$), NCNM and

C / N ratio ($r = -0.69$; $p \leq 0.05$), NCNM and LG / N ratio ($r = -0.68$; $p \leq 0.05$), NCNM and PP/N ratio ($r = -0.73$; $p \leq 0.05$) and NCNM and LG + PP / N ratio ($r = -0.70$; $p \leq 0.05$). The correlation between N mineralization and PP was nonsignificant with $p \leq 0.05$. The significant positive correlation between net rates of N mineralization and residue N concentration observed is consistent with other studies (Nourbakhsh and Dick, 2005; Vahdat et al., 2011). It has been reported that N availability may control the decomposition of plant residues, particularly those with low N content such as cereals, when the N requirements of the soil decomposers are not met by the residue or soil N contents (Vahdat et al., 2011). A negative correlation was also observed between net N mineralization and C / N ratio of the plant materials. Previously, total N contents and C / N ratio were considered adequate for predicting the net N mineralization of crop residue. However, the latest studies, including the present work, highlight the role of other quality characteristics, including LG and PP, that affect net mineralization of plant residues. The closer relationship between net mineralization and residue lignin contents ($r = -0.84$; $p \leq 0.01$) than that of the C / N ratio ($r = -0.69$; $p \leq 0.05$) recorded in this study was in accordance with previous findings (Vahdat et al., 2011). The highly significant positive correlation between net N mineralization and the residue N content ($r = 0.89$; $p \leq 0.01$) confirms the previous results (Nourbakhsh and Dick, 2005; Vahdat et al., 2011), indicating that residue N concentration can be considered a better tool to predict mineralization of added organic residues compared to the C / N ratio.

4   Conclusions

The experiment showed that soil amended with plant residues displayed wide variation of N mineralization depending on the plant species and plant components/organs. The decomposition and N-release potential of added materials were largely related to their biochemical composition. In addition to residue N concentration and C / N ratio, LG contents of plant residues also appeared to be an important factor in predicting the net N mineralization of plant residues. Shoots of *Glycine max and Trifolium repens* and leaves of *Robinia pseudoacacia* and *Elaeagnus umbellata* exhibited a substantial mineralization potential, demonstrating that legumes and trees of these two plant species can produce high-quality residues and thus have the potential to promote N cycling in agroecosystems. This study suggested that plant residues showing rapid mineralization can be used for early N demands of a crop, while residues with high C : N and LG contents immobilize N and thus can help to counter the N loss generally observed due to rapid ammonification–nitrification turnover. Use of such plant materials in our cropping systems, especially in the regions subjected to land degradation, may be a useful management strategy to restore these soils for agriculture production.

Acknowledgements. The authors express their appreciation to Nuclear Institute for Food and Agriculture, Peshawar, Pakistan for providing lab facilities to analyze biochemical characteristics of the plant materials used in the study.

Edited by: P. Pereira

References

Abbasi, M. K. and Khizar, A.: Microbial biomass carbon and nitrogen transformations in a loam soil amended with organic–inorganic N sources and their effect on growth and N-uptake in maize, Ecol. Eng., 39, 23–132, 2012.

Abbasi, M. K., Hina, M., and Tahir, M. M.: Effect of *Azadirachta indica* (neem), sodium thiosulphate and calcium chloride on changes in nitrogen transformations and inhibition of nitrification in soil incubated under laboratory conditions, Chemospher, 82, 1629–1635, 2011.

Abera, G., Wolde-meskel, E., and Bakken, L. R.: Carbon and nitrogen mineralization dynamics in different soils of the tropics amended with legume residues and contrasting soil moisture contents. Biol. Fertil. Soils, 48, 51–66, 2012.

Abiven, S., Recous, S., Reyes, V., and Oliver, R.: Mineralisation of C and N from root, stem and leaf residues in soil and role of their biochemical quality, Biol. Fertil. Soils, 42, 119–128, 2005.

Ali, B., Mohmand, H., and Muhammad, F.: Integrated land resource survey and evaluation of Azad Jammu & Kashmir area 2004, Soil Survey of Pakistan, Government of Pakistan, Ministry of Food, Agriculture & Livestock, 156–157, 2006.

Baldi, E. and Toselli, M.: Mineralization dynamics of different commercial organic fertilizers from agro-industry organic waste recycling: an incubation experiment, Plant Soil Environ., 60, 93–99, 2014.

Bremner, J. M. and Mulvaney, C. S.: Nitrogen–total, in: Methods of Soil Analysis Part 2 Chemical and Microbiological Properties, edited by: Page, A. L., Miller, R. H., and Keeney, D. R., SSSA Madison, WI, 595–624, 1982.

Campos, A. C., Etchevers, J. B., Oleschko, K. L., and Hidalgo, C. M.: Soil microbial biomass and nitrogen mineralization rates along an altitudinal gradient on the cofre de perote volcano (Mexico): the importance of landscape position and land use, Land Degrad. Dev., 25, 581–593, doi:10.1002/ldr.2185, 2013.

Cayuela, M. L., Sinicco, T., and Mondini, V.: Mineralization dynamics and biochemical properties during initial decomposition of plant and animal residues in soil, Appl. Soil Ecol., 48, 118–127, 2009.

Folin, O. and Denis, W.: A colorimetric estimation of phenol and phenol and derivatives in urine, J. Biol. Chem., 22, 305–308, 1915.

Frankenberger Jr., W. T., and Abdelmagid, H. M.: Kinetic parameters of nitrogen mineralization rates of leguminous crops incorporated into soil, Plant Soil, 87, 257–271, 1985.

Giménez Morera, A., Ruiz Sinoga, J. D., and Cerdà, A: The impact of cotton geotextiles on soil and water losses in Mediterranean rainfed agricultural land, Land Degrad. Dev., 210–217, doi:10.1002/ldr.971, 2010.

Hadas, A., Kautsky, L., Goek, M., and Kara, E. E.: Rates of decomposition of plant residues and available nitrogen in soil, related to residue composition through simulation of carbon and nitrogen turnover, Soil Biol. Biochem., 36, 255–266, 2004.

Handayanto, E., Cadish, G., and Giller, K. E.: Nitrogen release from prunings of hedgerow trees in relation to the quality of the pruning and incubation method, Plant Soil, 160, 237–248, 1994.

Huang, Y., Zou, J., Zheng, X., Wang, Y., and Xu, X.: Nitrous oxide emissions as influenced by amendment of plant residues with different C : N ratios, Soil Biol. Biochem., 36, 973–981, 2004.

Hueso-González, P., Martínez-Murillo, J. F., and Ruiz-Sinoga, J. D.: The impact of organic amendments on forest soil properties under Mediterranean climatic conditions, Land Degrad. Dev., 25, 604–612, doi:10.1002/ldr.2296, 2014.

Jansson, S. L. and Persson, J.: Mineralization and immobilization of soil nitrogen, in: Nitrogen in Agricultural Soils, edited by: Stevenson, F. J., ASA, SSSA Special Publication No. 22, Madison WI, 229–252, 1982.

Jiménez, M. N., Fernández-Ondoño, E., Ripoll, M. Á., Castro-Rodríguez, J., Huntsinger, L., and Navarro, F. B.: Stones and organic mulches improve the Quercus ilex L, afforestation success under Mediterranean climatic conditions, Land Degrad. Dev., online first, doi:10.1002/ldr.2250, 2013.

Keeny, D. R. and Nelson, D. W.: Nitrogen – inorganic forms, in: Methods of Soil Analysis Part 2 Chemical and Microbiological Properties, eddied by: Page, A. L., Miller, R. H., and Keeney D. R., SSSA Madison WI, 643–693, 1982.

Khalil, M. I., Hossain, M. B., and Schmidhalte, U.: Carbon and nitrogen mineralization in different upland soils of the subtropics treated with organic materials, Soil Biol. Biochem. ,37, 1507–1518, 2005.

Kumar, K. and Goh, K. M.: Nitrogen release from crop residues and organic amendments as affected by biochemical composition, Commun. Soil Sci. Plant Anal., 34, 2441–2460, 2003.

Mafongoya, P. L., Nair, P. K. R., and Dzowela, B. H.: Mineralization of nitrogen from decomposing leaves of multipurpose trees as affected by their chemical composition, Biol. Fertil. Soils, 27, 143–148, 1998.

Mohanty, M., Reddy, K. S., Probert, M. E., Dalal, R. C., Subba Rao, A., and Menzie, N. W.: Modeling N mineralization from green manure and farmyard manure from a laboratory incubation study, Ecol. Model., 222, 719–726, 2011.

MSTATC: A microcomputer program for the design, management, and analysis of agronomic research experiments, Michigan State University, Michigan, USA, 1990.

Mtambanengwe, F. and Kirchmann, H.,: Litter from a tropical savanna woodland (miombo): chemical composition and C and N mineralization, Soil Biol. Biochem., 27, 1639–1651, 1995.

Murungu, F. S., Chiduza, C., Muchaonyerwa, P., and Mnkeni, P. N. S.: Decomposition, nitrogen, and phosphorus mineralization from residues of summer-grown cover crops and suitability for a smallholder farming system in South Africa, Commun. Soil Sci. Plant Anal., 42, 2461–2472, 2011.

Nakhone, L. N. and Tabatabai, M. A.: Nitrogen mineralization of leguminous crops in soils, J. Plant Nutr. Soil Sci., 171, 231–241, 2008.

Nelson, D. N. and Sommer, L. E.: Total carbon, organic carbon and organic matter, in: Methods of Soil Analysis Part 2 Chemical and Microbiological Properties, edited by: Page, A. L., Miller, R. H., and Keeney, D. R., SSSA Madison, WI, 539–589, 1982.

Nourbakhsh, F., and Dick, R. P.: Net nitrogen mineralization or immobilization potential in a residue-amended calcareous soil, Arid. Land. Res. Manage., 19, 299–306, 2005.

Novara, A., Gristina, L., Rühl, J., Pasta, S., D'Angelo, G., La Mantia, T., and Pereira, P.: Grassland fire effect on soil organic carbon reservoirs in a semiarid environment, Solid Earth, 4, 381–385, doi:10.5194/se-4-381-2013, 2013.

Oliveira, S. P., Lacerda, N. B., Blum, S. C., Escobar, M. E. O., and Oliveira, T. S.: Organic carbon and nitrogen stocks in soils of northeastern brazil converted to irrigated agriculture, Land Degrad. Dev., 26, 9–21, doi:10.1002/ldr.2264, 2014.

Palm, C. A. and Sanchez, P. A.: Nitrogen release from the leaves of some tropical legumes as affected by their lignin and polyphenolic contents, Soil Biol. Biochem., 23, 83–88, 1991.

Prats, S. A., Malvar, M.C., Simões-Vieira, D. C., MacDonald, L., and Keizer, J. J.: Effectiveness of hydro- mulching to reduce runoff and erosion in a recently burnt pine plantation in central Portugal, Land Degrad. Dev., online first, doi:10.1002/ldr.2236, 2014.

Rasmussen, P. E. and Parton, W. J.: Long term effects of residue management in wheat-fallow: I. Inputs, yield, and soil organic matter, Soil Sci. Soc. Am. J., 58, 523–530, 1994.

Singh, K., Trivedi, P., Singh, G., Singh, B., and Patra, D. D.: Effect of different leaf litters on carbon, nitrogen and microbial activities of sodic soils, Land Degrad. Dev., online first, doi:10.1002/ldr.2313, 2014.

Sistani, K. R., Adeli, A., McGowen, S. L., Tewolde, H., and Brink, G. E.: Laboratory and field evaluation of broiler litter nitrogen mineralization, Bioresour. Technol., 99, 2603–2611, 2008.

Steel, R. G. D. and Torrie, J. H.: Principles and Procedure of Statistics, McGrraw Hill Book Co Inc, New York, 1980.

Tejada, M. and Benítez, C.: Effects of crushed maize straw residues on soil biological properties and soil restoration, Land Degrad. Dev., 25, 501–509, doi:10.1002/ldr.2316, 2014.

Thippayarugs, S., Toomsan, B., Vityakon, P., Limpinuntana, V., Patanothai, A., and Cadisch, G. G.: Interactions in decomposition and N mineralization between tropical legume residue components, Agr. Forest. Syst., 72, 137–148, 2008.

Trinsoutrot, I., Recous, S., Mary, B., and Nicolardot, B.: C and N flux of decomposing ^{13}C and ^{15}N Brassica napus L.: effect of residue composition and N content, Soil Biol. Biochem., 32, 1717–1730, 2000.

Vahdat, E., Nourbakhsh, F., and Basiri, M.: Lignin content of range plant residues controls N mineralization in soil, Eur. J. Soil Biol., 47, 243–246, 2011.

Van Soest, P. J., Robertson, J. B., and Lewis, B.,A.: Methods for dietary fiber, neutral detergent fiber and non-starch polysaccharides in relation to animal nutrition, J. Dairy Sci., 74, 3584–3597, 1991.

Zhao, G., Mu, X., Wen, Z., Wang, F., and Gao, P.: Soil erosion, conservation, and eco-environment changes in the loess plateau of China, Land Degrad. Dev., 24, 499–510, doi:10.1002/ldr.2246, 2013.

6

Kinetics of potassium release in sweet potato cropped soils: a case study in the highlands of Papua New Guinea

B. K. Rajashekhar Rao

Department of Agriculture, Papua New Guinea University of Technology, Lae 411, Papua New Guinea

Correspondence to: B. K. Rajashekhar Rao (rsraobk@rediffmail.com)

Abstract. The present study attempts to employ potassium (K) release parameters to identify soil-quality degradation due to changed land use patterns in sweet potato (*Ipomoea batatas* (L.) Lam) farms of the highlands of Papua New Guinea. Rapid population increase in the region increased pressure on the land to intensify subsistence production mainly by reducing fallow periods. Such continuous cropping practice coupled with lack of K fertilization practices could lead to a rapid loss of soil fertility and soil-resource degradation. The study aims to evaluate the effects of crop intensification on the K-release pattern and identify soil groups vulnerable to K depletion. Soils with widely differing exchangeable and non-exchangeable K contents were sequentially extracted for periods between 1 and 569 h in 0.01 M CaCl$_2$, and K-release data were fitted to four mathematical models: first order, power, parabolic diffusion and Elovich equations. Results showed two distinct parts in the K-release curves, and 58–80 % of total K was released to solution phase within 76 h (first five extractions) with 20–42 % K released in the later parts (after 76 h). Soils from older farms that were subjected to intensive and prolonged land use showed significantly ($P < 0.05$) lower cumulative K-release potential than the farms recently brought to cultivation (new farms). Among the four equations, first-order and power equations best described the K-release pattern; the constant b, an index of K-release rates, ranged from 0.005 to 0.008 mg kg^{-1} h^{-1} in the first-order model and was between 0.14 and 0.83 mg kg^{-1} h^{-1} in the power model for the soils. In the non-volcanic soils, model constant b values were significantly ($P < 0.05$) higher than the volcanic soils, thus indicating the vulnerability of volcanic soils to K deficiency. The volcanic soils cropped for several crop cycles need immediate management interventions either through improved

fallow management or through mineral fertilizers plus animal manures to sustain productivity.

1 Introduction

Sweet potato (*Ipomoea batatas* (L.) Lam) is the major staple food crop in highlands of Papua New Guinea (PNG) with production and consumption of tubers well over 1.5 million tonnes (Bourke, 2005). The vine tips of sweet potato are also an integral part of the human diet besides being a feed in traditional pig husbandry. In PNG, much of the sweet potato production is through subsistence agriculture with hardly any input of mineral fertilizers and little or no manure use. Traditionally, cropping areas are cleared of shrubs and other vegetation and the slashed vegetation is burnt to give a nutrient-rich ash (Bailey et al., 2008). The sweet potato farms may be old farms (cultivated over many seasons and about to be fallowed) or new farms (newly brought into cultivation). Over several years of continuous farming and in the absence of any mineral nutrient inputs, fertility of old farms generally decreases and farmers abandon such farm areas for fallow. The population of the highlands region, however, has been increasing by ~ 3 % each year, thus placing increasing pressure on the land resources to produce extra food for the growing populace, as observed in other parts of the world (Abu Hammad and Tumeizi, 2012). Simultaneously, crop productivity appears to be declining, which has been attributed to a degradation of soil fertility linked to the progressive shortening of the fallow rejuvenation periods (Allen et al., 1995; Sem, 1996; Bourke, 2005; Walter et al., 2011). Pressure on land resources has increased dramatically because of the population growth; fallow periods between cropping cycles have been shortened from several decades to less than 1 year in the recent past. Such decline in soil fertility and productivity induced by land use change has been reported in Africa, Mediterranean regions and Asia (Biro et al., 2013; Abu Hammad and Tumeizi, 2012; Liu et al., 2014)

Previous work conducted across four of the highland provinces (Southern Highlands, Eastern Highlands, Simbu and Enga) established potassium (K) deficiency as the major nutrient-related cause for the poor sweet potato productivity in almost a third of sweet potato farms (Bailey et al., 2009; Ramakrishna et al., 2009; Walter et al., 2011). These studies also reported that K deficiency was more of a problem in old farms (which have been cropped for several crop cycles) than in new farms (which are ready for cropping after fallow periods). Potassium requirement for the tuber crops such as sweet potatoes is larger than for other food crops. Sweet potato crops yielding $12\,Mg\,ha^{-1}$ tubers can mine ca. $100\,kg$ K in storage roots and vines, and more than $375\,kg$ K can be removed by sweet potatoes yielding $50\,Mg\,ha^{-1}$ (O'Sullivan et al., 1997). Negative K balances subsequent to several crop cycles have been reported globally. Depletion of K stocks in soil resources has been reported due to suboptimal application rates of K fertilizers and manures in India (Srinivasarao et al., 2013, 2014) and lack of fertilizer application in Africa (Hanao and Baanante, 1999). Changes in land use systems can also affect the status and form of K (Rezapour and Samadi, 2014).

In the absence of any external K inputs in PNG, crop production solely depends on native K supply potential of soils and their release rates to a soil solution from non-exchangeable pools. Non-exchangeable K from reserves makes an important contribution to plant K supply (Mengel and Uhlenbecker, 1993). For optimal nutrition of the crop, the replenishment of a K-depleted soil solution is affected predominately by the release of non-exchangeable K from clay minerals and organic matter. Under intensive cropping with tropical conditions of high rainfall and leaching, labile "K pool" may be rapidly depleted. In the absence of fertilization, how well it is replenished depends largely on the amount of K in non-exchangeable pools and their release rates (Steffens and Sparks, 1997). As many well-weathered tropical soils have predominantly kaolinitic clay and low K reserves (Malavolta, 1985), it is expected that their K solution would be rapidly depleted, especially under intensive cropping. From a sustainability perspective, it is essential to ascertain if soil reserves alone are sufficiently large and sufficiently accessible to sustain sweet potato production in the medium- to long-term (decades to centuries) in the absence of external inputs (fertilizers). Because plants use varying proportions of non-exchangeable K, measurement of exchangeable K (NH_4OAc-K) is not always a reliable measurement of plant availability or accessibility. Thus, more information is needed on the nature and rates of non-exchangeable K release in these soils. The K-release kinetics studies with different extractants including organic acids, nitric acid and

dilute salt solutions such as $CaCl_2$ could be used for generating such information (Lopez-Pineiro and Navarro, 1997). For long-term management of K under intensive or prolonged sweet potato cropping, knowledge of the release potential and release rates of K from soil mineral pools is vital.

Therefore, the present study was initiated with the objectives of (1) evaluating the K-supplying powers of sweet potato garden soils of the highlands region by the K-release kinetics approach and (2) elucidating the relationship between K-supply potentials and rates of K release in these soils with the soil types, soil sampling depth and garden types.

2   Material and methods

2.1   Study location and sampling sites

A range of soil samples used by Walter et al. (2011) from the four highland provinces (Western Highlands, Eastern Highlands, Simbu and Enga) of PNG, with widely differing K fractions were selected for the present study (Table 1). Sites were chosen with a range of available K statuses from optimum (NH_4OAc-K $> 125\,mg\,kg^{-1}$) to very deficient (NH_4OAc-K $< 50\,mg\,kg^{-1}$) and with an equal number of old and new farm sites, situated on a range of soil types or parent materials of volcanic and non-volcanic origin. The volcanic soils chosen in the study belonged to the great soil groups Hydrandepts and Endoaquepts in the Enga and Western Highlands provinces (Soil Survey Staff, 2014). Those derived from non-volcanic parent material (e.g., Dystropepts, Eutropepts and/or Tropaqualfs) were dominant in Simbu and Eastern Highlands. Eight soil samples were of volcanic and 17 of non-volcanic origin. In PNG, the majority of the soils belong to Inceptisols ($> 50\%$ of the land area), Entisols (25%) and Ultisols (14%), while Alfisols, Histosols and Mollisols occupy smaller parcels of land area (Bourke and Harwood, 2009). Andisols occupy 5.5% of the land area and yet are agriculturally quite significant because much of population cultivates these soils (Radcliffe and Kanua, 1998). Lithologically, PNG soils are derived from diverse parent material such as lava basalt, andesite, siliciclastic materials, siltstone, limestone, lithic sandstone, mudstone and shale, carbonaceous siltstone and sandstone, Triassic granodiorite, etc. (Davies, 2012).

In this study, older farms are those under continuous sweet potato cultivation without any fallow periods, while new gardens refer to those freshly brought in to cropping either after a fallow or native primary forest. The old farms sampled were under cropping for 3–4 years whilst most of the new farms were due for cropping after fallow for 1–5 years. Details of the site selection and soil sampling are provided in greater detail elsewhere (Walter et al., 2011). Briefly stated, at every farm site soil was sampled from one or two planting stations (at least 10 m apart) from surface (0–10 cm) and sub-

Table 1. Soil management history of the sweet potato farms and physico-chemical properties of the soil samples used in the study.

Sample no.	Province	Soil group	Soil depth	Soil type/origin	Farm type	pH	Total C (%)	NH$_4$OAc K (mg kg^{-1})	1 N HNO$_3$ K (mg kg^{-1})	Non-exchangeable K (mg kg^{-1})
1	EHP	Tropohumult	Surface	Non-volcanic	New	6.09	6.82	323	396	73.0
2	EHP	Dystropepts	Surface	Non-volcanic	Old	5.84	3.25	92.0	251	159
3	Enga	Hydrandept	Surface	Volcanic	Old	6.42	3.86	56.9	121	64.1
4	EHP	Tropoaqualf	Surface	Non-volcanic	New	6.46	2.33	88.8	493	404
5	WHP	Hydrandept	Surface	Volcanic	Old	6.13	3.82	68.1	124	55.9
6	EHP	Hydrandept	Surface	Non-volcanic	New	5.93	5.56	120	148	28.0
7	EHP	Hydrandept	Surface	Non-volcanic	Old	4.75	12.4	68.6	119	50.4
8	WHP	Tropoaqualf	Surface	Non-volcanic	Old	6.89	1.62	218	658	440
9	Enga	Hydrandept	Surface	Volcanic	Old	5.42	15.5	102	189	87.0
10	EHP	Tropoaqualf	Surface	Non-volcanic	New	6.29	4.18	166	419	253
11	EHP	Tropohumult	Subsurface	Non-volcanic	New	5.96	6.93	9.46	177	168
12	EHP	Dystropepts	Surface	Non-volcanic	Old	5.66	3.96	6.55	225	219
13	Enga	Hydrandept	Subsurface	Volcanic	Old	5.41	4.27	6.53	216	210
14	WHP	Hydrandept	Subsurface	Volcanic	Old	6.49	3.94	11.4	148	137
15	Enga	Hydrandept	Subsurface	Volcanic	Old	5.70	4.70	12.6	142	129
16	WHP	Tropoaqualf	Subsurface	Non-volcanic	New	6.36	2.55	5.12	642	637
17	Simbu	Hydrandept	Subsurface	Volcanic	New	6.62	2.93	16.8	333	316
18	EHP	Tropoaqualf	Subsurface	Non-volcanic	Old	6.83	1.55	17.1	601	584
19	Simbu	Hydrandept	Subsurface	Volcanic	Old	5.61	16.6	4.18	127	123
20	EHP	Tropepts	Surface	Non-volcanic	Old	6.00	3.27	110	117	7.00
21	EHP	Aqualfs	Surface	Non-volcanic	New	5.30	5.72	369	701	332
22	EHP	Aquepts	Surface	Non-volcanic	Old	6.50	3.96	86.0	113	27.0
23	Simbu	Andepts	Surface	Volcanic	New	5.40	5.79	215	318	103
24	Simbu	Aquepts	Surface	Non-volcanic	Old	6.50	1.74	161	328	167
25	Simbu	Aquepts	Surface	Non-volcanic	Old	5.80	2.58	168	217	49.0

EHP – Eastern Highlands province, WHP – Western Highlands province.

surface (10–20 cm) using a trowel. The air-dried soil samples were sieved (< 2 mm) and then analyzed for total carbon (C) by the dry combustion method (Nelson and Sommers, 1996) and for pH in a 1 : 5 soil : water extract. Water-soluble K was extracted with de-ionized water (1 : 5 w/v) after shaking for 30 min on a mechanical shaker. Non-exchangeable K was estimated as the difference between boiling 1N HNO$_3$-K and 1N NH$_4$OAc-K (Walter et al., 2011).

2.2 Potassium-release study

A sequential extraction of soil K reserves with a 0.01 M CaCl$_2$ solution was conducted according to Jalali (2005). About 2 g of a 2 mm sieved soil sample was treated with 20 mL of CaCl$_2$ solution in a 50 mL centrifuge tube. The soil suspension was equilibrated for 1 to 569 h at 25 °C. After the addition of CaCl$_2$ solution, the soil suspension was shaken in a rotary shaker for 15 min (200 rpm) and later centrifuged at 4000 rpm. K content in the supernatant solution was estimated by an inductively coupled plasma/optical emission spectrophotometer (Varian 700ES model). Sequential extractions were followed at 1, 4, 7, 21, 76, 165, 242, 333, 408 and 569 h. The K extracted over time was used to construct K-release curves. The K-release curves have two distinct parts: the initial part (1–76 h) was used to compute the amount of K in edge position and the latter part (76–569 h) was used to compute the amount of K in internal positions.

2.3 Mathematical and statistical analysis

The K-release data obtained from the analysis of K contents from the extracts were tested for the mathematical fit to different kinetic equations:

$$\text{power-function equation:} \ln q = \ln a + b \ln t, \tag{1}$$

$$\text{parabolic diffusion:} q = (a + b)t^{1/2}, \tag{2}$$

$$\text{first-order reaction:} \ln(q_0 - q_t) = (a - b)t, \tag{3}$$

$$\text{Elovich equation:} q = (a + b) \ln t, \tag{4}$$

where q_t is the cumulative K released (mg kg^{-1}) at time t (h), q_0 is the maximum cumulative K released (mg kg^{-1}) and a and b are constants. Four models were tested by least-square regression analysis to determine which equation describes the non-exchangeable K release in a better manner. Standard error of estimate (SE) was computed as SE $= [(q - q^*)^2 / (n - 2)]^{1/2}$, where q and q^* represent the measured and calculated amounts of non-exchangeable K in soil at time t, respectively, and n is the number of data points evaluated. Samples were grouped into old- and new-farm soil samples, volcanic and non-volcanic soil samples and surface and subsurface samples prior to statistical analysis. The K-release data at 76 and 569 h (representing K in edge positions and K in internal sites, respectively) and K-release constants (a and b) were analyzed by two-sample t test to reveal differences between means of the two independent groups of

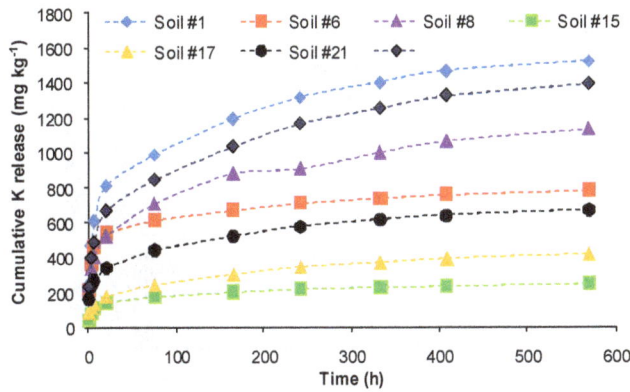

Figure 1. The cumulative K-release pattern in some representative sweet potato farm soils of PNG.

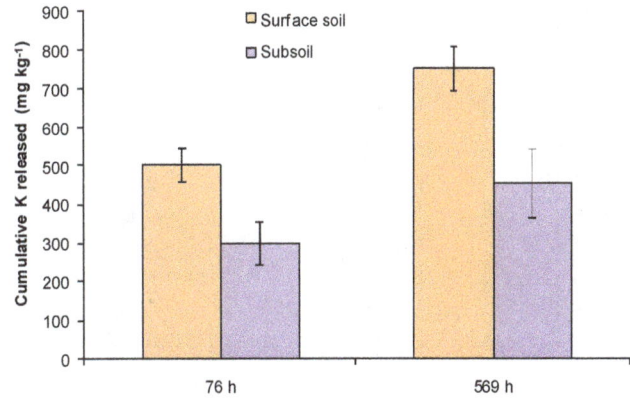

Figure 2. The cumulative K released at 76 and 569 h in surface and subsoils. At both study intervals, K release was significantly different ($P < 0.05$) between surface and subsoils. Error bars indicate standard errors for each group of soil.

samples. At $P < 0.05$, differences between groups were considered significantly different. Statistical analysis was carried out with Statistix 8 software for Windows.

3 Results and discussion

3.1 Potassium status of soils

Soils selected for this study varied with respect to geological origin and past management practices (Table 1). Soils selected were moderate to strongly acidic with pH values ranging from 4.75 to 6.62. The total C contents varied between 1.55 and 15.5 %. Two soil samples had surprisingly high total C contents of above 10 %, which is not unusual for soils of PNG (Ruxton, 2003). The NH_4OAc-K content varied widely and ranged from $2.3\,\mathrm{mg\,kg^{-1}}$ to as high as $369\,\mathrm{mg\,kg^{-1}}$. Non-exchangeable K in most of the samples ranged from low to medium (Srinivasarao et al., 2007; Walter et al., 2011). About 76 % of the samples were "low" (non-exchangeable K contents $<300\,\mathrm{mg\,kg^{-1}}$) in the non-exchangeable K supply, while 20 % of samples were "medium" (300–$600\,\mathrm{mg\,kg^{-1}}$) and only 4 % of samples were in the "high" ($> 600\,\mathrm{mg\,kg^{-1}}$) category. About 40 % of samples were "low" in plant-available K (exchangeable K content below $50\,\mathrm{mg\,kg^{-1}}$), 40 % of samples were "medium" (K content of 50–$125\,\mathrm{mg\,kg^{-1}}$) and only 20 % of samples were "high" (K content $> 125\,\mathrm{mg\,kg^{-1}}$) in exchangeable K content (Srinivasarao et al., 2007; Walter et al., 2011).

3.2 Potassium-release pattern

The cumulative amounts of K released during successive extractions from some representative soils are presented in Fig. 1. Cumulative K release was greatest ($1.53\,\mathrm{g\,kg^{-1}}$) at 569 h of incubation in soil #1 and smallest ($256\,\mathrm{mg\,kg^{-1}}$) in soil #15; thus, samples showed wide variation in total K release. Among the samples, maximum amounts of K (58–80 % of total K) were released to the solution phase within

76 h (first five extractions). Quantities ranging between 20 and 42 % were released in the later parts (after 76 h) of the study. K release greatly varied in the initial 4–5 extractions; later, they almost plateaued. To visualize the variations in K-release patterns with time in different parent materials, soil depths and farm management types, K released up to 76 h and between 76 and 569 h were separately subjected to a two-sample t test. This was necessary as the K-release curves had two distinct parts: the initial part (1–76 h) corresponding to K in edge position and the later part (76–569 h) representing the amount of K in internal positions. Because in these samples 58–80 % of cumulative K released before 76 h, it can be inferred that a major chunk of the plant-available K is present in edge positions. These soils may contain some illite and vermiculite minerals with surface, edge and interlayer sites that hold K (Jalali, 2005).

The surface soils had significantly ($P < 0.05$) greater cumulative K released both at edge positions (76 h) and K at interlayer positions (569 h) than that of subsoils, which is an indication of the exhaustion of soil K in the majority of the subsoils (Fig. 2). During crop plantings, the leftover residues, manures, wood ash and other inputs are generally spread on the soil surface and later covered with a thin layer of soil to form mounds. This practice probably leads to very little mixing of inputs and plant nutrients with the subsoil. Distinct absence of manure–soil mixing techniques, tillage and land preparation in PNG could also partly be the reason for K depletion in subsoils. Traditionally, farmers perform shallow manual digging with digging sticks and spades. Besides, during fallow periods substantial subsoil nutrients are mined by fallow vegetation species and added to topsoil. For example, a common fallow species *Piper aduncum* could add up to $377\,\mathrm{kg\,K\,ha^{-1}}$ through its root mass in the top 15 cm of soil (Hartemink, 2004). Besides, almost $300\,\mathrm{kg\,K\,ha^{-1}}$ could be added through the above-ground biomass by way of slash-

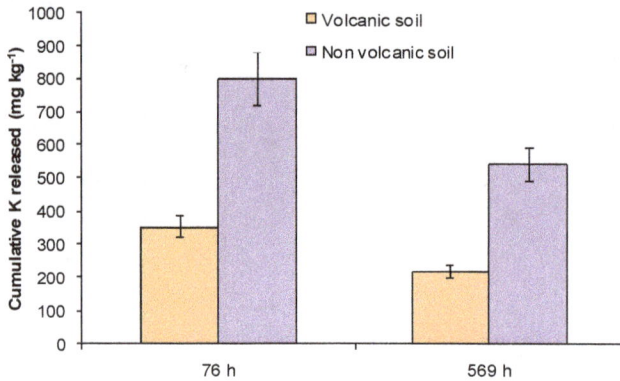

Figure 3. The cumulative K released at 76 and 569 h in volcanic and non-volcanic soils cropped in sweet potatoes. At both study intervals, K release was significantly different ($P < 0.05$) between volcanic and non-volcanic soils. Error bars indicate standard errors for each group of soil.

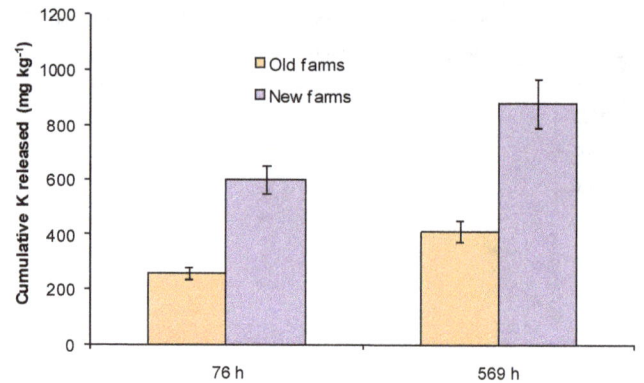

Figure 4. The cumulative K released at 76 and 569 h in old farms and new farms soils cropped in sweet potatoes. At both study intervals, K release was significantly different ($P < 0.05$) between old farms soils and new farms soils. Error bars indicate standard errors for each group of soil.

and-burn practices in the cultivation cycle, thus increasing K status and consequently K release.

Mean cumulative K release in soils of volcanic and non-volcanic origin were significantly ($P < 0.05$) different at 76 and 569 h (Fig. 3). The volcanic soils were poorer in cumulative releasable K compared to non-volcanic soils. Several of the volcanic soils are reported to have lower non-exchangeable K and low K-fixing abilities mainly due to the predominance of minerals such as volcanic glass, feldspars, pyroxenes and amphiboles (Moss and Coulter, 1964; Zharikova and Golognaya, 2009), and such minerals show inherently lower K-release potentials. The soils from older gardens had significantly ($P < 0.05$) lower quantities of K on edge (up to 76 h) and internal sites (76–569 h) compared to new garden soils (Fig. 4). Severe K depletion or exhaustion noted in older farms may be due to continuous crop mining with very few additions of fertilizers and manures. Possibly, the short fallow periods do not provide ample opportunity for revitalization of soil fertility with respect to K. Continuous crop cultivation is known to exhaust exchangeable and non-exchangeable K reserves in the sugarcane fields in Fiji (Gawander et al., 2002), the calcareous soils of sugar beet (Samadi et al., 2008) and the sweet potato gardens of PNG (Walter et al., 2011). Besides crop mining, an inevitable soil erosion followed by vegetation clearing and cropping are potential causes of land productivity decline when land covers are changed (Leh et al., 2013; Ziadat and Taimeh, 2013).

3.3 Modeling potassium release

The K-release data of some representative soils fitted to mathematical models in describing the release mechanism are shown in Fig. 5. The data fitted to first-order and parabolic diffusion models demonstrate two distinct parts representing two phases of K release, which corroborates with Rubio and Gil-Sotres (1997) and Jalali (2005). The

coefficient of determination (R^2) and standard error values showed that all equations could be fitted well to the observed K-release rates (Table 2). However, power and first-order equations were the best of the kinetic equations to describe the K-release pattern in 0.01 M CaCl$_2$. These two equations showed the overall highest values of R^2 and lowest values of SE. The order of application of various kinetics models to describe K-release data in 0.01 M CaCl$_2$ is power-function > first-order > parabolic diffusion > Elovich models. The constant b represents the slope and can be used as an index of ionic-K-release rates, ranged from 0.005 to 0.008 mg kg^{-1} h^{-1} in the first-order model. The constant a (the intercept value) ranged from 5.01 to 6.92 mg kg^{-1} in the first-order model, while the range was from 3.76 to 5.69 mg kg^{-1} in the power model. These ranges were lower than that observed in some Iranian soils (Jalali and Zarabi, 2006; Jalali, 2008). A successful description of K release with power equation in soils (Hosseinpur et al., 2012) and individual soil size fractions (sand, silt and clay) has been reported by Najafi-Ghiri and Jaberi (2013). K-release pattern was also successfully modeled through the Elovich equation to discriminate between K fertilizer management zones in oil palm plantations of Milne Bay, PNG (Steven, 2010).

In non-volcanic soils, model constant b (in first-order, parabolic diffusion and Elovich equations) values were significantly ($P < 0.05$) higher than the volcanic soils (Table 3). Interestingly, the power equation was not able to distinguish between the soils types with regard to the rate of K release. The values of constant a were also greater for the non-volcanic soils compared to volcanic soils in all four models in the study, thus confirming a greater K supply potential of non-volcanic soils. Volcanic ash soils in many sites of the highlands of PNG were known to be high in non-crystalline aluminosilicate mineral (allophone/imogolite), hydroxy interlayered vermiculite and volcanic glass (Bleeker and Sage-

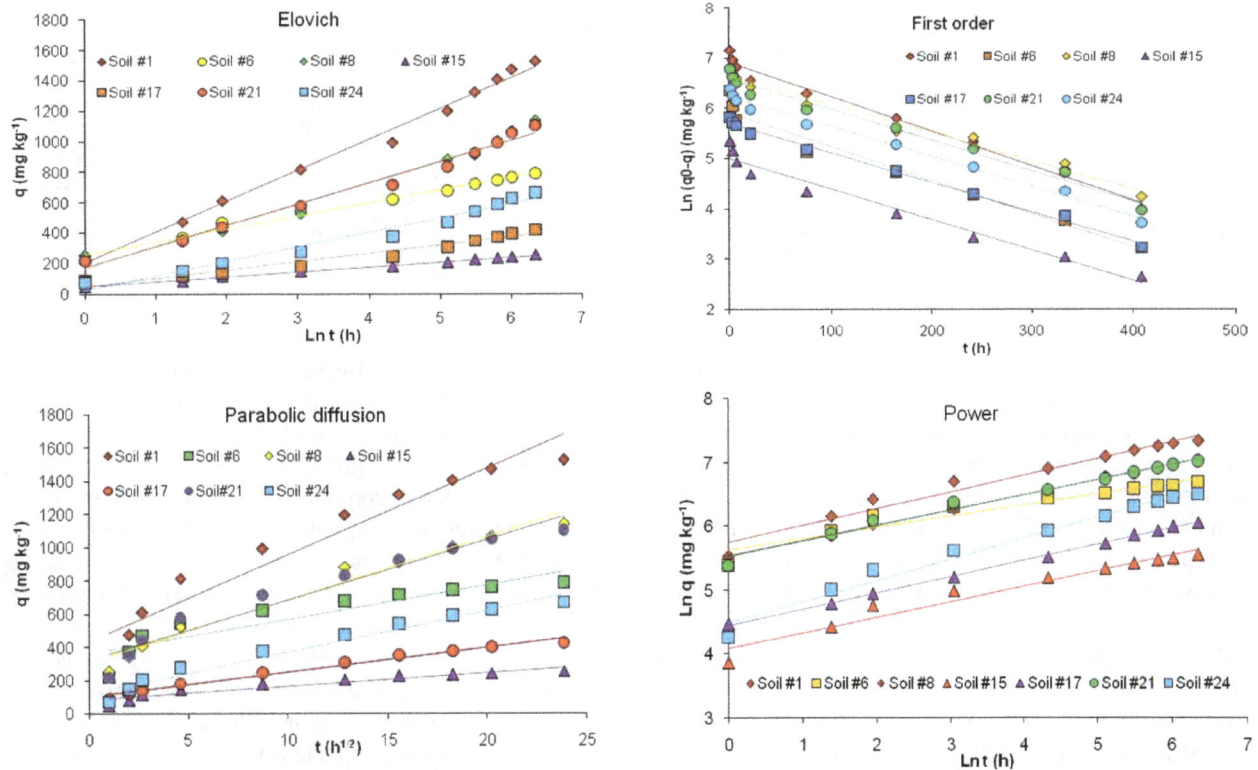

Figure 5. The K-release data fitted to four kinetic models in some sweet potato farm soils

Table 2. Descriptive statistics of model parameters (constant a, constant b, coefficient of determination, R^2, and standard error, SE) for the data fitted to different K-release equations ($N = 25$).

Release models	Model parameters	Minimum	Maximum	Mean
First order	constant a (mg kg^{-1})	5.01	6.92	5.88
	constant b (mg kg^{-1} h^{-1})	−0.005	−0.008	−0.006
	R^2	0.956***	0.991***	0.978***
	SE	0.04	0.11	0.068
Power	constant a (mg kg^{-1})	3.76	5.69	4.94
	constant b (mg kg^{-1} h^{-1})	0.14	0.83	0.26
	R^2	0.919***	0.997***	0.972***
	SE	0.01	0.12	0.057
Parabolic diffusion	constant a (mg kg^{-1})	58.50	432.40	210.10
	constant b (mg kg^{-1} h$^{-1/2}$)	8.23	52.24	20.99
	R^2	0.836***	0.979***	0.933***
	SE	8.19	69.3	26.0
Elovich	constant a (mg kg^{-1})	15.6	257	122
	constant b (mg kg^{-1} h^{-1})	31.4	203	80.7
	R^2	0.953***	0.999***	0.981***
	SE	2.86	44.1	16.1

*** indicates $P < 0.001$

Table 3. Comparison of means of model constants (a and b) in soils differing in farm type, soil type and soil depth.

		First order		Parabolic		Power		Elovich	
		a (mg kg^{-1})	b (mg kg^{-1} h^{-1})	a (mg kg^{-1})	b (mg kg^{-1} h$^{-1/2}$)	a (mg kg^{-1})	b (mg kg^{-1} h^{-1})	a (mg kg^{-1})	b (mg kg^{-1} h^{-1})
Farm type	New farms ($N = 17$)	5.96	−0.007	22.7	245	5.12	0.22	145	88.4
	Old farms ($N = 8$)	5.83	−0.006	20.0	191	4.84	0.21	108	76.4
	P value	0.638	0.151	0.597	0.249	0.263	0.397	0.258	0.536
Soil type	Volcanic soils ($N = 9$)	5.47	−0.006	12.6	116	4.38	0.24	64.0	47.9
	Non-volcanic soils ($N = 16$)	6.11	−0.007	126	263	5.25	0.25	154	99.2
	P value	0.001	0.016	0.005	0.000	0.000	0.862	0.003	0.004
Soil depth	Surface soils ($N = 9$)	6.03	−0.007	24.3	238	5.06	0.26	136	93.2
	Subsoils ($N = 16$)	5.56	−0.006	14.1	151	4.67	0.22	91.4	54.2
	P value	0.025	0.024	0.041	0.066	0.124	0.455	0.181	0.040

man, 1990; Rijkse and Trangmar, 1995). Soils with appreciable quantities of these minerals are likely to be inherently low in non-exchangeable K, owing to the absence of interlayer spaces in the clay-sized mineral particles compared with those with a preponderance of mica type minerals (Rubio and Gil-Sotres, 1997; Rezapour et al., 2009). The high non-exchangeable K status of non-volcanic soils may be an indication of the presence of comparable amounts of K-rich micaceous and other 2 : 1 minerals. Length of cropping (represented by old and new farms) did not have any significant effect ($P > 0.05$) on the K-release potential (constant a) or release rate (constant b). However, the older farms had lower K-release potential and K-release rates than new farms. Serious soil-fertility degradation (regarding available K contents) could occur in certain soils due to inherently lower nutrient status or lower rates of nutrient release from resistant soil parent materials (Hartemink and Bridges, 1995). Irrespective of cropping system, soils degrade with respect to K to an extent of 0.018 cmol kg^{-1} year^{-1}, which could also due to soil-degradation processes, such as soil erosion and nutrient runoff, as well as crop mining under tropical conditions (Lal, 1996). New land uses trigger soil erosion and degradation processes during and after land abandonment (Cerda et al., 2010). Besides affecting physical properties, long-term intensive land use can affect biochemical and microbial properties of the soil (Balota et al., 2014). Land-degradation processes are dominant in surface soil that is exposed due to clearing of vegetation and thus stores less soil moisture (Garcia-Orenes et al., 2009). Surface soils had the greater K-release potential and consequently had higher K-release rates when tested using four models. The first-order equation differentiated well the K-release constants between surface and subsoil, while the parabolic diffusion model could only make a distinction on K-release potential (constant a).

4   Conclusions

This study made use of the K-release kinetics approach to examine differences due to changes in land use pattern in the highlands of PNG. Results showed that sweet potato farm soils of the study region are inferior in potentially releasable K and K-release rates. The potential and rates of non-exchangeable K release varied greatly among the soil types, farm types and soil depth. The older farms in volcanic soils were particularly inferior in K-release potential and rate of K release. Due to continuous nutrient exploitation by the crops and fallow species, subsoils (10–20 cm) were markedly lower in K-release parameters. Soil-degradation processes such as vegetation loss and soil erosion due to intensive crop production resulted in a loss of soil fertility. Poor K-kinetics parameters warrant the application of mineral K fertilizers to compensate for mined nutrient K. Furthermore, improved fallow management practices need to be explored to meet the increased nutrient requirements as an inevitable consequence of crop intensification.

Acknowledgements. The author expresses his gratitude to the research committee of the PNG University of Technology for the funding awarded towards soil analysis and to the ACIAR project with PNG-NARI, during which the soil samples were collected.

Edited by: A. Cerdà

References

Abu Hammad, A. and Tumeizi, A.: Land degradation: socioeconomic and environmental causes and consequences in the eastern Mediterranean, Land Degrad. Develop., 23, 216–226. 2012.

Allen, B. J., Bourke, R. M., and Hide, R. L.: The sustainability of Papua New Guinea agricultural systems: the conceptual background, Global Environ. Change., 5, 297–312, 1995.

Bailey, J. S., Ramakrishna, A., and Kirchhof, G.: Relationship between important soil variables in moderately acidic soils (pH < 5.5) in the highlands of Papua New Guinea and management implications for subsistence farmers, Soil Use Manage., 24, 281–291, 2008.

Bailey, J. S., Ramakrishna, A., and Kirchhof, G.: An evaluation of nutritional constraints on sweet potato (*Ipomoea batatas*) production in the central highlands of Papua New Guinea, Plant Soil, 316, 9–7105, 2009.

Balota, E. L., Yada, I. F., Amaral, H., Nakatani, A. S., Dick, R. P., and Coyne, M. S.: Long-term land use influences soil microbial biomass P and S, phosphatase and arylsulfatase activities, and

S mineralization in a Brazilian Oxisol, Land Degrad. Dev., 25, 397–406, 2014.

Bleeker, P. and Sageman, R.: Surface charge characteristics and clay mineralogy of some variable charge soils in Papua New Guinea, Austral. J. Soil Res., 28, 901–917, 1990.

Biro, K., Pradhan, B., Buchroithner, M., and Makeschin, F.: Land use/land cover change analysis and its impact on soil properties in the Northern part of Gadarif region, Sudan, Land Degrad. Dev., 24, 90–102, 2013.

Bourke, R. M.: Sweet potato in Papua New Guinea: the plant and people, in: The sweet potato in Oceania: a reappraisal, edited by: Ballard, C., Brown, P., Bourke, R. M., and Harwood, T., The University of Sydney, Australia, 15–24, 2005.

Bourke, R. M. and Harwood, T.: Food and agriculture in Papua New Guinea, Australian National University E Press, Canberra, Australia, 2009.

Cerda, A., Hooke, J. Romero-Diaz, A., Montanarella, L., and Lavee, H.: Soil erosion on Mediterranean type-ecosystems, Land Degrad. Dev., 21, 71–74, 2010.

Davies, H. L.: The geology of New Guinea-the cordilleran margin of the Australian continent, Episodes, 35, 87–102, 2012.

Garcia-Orenes, F., Cerda, A., Mataix-Solera, J., Guerrero, C., Bodi, M. B., Arcenegui, V., Zornoza, R., and Sempere, J. G.: Effects of agricultural management on surface soil properties and soil-water losses in eastern Spain, Soil Tillage Res., 106, 117–123, 2009.

Gawander, J. S., Gangaiya, P., and Morrison, R. J.: Potassium studies on some sugarcane growing soils in Fiji, South Pacific J. Natural Sci., 20, 15–21, 2002.

Hartemink, A. E.: Nutrient stocks of short-term fallows on a high base status soil in the humid tropics of Papua New Guinea, Agrofor. Syst., 63, 33–43, 2004.

Hartemink, A. E. and Bridges, E. M.: The influence of parent material on soil fertility degradation in the coastal plain of Tanzania, Land Degrad. Dev., 6, 215–221, 1995.

Henao, J. and Baanante, C.: Estimating rates of nutrient depletion in soils of agricultural lands of Africa, Muscle Shoals, Alabama, USA, International Fertilizer Development Center, 1999.

Hosseinpur, A. R., Motaghian, H. R., and Salehi, M. H.: Potassium release kinetics and its correlation with pinto bean (*Phaseolus vulgaris*) plant indices, Plant Soil Environ., 58, 328–333, 2012.

Jalali, M.: Release kinetics of non-exchangeable potassium in calcareous soils, Commun. Soil Sci. Plant Anal., 36, 1903–1917, 2005.

Jalali, M.: Effect of sodium and magnesium on kinetics of K release in some calcareous soils of Western Iran, Geoderma, 145, 207–215, 2008.

Jalali, M. and Zarabi, M.: Kinetics of nonexchangeable-potassium release and plant response in some calcareous soils, J. Plant Nutr. Soil Sci., 169, 194–204, 2006.

Lal, R.: Deforestation and land-use effects on soil degradation and rehabilitation in western Nigeria, II. Soil chemical properties, Land Degrad. Dev., 7, 87–98, 1996.

Leh, M., Bajwa, S., and Chaubey, I.: Impact of land use change on erosion risk: and integrated remote sensing geographic information system and modeling methodology, Land Degrad. Dev., 24, 409–421, 2013.

Liu, Z., Yao, Z., Huang, H., Wu, S., and Liu, G.: Land use and climate changes and their impacts on runoff in the Yarlung Zangbo river basin, China, Land Degrad. Dev., 25, 203–215, 2014.

Lopez-Pineiro, A. and Navarro, A. G.: Potassium release kinetics and availability in unfertilized vertisols of south western Spain, Soil Sci., 162, 912–918, 1997.

Malavolta, E.: Potassium status of tropical and subtropical region soils, in: Potassium in agriculture, edited by: Munson, R. D., American Society of Agronomy, Madison, WI, USA, 163–200, 1985.

Mengel, K. and Uhlenbecker, K.: Determination of available interlayer potassium and its uptake by ryegrass, Soil Sci. Soc. Am. J., 57, 561–566, 1993.

Moss, P. and Coulter, J. K.: The potassium status of West Indian volcanic soils, J. Soil Sci., 15, 284–298, 1964.

Najafi-Ghiri, M. and Jaberi, H. R.: Effect of soil minerals on potassium release from soil fractions by different extractants, Arid Land Res. Manage., 27, 111–127, 2013.

Nelson, D. W. and Sommers, L. E.: Total carbon, organic carbon, and organic matter, in: Methods of Soil Analysis, Part 2, 2nd Edn., edited by: Page, A. L., Am. Soc. Agron. Inc. Madison, WI, USA, 961–1010, 1996.

O'Sullivan, J. N., Asher, C. J., and Blamey, F. P. C.: Nutrient disorders of sweet potato, ACIAR monograph, ACIAR, Canberra, Australia, 1997.

Radcliffe, D. J. and Kanua, M. B.: Properties and management of Andisols in the Highlands of Papua New Guinea, Papua New Guinea J. Agric. Forest. Fisheries, 41, 29–43, 1998.

Ramakrishna, A., Bailey, J. S., and Kirchhof, G.: A preliminary diagnosis and recommendation integrated system (DRIS) model for diagnosing the nutrient status of sweet potato (*Ipomoea batatas*), Plant Soil, 316, 107–113, 2009.

Rezapour, S. and Samadi, A.: The spatial distribution of potassium status and clay mineralogy in relation to different land-use types in a calcareous Mediterranean environment, Arabian J. Geosci., 7, 1037–1047, 2014.

Rezapour, S., Jafarzadeh, A. A., Samadi, A., and Oustan, S.: Impacts of clay mineralogy and physiographic units on the distribution of potassium forms in calcareous soils in Iran, Clay Min., 44, 327–337, 2009.

Rijkse, W. C. and Trangmar, B. B.: Soil- landscape models and soils of Eastern highlands, Papua New Guinea, Austral. J. Soil Res., 33, 735–755, 1995.

Rubio, B. and Gil-Sotres, F.: Distribution of four major forms of potassium in soils of Galicia (N.W Spain), Commun. Soil Sci. Plant Anal., 28, 1805–1816, 1997.

Ruxton, B. P.: Kinetics of organic matter in soils, in: Advances in Regolith, edited by: Roach, I. C., CRC Leme, Canberra, Australia, 369–372, 2003.

Samadi, A., Dovlati, B., and Barin, M.: Effect of continuous cropping on potassium forms and potassium adsorption characteristics in calcareous soils of Iran, Austral. J. Soil Res., 46, 265–272, 2008.

Sem, G.: Land-use change and population in Papua New Guinea, in: Population, Land Management, and Environmental Change, edited by: Uitto, J. I. and Akiko, O., The UNU global environmental forum IV, The United Nations University, Tokyo, Japan, 34–45, 1996.

Soil Survey Staff.: Keys to Soil Taxonomy, 12th Edn., USDA-Natural Resources Conservation Service. Washington, DC, 2014.

Srinivasarao, Ch., Vittal, K. P. R., Tiwari, K. N., Gajbhiye, P. N., and Kundu, S.: Categorisation of soils based on potassium reserves and production systems: implications in K management, Austral. J. Soil Res., 45, 438–447, 2007.

Srinivasarao, Ch., Kundu, S., Venkateswarlu, B., Lal, R., Singh, A. K., Balaguravaiah, G., Vijayasankarbabu, M., Vittal, K. P. R., Reddy, S., and Manideep, V. P.: Long-term effects of fertilization and manuring on groundnut yield and nutrient balance of Alfisols under rainfed farming in India, Nutr. Cycl. Agroecosyst., 96, 29–46, 2013.

Srinivasarao, Ch., Kundu, S., Ramachandrappa, B. K., Reddy, S., Lal, R., Venkateswarlu, B., Sahrawat, K. L., and Naik, R. P.: Potassium release characteristics, potassium balance, and finger-millet (*Eleusine coracana* G.) yield sustainability in a 27-year long experiment on an Alfisol in the semi-arid tropical India, Plant Soil, 374, 315–330, 2014.

Steffens, D. and Sparks, D. L.: Kinetics of non-exchangeable ammonium release from soils, Soil Sci. Soc. Am. J., 61, 455–462, 1997.

Steven, N.: Potassium fixation and release in alluvial clay soils of Milne Bay, Papua New Guinea: Effects of management under oil palm, MS (Research) thesis, James Cook University, available at: http://eprints.jcu.edu.au/16179 (last access: 20 September 2014) Cairn, Australia, 2010.

Walter, R., Rajashekhar Rao, B. K., and Bailey, J. S.: Distribution of potassium fractions in sweetpotato (*Ipomoea batatas*) garden soils in the Highlands of Papua New Guinea and management implications, Soil Use Manage., 27, 77–83, 2011.

Zharikova, E. A. and Golognaya, O. M.: Available potassium in volcanic soils of Kamchatka, Eurasian Soil Sci., 42, 850–860, 2009.

Ziadat, F. M. and Taimeh, A. Y.: Effect of rainfall intensity, slope and land use and antecedent soil moisture on soil erosion in an arid environment, Land Degrad. Dev., 24, 582–590, 2013.

7

Soil microbiological properties and enzymatic activities of long-term post-fire recovery in dry and semiarid Aleppo pine (*Pinus halepensis* M.) forest stands

J. Hedo[1], M. E. Lucas-Borja[2], C. Wic[2], M. Andrés-Abellán[2], and J. de Las Heras[1]

[1]Department of Plant Production and Agricultural Technology, School of Advanced Agricultural Engineering, Castilla La Mancha University, Campus Universitario s/n, CP 02071, Albacete, Spain
[2]Department of Agroforestry Technology and Science and Genetics, School of Advanced Agricultural Engineering, Castilla La Mancha University, Campus Universitario s/n, CP 02071, Albacete, Spain

Correspondence to: J. Hedo (javier.hedo@gmail.com)

Abstract. Wildfires affecting forest ecosystems and post-fire silvicultural treatments may cause considerable changes in soil properties. The capacity of different microbial groups to recolonise soil after disturbances is crucial for proper soil functioning. The aim of this work was to investigate some microbial soil properties and enzyme activities in semiarid and dry Aleppo pine (*Pinus halepensis* M.) forest stands. Different plots affected by a wildfire event 17 years ago without or with post-fire silvicultural treatments 5 years after the fire event were selected. A mature Aleppo pine stand, unaffected by wildfire and not thinned was used as a control. Physicochemical soil properties (soil texture, pH, carbonates, organic matter, electrical conductivity, total N and P), soil enzymes (urease, phosphatase, β-glucosidase and dehydrogenase activities), soil respiration and soil microbial biomass carbon were analysed in the selected forests areas and plots. The main finding was that long time after this fire event produces no differences in the microbiological soil properties and enzyme activities of soil after comparing burned and thinned, burned and not thinned, and mature plots. Moreover, significant site variation was generally seen in soil enzyme activities and microbiological parameters. We conclude that total vegetation recovery normalises post-fire soil microbial parameters, and that wildfire and post-fire silvicultural treatments are not significant factors affecting soil properties after 17 years.

1 Introduction

Fire is one of the most important disturbances in the Mediterranean region, as it shapes and structures many plant communities, forest ecosystems and landscapes (Boydak et al., 2006). After a fire event, forest functions, nutrient cycling, and the physical, chemical and biological properties of soils are significantly affected (Wic-Baena et al., 2013), and runoff and surface erosion rates can greatly increase (Prats et al., 2013). Moreover, global change is affecting fire regime, increasing fire frequency, area burned, and its destructiveness to Mediterranean ecosystems (Pausas, 2004). In this context, post-fire forest management is useful to accelerate the recovery of soil forest functions, and to improve health, growth and reproductive processes (Moya et al., 2008). For fire-adapted pines, such as *Pinus pinaster* Ait. (Maritime pine) and *Pinus halepensis* Mill. (Aleppo pine), three main forest management guidelines have been proposed as proper post-fire silvicultural treatments. The guidelines are in accordance with the success of natural regeneration: (1) no treatments if natural regeneration is achieved after the fire event; (2) assisted natural regeneration or (3) active restoration (De las Heras et al., 2012). Moreover, several studies have shown that early thinning reduces both intra-specific competition and fire recurrence events (Espelta et al., 2008).

Soil plays an essential role in the forest ecosystem's fertility and stability (Smith and Papendick, 1993). Also, soil microorganisms accomplish reactions to release soil nutrients for vegetation development (Hannam et al., 2006). As

Rutigliano et al. (2004) reported, microbial biomass and activity increased from younger to later stages of ecological succession and the introduction of pine into Mediterranean areas retards soil development. However, soil properties and plant cover relationships can be in various ways at various rates, and since different studies were not replicated across a range of site types, conclusions cannot be generalized (Muscolo et al., 2007). Forest fires and post-fire silvicultural treatments may significantly change forest and soil properties (Grady and Hart, 2006; Wic-Baena et al., 2013). After forest fires, changes in vegetation dynamics and soil properties are expected to occur due to the plant–soil feedback (Van der Putten et al., 2013; Brandt et al., 2013). Soil erosion is a key process in redistributing the soil particles, the seeds, and the nutrients (Cerdà and Lasanta, 2005; Lasanta and Cerdà, 2005). Fire may alter physicochemical soil properties (i.e., soil organic matter content and structure, hydrophobicity, pH and nutrient cycles) and microbiological or biochemical soil properties (i.e., microbial biomass, microbial activity, soil enzymes activities) (Mataix-Solera et al., 2009). These changes mostly occur below 5 cm of the surface, where the soil temperature rarely overtakes 100 °C (Úbeda and Outeiro, 2009; Aznar et al., 2013). Post-fire silvicultural treatments may also modify the soil microbiological and biochemical variables, such as belowground biological activity and soil nutrients' availability (Grady and Hart, 2006) or enzyme activities (Wic-Baena et al., 2013). Tree felling or shrub clearing modifies microclimatic conditions at the ground level, as well as the amount and quality of potential organic inputs to soil (Grady and Hart, 2006). The magnitude of the changes occurring after wildfire events or post-fire silvicultural treatments depends on forest characteristics, such as the recovery capacity of vegetation (Irvine et al., 2007), climatic factors (Almagro et al., 2009) and post-fire soil rehabilitation management (Fernández et al., 2012; Prats et al., 2013).

Given the fundamental importance of soil microbial communities in soil ecosystem sustainability, information on how microbial functionality is affected by fire or post-fire silvicultural treatments under semiarid climatic conditions is required. Estimation of microbiota and soil status are necessary to determine optimal management strategies (Mabuhay et al., 2003; Mataix-Solera et al., 2009). In this context, the use of one parameter is not consistent because soil quality depends on a wide range of chemical, physical, biochemical and microbiological variables (Nannipieri et al., 1990; Bastida et al., 2008b). Thus, many authors have proposed using a combination of several variables as indicators of soil status (Dick et al., 1996). Specific indicators of microbial activity, such as variables relating to nutrient cycles (nitrogen, carbon and phosphorus) and enzymatic activities (urease, β-glucosidase and phosphatase), have been proposed to evaluate soil status (Trasar-Cepeda et al., 1998). Moreover, general indicators of microbial activity have been extensively used in forest and agricultural soil status characterisation (Armas et al., 2007; García-Orenes et al., 2010; Fterich et al., 2014; Ferreira et al., 2014).

Long-term studies into soil quality or those that evaluate soil recovery capacity are scarce. However, long term studies are necessary to reach reasonable conclusions on the impacts that fire events and post-fire silvicultural treatments have on soil properties, particularly in Mediterranean ecosystems (Wic-Baena et al., 2013). Some long-term studies appreciated that soil organic matter and microbial communities can recover to the pre-fire levels in the Mediterranean region, taking into account study areas dominated by *Quercus ilex* L., *Quercus suber* L. and *Pinus pinaster* Aiton subsp. *pinaster* (Guénon et al., 2013). The aim of this study is to investigate soil microbiological and soil enzymatic activities in different semiarid and dry Aleppo pine forest ecosystems affected by the following: (i) a wildfire event occurred 17 years ago; (ii) a wildfire event 17 years ago and treated with early thinning 12 years earlier; (iii) an Aleppo pine in a mature stand not affected by wildfire with no silvicultural treatments. We hypothesised the following: (1) that microbiological soil properties and enzymatic activities are influenced by the climatic conditions recorded at each semiarid and dry location; (2) that microbiological soil properties and enzymatic activities recover after the wildfire and the thinning at the mid-term.

2 Material and methods

2.1 Study area

The study was conducted at two sites burnt in the summer of 1994, Yeste and Calasparra (in the provinces of Albacete and Murcia, respectively) in SE Spain. The total burnt area covered about 44 000 ha in both provinces. The forest tree composition in the study area was dominated by mature even-aged Aleppo pine stands, with shrubs and herbaceous vegetation in the understory (Table 1). Natural post-fire regeneration took place at both sites (45 000 saplings ha^{-1} in Calasparra and 7000 saplings ha^{-1} in Yeste) (Table 1). The climate of both experimental areas is classified as Mediterranean (Allué, 1990), with Yeste and Calasparra classed as a dry site and a semi-arid ombroclimate site, respectively (Rivas Martínez, 1987). Average annual rainfall and temperature for the last 30 years were respectively 503 mm and 13.5 °C in Yeste as compared to 282 mm and 16.3 °C in Calasparra. According to the Spanish Soil Map (Guerra Delgado, 1968), Yeste and Calasparra soils are classified as Inceptisols and Aridisols, respectively (Table 1). Soil texture at both sites is classified as loam/clay loam (Table 1).

2.2 Experimental design

Two experimental sites of 3 ha were selected in both Yeste (2°20' W 38°21' S) and Calasparra (1°38' W 38°16' S). Three plots were set up inside each site, one of which (1 ha) was naturally burnt in summer 1994 and was then occupied by

Table 1. Soil, climatic and stand characteristics of each experimental site[a].

Site	Exp. condition	Altitude (m a.s.l.)	Vegetation cover[a]	Aleppo pine density (trees ha^{-1})	T (°C)	H (%)	Mean age (years)	Shrub and herbaceous vegetation	Soil order (suborder)/texture[b]
Calasparra	MAT	330	90% Ph 10% Shrub and herbaceous	400	12.0 ± 1.1	5.9 ± 2.0	70–80	*Macrochloa tenacissima* (L.) Kunth; *Rosmarinus officinalis*	Aridisol (Orthid) loam
	BNOT	430	80% Ph 20% Shrub and herbaceous	45 000	9.2 ± 1.8	7.5 ± 1.1	17	*Macrochloa tenacissima* (L.) Kunth; *Rosmarinus officinalis*; *Brachypodium retusum*; *Thymus vulgaris* L.	Aridisol (Orthid) loam
	BT	330	70% Ph 30% Shrub and herbaceous	1600	9.5 ± 1.4	5.2 ± 0.9	17	*Macrochloa tenacissima* (L.) Kunth; *Rosmarinus officinalis* *Brachypodium retusum*; *Thymus vulgaris* L.	Aridisol (Orthid) loam
Yeste	MAT	1010	90% Ph 10% Shrub and herbaceous	500	8.0 ± 1.2	10.6 ± 1.8	70–80	*Rosmarinus officinalis* L., *Brachypodium retusum*	Inceptisol (Ochrept) loam
	BNOT	860	80% Ph 20% Shrub and herbaceous	7000	7.1 ± 0.9	14.6 ± 3.1	17	*Rosmarinus officinalis* L., *Brachypodium retusum*	Inceptisol (Ochrept) loam
	BT	1010	70% Ph 30% Shrub and herbaceous	1600	7.5 ± 1.3	12.4 ± 2.6	17	*Rosmarinus officinalis* L., *Brachypodium retusum*	Inceptisol (Ochrept) clay loam

[a] Ph: Aleppo pine; T: soil temperature (mean ± standard error) during the sampling period; H: soil and stand characteristics of each experimental site (mean ± standard error) during the season of sampling.
[b] Soil taxonomy (USDA).

high Aleppo pine post-fire natural. The second plot of 1 ha was naturally burnt in summer 1994 and then thinned in 1999. The post-fire silvicultural treatment and thinning operations left 1600 saplings ha^{-1} at both the Calasparra and Yeste sites. The third plot was a mature stand of 1 ha used as a control. The mature Aleppo pine stand was located adjacent to the fire perimeter at both the Calasparra and Yeste sites and has not been affected by either forest-fire or silvicultural treatments in the last 20 years. It is noteworthy that we define recovery as a scenario which returns to the same soil functioning activity levels between the burnt or thinned and mature plots. All the plots were selected in areas with a low slope (< 5%). Sampling was carried out in early winter as Ferguson et al. (2007) recommended in their guidelines for soil sampling.

In December 2011, six soil samples (1000 g) were randomly taken from each plot: (i) the plot affected by a wildfire event and post-fire silvicultural treatments 17 and 12 years earlier, respectively (burned and thinned, hereafter named "BT"); (ii) the plot affected by a wildfire event 17 years earlier with no post-fire silvicultural treatments (burned and not thinner, hereafter named "BNOT"); (iii) the plot occupied by a mature Aleppo pine stand (hereafter named "MAT"). Each soil sample was composed of six subsamples collected in a 5 m × 5 m subplot area, which were thoroughly mixed to obtain a composite sample (Andrés Abellan et al., 2011). Finally 36 samples were obtained: 2 experimental sites × 3 treatments × 6 replicates. The results shown are the average of the samples taken at each subplot. Soil samples were

taken from the uppermost mineral layer (0–15 cm) after removing litter. Samples were passed through a 2 mm sieve and were kept at 4 °C for 1 month to avoid any influence on the parameters analysed in the laboratory (Andrés Abellan et al., 2011).

2.3 Physical and chemical variables

A total of 500 g of the collected soil samples were used to analyse some physical and chemical soil properties. pH and electrical conductivity (EC) were measured in a $1/5$ (w/v) aqueous solution using a pH-meter (Navi Horiba model). Total organic carbon (TOC) was determined by wet oxidation with K_2CrO_7 and titration of dichromate excess with Mohr's salt (Yeomans and Bremner, 1989), while organic matter (OM) was inferred by multiplying the TOC content by 1.728 (Lucas-Borja et al., 2010b). Total carbonates (CO_3^{2-}) were measured in a Bernard calcimeter according to the method of Guitián and Carballas (1976). Bioavailable phosphorus (P) was determined using the method described by Olsen and Sommers (1982). Total nitrogen (total N) was measured following Kjeldhal's method modified by Bremner (1965). The texture analysis was performed using the method of Guitián and Carballas (1976). Soil moisture and temperatures were recorded during the sampling season (winter 2011) using a soil moisture sensor (ECHO EC-10 model), a soil temperature sensor (TMC6-HD model) and a data-logger (Hobo U12-006 model). Soil temperature and humidity sensors were installed at a depth of 10 cm in each plot.

2.4 Biochemical and microbiological variables

Soil dehydrogenase activity (DHA) was determined by using 1 g of soil, and the reduction of p-iodonitrotetrazolium chloride (INT) to p-iodonitrotetrazolium formazan was measured by a modified version of the method reported by García et al. (1993). Soil dehydrogenase activity was expressed as μmol INTF g^{-1} soil h^{-1}. Urease activity (UA) was determined as the NH$_4^+$ released in the hydrolysis reaction (Kandeler et al., 1999). Alkaline phosphatase (PA) and β-glucosidase (BA) activities were measured following the methods reported by Tabatabai and Bremner (1969) and Tabatabai (1982), respectively. Basal soil respiration (RESP) was analysed by placing 50 of soil moistened to 40–50 % of its water-holding capacity (water potential: 0.055 MPa) in hermetically sealed flasks and by incubating for 20 days at 28 °C. Released CO$_2$ was periodically measured (daily for the first 4 days and then weekly) using an infrared gas analyzer (Toray PG-100, Toray Engineering Co. Ltd., Japan). The data were summed to give a cumulative amount of released CO$_2$ after a 20-day incubation. Basal soil respiration was expressed as mg CO$_2$-C kg^{-1} soil per day. Microbial biomass carbon (CB) was determined by Vance et al. (1987) following the method adapted by García et al. (2003).

2.5 Statistical analysis

Data were analysed by a two-way analysis of variance (ANOVA) at which site level (Yeste and Calasparra) and the silvicultural management level ("BT", "MAT" and "BNOT") were selected as the factors. All the subplots were assumed to be spatially independent. The post hoc test applied was Fisher's least significant difference. A $P < 0.05$ level of significance was adopted throughout, unless otherwise stated.

Moreover, a multivariate statistical method using a principal component analysis (PCA) was carried out to study the structure of the dependence and correlation between the physicochemical and microbiological soil properties at the different sites and for the various treatments. Another multivariate statistical method (correlation analysis) was carried out. To satisfy the assumptions of the statistical test (equality of variance and normal distribution), variables were square-root transformed whenever necessary. The statistical analyses were done with the Statgraphics Centurion software.

3 Results

3.1 Physical and chemical variables

Soil temperature and soil moisture differed significantly ($P < 0.05$) between both experimental sites (Yeste and Calasparra), but not between different treatments ("BT", "MAT" and "BNOT") (Table 1). Soil texture (Table 1) and electrical conductivity (Table 2) were similar for both study sites and for the different treatments. The percentage of carbonates, organic matter, phosphorus and total nitrogen differed between sites, with higher values recorded for Yeste. Significant differences were also observed ($P < 0.05$) in the pH values and C/N ratio between sites, with Yeste obtaining lower values. Under the experimental conditions ("BT", "MAT" and "BNOT"), the physical and chemical variables showed a different behaviour depending on the site (Yeste and Calasparra; Table 2). There were not significant differences in any of the studied parameters taking into account the interaction.

3.2 Biochemical and microbiological variables

The experimental treatments considered in this study and the interaction between sites and experimental treatments did not significantly ($P < 0.05$) influence the microbiological properties and enzyme activities (Table 3, Figs. 1 and 2). The experimental site was the only influential factor ($P < 0.05$) found for microbial biomass carbon, soil respiration and enzymatic activities (Table 3). Urease activity showed higher values in Calasparra than in Yeste, whereas β-glucosidase and dehydrogenase activities displayed higher values in Yeste than in Calasparra (Fig. 1). No significant differences for phosphatase activity were found (Fig. 1).

3.3 Correlation analysis

Positive and significant correlation coefficients were found between organic matter and some microbiological and biochemical variables (dehydrogenase, β-glucosidase and soil respiration). Negative and significant correlation coefficients were observed between organic matter and the physical-chemical variables, such as pH and C/N ratio and with the urease activity (Table 4). pH also showed a positive correlation and a significant coefficient with urease activity. pH negatively and significantly correlated with soil respiration, dehydrogenase and β-glucosidase activity. Urease activity presented different correlation coefficients, and positively and significantly correlated with phosphatase activity, pH and C/N ratio, while a negative and significant correlation was observed with dehydrogenase and β-glucosidase activities and total carbonates, phosphorus and total nitrogen. Conversely, a positive and significant correlation was seen between dehydrogenase and β-glucosidase activity. pH and C/N ratio correlated significantly and negatively with dehydrogenase and β-glucosidase activities (Table 4).

3.4 PCA analysis

The multivariate PCA analysis showed differences between the two study sites by separating into homogeneous groups (Fig. 3). Conversely, the PCA did not separate among different treatments. The PCA analysis clustered the plots located in Yeste on the negative axis of PC 2 (Fig. 3), which explained about 13.81 % of variability. PC 1 explained around 42.22 % of variability. The plots located in Calasparra were

Table 2. Soil physicochemical parameters for each site and experimental condition.

Site	Exp. condition	pH	Electrical conductivity (μS cm^{-1})	Organic matter (%)	Total carbonates (%)	P (mg kg^{-1})	Total N (%)	C/N
Calasparra	BT	8.66 (0.07) aA	21.15 (0.78) aA	6.73 (0.66) aB	2.72 (0.07) aB	11.32 (1.35) aB	0.18 (0.00) aB	53.5 (4.26) bA
	BNOT	8.75 (0.06) aA	20.28 (1.03) aA	5.87 (0.38) bB	2.13 (0.02) bB	12.74 (3.84) aB	0.11 (0.00) bB	83 (5.95) aA
	MAT	8.39 (0.02) bA	23.48 (2.69) aA	5.35 (0.68) bB	1.92 (0.20) bB	16.95 (1.12) aB	0.20 (0.03) aB	44 (3.73) bA
Yeste	BT	8.30 (0.17) aB	20.85 (0.02) aA	8.24 (0.60) aA	2.94 (0.01) aA	27.99 (0.57) aA	0.98 (0.22) aA	16.5 (4.22) aB
	BNOT	7.83 (0.17) aB	21.15 (0.73) aA	9.17 (0.19) aA	2.94 (0.01) aA	14.24 (2.38) cA	1.09 (0.26) aA	15 (3.71) aB
	MAT	8.37 (0.22) aB	21.89 (2.28) aA	6.42 (0.22) bA	2.88 (0.02) bA	20.63 (2.67) bA	0.76 (0.27) aA	33 (12.07) aB

For each parameter, values represent mean (standard error). Data followed by the same small letter are not significantly different according to the Fisher's Least Significant Difference (LSD) test ($P < 0.05$) for each experimental condition. For each experimental site, data followed by the same capital letter are not significantly different according to the LSD test ($P < 0.05$).

Table 3. Result of the two-factor ANOVA (site and experimental condition) for the microbiological properties and enzymatic activities analysis.

Factors	Dehydrogenase activity		Urease activity		Phosphatase activity		β-glucosidase activity		Basal respiration		Microbial biomass carbon	
	F ratio	p value	F ratio	p value	F ratio	p value	F ratio	p value	F ratio	p value	F ratio	p value
S	170.21	0.0001	45.15	0.0001	0.37	0.5486	65.14	0.0001	14.88	0.0006	4.61	0.0399
T	0.34	0.7137	0.01	0.9932	0.29	0.754	1.70	0.1993	1.35	0.274	0.20	0.8202
S × T	2.16	0.1334	0.02	0.9819	0.05	0.9519	0.72	0.4948	0.01	0.9885	0.07	0.9363

S: site; T: experimental treatment; S × T: interaction between S and T.

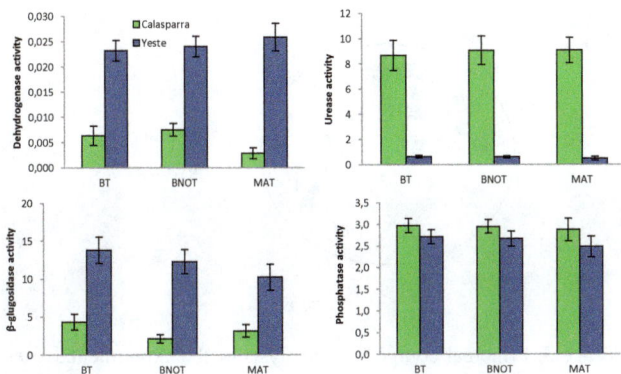

Figure 1. Dehydrogenase activity (μg (INTF) g^{-1} soil h^{-1}), β-glucosidase activity (μmol PNP g^{-1} dry soil h^{-1}), phosphatase activity (μmol PNP g^{-1} dry soil h^{-1}) and urease activity (μmol N-NH$_4^+$ g^{-1} dry soil h^{-1}) in relation to the experimental site. Error bars are the LSD intervals at $P < 0.05$.

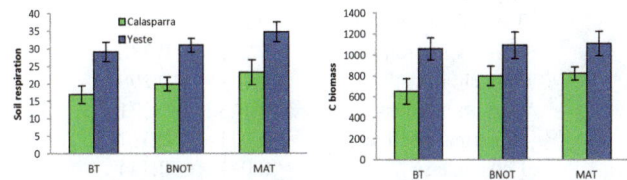

Figure 2. Basal soil respiration (mg CO$_2$ kg^{-1} soil) and microbial biomass carbon (mg kg^{-1}) in relation to the experimental site. Error bars are the LSD intervals at $P < 0.05$.

4   Discussion

Vegetation and soil type are key factors that can modify soil characteristics and are responsible for maintaining a stable microbial community (Bastida et al., 2008a). Since Aleppo pine forest dominates both experimental sites, variations in soil properties can be related mainly to site-specific differences, such as soil temperature and moisture and soil type (soil organic matter, C/N ratio, pH and P, soil texture). Micro-climatic factors influence microbial enzymes, and also change the quality and quantity of the substrate upon which they act (Kumar et al., 1992). Different authors have demonstrated that scarce soil moisture generate lower soil respiration rates, microbial biomass carbon values and dehydrogenase, phosphatase and β-glucosidase enzymatic activities (Criquet et al., 2004; Sardans and Peñuelas, 2005; Baldrian et al., 2010; Lucas-Borja et al., 2012). Our results coincide with these trends since Calasparra (higher temperatures at lower soil moisture values) obtained lower values of microbiological parameters, β-glucosidase and dehydrogenase activities, but higher values for urease and phosphatase enzymes. The

clustered on the positive axis of PC 2. Urease activity, C/N ratio and pH had a positive weight on PC 1, whereas dehydrogenase, β-glucosidase and organic matter had a negative weight (Table 5). Moreover, respiration, phosphatase and electrical conductivity had a positive weight on PC 2, while phosphorus and biomass carbon had a negative weight. The other loading factors of the different variables appear in Table 5.

Table 4. Correlation matrix between the different variables determined.

	UA	PA	DHA	BA	BC	RESP	H	OM	P	pH	EC	Total N	CO_3^{2-}
PA	0.38*	–	–	–	–	–	–	–	–	–	–	–	–
DHA	−0.66***	−0.14ns	–	–	–	–	–	–	–	–	–	–	–
BA	−0.58***	−0.06ns	0.67***	–	–	–	–	–	–	–	–	–	–
CB	−0.26ns	−0.11ns	0.36*	0.28ns	–	–	–	–	–	–	–	–	–
RESP	−0.12ns	0.18ns	0.61***	0.42*	0.10ns	–	–	–	–	–	–	–	–
H	−0.62***	0.01ns	0.77***	0.76***	0.28ns	0.41*	–	–	–	–	–	–	–
OM	−0.38*	0.18ns	0.50**	0.56***	0.02ns	0.54***	0.47**	–	–	–	–	–	–
P	−0.41*	−0.13ns	0.26ns	0.52**	0.19ns	0.10ns	0.43**	0.11ns	–	–	–	–	–
pH	0.39*	−0.15ns	−0.39*	−0.57***	−0.19ns	−0.40*	−0.50*	−0.63***	−0.11ns	–	–	–	–
EC	0.18ns	0.10ns	0.10ns	−0.13ns	−0.03ns	0.43**	−0.09ns	0.07ns	−0.16ns	0.16ns	–	–	–
Total N	−0.51**	−0.07ns	0.73***	0.54***	0.35*	0.64*	0.62***	0.27ns	0.06ns	−0.21ns	0.33*	–	–
CO_3^{2-}	−0.58***	−0.06ns	0.61***	0.58***	0.27ns	0.45*	0.46**	0.38*	0.30ns	−0.63***	−0.26ns	0.20ns	–
C/N	0.50**	0.12ns	−0.67***	−0.55***	−0.27ns	−0.41*	−0.57***	−0.33*	−0.27ns	0.25ns	−0.40*	−0.82***	−0.26ns

UA: urease activity; PA: phosphatase activity; DHA: Dehydrogenase activity; BA: β-Glucosidase activity; CB: biomass carbon; RESP: soil respiration; H: soil moisture; OM: organic matter; P: phosphorus, pH; CE: electrical conductivity; Total N: total nitrogen; CO_3^{2-}: total carbonates; C/N: carbon nitrogen ratio. Significant correlations: * $P \leq 0.05$; ** $P \leq 0.01$; *** $P \leq 0.001$; ns: non-significant.

Table 5. Weights of principal components analysis.

	PC 1	PC 2
Dehydrogenase	−0.351	−0.072
β-glucosidase	−0.341	−0.092
Moisture	−0.334	−0.103
Organic matter	−0.283	0.359
Total nitrogen	−0.273	−0.079
Soil respiration	−0.247	0.338
Phosphorus	−0.161	−0.298
Carbon biomass	−0.143	−0.271
Electrical conductivity	−0.021	0.317
Phosphatase	0.017	0.444
pH	0.264	−0.168
Total carbonates	−0.271	−0.093
C/N	0.284	0.058
Urease	0.288	0.312

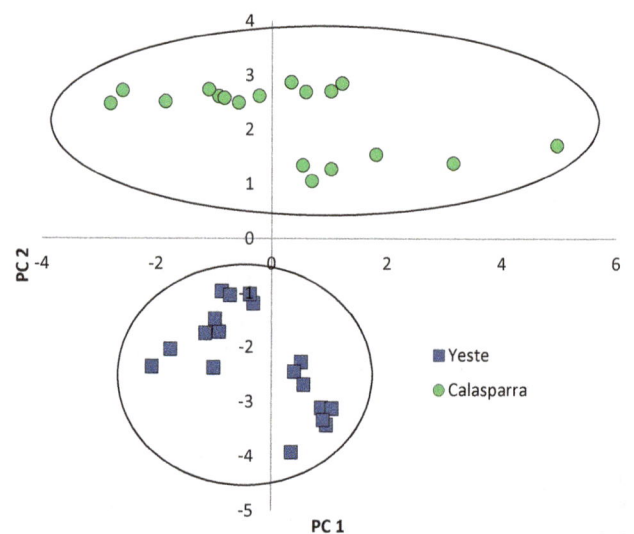

Figure 3. Principal components analysis of the experimental sites Yeste and Calasparra.

latter may be explained by quantity of total N and P present at each site. Given the lower total N and P values found in Calasparra, greater urease and phosphatase activity may be required to produce inorganic N and P ready for plant development. Gutknecht et al. (2010) showed that decreased N and P results in greater urease and phosphatase activity and higher enzyme production through soil microorganisms. On the other hand, Bastida et al. (2008a) indicated that seasonality affects enzymatic activities or microbial biomass, and in this work only we sampled in early winter, so it would be suitable to conduct sampling in different seasons.

In relation to fire and post-fire silvicultural treatments, soil moisture and temperature showed no significant differences between the "BT", "MAT" and "BNOT" plots, thus may largely explain similar microbiological parameters values and enzymatic activities. Moreover, Aleppo pine is a pyrophyte species that exhibits good post-fire natural regeneration; good post-fire seedling recruitment during the first growth season after the wildfire event (Leone et al., 2000)

was observed. Thus, initial vegetation recovery is promptly ensured after a wildfire event (De las Heras et al., 2012). In this context, temporary plant-cover loss and subsequent plant recruitment after a fire event may enhance the recovery of microbiological soil properties. According to our results, the microbiological soil properties and enzymatic activities capacity recovery should be achieved 17 years after the wildfire event and the post-fire silvicultural treatment. This longterm study demonstrated that soil parameters might recover, at least, to the pre-fire levels 17 years after the fire event and thinning operations. Wic-Baena et al. (2013) have recently shown that soil enzymatic activities recovered 6 years after thinning. The same authors stated that the time period since the silvicultural treatment was applied seemed to significantly affect soil properties. It may be explained because long-term effects on soil processes are likely driven by changes in the quality of the organic matter inputs (Hart et al., 2005), and the relationship between post-fire recovery of Aleppo pine (dominant species in both experimental sites) and soil properties.

The organic matter greatly differed, obtaining higher values for Yeste than for Calasparra. Higher values for the general soil microbial activity indicators (i.e., soil respiration and dehydrogenase activity) and for β-glucosidase and phosphatase activity have been reported by Lucas-Borja et al. (2010a, 2011) in forest soil at a higher organic matter concentration. Some organic matter fractions contain readily metabolisable compounds, which can act as energy sources for microorganisms. In relation to fire and post-fire silvicultural treatments, the organic matter content was similar when comparing "BT", "MAT" and "BNOT" plots, which may be explained by the Aleppo pine post-fire initial recruitment. The organic matter derived from new trees may be responsible for the similarities comparing "BT", "MAT" and "BNOT" plots. We found significant positive correlations between microbiological measurements (soil respiration) and enzymatic activities (dehydrogenase and β-glucosidase activities) and organic matter content. Our results also indicate lower C/N values at Yeste, but no significant differences among treatments, and we found significant negative correlations between microbiological measurements and enzymatic activities (except urease enzyme) with the C/N ratio. As Merilä et al. (2002) have shown, substrate quality, as determined by C/N , generally influences microbial biomass and respiration, so the main substrate of the litterfall were pine needles, which have high content of lignin. Berg (1986) stated that higher C/N ratios may be an indicator of the more recalcitrant nature of the soil organic matter. In Yeste and Calasparra, the Aleppo pine was the dominant species, so the main explanation of this different behaviour in each site may be the contrasting climatic conditions, which let the litterfall degrade faster. On the other hand, the experimental treatments showed the same climatic conditions and the same dominant tree species at each site, which can explain the absence of differences on C/N values depending on the post-fire treatment. Moreover, lower C/N rates have been associated with higher respiration rates and microbiological properties (Schmitz et al., 1998).

Regarding pH, some authors have denoted its influence on soil microbial biomass properties (Bååth and Anderson, 2003). According to Sinsabaugh et al. (2008), soil pH has direct biochemical effects on the activity of the extracellular enzymes immobilised in the soil matrix. The same author has also argued that soil pH reflects climatic controls in soil and plant community composition, which may affect the large-scale distribution of extracellular enzymatic activities through changes in nutrient availability, soil organic composition and microbial community composition. Our results agree with this trend and indicate that pH correlates negatively with soil enzyme activities (except urease activity), soil respiration and organic matter.

Finally, the PCA results reveal that the sites were significantly discriminated. The higher soil temperatures and lower soil moisture values recorded at Calasparra provide unfavourable conditions for balanced soil functional diversity, as reflected by poorer enzyme activities, soil respiration and biomass carbon if compared with Yeste. On the contrary, treatments were not significantly discriminated, which reflects that vegetation recovery after a wild-fire event and the time elapsed since the post-silvicultural treatments applied were enough to achieve the initial soil property values found in mature and unaffected plots.

5   Conclusions

Biochemical, microbiological and physicochemical variables are affected by site, but not by post-fire silvicultural treatment, under dry and semiarid conditions. A total of 17 years after the wildfire event and the post-fire silvicultural treatment, microbiological soil properties may recover the initial status and values shown for mature and undisturbed Aleppo pine forest stands. The micro-climatic conditions, higher soil temperature and lower soil moisture values obtained at Calasparra indicate unfavourable conditions for microbiological properties and enzyme activities if compared with Yeste. Our results provide data on the long-term recovery pattern of microbiological and enzymatic activities, and clearly distinguish between sites with different microclimatic conditions (temperature and moisture), but not among burnt/unburnt or post-fire thinned/unthinned Aleppo pine forests stands for more than 17 years after the wildfire and silvicultural treatment. Forest management guidelines should consider that forest site plays an important role in forest recovery after wildfire, and therefore in soil quality. Thus, forest management policies should take these aspects into account when designing (and budgeting) restoration plans.

Acknowledgements. The authors thank the Agencia Estatal de Meteorología (AEMET) for climatic data. We thank Helen Warburton for reviewing the English. The study has been funded by CONSOLIDER-INGENIO 2010: MONTES (CSD 2008-00040) of the Spanish Ministry of Science and Innovation.

Edited by: A. Jordán

References

Andrés Abellan, M., Wic Baena, C., García Morote, F., Picazo Córdoba, M., Candel Pérez, D., and Lucas-Borja, M.: Influence of the soil storage method on soil enzymatic activities, Forest Systems, 20, 379–388, 2011.

Allué, J. L.: Atlas Fitoclimático de España, Taxonomías, Instituto Nacional de Investigaciones Agrarias, Ministerio de Agricultura, Pesca y Alimentación, Madrid, 1990 (in Spanish).

Almagro, M., López, J., Querejeta, J. I., and Martínez-Mena, M.: Temperature dependence of soil CO_2 efflux is strongly modulated by seasonal patterns of moisture availability in a Mediterranean ecosystem, Soil Biol. Biochem., 41, 594–605, 2009.

Armas, C. M., Santana, B., Mora, J. L., Notario, J. S., Arbelo, C. D., and Rodríguez-Rodríguez, A.: A biological quality index for volcanic Andisols and Aridisols (Canary Islands, Spain): variations related to the ecosystem development, Sci. Total Environ., 378, 238–244, 2007.

Aznar, J. M., González-Pérez, J. A., Badía, D., and Martí, C.: At what depth are the properties of a gypseous forest topsoil affected by burning?, Land Degrad. Dev., doi:10.1002/ldr.2258, online first, 2013.

Bååth, E. and Anderson, T. H.: Comparison of soil fungal/bacterial ratios in a pH gradient using physiological and PLFA-based techniques, Soil Biol. Biochem., 35, 955–963, 2003.

Baldrian, P., Merhautová, V., Petránková, M., Cajthaml, T., Šnajdr, J.: Distribution of microbial biomass and activity of extracellular enzymes in a hardwood forest soil reflect soil moisture content, Appl. Soil Ecol., 46, 177–182, 2010.

Bastida, F., Barberá, G. G., García, C., and Hernández, T.: Influence of orientation, vegetation and season on soil microbial and biochemical characteristics under semiarid conditions, Appl. Soil Ecol., 38, 62–70, 2008a.

Bastida, F., Zsolnay, A., Hernández, T., and García, C.: Past, present and future of soil quality indices: A biological perspective, Geoderma, 147, 159–171, 2008b.

Berg, B.: Nutrient release from litter and humus in coniferous forest soils: a mini review, Scand. J. Forest Res., 1, 359–369, 1986.

Boydak, M., Dirik, H., and Çalıkoğlu, M.: Biology and silviculture of Turkish red pine Pinus brutia Ten., Laser Ofset Matbaa Tesisleri San. Tic. Ltd. Sti., Ankara, 2006.

Brandt, A. J., de Kroon, H., Reynolds, H. L., and Burns, J. H.: Soil heterogeneity generated by plant-soil feedbacks has implications for species recruitment and coexistence, J. Ecol., 101, 277–286, 2013.

Bremner, J. M.: Nitrogen availability indexes, in: Methods of Soil Analysis, Part 2, edited by: Black, C. A., American Society of Agronomy, Agronomy, 9, 1324–1345, 1965.

Cerdà, A. and Lasanta, A.: Long-term erosional responses after fire in the Central Spanish Pyrenees: 1. Water and sediment yield, Catena, 60, 59–80, 2005.

Criquet, S., Ferre, E., Farnet, A. M., and Le Petit, J.: Annual dynamics of phosphatase activities in an evergreen oak litter: Influence of biotic and abiotic factors, Soil Biol. Biochem., 36, 1111–1118, 2004.

De las Heras, J., Moya, D., Vega, J. A., Daskalakou, E., Vallejo, R., Grigoriadis, N., Tsitsoni, T., Baeza, J., Valdecantos, A., Fernandez, C., Espelta, J., and Fernandes, P.: Post-Fire Management of Serotinous Pine Forests, in: Post-Fire Management and Restoration of Southern European Forests, edited by: Moreira, F., Arianotsou, M., Corona, P., and De las Heras, J., Managing Forest Ecosystems, Springer, 24, 151–170, 2012.

Dick, R. P., Breakwell, D. P., and Turco, R. F.: Soil enzyme activities and biodiversity measurements as integrative microbiological indicators, in: Methods for assessing soil quality, edited by: Doran, J. W. and Jones, A. J., SSSA Special Publication 49, 247–271, 1996.

Espelta, J. M., Verkaik, I., Eugenio, M., and Lloret, F.: Recurrent wildfires constrain long-term reproduction ability in Pinus halepensis Mill., Int. J. Wildland. Fire, 17, 579–585, 2008.

Ferguson, R. B., Hergert, G. W., Shapiro, C. S., and Wortmann, C. S.: Guidelines for Soil Sampling, NebGuide G1740, University of Nebraska–Lincoln, USA, 2007.

Fernández, C., Vega, J. A., Jiménez, E., Vieira, D. C. S., Merino, A., Ferreiro, A., and Fonturbel, T.: Seeding and mulching + seeding effects on post-fire runoff, soil erosion and species diversity in Galicia (NW Spain), Land Degrad. Dev., 23, 150–156, doi:10.1002/ldr.1064, 2012.

Ferreira, A. C. C., Leite, L. F. C., de Araújo, A. S. F., and Eisenhauer, N.: Land-use type effects on soil organic carbon and microbial properties in a semi-arid region of northeast Brazil, Land Degrad. Dev., doi:10.1002/ldr.2282, online first, 2014.

Fterich, A., Mahdhi, M., and Mars, M.: The effects of Acacia tortilis subsp. raddiana, soil texture and soil depth on soil microbial and biochemical characteristics in arid zones of Tunisia, Land Degrad. Dev., 25, 143–152, doi:10.1002/ldr.1154, 2014.

García, C., Hernández, T., Costa, F., Ceccanti, B., Masciandaro, G.: The dehydrogenase activity of soil as an ecological marker in processes of perturbed system regeneration, in: Proceedings of the XI International Symposium of Environmental Biochemistry, edited by: Gallardo-Lancho, J., CSIC, Salamanca, España, 89–100, 1993.

García, C., Gil, F., Hernández, M. T., and Trasar, C.: Técnicas de Análisis de Parámetros Bioquímicos en Suelos, Mundi-Prensa, Madrid, 2003 (in Spanish).

García-Orenes, F., Guerrero, C., Roldán, A., Mataix-Solera, J., Cerdà, A., Campoy, M., Zornoza, R., Bárcenas, G., and Caravaca, F.: Soil microbial biomass and activity under different agricultural management systems in a semiarid Mediterranean agroecosystem, Soil and Tillage Research, 109, 110–115, doi:10.1016/j.still.2010.05.005, 2010.

Grady, K. C. and Hart, S. C.: Influences of thinning, prescribed burning, and wildfire on soil processes and properties in southwestern ponderosa pine forests: a retrospective study, Forest Ecol. Manag., 234, 123–135, 2006.

Guénon, R., Vennetier, M., Dupuy, N., Roussos, S., Pailler, A., and Gros, R.: Trends in recovery of Mediterranean soil chemical properties and microbial activities after infrequent and frequent wildfires, Land Degrad. Dev., 24, 115–128, doi:10.1002/ldr.1109, 2013.

Guerra Delgado, A.: Leyenda del Mapa de Suelos de España (1 : 1.000.000), Instituto Nacional de Edafología, The Consejo Superior de Investigaciones Científicas (CSIC), Madrid, 1968 (in Spanish).

Guitián, F. and Carballas, T.: Técnicas de análisis de suelos, Pico Sacro, Santiago de Compostela, 1976 (in Spanish).

Gutknecht, J. L. M., Henry, H. A. L., and Balser, T. C.: Inter-annual variation in soil extra-cellular enzyme activity in response to simulated global change and fire disturbance, Pedobiologia, 53, 283–293, 2010.

Hannam, K. D., Quideau, S. A., and Kishchuk, B. E.: Forest floor microbial communities in relation to stand composition and timber harvesting in northern Alberta, Soil Biol. Biochem., 38, 2565–2575, 2006.

Hart, S. C., DeLuca, T. H., Newman, G. S., MacKenzie, M. D., and Boyle, S. I.: Post-fire vegetative dynamics as drivers of microbial community structure and function in forest soils, Forest Ecol. Manag., 220, 166–184, 2005.

Irvine, J., Law, B. E., and Hibbard, K. A.: Postfire carbon pools and fluxes in semiarid ponderosa pine in Central Oregon, Glob. Change Biol., 13, 1748–1760, 2007.

Kandeler, E., Stemmer, M., and Klimanek, E.: Response of soil microbial biomass, urease and xylanase within particle size fractions to long-term soil management, Soil Biol. Biochem., 31, 261–273, 1999.

Kumar, J. D., Sharma, G. D., and Mishra, R. R.: Soil microbial population numbers and enzyme activities in relation to altitude and forest degradation, Soil Biol. Biochem., 24, 761–767, 1992.

Lasanta, A. and Cerdà, A.: Long-term erosional responses after fire in the Central Spanish Pyrenees: 2. Solute release, Catena, 60, 80–101, 2005.

Leone, V., Borghetti, M., and Saracino, A.: Ecology of post-fire recovery in *Pinus halepensis* in southern Italy, in: Life and Environment in Mediterranean Ecosystems, edited by: Trabaud, L., WIT Press, Southampton, 129–154, 2000.

Lucas-Borja, M. E., Bastida, F., Moreno, J. L., Nicolás, C., Andrés, M., and López, F. R.: The effects of human trampling on the microbiological properties of soil and vegetation in Mediterranean Mountain areas, Land Degrad. Dev., 22, 383–394, 2010a.

Lucas-Borja, M. E., Bastida, F., Nicolás, C., Moreno, J. L., Del Cerro, A., and Andrés, M.: Influence of forest cover and herbaceous vegetation on the microbiological and biochemical properties of soil under Mediterranean humid climate, Eur. J. Soil Biol., 46, 273–279, 2010b

Lucas-Borja, M. E., Candel, D., Jindo, K., Moreno, J. L., Andrés, M., and Bastida, F.: Soil microbial community structure and activity in monospecific and mixed forest stands, under Mediterranean humid conditions, Plant Soil, 354, 359–370, 2011.

Lucas-Borja, M. E., Candel, D., López-Serrano, F. R., Andrés, M., and Bastida, F.: Altitude-related factors but not *Pinus* community exert a dominant role over chemical and microbiological properties of a Mediterranean humid soil, Eur. J. Soil Sci., 63, 541–549, 2012.

Mabuhay, J. A., Nobukazu, N., and Horikoshi, T.: Microbial biomass and abundance after forest fire in pine forests in Japan, Ecol. Res., 18, 431–441, 2003.

Mataix-Solera, J., Guerrero, C., García-Orenes, F., Bárcenas, G. M., Torres, M. P., and Bárcenas, M.: Forest Fire Effects on Soil Microbiology, in: Fire Effects on Soils and Restoration Strategies Enfield, edited by: Cerdá, A. and Robichaud, P. R., Science Publishers, New Hampshire, 133–175, 2009.

Merilä, P., Smolander, A., and Strömmer, R.: Soil nitrogen transformations along a primary succession transect on the land-uplift coast in western Finland, Soil Biol. Biochem., 34, 373–385, 2002.

Moya, D., De las Heras, J., López-Serrano, F., and Leone, V.: A post-fire management model to improve Aleppo pine forest resilience, in: Modelling, monitoring and management of forest fires I, edited by: De las Heras, J., Brebbia, C. A., Viegas, D., and Leone, V., WittPress, Southhampton, 311–319, 2008.

Muscolo, A., Sidari, M., and Mercurio, R.: Influence of gap size on organic matter decomposition, microbial biomass and nutrient

cycle in Calabrian pine (*Pinus laricio*, Poiret) stands, Forest Ecol. Manag., 242, 412–418, 2007.

Nannipieri, P., Grego, S., and Ceccanti, B.: Ecological significance of the biological activity in soil, in: Soil Biochemistry, edited by: Bollag, J. M. and Stotzky, G., Marcel Dekker, New York, 6, 293–355, 1990.

Olsen, S. R. and Sommers, L. E.: Phosphorus, in: Methods of Soil Analysis. Chemical and Microbiological Properties, 2nd Edn., edited by: Page, A. L., Miller, R. H., and Keeney, D. R., American Society of Agronomy, Madison, 403–427, 1982.

Pausas, J. G.: Changes in fire and climate in the eastern Iberian Peninsula (Mediterranean basin), Climatic Change, 63, 337–350, 2004.

Prats, S. A., Cortizo Malvar, M., Simões Vieira, D. C., MacDonald, L., and Keizer, J. J.: Effectiveness of hydromulching to reduce runoff and erosion in a recently burnt pine plantation in central Portugal, Land Degrad. Dev., doi:10.1002/ldr.2236, online first, 2013.

Rivas Martínez S.: Memoria del mapa de series de vegetación de España (1 : 400.000), ICONA, Ministerio de Agricultura, Pesca y Alimentación, Madrid, 1987 (in Spanish).

Rutigliano, F. A., Ascoli, R. D., and De Santo, A. V.: Soil microbial metabolism and nutrient status in a Mediterranean area as affected by plant cover, Soil Biol. Biochem., 36, 1719–1729, 2004.

Sardans, J. and Peñuelas, J.: Drought decreases soil enzyme activity in a Mediterranean holm oak forest, Soil Biol. Biochem., 37, 455–461, 2005.

Schmitz, M. F., Atauri, J. A., de Pablo, C. L., Martín de Agar, P., Rescia, A. J., and Pineda, F. D.: Changes in land use in Northern Spain: effects of forestry management on soil conservation, Forest Ecol. Manag., 109, 137–150, 1998.

Sinsabaugh, R. L., Lauber, C. L., Weintraub, M. N., Ahmed, B., Allison, S. D., Crenshaw, C., Contosta, A. R., Cusack, D., Frey, S., Gallo, M. E., Gartner, T .B., Hobbie, S. E., Holland, K., Keeler, B. L., Powers, J. S., Stursova, M., Takacs-Vesbach, C., Waldrop, M. P., Wallenstein, M. D., Zak, D. R., and Zeglin, L. H.: Stoichiometry of soil enzyme activity at global scale, Ecol. Lett., 11, 1252–1264, 2008.

Smith, L. J. and Papendick, R. I.: Soil organic matter dynamics and crop residue management, in: Soil Microbial Ecology, edited by: Metting, B., Marcel Dekker, New York, 65–94, 1993.

Tabatabai, M. A.: Soil enzymes, in: Methods of analysis, part 2, 2nd Edn., edited by: Page, A. L., Miller, E. M., and Keeney, D. R., Agronomy, 9, 389–396, 1982.

Tabatabai, M. A. and Bremner, J. M.: Use of p-nitrophenyl phosphate for assay of soil phosphatase activity, Soil Biol. Biochem, 1, 301–307, 1969.

Trasar-Cepeda, C., Leirós, C., Gil-Sotres, F., and Seoane, S.: Towards a biochemical quality index for soils: an expression relating several biological and biochemical properties, Biol. Fert. Soils, 26, 100–106, 1998.

Úbeda, X. and Outeiro, L.: Physical and chemical effects offire on soil, in: Fire Effects on Soils and Restoration Strategies, edited by: Cerdà, A. and Robichaud, P. R., Science Publishers, CRC Press, Boca Raton, FL, 105–133, 2009.

Vance, E. D., Brookes, P. C., and Jenkinson, D. S.: An extraction method for measuring soil microbial biomass, Soil Biol. Biochem., 19, 703–707, 1987.

Van der Putten, W. H., Bardgett, R. D., Bever, J. D., Bezemer, T. M., Casper, B. B., Fukami, T., Kardol, P., Klironomos, J. N., Kulma-tiski, A., Schweitzer, J. A., Suding, K. N., Van de Voorde, T. F. J., and Wardle, D. A.: Plant-soil feedback: the past, the present and future challenges, J. Ecol., 101, 265–276, 2013.

Wic-Baena, C., Andrés-Abellán, M., Lucas-Borja, M. E., Martínez-García, E., García-Morote, F. A., Rubio, E., and López-Serrano, F. R.: Thinning and recovery effects on soil properties in two sites of a Mediterranean forest, in Cuenca Mountain (South-eastern of Spain), Forest Ecol. Manag., 308, 223–230, 2013.

Yeomans, J. and Bremner, J. M.: A rapid and precise method for routine determination of organic carbon in soil, Commun. Soil Sci. Plan., 19, 1467–1476, 1989.

Effects of rodent-induced land degradation on ecosystem carbon fluxes in an alpine meadow in the Qinghai–Tibet Plateau, China

F. Peng, Y. Quangang, X. Xue, J. Guo, and T. Wang

Key laboratory of desert and desertification, Cold and Arid Regions Environmental and Engineering Research Institute (CAREERI), Chinese Academy of Sciences (CAS), Lanzhou, China

Correspondence to: X. Xue (xianxue@lzb.ac.cn)

Abstract. The widespread land degradation in an alpine meadow ecosystem would affect ecosystem carbon (C) balance. Biomass, soil chemical properties and carbon dioxide (CO_2) of six levels of degraded lands (D1–D6, according to the number of rodent holes and coverage) were investigated to examine the effects of rodent-induced land degradation on an alpine meadow ecosystem. Soil organic carbon (SOC), labile soil carbon (LC), total nitrogen (TN) and inorganic nitrogen (N) were obtained by chemical analysis. Soil respiration (R_s), net ecosystem exchange (NEE) and ecosystem respiration (ER) were measured by a Li-Cor 6400XT. Gross ecosystem production (GEP) was the sum of NEE and ER. Aboveground biomass (AGB) was based on a linear regression with coverage and plant height as independent variables. Root biomass (RB) was obtained by using a core method. Soil respiration, ER, GEP and AGB were significantly higher in slightly degraded (D3 and D6, group I) than in severely degraded land (D1, D2, D4 and D5, group II). Positive values of NEE average indicate that the alpine meadow ecosystem is a weak C sink during the growing season. The only significant difference was in ER among different degradation levels. R_s, ER and GEP were 38.2, 44.3 and 46.5 % higher in group I than in group II, respectively. Similar difference of ER and GEP between the two groups resulted in an insignificant difference of NEE. Positive correlations of AGB with ER, NEE and GEP, and relatively small AGB and lower CO_2 fluxes in group II, suggest the control of AGB on ecosystem CO_2 fluxes. Correlations of RB with SOC, LC, TN and inorganic N indicate the regulation of RB on soil C and N with increasing number of rodent holes in an alpine meadow ecosystem in the permafrost region of the Qinghai–Tibet Plateau (QTP).

1 Introduction

Soil contains the largest ecosystem carbon (C) stock (Batjes, 1996). Widespread land degradation (Dregne, 2002), including land use change and soil and vegetation degradation, has resulted in severe soil C and nitrogen (N) loss (Wang et al., 2009; Parras-Alcántara et al., 2013), which is estimated to be 19–29 Pg C worldwide (Lal, 2001). Restoration of the degraded ecosystems, therefore, has a great potential to sequestrate C from the atmosphere (Lal, 2004) at an annual rate of 0.9–1.9 Pg C for a 25- to 50-year period in drylands (Lal, 2001).

Grassland stores about 15.2 % of the terrestrial ecosystem vegetation and soil C stock (Ajtay, 1979). Either the aboveground vegetation (Fan et al., 2007) or the top 1 m of soil and root C stock (Yang et al., 2008) in an alpine meadow in the Qinghai–Tibet Plateau (QTP) account for a large proportion of those in grassland ecosystems in China (Ni, 2002). However, over one-third of the grassland in the QTP has been severely degraded due to climate change, grazing and road constructing since the 1990s (Ma et al., 1999), which has led to 1.8 Gg C loss in aboveground C stock from 1986 to 2000 (Wang et al., 2008). In addition to the vegetation C loss, land degradation could also result in decline in soil C and N (Wang et al., 2008; Wen et al., 2013), and consequently might alter net C balance in the alpine meadow (Li et al., 2011).

The primary factor causing "black soil type" degradation over the QTP is rodent grazing and burrowing (Ma et al., 1999). Rodent grazing activities trigger decline in biomass; lead to change in belowground biomass distribution, soil structure and microclimate; cause soil erosion and nutrient loss; and finally affect the ecosystem C balance (Li et al., 2011). Current studies about ecosystem C balance in alpine

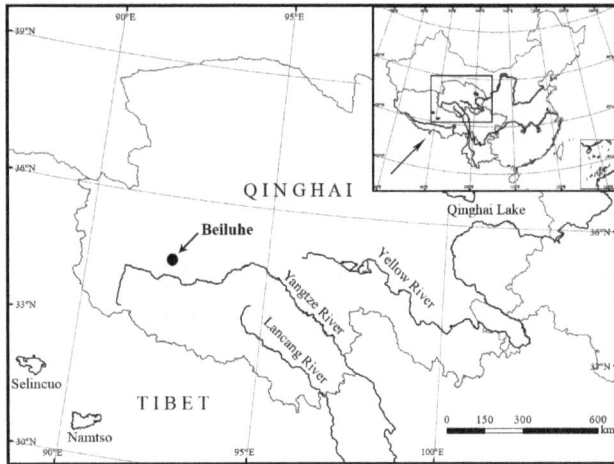

Figure 1. Location of the study area.

meadows focus on net ecosystem exchange (NEE) (Kato et al., 2004), inter-annual variation in NEE (Kato et al., 2006), soil respiration (R_s) and ecosystem respiration (ER) responses to experimental warming (Peng et al., 2014b; Luo et al., 2010; Lin et al., 2011). Effects of rodent-induced land degradation on ecosystem CO_2 fluxes have rarely been investigated. Studies only examining the responses of R_s to land degradation (Zhang et al., 2010; J. Wang et al., 2007) cannot provide solid evidence for determining the response of ecosystem C balance. No field experiment has been conducted in the permafrost region of the QTP to investigate the effect of land degradation on NEE, a direct measure of the ecosystem C balance, and on its components: ER and gross ecosystem production (GEP). We conducted a field study to investigate (1) how the NEE and its components respond to rodent-induced land degradation, and (2) how biotic and abiotic factors affect those CO_2 fluxes with land degradation processes in a *Kobresia pygmaea*-dominated alpine meadow in a permafrost area of the QTP.

2   Materials and methods

2.1   Site description

The study site is situated in the source region of the Yangtze River and in the middle of the QTP (Fig. 1, 92°56′ E, 34°49′ N) with a mean altitude of 4635 m a.s.l. and a typical alpine climate. Mean annual temperature is −3.8 °C (2000–2010) with minimum mean monthly temperature of −27.9 °C in January and maximum mean of 19.2 °C in July. Mean annual precipitation is 290.9 mm with 95 % falling from May to October. Mean annual potential evaporation is 1316.9 mm, mean annual relative humidity is 57 % and mean annual wind velocity is 4.1 m s^{-1} (Lu et al., 2006). The study site is a winter-grazed range, dominated by alpine meadow vegetation: *Kobresia capillifolia*, *Kobresia pygmaea*

and *Carex moorcroftii*, with a mean plant height of 5 cm. Plant roots are mainly within the 0–20 cm soil layer, with average soil organic carbon (SOC) of 1.5 %. Soil development is weak and is alpine meadow soil (soil taxonomy in China, and Cryosols in World Reference Base (WRB) taxonomy; IUSS, 2006) with a mattic epipedon at approximately 0–10 cm depth and an organic-rich layer at 20–30 cm (G. Wang et al., 2007). The parent soil material is of fluvio-glacial origin, and sand (> 0.05 mm) content is about 95 %. Permafrost thickness observed near the experimental site is 30–70 m, and the depth of the active layer is 1.5–3.5 m (Wu and Liu, 2004). However, the thickness of the active layer has been increasing at a rate of 3.1 cm yr^{-1} since 1995 due to climatic warming (Wu and Liu, 2004).

2.2   Experimental design and measurement protocol

2.2.1   Experimental design

We selected six habitats with different number of rodent holes (NRHs) and community coverage in a mountain slope based on our filed observation. The habitats were sequenced D1–D6 from east to northeast. The distance between each habitat was about 200–300 m. Two subplots (2 m × 4 m) were set up in each habitat. The NRHs, coverage, plant height and major species in D1–D6 were shown in Table 1.

2.2.2   Measurement protocol

Soil temperature

Soil temperature at the depth of 5 cm was monitored by a thermo-probe attached to a Li-Cor 6400 (Lincoln, NE, USA) when measurements of R_s, NEE and ER were conducted.

CO_2 fluxes: a PVC collar (80 cm^2 in area and 5 cm in height) was inserted 2–3 cm into soil permanently at the centre of each plot for measuring R_s. The measuring procedure of R_s was similar to that reported in former studies (Peng et al., 2014b; Zhou et al., 2007). Ecosystem respiration and NEE were measured with a transparent chamber (0.5 × 0.5 × 0.5 m) attached to an infrared gas analyser (Li-Cor 6400, Lincoln, NE, USA). The method used was similar to that reported by Steduto et al. (2002) and Niu et al. (2008). Gross ecosystem production was the calculated as the sum of NEE and ER. Ecosystem CO_2 fluxes were measured once a month from June to September in each plot.

Soil sampling

One soil sample was collected at the soil depth of 0–30 cm in each plot in June 2012.

AGB and RB

Aboveground biomass (AGB) was obtained from a step-wise linear regression, with AGB as the dependent variable, and

Table 1. Features of different habitats, which are represented by different number of rodents holes (NRHs, deep and shallow), coverage, plant height (H) and major plant species.

DD	NRHs (deep)	NRHs (shallow)	Coverage (%)	$H \pm SE$ (cm)	Major species
D1	19	7	18	9.5 ± 1.2	*Carex moorcroftii*
D2	5	13	35	7 ± 0.7	*Kobresia humilis, Kobresia pygmaea*
D3	0	3	80	6.5 ± 0.2	*Kobresia pygmaea*
D4	12	15	42	8 ± 0.4	*Carex moorcroftii, Kobresia pygmaea*
D5	17	13	30	7.5 ± 0.5	*Carex moorcroftii, Kobresia pygmaea*
D6	2	0	60	12 ± 0.3	*Carex moorcroftii*

Table 2. Major devices, measuring procedure, specific feature of methods and equipments to conduct the measurement of soil chemical properties and ecosystem CO_2 fluxes.

Items	Devices or procedure	Specific feature	Literature
T	6000-09TC, Li-Cor, Utah, USA	A thermo-probe	
R_s	6400-09, Li-Cor, Utah, USA	A collar 5 cm in depth	Zhou et al. (2007)
ER	6400XT, Li-Cor, Utah, USA	A collar 50 cm in depth	Steduo et al. (2002), Niu et al. (2008)
NEE	6400XT, Li-Cor, Utah, USA	A transparent chamber CO_2 gradient	Steduo et al. (2002), Niu et al. (2008)
AGB	A frame and a ruler	Linear regression	Xu et al. (2015)
RB	An auger		Xu et al. (2015)
SOC	Walkley–Black method		Walkley et al. (1947)
TN	Kjeldahl nitrogen method		
NH_4^+, NO_3^-	Spectrometer		
LC	Spectrometer		Blair et al. (1995)

T is the soil temperature at 5cm depth; R_s, soil respiration; ER, ecosystem respiration; NEE, net ecosystem exchange; AGB, aboveground biomass; RB, root biomass; SOC, soil organic carbon; TN, total nitrogen; NH_4^+ and NO_3^-, ammonia and nitrate nitrogen, respectively; LC, labile carbon.

coverage and plant height as independent variables (Peng et al., 2014b; Xu et al., 2015). Coverage of each plot was measured using a 10 cm × 10 cm frame in four diagonally divided subplots replicated eight times in D1–D6 in June 2012. Plant height was measured 40 times by a ruler and averaged for each plot. Root biomass (RB) was obtained from soil samples that were air-dried for 1 week and passed through a sieve ($\Phi = 2$ mm) to remove large particles. Roots were separated from the soil by washing, and fine roots was retrieved by sieve ($\Phi = 0.25$ mm). Living roots were separated from dead roots by their colour and consistency (Yang et al., 2007), and the separated roots were dried at 75 °C for 48 h.

Chemical analysis

Soil organic carbon was analysed using the Walkley–Black method (Walkley, 1947). Total nitrogen was measured via the Kjeldahl method. Ammonia and nitrate N were measured colorimetrically through a spectrometer. Labile soil carbon (LC) measurement was carried out by the procedure by Moscatelli et al. (2007). Devices or procedure used in measuring above parameters, specific feature of the measurement were inclued in Table 2.

2.3 Data analysis

The statistical differences of soil temperature, R_s, NEE, ER, GEP, SOC, total nitrogen (TN), inorganic N, C : N and biomass in D1–D6 were tested by the one-way ANOVA analysis at the $P < 0.05$, and the Tukey test was used in doing the post hoc analysis for SOC, TN, inorganic N, C : N and biomass. Previously to conduct the ANOVA, the Kolmogorov–Smirnov test and Levene test were used to test the normality and homogeneity of variance of the parameters. Monthly data of soil temperature, R_s, NEE and GEP measured in each subplot from June to September were used in the analysis. Plots in D3 and D6 were ranked as group I, and those in D1, D2, D4 and D5 were considered group II because the total NRHs in D3 and D6 was much lower than that in D1, D2, D4 and D5. The statistical significance of CO_2 fluxes between the two groups was also tested by one-way ANOVA analysis. The monthly differences in CO_2 fluxes were analysed by repeated ANOVA. Relationships of R_s, NEE and ER with soil temperature, ABG, RB and TN or inorganic N were analysed by linear regression analyses. Pearson correlation analysis was used to investigate the relationships of NRHs with soil chemical properties and biomass. The linear regression and Pearson correlation were considered significant with $P < 0.05$. Soil respiration, NEE and ER data were the averages of 4 months in D1–D6 when conducting the correlation analysis. All the analyses were conducted in SPSS 16.0 for Windows.

3 Results

3.1 Soil temperature

Soil temperature at 5 cm depth maximized in July (Fig. 2) and average soil temperature of the 4 months had no significant difference ($P > 0.05$) among treatments. The monthly average soil temperature was about 9.6–12.4 °C from D1 to D6. Soil temperature also had no significant difference between group I and II.

3.2 Soil chemical properties and biomass

Soil organic carbon, LC, TN, ammonia N and RB were higher in D2 than in other habitats (Table 3). AGB was higher in D3 and D6 than in others. Soil organic carbon, LC, TN and inorganic N (NH_4^+-N and NO_3^--N) had no obvious trend with the increasing NRHs, whereas AGB was negatively correlated with the NRHs ($r = -0.89$, $P < 0.05$).

3.3 Soil respiration, net ecosystem exchange, ecosystem respiration and gross ecosystem production

Repeated one-way ANOVA showed the significant seasonal change in R_s ($P < 0.01$), ER ($P < 0.05$), NEE ($P < 0.01$) and GEP ($P < 0.01$). The maximum R_s and ER were in July (Fig. 3a, b), whereas the maximum NEE and GEP were in June (Fig. 3c, d). Growing season average R_s and NEE had no significant difference in D1–D6, while ER and GEP were marginally higher in D3 and D6 than in others (Table 4). Soil respiration, ER and GEP were significantly higher in group I than in group II ($P < 0.05$).

3.4 Relationship of R_s, ER and NEE with soil temperature, soil nitrogen and biomass

Ecosystem CO_2 fluxes had no obvious relationship with soil temperature (Fig. 4a), soil inorganic N (Fig. 4b) and RB (Fig. 4d), while they correlated positively with AGB, with the steepest regression slope in R_s, followed by ER and NEE (Fig. 4c).

4 Discussion

4.1 C and N loss

Soil organic carbon, LC, TN and inorganic N were only significantly higher in D2 in our study (Table 3). The results indicate that C and N loss induced by rodent activities were different, with nutrient loss associated with desertification and wind erosion in temperate grassland (Zhang et al., 2010) or

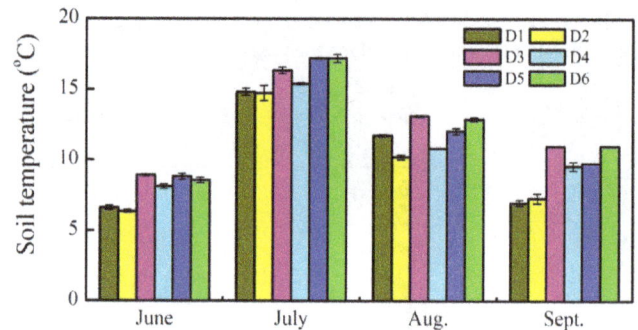

Figure 2. Soil temperature in each degradation level (D1–D6) from June to September. Error bars represent the standard error for D1–D6 in each month.

in the alpine meadow ecosystem (Xue et al., 2009). Soil C and N loss can occur in degraded land by (1) reducing vegetative growth and exposing the soil surface to wind and water erosion, and by (2) reducing the return of litter to soil (Nunes et al., 2012). Higher AGB in D3 and D6 (Table 3) suggests more litter returning to the soil, but more ecosystem CO_2 emission from soil in terms of higher R_s could be the reason for lower SOC, LC and TN in D3 and D6. Positive correlation between AGB and R_s indicates that decomposition of fresh litter from AGB might be the major component of R_s in the alpine meadow. The highest RB in D2 in spite of lower AGB compared with that in D3 and D6 (Table 3) provides evidence that RB is the major source of soil C and N in the alpine meadow ecosystem. The results was similar to a study where soil C and N storage increases were positively correlated with the increase of belowground biomass allocation with grazing pressure (Li et al., 2011).

4.2 CO_2 fluxes

Soil temperature explains most of the temporal variation (Peng et al., 2014a), but RB determines the spatial variation in R_s over the QTP (Geng et al., 2012). The lack of an obvious relationship between R_s and soil temperature (Fig. 4a) suggests other factors might be involved in controlling the temporal variation in R_s in degraded land. Significant reduction of R_s only appears on the severely degraded alpine meadow level (Zhang et al., 2010). R_s being lower in group I than in group II (1) supports the above finding because community coverage in D1, D2, D4 and D5 (Table 1) conforms to the standard of the severely degraded alpine meadow (Xue et al., 2009) and (2) indicates the controlling effect of biomass on R_s in degraded land induced by rodent activities. Soil respiration is composed of autotrophic respiration from plant roots and their symbionts, and heterotrophic respiration from litter and SOC decomposition (Hanson et al., 2000). Aboveground biomass and dead roots are the major sources of alpine meadow litter (Sun and Wang, 2008), and SOC abates due to the decreasing litter input into soil as a result of lower

Table 3. Soil organic carbon (SOC), labile soil carbon (LC), total nitrogen (TN), inorganic nitrogen (NH_4^+-N and NO_3^--N), aboveground biomass (AGB) and root biomass (RB) in different sites (D1–D6) and results (F values) of one-way ANOVA analysis.

DD	SOC $(g\,kg^{-1})$	LC $(g\,kg^{-1})$	TN $(mg\,kg^{-1})$	NH_4^+_N $(mg\,kg^{-1})$	NO_3^-_N $(mg\,kg^{-1})$	AGB $(g\,m^{-2})$	RB $(kg\,m^{-2})$	C:N
D1	4.91 ± 0.13b	1.21 ± 0.13b	0.44 ± 0.02b	8.21 ± 0.32b	4.16 ± 0.62a	149	3.8 ± 0.06c	11.2 ± 0.7ab
D2	8.70 ± 1.19a	2.12 ± 0.31a	0.75 ± 0.10a	13.11 ± 1.23a	3.81 ± 0.51ab	145	13.8 ± 3.5a	11.5 ± 0.2ab
D3	5.02 ± 1.01b	1.23 ± 0.29b	0.46 ± 0.11ab	8.54 ± 1.00b	2.31 ± 0.38bc	272	11.3 ± 1.3a	10.9 ± 0.3a
D4	3.95 ± 0.62b	1.28 ± 0.34b	0.36 ± 0.05ab	7.56 ± 1.39b	2.62 ± 0.24bc	189	6.0 ± 1.5b	10.8 ± 0.3b
D5	3.77 ± 0.32b	0.9 ± 0.09b	0.38 ± 0.03b	9.38 ± 1.33b	1.98 ± 0.21c	141	6.1 ± 0.9b	10.1 ± 0.2b
D6	3.41 ± 0.35b	0.83 ± 0.04b	0.34 ± 0.02b	8.08 ± 0.76b	2.64 ± 0.10bc	336	5.7 ± 0.3b	9.9 ± 0.4b
F, P value	$F = 11.05, P < 0.01$	$F = 7.9, P < 0.01$	$F = 5.9, P < 0.01$	$F = 5.5, P < 0.01$	$F = 7.5, P < 0.01$		$F = 15.7, P < 0.01$	$F = 4.8, P = 0.012$

The values in the table were the average and standard error of soil samples at each site. Different letters in each column stands for significant difference of at $P < 0.05$ level, $n = 4$.

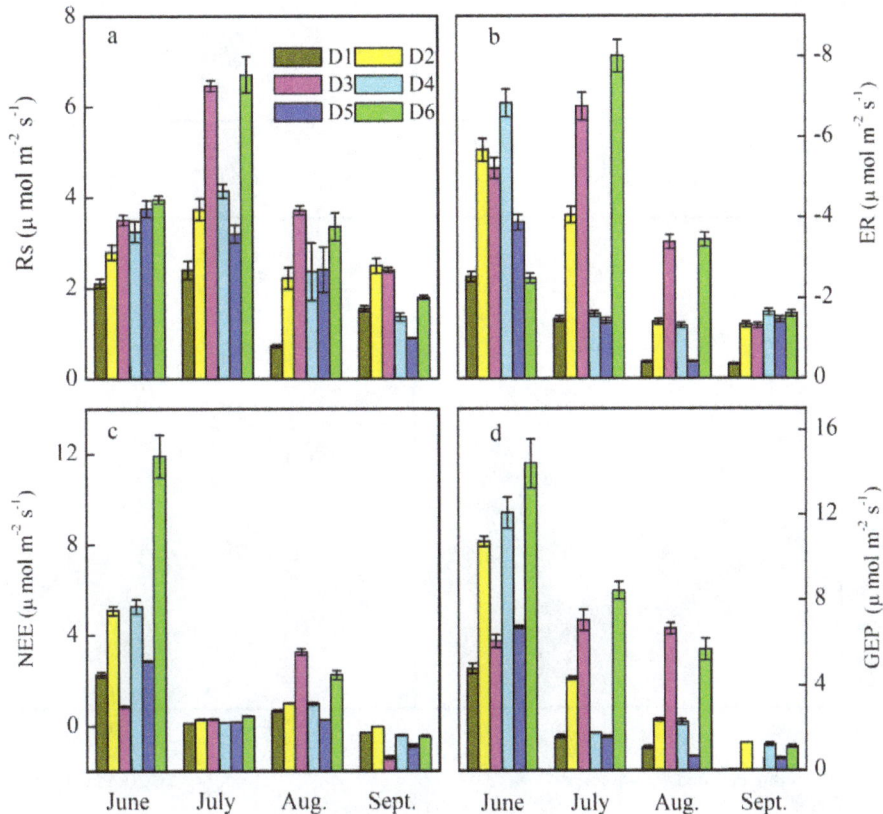

Figure 3. Monthly soil respiration (R_s, **a**), ecosystem respiration (ER, **b**), net ecosystem exchange (NEE, **c**) and gross ecosystem production (GEP, **d**) among different degradation levels from June to September. Values in the bars were the average of four replicates (two replicates in two subplots), and error bars are standard errors.

AGB and plant detritus (Wang et al., 2009; Wen et al., 2013). RB, SOC and LC being lower yet AGB being higher in D3 and D6 than in D1 and D2 (Table 3) implies that AGB is the major controlling factor of R_s with the development of land degradation, which is proved by the positive correlation between R_s and AGB (Fig. 4c). In disturbed ecosystems, competition among microorganisms induces the microbes to use more C energy for cell integrity and maintenance (Moscatelli et al., 2007), and the consequently higher respiration quotient (Nunes et al., 2012) could contribute to the insignificant change in R_s with development of land degradation.

Ecosystem respiration comprises of respiration of AGB and R_s (Zhang et al., 2009). Higher R_s therefore could be one reason for the higher ER in D3 and D6. Lower relative difference in R_s (38.2 %) than in ER (44.5 %) between the two groups suggests the influence of other factors like AGB on ER difference, which is supported by the positive correlation between ER and AGB (Fig. 4c).

The highest net photosynthesis in June in the alpine meadow ecosystem (Yi et al., 2000) justifies the maximum GEP in June (Fig. 3d). Sedge percentage will decrease and forb percentage increase with the development of land degra-

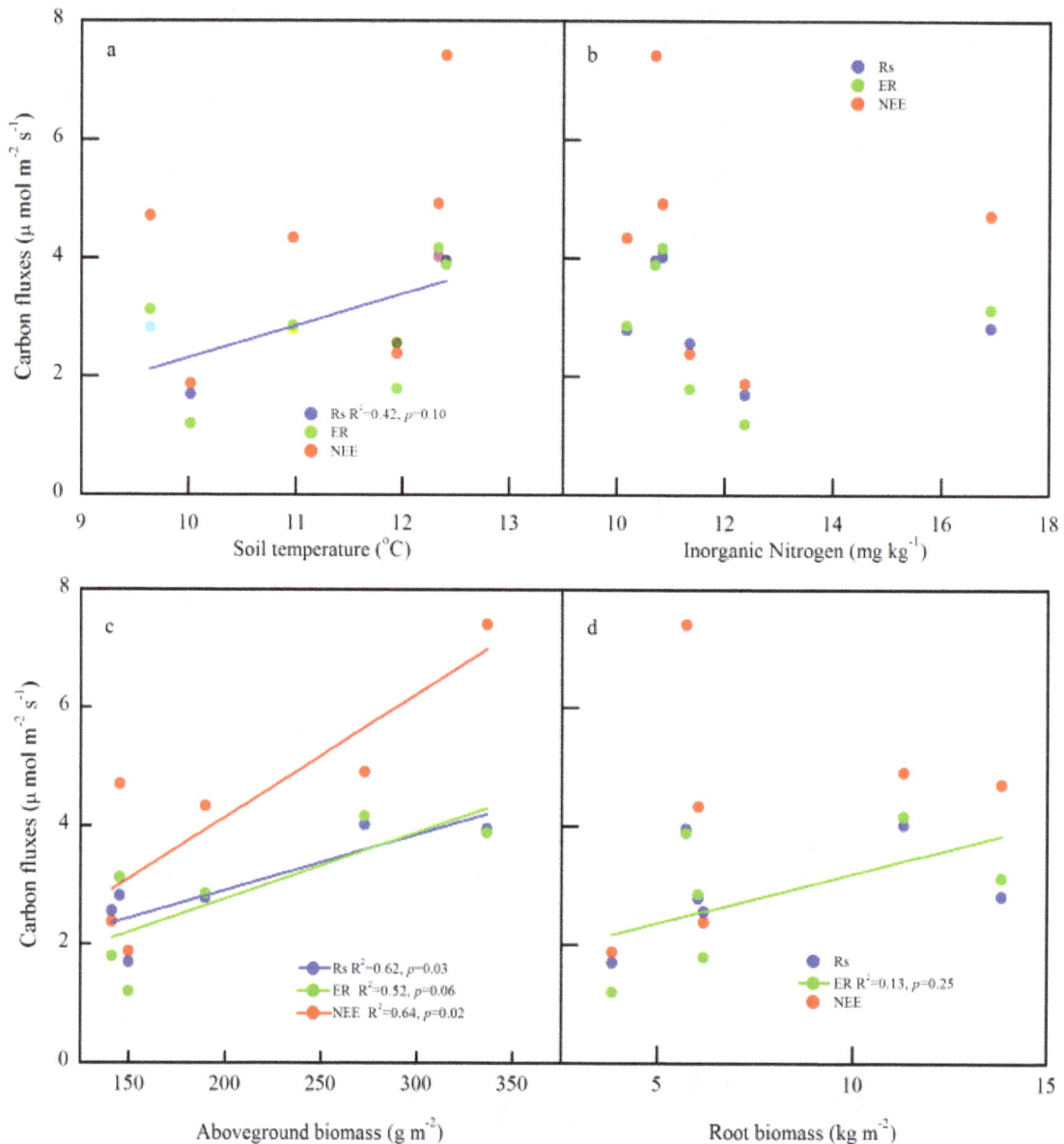

Figure 4. Linear regressions of CO_2 fluxes (soil respiration, R_s; ecosystem respiration, ER; net ecosystem exchange, NEE) with soil temperature **(a)**, inorganic nitrogen **(b)**, aboveground biomass **(c)** and root biomass **(d)**. R_s, ER and NEE data were the average of four measurements from June to September within two subplots; inorganic nitrogen and root biomass (0–30 cm) were derived from soil samples at 0–30 cm depth in June.

dation (Liu et al., 2008). The relatively higher net photosynthetic rate of forb species (*Polygonum viviparum* Linn.) than that of sedge species (*Carex atrofusca* Schkuhr, unpublished data) and higher AGB might compensate for the effect of species composition change on GEP due to the positive correlation between GEP and AGB ($r = 0.84$, $P < 0.05$) in the current study.

The maximum NEE in June is a result of the highest GEP and lower ER in this time (Fig. 3). Positive average NEE (Fig. 3c) indicates alpine meadow is weak C sink in the growing season. The insignificant difference of NEE in the two

groups (Table 4) might be the result of the corresponding change of ER (44.5 % higher in group I than in group II) and GEP (46.5 % higher in group I than in group II).

4.3 Implication of the soil C dynamics

The insignificant difference in NEE among different degradation levels suggests that SOC loss (in D1, D4 and D5) with land degradation is not a direct result of changes in net C uptake and emission. The higher SOC, LC and TN in D2 with more NRHs, and the positive correlation between RB and

Table 4. Results (F values) of ANOVA on the effect of land degradation on soil respiration (R_s), ER (ecosystem respiration), NEE (net ecosystem exchange) and GEP (gross ecosystem respiration).

	\multicolumn{4}{	}{D1–D6}	\multicolumn{4}{	}{Group I and group II}
	R_s	ER	NEE	GEP	R_s	ER	NEE	GEP
F	1.69	2.64	1.35	2.27	7.41	8.21	1.59	6.01
P	0.12	**0.04**	0.26	0.06	**0.01**	**0.006**	0.21	**0.02**

Group I includes D3 and D6, while group II includes D1, D2, D4 and D5. Numbers in bold stand for the statistical significance at the $P < 0.05$ level.

SOC suggest that other dynamics associated with land degradation, like species composition (W. Li et al., 2011) and C allocation between AGB and RB change (G. Li et al., 2011), might be involved in the soil C and N dynamics in degraded land in the alpine meadow ecosystem (Zhang et al., 2010).

5 Conclusions

Soil respiration, ER and GEP all decreased with increasing NRHs. The corresponding change in ER and GEP leads to insignificant change in NEE. All the ecosystem CO_2 fluxes are primarily affected by AGB. SOC and soil nutrient change in degraded land is not a direct result of the response of net ecosystem C balance to land degradation. Other processes like species composition and above- and belowground biomass allocation might play a role in the soil C dynamic with development of land degradation.

Acknowledgements. The authors thank Yongzhi Liu, Hanbo Yun, Guilong Wu and Yuanwu Yang for their help in setting up the field experiment. Financial support came from the National Natural Science Foundation of China (41301211, 41201195 and 41301210); the Foundation for Excellent Youth Scholars of CAREERI, CAS (Y351191001); and the Chinese Academy of Sciences (Hundred Talents Program).

Edited by: P. Pereira

References

Ajtay, G. L.: Terrestrial primary production and phytomass: The global carbon cycle, John wiley sons, Chichester, 1979.

Batjes, N. H.: Total carbon and nitrogen in the soils of the world, European J. Soil Sci., 47, 151–163, 1996.

Dregne, H. E.: Land degradation in the drylands, Arid Land Res. Manage., 16, 99–132, 2002.

Fan, J. W., Zhong, H. P., Harris, W., Yu, G. R., Wang, S. Q., Hu, Z. M., and Yue, Y. Z.: Carbon storage in the grasslands of China based on field measurements of above- and below-ground biomass, Clim. Change, 86, 375–396, doi:10.1007/s10584-007-9316-6, 2008.

Geng, Y., Wang, Y., Yang, K., Wang, S., Zeng, H., Baumann, F., Kuehn, P., Scholten, T., and He, J. S.: Soil Respiration in Tibetan Alpine Grasslands: Belowground Biomass and Soil Moisture, but Not Soil Temperature, Best Explain the Large-Scale Patterns, PLoS ONE, 7, e34968, doi:10.1371/journal.pone.0034968, 2012.

Hanson, P. J., Edwards, N. T., Garten, C. T., and Andrews, J. A.: Separating root and soil microbial contributions to soil respiratin: a review of methods and observations, Biogeochemistry, 48, 115–146, 2000.

Kato, T., Tang, Y., Gu, S., Cui, X., Hirota, M., Du, M., Li, Y., Zhao, X., and Oikawa, T.: Carbon dioxide exchange between the atmosphere and an alpine meadow ecosystem on the Qinghai–Tibetan Plateau, China, Agr. Forest Meteorol., 124, 121–134, 2004.

Kato, T., Tang, Y., Gu, S., Hirota, M., Du, M., Li, Y., and Zhao, X.: Temperature and biomass influences on interannual changes in CO_2 exchange in an alpine meadow on the Qinghai-Tibetan Plateau, Global Change Biol., 12, 1285–1298, 2006.

Lal, R.: Potential of desertification control to sequester carbon and mitigate the greehouse effect, Clim. Change, 51, 35–72, 2001.

Lal, R.: Soil carbon sequestration impacts on global climate change and food security, Science, 304, 1623–1626, 2004.

Li, G., Liu, Y., Frelich, L. E., and Sun, S.: Experimental warming induces degradation of a Tibetan alpine meadow through trophic interactions, J. Appl. Ecol., 48, 659–667, 2011.

Li, W., Huang, H. Z., Zhang, Z. N., and Wu, G. L.: Effects of grazing on the soil properties and C and N storage in relation to biomass allocation in an alpine meadow, J. Soil Sci. Plant Nutr., 11, 27–39, 2011.

Lin, X. W., Zhang, Z. H., Wang, S. P., Hu, Y. G., Xu, G. P., Luo, C. Y., Chang, X. F., Duan, J. C., Lin, Q. Y., Xu, B., Wang, Y. F., Zhao, X. Q., and Xie, Z. B.: Response of ecosystem respiration to warming and grazing during the growing seasons in the alpine meadow on the Tibetan plateau, Agr. Forest Meteorol., 151, 792–802, 2011.

Liu, X. N., Sun, J. L., Sun, D. G., Pu, X. P., and Xu, G. P.: A study on the community structure and plant diversity of alpine meadow under differetn degrees of degradation in the Eastern Qilian Mountains, Ac. Pratacult. Sinica, 17, 1–11, 2008.

Lu, Z., Wu, Q., Yu, S., and Zhang, L.: Heat and water difference of active layers beneath different surface conditions near Beiluhe in Qinghai-Xizang Plateau, J. Glaciol. Geogryol., 28, 642–647, 2006.

Luo, C. Y., Xu, G. P., Chao, Z. G., Wang, S. P., Lin, X. W., Hu, Y. G., Zhang, Z. H., Duan, J. C., Chang, X. F., Su, A. L., Li, Y. N., Zhao, X. Q., Du, M. Y., Tang, Y. H., and Kimball, B. A.: Effect of warming and grazing on litter mass loss and temperature sensitivity of litter and dung mass loss on the Tibetan plateau, Glob. Change Biol., 16, 1606–1617, 2010.

Ma, Y. S., Lang, B. N., and Wang, Q. J.: Review and prospect of the study on black soil type deteriorated grassland, Pratacul. Sci., 16, 5–9, 1999.

Moscatelli, M. C., Tizio, D., Marinari, A., and Grego, S.: Microbial indicators related to soil carbon in Mediterranean land use systems, Soil Till. Res., 97, 51–59, 2007.

Ni, J.: Carbon storage in grasslands of China, J. Arid Environ., 20, 205–211, 2002.

Niu, S., Wu, M., Han, Y., Xia, J., LI, L., and Wan, S.: Water-mediated responses of ecosystem carbon fluxes to climatic change in a temperate steppe, New Phytol., 177, 209–219, 2008.

Nunes, J. S., Araujo, A. S. F., Nunes, L. A. P. L., Lima, L. M., Carneiro, R. F. V., Salviano, A. A. C., and Tsai, S. M.: Impact of land degradation on soil microbial biomass and activity in Northeast Brazil, Pedosphere, 22, 88–95, 2012.

Parras-Alcántara, L., Martín-Carrillo, M., and Lozano-García, B.: Impacts of land use change in soil carbon and nitrogen in a Mediterranean agricultural area (Southern Spain), Solid Earth, 4, 167–177, doi:10.5194/se-4-167-2013, 2013.

Peng, F., Xue, X., You, Q. G., Zhou, X. H., and Wang, T.: Warming effects on carbon release in a permafrost area of Qinghai-Tibet Plateau, Environ. Earth Sci., 73, 57–66, doi:10.1007/s12665-014-3394-3, 2014a.

Peng, F., You, Q., Xu, M., Guo, J., Wang, T., and Xue, X.: Effects of Warming and Clipping on Ecosystem Carbon Fluxes across Two Hydrologically Contrasting Years in an Alpine Meadow of the Qinghai-Tibet Plateau, PLoS ONE, 9, e109319, doi:10.1371/journal.pone.0109319, 2014b.

Steduto, P., Çetinkökü, Ö., Albrizio, R., and Kanber, R.: Automated closed-system canopy-chamber for continuous field-crop monitoring of CO_2 and H_2O fluxes, Agr. Forest Meteorol., 111, 171–186, 2002.

Sun, X. D. and Wang, Y. L.: Difference of biomass and soil nutrition in alpine cold meadow of different degraded degree, Chin. Qinghai J. Animal Veterin. Sci., 38, 6–8, 2008.

Walkley, A.: A critical examination of a rapid method for determining organic carbon in soils-effect of variations in digestion conditions and of inorganic soil constituents, Soil Science, 63, 251–264, 1947.

Wang, G., Wang, Y., Li, Y., and Cheng, H.: Influences of alpine ecosystem responses to climatic change on soil properties on the Qinghai–Tibet Plateau, China, CATENA, 70, 506–514, 2007.

Wang, G., Li, Y., Wang, Y., and Wu, Q.: Effects of permafrost thawing on vegetation and soil carbon losses on the Qinghai-Tibet Plateau, China, Geoderma, 143, 143–152, 2008.

Wang, J., Wang, G., Wang, Y., and Li, Y.: Influences of the degradation of swamp and alpine meadows on CO_2 emission during growing season on the Qinghai-TIbet plateau, Chinese Sci. Bull., 52, 2565–2574, 2007.

Wang, W. Y., Wang, Q. J., and Lu, Z. Y.: Soil organic carbon and nitrogen content of density fractions and effect of meadow degradation to soil carbon and nitrogen of fractions in alpine Kobresia meadow, Sci. China Series D-Earth Sci., 52, 660–668, 2009.

Wen, L., Dong, S., Li, Y., Wang, X., Li, X., Shi, J., and Dong, Q.: The impact of land degradation on the C pools in alpine grasslands of the Qinghai-Tibet Plateau, Plant Soil, 368, 329–340, 2013.

Wu, Q. B. and Liu, Y. Z.: Ground temperature monitoring and its recent change in Qinghai–Tibet Plateau, Cold Reg. Sci. Technol., 38, 85–92, 2004.

Xu, M. H., Peng, F., You, Q. G., Guo, J., Tian, X. F., Xue, X., and Liu, M.: Year-round warming and autumnal clipping lead to downward transport of root biomass, carbon and total nitrogen in soil of an alpine meadow, Environ. Experiment. Bot., 109, 54–62, 2015.

Xue, X., Guo, J., Han, B. S., Sun, Q. W., and Liu, L. C.: The effect of warming and permafrost thaw on desertification in the Qinghai-Tibet Plateau, Geomorphology, 108, 182–190, 2009.

Yang, Y., Ma, W., Mohammat, A., and Fang, J.: Storage,patterns and controls of soil nitrogen in China, Pedosphere, 17, 776–785, 2007.

Yang, Y. H., Fang, J. Y., Tang, Y. H., Ji, C. J., Zheng, C. Y., He, J. S., and Zhu, B.: Storage, patterns and controls of soil organic carbon in the Tibetan grasslands, Glob. Change Biol., 14, 1592–1599, 2008.

Yi, X. F., Pen, G. Y., Shi, S. B., and Han, F.: Seasonal variation in photosynthesis of Kobresia humilis population and community at Haibei alpine meadow, Grassland China, 1, 12–15, 2000.

Zhang, F., Wang, T., Xue, X., Han, B. S., Peng, F., and You, Q. G.: The responses of soil CO_2 to desertification on alpine meadow in the Qinghai-Tibet Plateau, Environ. Earth Sci., 60, 349–358, 2010.

Zhou, X. H., Wan, S. Q., and Luo, Y. Q.: Source components and interannual variability of soil CO_2 efflux under experimental warming and clipping in a grassland ecosystem, Glob. Change Biol., 13, 761–775, 2007.

9

Aggregate breakdown and surface seal development influenced by rain intensity, slope gradient and soil particle size

S. Arjmand Sajjadi and M. Mahmoodabadi

Department of Soil Science, Agriculture Faculty, Shahid Bahonar University of Kerman, Kerman, Iran

Correspondence to: M. Mahmoodabadi (mahmoodabadi@uk.ac.ir)

Abstract. Aggregate breakdown is an important process which controls infiltration rate (IR) and the availability of fine materials necessary for structural sealing under rainfall. The purpose of this study was to investigate the effects of different slope gradients, rain intensities and particle size distributions on aggregate breakdown and IR to describe the formation of surface seal. To address this issue, 60 experiments were carried out in a $35 \times 30 \times 10$ cm detachment tray using a rainfall simulator. By sieving a sandy loam soil, two sub-samples with different maximum aggregate sizes of 2 mm ($D_{max}2$ mm) and 4.75 mm ($D_{max}4.75$ mm) were prepared. The soils were exposed to two different rain intensities (57 and 80 mm h^{-1}) on several slopes (0.5, 2.5, 5, 10 and 20 %) each at three replicates. The result showed that for all slope gradients and rain intensities, the most fraction percentages in soils $D_{max}2$ and $D_{max}4.75$ mm were in the finest size classes of 0.02 and 0.043 mm, respectively. The soil containing finer aggregates exhibited higher transportability of pre-detached material than the soil containing larger aggregates. Also, IR increased with increasing slope gradient, rain intensity and aggregate size under unsteady state conditions because of less development of surface seal. However, under steady state conditions, no significant relationship was found between slope and IR. The findings of this study revealed the importance of rain intensity, slope steepness and soil aggregate size on aggregate breakdown and seal formation, which can control infiltration rate and the consequent runoff and erosion rates.

1 Introduction

Soil erosion is one of the most serious environmental problems in the world (Leh et al., 2013; Lieskovský and Kenderessy, 2014). Soil erosion affects forests and agricultural lands and is a key factor for land degradation (Cerdà et al., 2009; Mahmoodabadi, 2011; Mandal and Sharda, 2013); it also explains the changes in landforms, soil and water resources and the recovery of vegetation (García Orenes et al., 2009; García Fayos et al., 2010; Zhao et al., 2013). To improve the accuracy and precision of erosion models and develop more rationally based soil erosion control techniques, the development of process-based models is very important (Romkens et al., 2001; Haregeweyn et al., 2013). Raindrops that impact soil surface can influence erosion rate and change the structure of soil in various ways (Kinnell, 2005), although the size of the drops is a key factor (Cerdà, 1997). In this regard, surface seal is formed by raindrop impact, which further leads to slaking and breakdown of soil aggregates (Assouline, 2004). The development of surface seal depends on the extent of the breakdown of surface aggregates, which depends on soil structure stability (Pulido Moncada et al., 2013; Wick et al., 2014; Gelaw et al., 2015). This is directly related to the kinetic energy of raindrops, the rain intensity and the duration of the rainstorm as well as the stability of aggregates to resist such breakdown. However, vegetation cover is the key factor in reducing soil erosion through the reduction of crusting in the soil surface and the enhancement of infiltration (Cerdà et al., 1998; Gabarrón-Galeote et al., 2013; Brevik et al., 2015). Reduction of infiltration rate (IR), intensification of runoff and interference with seed germination are some of the consequences of surface sealing (Mermut et al., 1997).

Some studies have shown that seal formation is a key factor in soil erosion processes, because it can reduce the surface roughness as well as IR and soil loss by splash (Assouline and Mualem, 2000; Robinson and Phillips, 2001; Assouline, 2004; Assoualine and Ben-Hur, 2006). In general, aggregate breakdown occurs when its strength is reduced by wetting to a level where the stress imposed by raindrops is sufficient to disrupt the aggregate (Assouline, 2004). The main mechanisms of aggregate breakdown during water erosion processes are slaking by fast wetting and mechanical breakdown due to raindrop impact (Le Bissonnais, 1996; Legout et al., 2005; Shi et al., 2010). Therefore, a certain threshold kinetic energy is needed to start detachment (Lujan, 2003). Consequently, when aggregates are broken down by raindrops impact and/or slaking, the disaggregated particles are deposited within the upper soil pore spaces, forming a thin, dense and low-permeable layer, namely surface seal (Assouline, 2004).

Some studies have shown that when rainfall detachment is the dominant erosion process, the size distribution of the eroded soil differs from the original soil from which it was derived (Proffitt et al., 1993; Slattery and Burt, 1997). Also, aggregate breakdown due to the raindrop impact is likely to be a major factor affecting sediment size distribution in soil erosion experiments (Hairsine et al., 1999). Aggregate breakdown produces smaller particles than the original soil, which may then be displaced and reoriented into a more continuous structure. They clog conducting pores and, consequently, a surface seal is developed (Ramos et al., 2003). The particle size distribution of the eroded soil can be influenced by the particle size distribution of the original soil, the aggregate breakdown during erosion event and the settling velocity of different size classes of particles (Rose et al., 2007; Mahmoodabadi et al., 2014a). The particle size distribution of eroded soil also seems to be dependent on the erosive agent of rainfall and or runoff, flow hydraulic characteristics and slope gradient (Ruff et al., 2003; Sirjani and Mahmoodabadi, 2012).

Soil infiltration during a rainstorm is closely related to the intensity and kinetic energy of the rainfall, surface conditions and soil properties such as those related to aggregate stability (Hawke et al., 2006; Mazaheri and Mahmoodabadi, 2012). These can affect IR through the surface seal formation, which results from physico-chemical compaction and dispersion due to raindrop impact (Assouline, 2004). In addition, slope gradient is considered to play a key role in controlling IR and erosion rate (Essig et al., 2009; Mahmoodabadi and Cerdà, 2013). Ekwue et al. (2009) and Sirjani and Mahmoodabadi (2014) reported that soil erosion increased with increasing slope gradient as a result of reduced IR and greater runoff rate. Janeau et al. (2003) observed a reduction in IR when slope gradient increased. Poesen (1987) noted contradictory results dealing with the relationship between slope gradient and IR: on soils susceptible to surface seal formation, a decrease in IR with increasing slope gradient was found.

Soil infiltration is also highly dependent on rainfall intensity and the relationship between these two parameters has been studied (Foley and Silburn, 2002; Hawke et al., 2006). Foley and Silburn (2002) found that higher IR often occurred with greater rainfall intensities. Romkens et al. (1985) reported that raindrops can destroy or deform the arrangement of soil particles; therefore, the detached particles can clog the soil pores, again reducing the IR. Ribolzi et al. (2011) concluded that the kinetic energy of raindrops and associated risks of soil crusting also decrease on steeper slopes, which might lead to increasing IR. The soils of arid and semiarid regions due to low content of organic carbon are generally susceptible to surface sealing and erosion (Cerdà, 2000; Mahmoodabadi and Cerdà, 2013). Under these conditions, only a few studies have investigated aggregate breakdown and surface sealing. The objective of this study was to evaluate aggregate breakdown under different rain intensities, slope gradients and soil aggregate sizes by the determination of aggregate size distribution and to assess the formation and development of surface seal on the basis of obtained data of IR.

2 Material and methods

2.1 Soil preparation and characteristics

In this study, a soil sample was taken from the upper 20 cm of agricultural land. It was air dried and then passed separately through 2 and 4.75 mm sieves. Two soils with different maximum aggregate sizes were provided (Zamani and Mahmoodabadi, 2013), named $D_{max}2$ mm and $D_{max}4.75$ mm. Note there were no primary particles coarser than 2 mm in the soils because the original soil was collected from agricultural land. Some physical and chemical properties were measured for both sub-samples separately. Texture of the soils was determined using the hydrometer method (Gee and Or, 2002). Aggregate size distribution was determined by wet and dry sieving (Kemper and Rosenau, 1986). Also, some chemical properties of the soils including pH and EC were measured in a soil : water suspension with a ratio of 1 : 5. Organic carbon content was determined as described by Walkley and Black (1934), and the percentage of $CaCO_3$ equivalent was measured using the titration method (Pansu and Gautheyrou, 2006). The measured physical and chemical properties of the soils are listed in Table 1. The obtained results showed that the mean weight diameter in terms of dry and wet for soil $D_{max}4.75$ mm was 0.78 and 0.3 mm, respectively, while these parameters for soil $D_{max}2$ mm had lower values. Both soils showed a very low organic carbon content (< 1 %), whereas the content of $CaCO_3$ equivalent, which is dominant in arid and semiarid region soils, was higher than 10 % (Mazaheri and Mahmoodabadi, 2012). The fraction percentage of aggregates for the soils is also shown in Fig. 1. For both soils $D_{max}2$ and $D_{max}4.75$ mm the most frequent size classes were found to be in the range of 0.063 to 0.5 mm with 75.9 and 79.9 %, respectively, while larger and finer size classes were lower.

Table 1. Some physical and chemical properties of the soils used in the experiments.

Soil properties	Soil containing particles finer than 2 mm (D_{max} 2 mm)	Soil containing particles finer than 4.75 mm (D_{max} 4.75 mm)
Sand (%)	58.8	56.6
Silt (%)	23.4	31.3
Clay (%)	17.8	12.1
Dry MWD (mm)	0.46	0.78
Wet MWD (mm)	0.26	0.3
OC (%)	0.9	0.75
pH	7.13	7.47
EC (dS m^{-1})	3.11	3.31
CaCO$_3$ (%)	17.4	21

MWD: mean weight diameter, EC: electrical conductivity, OC: organic carbon.

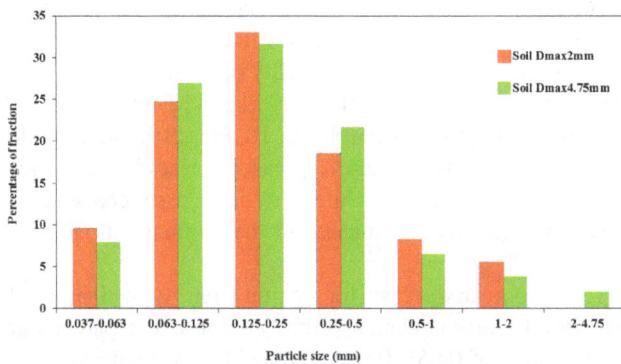

Figure 1. The fraction percentage obtained by the wet sieving procedure.

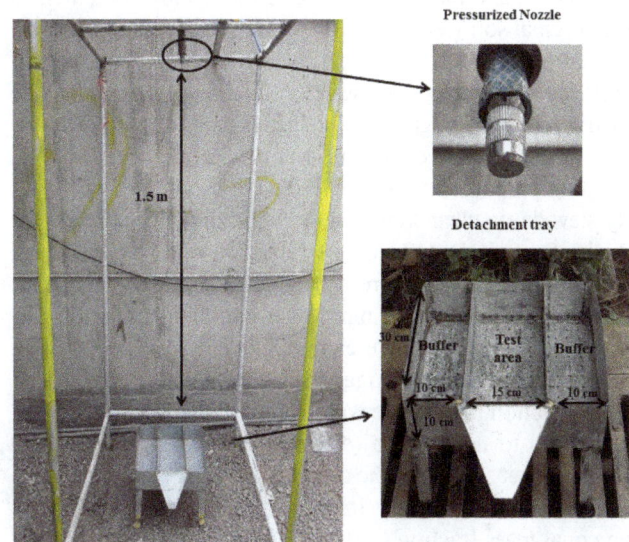

Figure 2. The rainfall simulator and detachment tray used in the experiments.

2.2 Treatments and experimental setup

In total, 60 experiments were carried out using the prepared soil samples under different rain intensities of 57 and 80 mm h^{-1} and several slopes (0.5, 2.5, 5, 10 and 20 %), each at three replicates. For this purpose, an experiment was done with a rainfall simulator to generate different rain intensities (Bodí et al., 2012; Mahmoodabadi and Cerdà, 2013; Moreno-Ramón et al., 2014). The nozzle used in the rainfall simulator was a pressurized one which was placed 1.5 m above the soil surface (Fig. 2). In order to measure rain intensity, 16 containers (6.8 cm diameter) were placed at regular distances under the simulated rains (Mahmoodabadi et al., 2007). To assess the uniformity of rain intensity, Christiansen's coefficient was calculated (Grierson and Oades, 1977):

$$CC = \left[1 - \frac{\sum |x_i - m|}{m \cdot n}\right] \times 100, \qquad (1)$$

where x_i is the measured intensity in each container, m is the average rain intensity and n is the number of containers. Also, the measurement of average drop size was done using the stain method (Arjmand Sajjadi and Mahmoodabadi, 2015). The average ($\pm$ standard deviation) drop sizes for the

rain intensities of 57 and 80 mm h^{-1} were 2.2 ± 0.08 and 2.5 ± 0.09 mm with the coefficient of uniformity of 86 and 80 %, respectively.

A 35×30 cm drainable tray with 10 cm depth was used in the experiments (Fig. 2). The washed sediment was collected from the central test area of the tray. On two sides of the test area a buffer section was provided so the soil was not only lost by splash but could also be returned from the buffer area (Arjmand Sajjadi and Mahmoodabadi, 2015). Different parts of the applied detachment tray are shown in Fig. 2.

2.3 Rainfall simulation experiments

Before every experiment, each soil sample was saturated for 24 h. Afterward, the drainage water was removed out of the tray. Simulated rainfall lasted until a constant runoff rate was reached (40–45 min). For each rainfall event, the sediment-laden overland flow was sampled at time intervals (2, 5, 15, 20, 30 and 40 min) and volumetrically measured. Collected samples were deposited, separated from the water and dried in an oven at 105 °C for 24 h. In addition, stream power as one of the hydraulic parameters was used, as defined by Mahmoodabadi et al. (2014b):

$$\Omega = \rho g q S, \tag{2}$$

where Ω is stream power (W m^{-2}), ρ is water mass density (kg m^{-3}), g is the gravitational acceleration (m s^{-2}), q (m^{-2} s) is volumetric flux per unit width and S is the gradient of bed slope (m m^{-1}).

During each experiment, infiltrated water was collected from the bottom of the detachment tray at different time intervals. Since the soil was saturated during each run, aggregate breakdown and the resultant size redistribution compared to the original soil were attributed to the seal formation. Therefore, at the end of each experiment the upper 5 mm of soil surface was sampled for the determination of aggregate size distribution. Aggregate size distribution of the eroded soil was measured by wet sieving (Kemper and Rosenau, 1986). For this purpose, soil aggregates were submerged and gently sieved into clear water, while each sample was sieved for 2 min. For soil D_{max}2 mm, six sieves of 1, 0.5, 0.25, 0.125, 0.063 and 0.037 mm were used, and for soil D_{max}4.75 mm, one additional sieve of 2 mm was used. Then, remaining aggregates on each sieve were dried in oven at 105 °C for 24 h.

For quantification of aggregate breakdown of the eroded soils, fraction percentage was determined for each size class compared to non-eroded (original) soil. The obtained data from the wet sieving of the original soil were subdivided into 10 size classes using the interpolation method, each having an equal mass fraction (10 %). Also, both soil samples were subdivided 10 size classes. Finally, the fraction of each size class was obtained using the subdivision of equal classes obtained from the original soil as described in Mahmoodabadi and Sirjani (2012). Thereupon, the fraction of the eroded soils for each experiment was calculated based on the size classes of the original soil. All statistical analyses were performed in the SAS statistical framework; to obtain the main differences between the treatments, the Duncan's ($\alpha = 0.05$) test was applied.

3 Results and discussion

3.1 Rain-induced particle size redistribution

The fraction percentages of 10 size classes of soil D_{max}2 mm created by different rain intensities and slope gradients, are compared to the original soil in Fig. 3. The fraction percentage of the original soil was indicated in Fig. 3 by uniform fraction of 10 % in each size class. When the fraction percentage of each size class (10 size classes of eroded soil) was greater than 10 %, the size class increased on the soil sur-

face. Generally, the fraction percentage of the size class of 0.02 mm was the highest among all the rain intensities and slope gradients. This size class was affected by decreasing in the fraction percentage of coarser size classes. Therefore, the fraction percentage in coarser size classes decreased while the opposite was found in finer size classes.

For the rain intensity of 57 mm h^{-1} and 0.5 % slope gradient, the fraction percentage of eroded soil in the range of 0.055–0.092 mm was slightly greater than that of the original soil (Fig. 3a). The fraction percentage in the range of 0.121–0.411 mm decreased and was slightly higher in the coarsest size class (1.5 mm) than the original soil. At 2.5 % slope gradient, the fraction percentage in the size class of 0.055 mm was higher than the original soil; the fraction percentages decreased in size classes coarser than 0.073 mm (Fig. 3b). At 5 % slope gradient, the fraction percentage in the range of 0.055–0.092 mm was higher, whereas it was less than the original soil in the size classes from 0.121 to 0.411 mm (Fig. 3c). However, for rain intensity of 57 mm h^{-1}, the fraction percentage of the coarsest size class (1.5 mm) increased compared to the original soil. At 10 and 20 % slope gradients, the fraction percentages increased in size classes ranged from 0.055 to 0.092 mm, while those size classes coarser than 0.121 mm decreased compared to the original soil (Fig. 3d and e).

In the comparison case, for the rain intensity of 80 mm h^{-1} and in all slope gradients (Fig. 3) the fraction percentage in the range of 0.055–0.092 mm was higher than the original soil (except 5 % slope gradient). In contrast, in the size classes coarser than 0.121 mm, the fraction percentage decreased compared to the original soil for all slope gradients (except 5 % slope gradient). At 5 % slope gradient, the fraction percentage in the range of 0.055–0.073 mm was higher; in size classes coarser than 0.092 mm, it was less than the original soil.

The obtained results for soil D_{max}2 mm exhibited some differences in the two applied rain intensities. The first difference refers to the fraction percentage in the size class of 0.02 mm, which was higher in rain intensity of 57 mm h^{-1} than that obtained in rain intensity of 80 mm h^{-1}. This means that in rain intensity of 57 mm h^{-1}, however, the aggregates were broken down by raindrop impact during the rainfall event and produced finer particles; the resultant surface flow did not have enough transportability to carry detached particles way out of the test area. Therefore, the fraction percentage of the finest size class (0.02 mm) was enhanced in the eroded soil under the lower rain intensity (57 mm h^{-1}). In contrast, the higher rain intensity of 80 mm h^{-1} caused more detachability of soil aggregates and higher flow rates, which intensified transportability of finer pre-detached materials as well. Asadi et al. (2011) reported that with increasing flow stream power, sediment size distribution became coarser, finally becoming similar to or even coarser than the original soil; therefore, finer sediment remained on the soil surface.

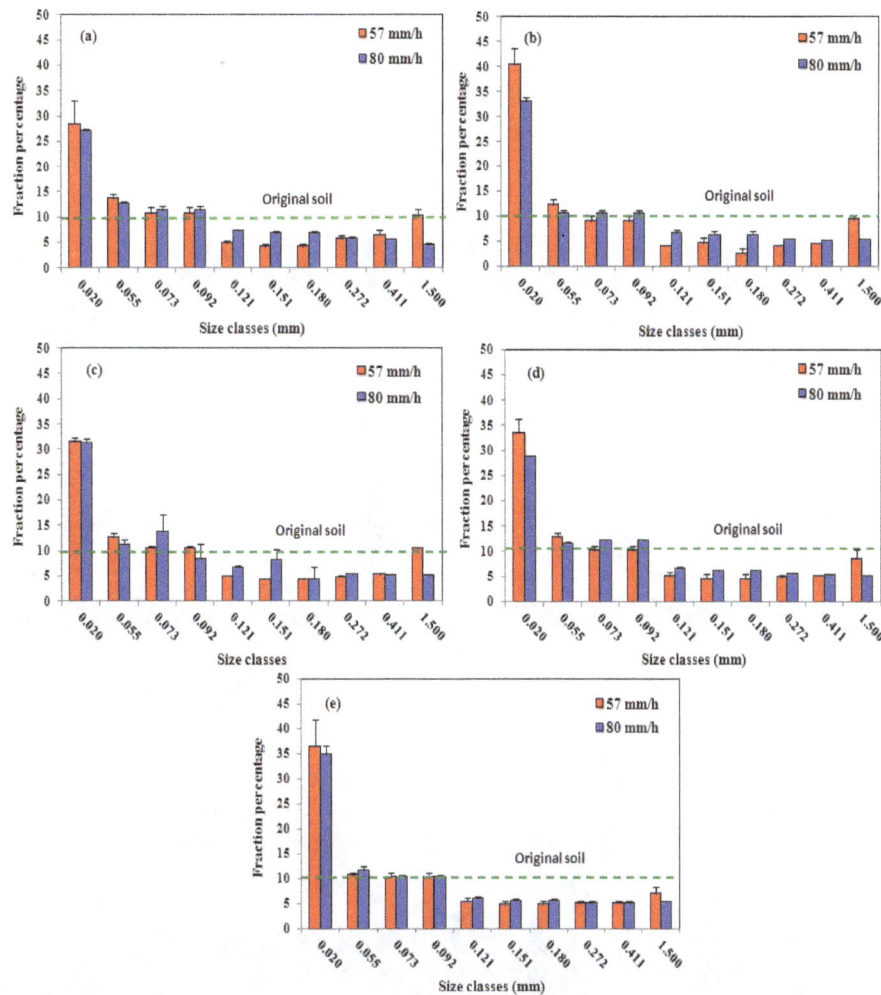

Figure 3. Comparison of particle size distribution in eroded soil D_{max} 2 mm compared to the original soil for different slopes of (**a**) 0.5, (**b**) 2.5, (**c**) 5, (**d**) 10 and (**e**) 20 %. Error bars represent standard errors of the means.

The second difference can be related to the coarsest size class (1.5 mm), which showed higher fraction percentage in rain intensity of $57\,mm\,h^{-1}$ than that observed in $80\,mm\,h^{-1}$. Since the erosive force of raindrops in the higher rain intensity ($80\,mm\,h^{-1}$) was higher than rain intensity of $57\,mm\,h^{-1}$, much larger aggregates were broken down. Consequently, the coarser particles size percentage was reduced under $80\,mm\,h^{-1}$ rain intensity compared to the lower rain intensity. In addition, the rain intensity of $80\,mm\,h^{-1}$ generated higher flow rates leading to higher transportability of aggregates. Meyer et al. (1980) found that the percentage of coarser particles in eroded sediment was higher for more intense rainstorms. Beuselink et al. (2000) reported that in lower stream powers, finer particles were transported selectively and large particles remained on soil surface; however, with increasing stream power, larger particles were also transported.

The obtained result for soil D_{max} 4.75 mm and rain intensity of $57\,mm\,h^{-1}$ showed that the fraction percentage for

size class of 0.043 mm were the highest, which implied a considerable increase compared to the original soil in all slope gradients (Fig. 4). For this lower rain intensity, the fraction percentage in the coarsest size class (3.375 mm) was more than the original soil for all slope gradients. Also, a reduction trend in the fraction percentage was found in the size class of 0.064 to 0.433 mm. Similarly, for the rain intensity of $80\,mm\,h^{-1}$, the most fraction percentage was placed at the finest size class (0.043 mm) and the size classes coarser than 0.064 mm showed less fraction percentages than the original soil in all the slopes (Fig. 4).

A comparison of the fraction percentages for soil D_{max} 4.75 mm under different rain intensities (Fig. 4) showed that in both rain intensities, the most fraction percentage compared to the original was the finest size class (0.043 mm). However, for the rain intensity of $80\,mm\,h^{-1}$, the fraction percentage of the finest size class was higher than that obtained for the intensity of $57\,mm\,h^{-1}$. In contrast, the fraction percentage of the coarsest size (3.375 mm) was reduced

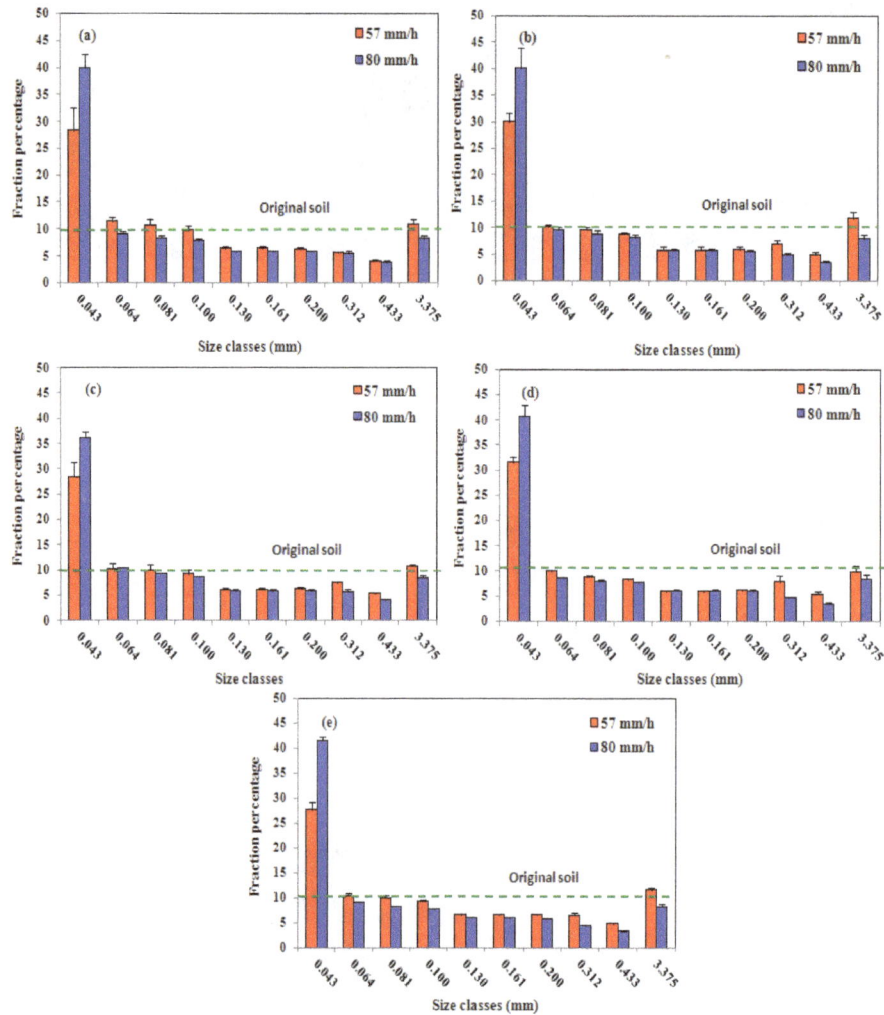

Figure 4. Comparison of particle size distribution in eroded soil D_{max}4.75 mm compared to the original soil and for different slopes of (**a**) 0.5, (**b**) 2.5, (**c**) 5, (**d**) 10 and (**e**) 20 %. Error bars represent standard errors of the means.

at higher rain intensity (80 mm h^{-1}) compared to the lower intensity (57 mm h^{-1}). The result for the finest size class is contradictory to soil D_{max}2 mm may be partly due to the fact that the soil containing larger aggregates exhibited higher infiltration rate and lower flow rates (Mazaheri and Mahmoodabadi, 2012). The result showed that the flow stream power generated on soil D_{max}2 mm and soil D_{max}4.75 mm ranged from 0.0007 to 0.0346 and from 0.0004 to 0.0313 W m^{-2}, respectively. In other words, the higher the rain intensity introduced on soil D_{max}4.75 mm, the greater amounts of finer particles were produced. Nevertheless, because of higher infiltration rate of this soil, the stream power of generated flow seems not to be enough to transport and move out all the pre-detached materials from the test area (Arjmand Sajjadi and Mahmoodabadi, 2015). This finding implies that the redistribution of particles or aggregates on the surface of eroding soil depends on aggregate size distribution as well as rain intensity and the resultant flow stream power.

3.2 Time changes of infiltration rate

Time changes of IR for soil D_{max}2 mm under different rain intensities and slope gradients is presented in Fig. 5. For both rain intensities at the beginning of event, infiltration values were at the highest rates; meanwhile, the fluctuations of IR for different slope gradients were relatively high. Due to the time changes of IR in these first minutes, this period can be considered as unsteady state conditions. Under these conditions, higher IR values were obtained for the steepest slope (20 %). Towards the end of the event, the variations of IR were minimal. Also, it was reduced to reach steady state conditions as the changes of IR found to be negligible with time. The highest fluctuation of IR with time was found when IR was at the maximum value; therefore for each experiment, this value was assumed to be an unsteady IR. To compare these two conditions, results of variance analysis for measured IR under unsteady and steady state conditions are pre-

Table 2. Analysis of variance for the applied treatments on measured infiltration rate under unsteady and steady state conditions.

Source of Variation	D.F.	Mean square for unsteady state conditions	Mean square for steady state conditions
Slope (A)	4	116.2**	4.2[ns]
Rain intensity(B)	1	3207.8**	57.4**
Particle size distribution (C)	1	69.4**	199.3**
A × B	4	63.8**	3.9[ns]
A × C	4	209.8[ns]	3.9[ns]
B × C	1	3431.1**	3.8[ns]
A × B × C	4	205.6[ns]	0.2[ns]
Error	40	4.1	3
Coefficient Variation	–	6.3	19.3

* significant at 0.05 probability; ** significant at 0.01 probability level; ns: non significant.

sented in Table 2. As is obvious, the single effects of rain intensity and soil particle size distribution on IR were significant under both unsteady and steady conditions. In contrast, the influence of slope gradient on IR was just significant in an unsteady state, whereas no significant effect was found under steady state conditions.

Since the studied soils remained saturated during the rainfall, the time changes of IR can only be attributed to seal formation. The results indicated that the surface seal was less-developed during the first minutes and become more developed with the progress of time. This explanation can be applied for the effects of slope gradient on IR under two different steady and unsteady state conditions. Under unsteady state conditions and at steeper slopes, higher values of IR were observed. This means that surface sealing could not be developed at steeper slopes due to the depletion of pre-detached materials by sheet flow. Poesen (1986) inferred that increased IR on steeper slopes can result from reduced surface sealing. In some studies, no significant relationship was found between slope gradient and IR (e.g., Singer and Blackard, 1982; Mah et al., 1992; Martínez-Murillo et al., 2013), whereas in others, a reduction in IR with increasing slope gradient was reported (e.g., Chaplot and Le Bissonnais, 2000; Essig et al., 2009). Fox et al. (1997) observed a reduction in IR with increasing slope gradient until a critical threshold was reached; thereafter, IR was found to be irrelevant to slope gradient. More counterintuitive are the studies that showed an increase in IR with increasing slope gradient (e.g., Poesen, 1986; Cerdà, 1999; Assouline and Ben-Hur, 2006).

In a steady state, lower rates of infiltration were observed compared to the unsteady state conditions. In addition, the effect of slope gradient on steady IR was insignificant (Table 2). According to Fig. 3, the aggregate breakdown due to raindrop impact produced finer aggregates, which were used to form a surface seal with lower hydraulic conductivity than the original soil. Freebairn et al. (1989) attributed the reduction in IR during rainfall in both laboratory and field

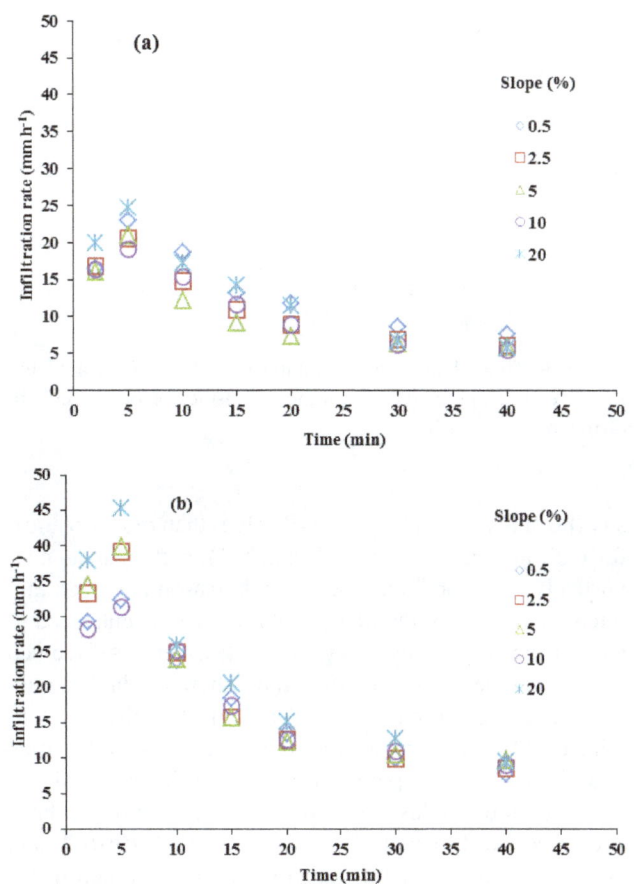

Figure 5. Time changes of infiltration rate in soil $D_{max}2$ mm for different slope gradients and rain intensities of **(a)** 57 and **(b)** 80 mm h^{-1}.

conditions to the formation of surface seal. Similarly, Moss and Watson (1991) reported that the reduction of IR is likely related to the obstruction of surface pores due to aggregate breakdown and seal formation.

A comparison of IR of the simulated rain intensities for soil $D_{max}2$ mm (Fig. 5) implied that the higher rain inten-

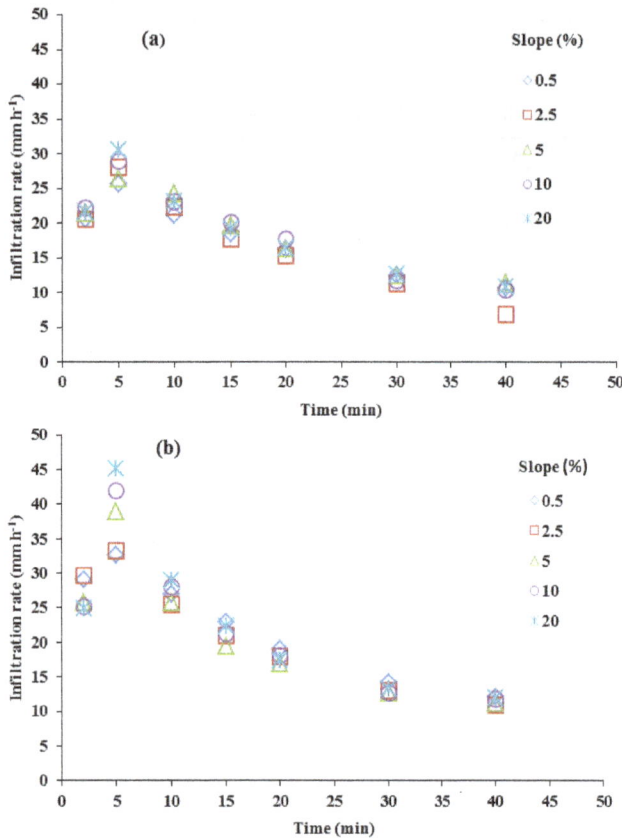

Figure 6. Time changes of infiltration rate in soil $D_{max}4.75$ mm for different slope gradients and rain intensities of (**a**) 57 and (**b**) 80 mm h^{-1}.

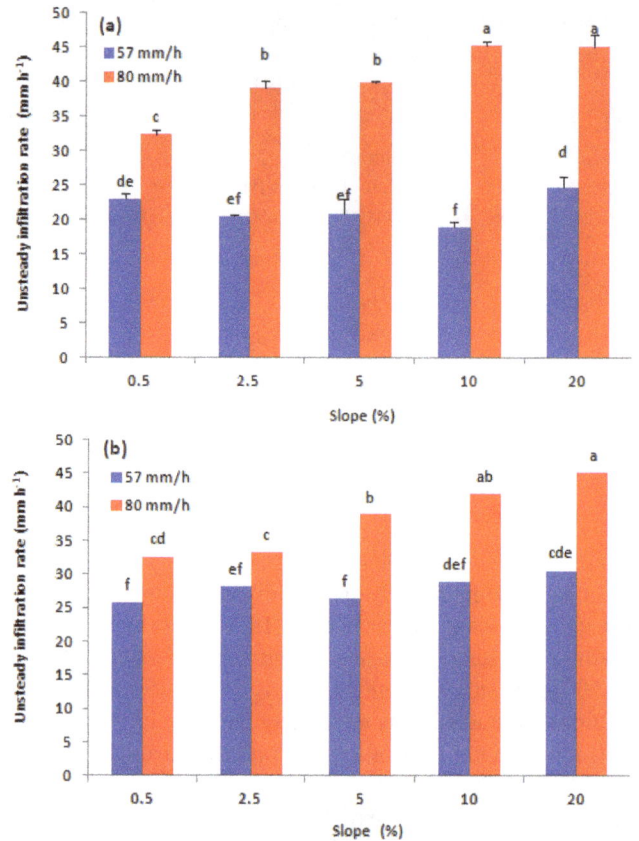

Figure 7. Comparison of the unsteady infiltration rate for soil samples with the maximum particles size of (**a**) 2 and (**b**) 4.75 mm (error bars represent standard errors of the means and mean comparison using Duncan's test; $\alpha = 0.05$).

sity (80 mm h^{-1}) led to greater IR values than those obtained for the lower rain intensity (57 mm h^{-1}), particularly under unsteady state conditions. A plausible reason is that as rain intensity increased, the transportability of flow enhanced to carry detached particles way out of the test area. As discussed above, the finest size class (0.02 mm) showed a higher fraction percentage in rain intensity of 57 mm h^{-1} than that obtained in 80 mm h^{-1}. However, some researchers (e.g., Foley and Silburn, 2002) reported an increase in IR due to higher rain intensities. In this regard, some inconsistent results have been reported (Liu et al., 2011; Schmidt et al., 2010). Liu et al. (2011) believed that the relationship between rain intensity and IR is reversed. Schmidt (2010) verified that higher rain intensities with more erosive impacts can increase the amount of runoff as a result of IR reduction. In our study, we show that in spite of the higher erosivity of more intense rain, the surface seal did not develop completely under unsteady state conditions because of washing out and removing fine soil particles.

Figure 6 shows the changes of IR with time for different rain intensities and slope gradients for soil $D_{max}4.75$ mm. The results of this soil are similar to those obtained for soil $D_{max}2$ mm. At the start of rain event, the unsteady IR fluctu-

ated highly among different slope gradients, while over time it approached a nearly constant value for all slopes. The result indicated that the unsteady IR increased with increasing slope gradient. Also, increasing rain intensity increased IR under unsteady state conditions.

A considerable point observed in both soils (Figs. 5 and 6) is that the measured IR in soil $D_{max}4.75$ mm was higher than in soil $D_{max}2$ mm. The reason for higher IR values in soil $D_{max}4.75$ mm can be attributed to the existence of larger aggregate sizes and the subsequent larger pores. In addition, larger aggregate create a relatively rough surface; therefore, the generated runoff have enough time to infiltrate into the soil.

3.3 Unsteady IR

The results of Table 2 indicate that the influence of slope on IR was significant just under unsteady state conditions. The effect of slope gradient and rain intensity on the unsteady IR for soil $D_{max}2$ mm and $D_{max}4.75$ mm is shown in Fig. 7. In general, the obtained unsteady IR increased as slope steepness increased, especially under the higher rain intensity. For soil $D_{max}2$ mm, the unsteady IR ranged from $19\,\mathrm{mm\,h^{-1}}$ at 10 % slope to $24.7\,\mathrm{mm\,h^{-1}}$ at 20 % slope under $57\,\mathrm{mm\,h^{-1}}$ rain intensity. In higher rain intensity ($80\,\mathrm{mm\,h^{-1}}$), it varied from 32.4 to $45.2\,\mathrm{mm\,h^{-1}}$ as slope gradient increased from 0.5 to 20 %. Therefore, the unsteady IR under $80\,\mathrm{mm\,h^{-1}}$ was higher than $57\,\mathrm{mm\,h^{-1}}$ rain intensity. This finding was consistent with the results of Assouline and Ben-Hur (2006), who reported that infiltration rate and soil loss increased at higher rain intensities. This was attributed to a thinner and less developed seal layer resulting from higher erosion of the soil surface and lower component of drop impact. Thus, the probable reason for the difference between the applied rain intensities in the present study may be partly as a consequence of greater stream power due to the higher rain intensity of $80\,\mathrm{mm\,h^{-1}}$ in removing fine soil particles and underdevelopment of surface seal.

For soil $D_{max}4.75$ mm, as slope gradient increased from 0.5 to 20 % the unsteady IR values due to rain intensities of 57 and $80\,\mathrm{mm\,h^{-1}}$ ranged from 25.7 to $30.6\,\mathrm{mm\,h^{-1}}$ and from 32.6 to $45.1\,\mathrm{mm\,h^{-1}}$, respectively. Therefore, for soil $D_{max}4.75$ mm similar to soil $D_{max}2$ mm, the unsteady IR was higher under rain intensity of $80\,\mathrm{mm\,h^{-1}}$ than that under $57\,\mathrm{mm\,h^{-1}}$. In both rain intensities, the unsteady IR values were higher at steeper slopes for both soils. This means that at steeper slopes and under unsteady state conditions due to faster depletion of pre-detached soil particles seal layer was less developed, which enhanced the infiltration of water into the soil.

4 Conclusions

Considering the obtained fraction percentage in size classes for both eroded soils, the percentage of the finest particles was found to increase compared to the original soil, whereas the reverse result was found for larger aggregates. Also, an increase in rain intensity led to an intensification of aggregate breakdown; however, the effect of rain intensity on the contribution of fraction percentage in size classes depended on the aggregate size. In addition, the soil containing finer aggregates exhibited relatively easy transportability of the predetached material in comparison to the soil containing larger aggregates. Since the studied soils remained saturated during the rainfall event, the change of infiltration rate with time was only attributed to seal formation. The surface seal was found to be less developed during the first minutes, while it formed a more developed seal layer with the progress of time. Fur-

thermore, the result showed that the measured infiltration rate increased with increasing rain intensity, aggregate size and slope under unsteady state conditions because of less development of the surface seal. However, under steady state conditions, no significant relationship was found between slope and the measured infiltration rate, which was attributed to the development of surface seal. In a steady state, lower rates of infiltration were observed compared to the unsteady state conditions. In addition, the soil containing larger aggregates exhibited higher rates of infiltration as this soil was less sensitive to raindrop impact and seal formation. The findings of this study highlight the importance of rain intensity, slope steepness and soil aggregate size on aggregate breakdown and seal formation that can control infiltration rate and the consequent runoff and erosion rates.

Acknowledgements. This work was supported by the Soil Science Department, Faculty of Agriculture, Shahid Bahonar University of Kerman, which is greatly acknowledged.

Edited by: A. Cerdà

References

Arjmand Sajjadi, S. and Mahmoodabadi, M.: Sediment concentration and hydraulic characteristics of rain-induced overland flows in arid land soils, J. Soil. Sedim., 15, 710–721, doi:10.1007/s11368-015-1072-z, 2015.

Asadi, H., Moussavi, A., Ghadiri, H., and Rose, C. W.: Flow-driven soil erosion processes and the size selectivity of sediment, J. Hydrol., 406, 73–81, 2011.

Assouline, S.: Rainfall-induced soil surface sealing: a critical review of observations, conceptual models, and solutions, Vadose Zone J., 3, 570–591, 2004.

Assouline, S. and Ben-Hur, M.: Effect of rainfall intensity and slope gradient on the dynamics of interrill erosion during soil surface sealing, Catena, 66, 211–220, 2006.

Assouline, S. and Mualem, Y.: Modeling the dynamics of seal formation: analysis of the effect of soil and rainfall properties, Water Resour. Res., 36, 2341–2349, 2000.

Beuselinck, L., Govers, G., Steegen, A., and Quine, T. A.: Sediment transport by overland flow over an area of net deposition, Hydrol. Proc., 13, 2769–2782, 1999.

Bodí, M. B., Doerr, S. H., Cerdà, A., and Mataix-Solera, J.: Hydrological effects of a layer of vegetation ash on underlying wettable and water repellent soils, Geoderma, 191, 14–23, 2012.

Brevik, E. C., Cerdà, A., Mataix-Solera, J., Pereg, L., Quinton, J. N., Six, J., and Van Oost, K.: The interdisciplinary nature of soil, Soil, 1, 117–129, 2015.

Cerdà, A.: Rainfall drop size distribution in Western Mediterranean Basin, València, Spain, Catena, 31, 23–38, 1997.

Cerdà, A.: The influence of aspect and vegetation on seasonal changes in erosion under rainfall simulation on a clay soil in Spain, Can. J. Soil Sci., 78, 321–330, 1998.

Cerdà, A.: Seasonal and spatial variations in infiltration rates in badland surfaces under Mediterranean climatic conditions, Water Resour. Res., 35, 319–328, 1999.

Cerdà, A.: Aggregate stability against water forces under different climates on agriculture land and scrubland in southern Bolivia, Soil Till. Res., 36, 1–8, 2000.

Cerdà, A., Giménez-Morera, A. Y., and Bodí, M. B.: Soil and water losses from new citrus orchards growing on sloped soils in the western Mediterranean basin, Earth Surf. Proc. Land., 34, 1822–1830, 2009.

Chaplot, V. and Le Bissonnais, Y.: Field measurements of interrill erosion under different slopes and plot sizes, Earth Surf. Proc. Land., 25, 145–153, 2000.

Ekwue, E. I., Bharat, C., and Samaroo, K.: Effect of soil type, peat and farmyard manure addition, slope and their interactions on wash erosion by overland flow of some Trinidadian soils, Biosyst. Engin., 102, 236–243, 2009.

Essig, E. T., Corradini, C., Morbidelli, R., and Govindaraju, R. S.: Infiltration and deep flow over sloping surfaces: Comparison of numerical and experimental results, J. Hydrol., 374, 30–42, 2009.

Foley, J. L. and Silburn, D. M.: Hydraulic properties of rain impact surface seals on three clay soils influence of raindrop impact frequency and rainfall intensity during steady state, Austr. J. Soil Res., 40, 1069–1083, 2002.

Fox, D. M., Bryan, R. B., and Price, A. G.: The influence of slope angle on final infiltration rate for interrill conditions, Geoderma, 80, 181–194, 1997.

Freebairn, D. M., Gupta, S. C., and Rawls, W. J.: Influence of aggregate size and microrelief on development of surface soil crusts, Soil Sci. Soc. Am. J., 55, 188–195, 1991.

Gabarrón-Galeote, M. A., Martínez-Murillo, J. F., Quesada, M. A., and Ruiz-Sinoga, J. D.: Seasonal changes in the soil hydrological and erosive response depending on aspect, vegetation type and soil water repellency in different Mediterranean micro environments, Solid Earth, 4, 497–509, doi:10.5194/se-4-497-2013, 2013.

García-Orenes, F., Cerdà, A., Mataix-Solera, J., Guerrero, C., Bodí, M. B., Arcenegui, V., Zornoza, R., and Sempere, J. G.: Effects of agricultural management on surface soil properties and soil-water losses in eastern Spain, Soil Till. Res., 106, 117–123, 2009.

García-Fayos, P., Bochet, E., and Cerdà, A.: Seed removal susceptibility through soil erosion shapes vegetation composition, Plant Soil, 334, 289–297, 2010.

Gee, G. W. and Or, D.: Particle size analysis. In: Dane, J. H., and Topp, G. C. (Eds.): Methods of Soil Analysis, Part 4. Physical Mmethods, Soil Science Society of America, Book Series, No. 5, ASA and SSA Madison, WI, 255–293, 2002.

Gelaw, A. M., Singh, B. R., and Lal, R.: Organic carbon and nitrogen associated with soil aggregates and particle sizes under different land uses in Tigray, northern Ethiopia, Land Degrad. Dev., doi:10.1002/ldr.2261, 2015.

Grierson, I. T. Y. and Oades, J. M.: A rainfall simulator for field studies of runoff and erosion, J. Agr. Engin. Res., 22, 37–44, 1977.

Hairsine, P. B., Sander, G. C., Rose, C. W., Parlange, J. Y., Hogarth, W. L., Lisle, I., and Rouhipour, H.: Unsteady soil erosion due to rainfall impact: a model of sediment sorting on the hillslope, J. Hydrol., 199, 115–128, 1999.

Haregeweyn, N., Poesen, J., Verstraeten, G., Govers, G., de Vente, J., Nyssen, J., Deckers, J., and Moeyersons, J.: Assessing the performance of a spatially distributed soil erosion and sediment delivery model (WATEM/SEDEM) in Northern Ethiopia, Land Degrad. Dev., 24, 188–204, 2013.

Hawke, R. M., Price, A. G., and Bryan, R. B.: The effect of initial soil water content and rainfall intensity on near-surface soil hydrologic conductivity: a laboratory investigation, Catena, 65, 237–346, 2006.

Janeau, J. L., Briquet, J. P., Planchon, O., and Valentin, C.: Soil crusting and infiltration on steep slopes in northern Thailand, Europ. J. Soil Sci., 54, 543–553, 2003.

Kemper, W. D. and Rosenau, R. C.: Aggregate stability and size distribution, in: Methods of Soil Analysis, Part 1. Physical and Mineralogical Methods, edited by: Klute, A., American Society of Agronomy, Madison, Wisconsin, 425–442, 1986.

Kinnell, P. I. A.: Raindrop impact induced erosion processes and prediction: A review, Hydrol. Proc., 19, 2815–2844, 2005.

Le Bissonnais, Y.: Aggregate stability and assessment of soil crustability and erodibility: I. Theory and methodology, Europ. J. Soil Sci., 47, 425–437, 1996.

Legout, C., Leguédois, S., Le Bissonnais, Y., and Issa, O. M.: Splash distance and size distributions for various soils, Geoderma, 124, 279–292, 2005.

Leh, M., Bajwa, S., and Chaubey, I.: Impact of land use change on erosion risk: an integrated remote sensing geographic information system and modeling methodology, Land Degrad. Dev., 24, 409–421, 2013.

Lieskovský, J. and Kenderessy, P.: Modelling the effect of vegetation cover and different tillage practices on soil erosion in vineyards: a case study in Vrable (Slovakia) using WATEM/SEDEM, Land Degrad. Dev., 25, 288–296, 2014.

Liu, H., Lei, T. W., Zhao, J., Yuan, C. P., Fan, Y. T., and Qu, L. Q.: Effects of rainfall intensity and antecedent soil water content on soil infiltrability under rainfall conditions using the run off-on-out method, J. Hydrol., 396, 24–36, 2011.

Lujan, D. L.: Soil physical properties affecting soil erosion in tropical soils, Lecture given at the College on Soil Physics, Trieste, 3–21 March, 2003.

Mah, M. G. C., Douglas, L. A., and Ringrose-Voase, A. J.: Effects of crust development and surface slope on erosion by rainfall, Soil Sci., 154, 37–43, 1992.

Mahmoodabadi, M.: Sediment yield estimation using a semi-quantitative model and GIS- remote sensing data, Internat. Agrophys., 25, 241–247, 2011.

Mahmoodabadi, M. and Cerdà, A.: WEPP calibration for improved predictions of interrill erosion in semi-arid to arid environments, Geoderma, 204/205, 75–83, 2013.

Mahmoodabadi, M. and Sirjani, E.: Study on sediment transport mechanisms due to sheet erosion using flume experiment, J. Watershed Engin. Manage., 4, 1–11, 2012 (in Persian).

Mahmoodabadi, M., Rouhipour, H., Arabkhedri, M., and Rafahi, H. G.: Intensity calibration of SCWMRI rainfall and erosion simulator, J. Watershed Manage. Sci. Engin., 1, 39–50, 2007 (in Persian).

Mahmoodabadi, M., Ghadiri, H., Bofu, Y., and Rose, C.: Morphodynamic quantification of flow-driven rill erosion parameters based on physical principles, J. Hydrol., 514, 328–336, 2014a.

Mahmoodabadi, M., Ghadiri, H., Rose, C., Bofu, Y., Rafahi, H., and Rouhipour, H.: Evaluation of GUEST and WEPP with a new approach for the determination of sediment transport capacity, J. Hydrol., 513, 413–421, 2014b.

Mandal, D. and Sharda, V. N.: Appraisal of soil erosion risk in the Eastern Himalayan region of India for soil conservation planning, Land Degrad. Dev., 24, 430–437, 2013.

Martínez-Murillo, J. F., Nadal-Romero, E., Regüés, D., Cerdà, A., and Poesen, J.: Soil erosion and hydrology of the western Mediterranean badlands throughout rainfall simulation experiments, Catena, 106, 101–112, 2013.

Mazaheri, M. R. and Mahmoodabadi, M.: Study on infiltration rate based on primary particle size distribution data in arid and semi arid region soils, Arab. J. Geosci., 5, 1039–1046, 2012.

Mermut, A. R., Luk, S. H., Romkens, M. J. M., and Poesen, J. W. A.: Soil loss by splash and wash during rainfall from two loess soils, Geoderma, 75, 203–214, 1997.

Meyer, L. D., Harmon, W. C., and McDowell, L. L.: Sediment size eroded from crop row sideslopes, Trans. ASAE, 23, 891–898, 1980.

Moreno-Ramón, H., Quizembe, S. J., and Ibáñez-Asensio, S.: Coffee husk mulch on soil erosion and runoff: experiences under rainfall simulation experiment, Solid Earth, 5, 851–862, doi:10.5194/se-5-851-2014, 2014.

Moss, A. J. and Watson, C. L.: Rain-impact soil crust III. Effects of continuous and flawed crusts on infiltration and the ability of plant cover to maintain crustal flaws, Austr. J. Soil Res., 29, 311–330, 1991.

Pansu, M. and Gautheyrou, J.: Handbook of soil analysis, mineralogical, organic and inorganic methods, Springer, Heidelberg, p. 993, 2006.

Poesen, J.: Surface sealing as influenced by slope angle and position of simulated stone sin the top layer of loose sediments, Earth Surf. Proc. Land., 11, 1–10, 1986.

Poesen, J.: The role of slope angle in surface seal formation, in: Proc. 1st International Conference on Geomorphology: Geomorphology, Resource Environment and Developing World, edited by: Gardner, V., John Wiley and Sons, New York, 437–448, 1987.

Proffitt, A. P. B., Hairsine, P. B., and Rose, C. W.: Modelling soil erosion by overland flow: application over a range of hydraulic conditions, Trans. ASAE, 36, 1743–1753, 1993.

Pulido Moncada, M., Gabriels, D., Cornelis, W., and Lobo, D.: Comparing aggregate stability tests for soil physical quality indicators, Land Degrad. Dev., doi:10.1002/ldr.2225, 2013.

Ramos, M. C., Nacci, S., and Pla, I.: Effect of raindrop impact and its relationship with aggregate stability to different disaggregation forces, Catena, 53, 365–376, 2003.

Ribolzi, O., Patin, J., Bresson, L. M., Latsachack, K. O., Mouche, E., Sengtaheuanghoung, O., Silvera, N., Thiebaux, J. P., and Valentin, C.: Impact of slope gradient on soil surface features and infiltration on steep slopes in northern Laos, Geomorphology, 127, 53–63, 2011.

Robinson, D. A. and Phillips, C. P.: Crust development in relation to vegetation and agricultural practice on erosion susceptible, dispersive clay soils from central and southern Italy, Soil Till. Res., 60, 1–9, 2001.

Romkens, M., Baumhardt, R., Parlange, J., Whistler, F., Parlange, M., and Prasad, S.: Rain-induced surface seals: their effect on ponding and infiltration, Ann. Geophys. (Series B), 4, 17–424, 1985.

Romkens, M. J. M., Helming, K., and Prasad, S. N.: Soil erosion under different rainfall intensities, surface roughness and soil water regimes, Catena, 46, 103–123, 2001.

Rose, C. W., Yu, B., Ghadiri, H., Asadi, H., Parlange, J. Y., Hogarth, W. L., and Hussein, J.: Dynamic erosion of soil in steady sheet flow, J. Hydrol., 333, 449–458, 2007.

Schmidt, J.: Effects of soil slaking and sealing on infiltration-experiments and model approach, Proceedings of the 19th World Congress of Soil Science: Soil solutions for a changing world, Brisbane, Australia, 1–6 August 2010, the physics of soil pore structure dynamics, 29–32, 2010.

Shi, Z. H., Yan, F. L., Li, L., Li, Z. X., and Cai, C. F.: Interrill erosion from disturbed and undisturbed samples in relation to topsoil aggregate stability in red soils from subtropical China, Catena, 81, 240–248, 2010.

Singer, M. J. and Blackard, J.: Slope angle-interrill soil loss relationships for slopes up to 50 %, Soil Sci. Soc. Am. J., 46, 1270–1273, 1982.

Sirjani, E. and Mahmoodabadi, M.: Study on flow erosivity indicators for predicting soil detachment rate at low slopes, Internat. J. Agr. Sci., Res. Technol., 2, 55–61, 2012.

Sirjani, E. and Mahmoodabadi, M.: Effects of sheet flow rate and slope gradient on sediment load, Arab. J. Geosci., 7, 203–210, 2014.

Slattery, M. C. and Burt, T. P.: Particle size characteristics of suspended sediment in hillslope runoff and stream flow, Earth Surf. Proc. Land., 22, 705–719, 1997.

Walkley, A. and Black, I. A.: An examination of the degtjareff method for determining soil organic matter, and proposed modification of the chromic acid titration method, Soil Sci., 37, 29–38, 1934.

Wick, A. F., Daniels, W. L. Nash, W. L., and Burger, J. A.: Aggregate recovery in reclaimed coal mine soils of SW Virginia, Land Degrad. Dev., doi:10.1002/ldr.2309, 2014.

Zamani, S. and Mahmoodabadi, M.: Effect of particle-size distribution on wind erosion rate and soil erodibility, Arch. Agr. Soil Sci., 59, 1743–1753, 2013.

Zhao, G., Mu, X., Wen, Z., Wang, F., and Gao, P.: Soil erosion, conservation, and Eco-environment changes in the Loess Plateau of China, Land Degrad. Dev., 24, 499–510, 2013.

10

Cobalt, chromium and nickel contents in soils and plants from a serpentinite quarry

M. Lago-Vila, D. Arenas-Lago, A. Rodríguez-Seijo, M. L. Andrade Couce, and F. A. Vega

Department of Plant Biology and Soil Science, University of Vigo, 36310 Vigo, Spain

Correspondence to: M. L. Andrade Couce (mandrade@uvigo.es)

Abstract. The former serpentinite quarry of Penas Albas (Moeche, Galicia, NW Spain) left behind a large amount of waste material scattered over the surrounding area, as well as tailing areas. In this area several soils were studied together with the vegetation growing spontaneously over them with the aim of identifying the bioavailability of heavy metals. The potential of spontaneous vegetation for phytoremediation and/or phytostabilization was evaluated. The pH of the soils ranges from neutral to basic, with very low organic matter and nitrogen contents. There are imbalances between exchangeable cations like potassium (K) and calcium (Ca), mainly due to high magnesium (Mg) content that can strongly limit plant production. Moreover, in all of the studied soils there are high levels of cobalt (Co), chromium (Cr) and nickel (Ni) (>70, >1300 and $>1300\,\mathrm{mg\,kg^{-1}}$, respectively). They exceed the intervention limits indicated by soil guideline values. Different soil extractions were performed in order to evaluate bioavailability. $CaCl_2$ 0.01 M is the most effective extraction reagent, although the reagent that best predicts plant availability is a mixture of low molecular weight organic acids. *Festuca rubra*, L. is the spontaneous plant growing in the soils that accumulates the highest amount of the metals, both in shoot and roots. *Festuca* also has the highest translocation factor values, although they are only >1 for Cr. The bioconcentration factor is >1 in all of the cases, except in the shoot of *Juncus* sp. for Co and Ni. The results indicate that *Festuca* is a phytostabilizer of Co and Ni and an accumulator of Cr, while *Juncus* sp. is suitable for phytostabilization.

1 Introduction

Environmental pollution is a global threat of increasing severity due to urban growth, industrialization and changing lifestyles (Liu et al., 2014). According to the "EC Guidance on Undertaking Non-Energy Extractive Activities in Accordance with Natura 2000 Requirements" and the COMG (Cámara Oficial Mineira de Galicia) it is possible to make extractive activities compatible with preservation of the natural environment (COMG, 2013). Land degradation is taking place in the world due to soil erosion, deforestation (Biro et al., 2013; de Souza et al., 2013; Mandal and Sharda, 2013; Milder et al., 2013; Zhao et al., 2013) and soil pollution (Fernández Calviño et al., 2013; Vacca et al., 2012; Yang et al, 2012). Serpentinite soils are stressful environment for plants and also for other living organisms, with low calcium to magnesium ratio, deficiencies of essential macronutrients, high concentrations of heavy metals and low water-holding capacity (Doubková et al., 2012).

Approximately 5 % of Galicia (Spain) is covered by serpentinitic areas; these sites were formerly quarries from which materials for roads, ballast for railway, and ornamental rock were extracted (Pereira et al., 2007). The tailings left behind are often a source of contamination. The soils formed on these tailings (Spolic Technosols; FAO, 2006) must be rehabilitated, as they provide an unsuitable environment for plant growth. They are susceptible to weathering and can cause environmental degradation mainly due to their high content of heavy metals and low organic matter and nutrient content (Asensio et al., 2013).

Spolic Technosols are very young soils that form over unstable materials with low cohesion and physical, chemical and biological deficiencies. These facts are due to their low nutrient and organic matter content, and a high content of heavy metals, which limits the development of bacteria, plants and animals (Deng et al., 2006; Ali et al., 2013).

The lack of nutrients and anomalous physicochemical properties means that the establishment of plant cover is strongly limited in these areas, favouring the accelerated weathering of the soil (Mendez and Maier, 2008). In addition, the limited plant cover contributes to the migration of heavy metals that contaminate surface and underground waters (Bidar et al., 2009).

In the soils from these types of tailings, the levels of copper (Cu), cobalt (Co), chromium (Cr) and nickel (Ni) are usually high (Brooks, 1987; Brooks et al., 1992; Gambi, 1992; Gough et al., 1989; Oze et al., 2004a, b; Rabenhorst et al., 1982; Schwertmann and Latham, 1986), and there is also a deficiency of essential nutrients for plants, such as nitrogen (N), phosphorus (P) and potassium (K) (Turitzin, 1991; Proctor and Woodell, 1975; Walker, 1954). Therefore, the recovery of serpentinite quarry soils must not only consist of eliminating or immobilizing the contaminants, but also of improving the quality and fertility of the soils.

The total heavy metal concentration of soils includes all of the chemical forms that are in there. Therefore this total content does not provide reliable information on the mobility, availability and toxicity of the metals (Adamo et al., 2002; Pueyo et al., 2004).

This means that it is essential to know the available content of heavy metals in soils, the one that can interact with an organism and become incorporated into its structure. This content depends on a large number of factors, which include the properties of the contaminating element (ionic radii, charge...) and the soil (pH, ionic strength, organic C and clay contents...) (Naidu et al., 2008).

Physicochemical and biological methods, such as precipitation–flocculation coupled with pre/post-oxidation, reduction and concentration, have all been studied (Agrawal et al., 2006) in order to decontaminate soils with a high content of heavy metals and to preserve the environment; they are also often employed to control environmental pollution. These techniques, known as "removal–disposal", have numerous drawbacks, such as their high cost, low efficiency, lengthy and complex treatments for a wide variety of metals, and the formation of large amounts of toxic sub-products (Adki et al., 2013). Consequently, processes based on "recovery–reuse" are now being increasingly projected and used (Agrawal et al., 2006). Phytoremediation could avoid some of the problems of the aforementioned treatments, as it is a harmless procedure that respects the environment (Adki et al., 2013; Ali et al., 2013; Paz-Ferreiro et al., 2014).

Therefore, it is of great interest to study and analyse the plants that grow spontaneously in these zones. Their adaptation to the high concentrations of certain metals present in these soils, together with other limiting factors for plant growth, may provide an indication of the procedure to apply in the restoration process.

Hyperaccumulator plants are able to grow in these soils, as they have an extraordinary ability to absorb metals; but their efficiency may be limited due to the low bioavailability of the metals in the soils (Knight et al., 1997; Ali et al., 2013). These plants have unique characteristics, such as the ability to absorb and translocate metals from their roots to their shoot, and a high tolerance. Hyperaccumulators normally have little biomass, because they need a great deal of energy for the mechanisms required to adapt to the high concentrations of metals in their tissues (Garbisu and Alkorta, 2001).

The ideal plant for phytoextraction should be capable of growing in soils with large amounts of metals. It should also have a large radicular system and high levels of biomass production based on optimum growth and development, and be able to accumulate high concentrations of metals in its shoot, store several different metals at the same time and be resistant to pests and diseases (Garbisu and Alkorta, 2001).

Phytoextraction reduces the metal content of the bioavailable fraction of soils, and so this technique is used to reduce the damage caused to the environment (Martin and Ruby, 2004; Ali et al., 2013). When phytoextraction is not possible, phytostabilization should be carried out. This consists of fixing the metals in the soil, stabilizing contaminated soils and reducing the flow of contaminants into the environment. Plant cover also protects against weathering, thus reducing the risk of water infiltration and metals reaching aquifers. In the phytostabilization process, plants do not accumulate metals in their shoots, limiting the risk in terms of food safety (Garbisu and Alkorta, 2001; Ali et al., 2013).

In light of these issues, the aim of this study is: (a) to verify which is the ideal extractant to determine the phytoavailability of the heavy metals contained in soils from a former serpentine mine and (b) to evaluate the phytoremediation/phytostabilization capacity of the spontaneous vegetation growing in these soils.

2 Material and methods

2.1 Material

The study area is located in the Penas Albas serpentinite quarry (43°31′42.46″ N, 8°0′35.61″ W) (Moeche, Coruña, Spain). It is located at the ultramafic complex of Ortegal Cape, a group of dominant, acidic and intermediate basic rocks with sedimentary insertions and discontinuous serpentinite bodies, separated from other geological units (Castroviejo et al., 2004).

This serpentinite quarry (formed by the metamorphism of ultrabasic rocks) operated between the 1960s and mid-1990s. It left behind a large amount of waste material scattered over the surrounding area, as well as tailing areas.

The quarry produced around 50 000 Mg year^{-1} of serpentinite needed in the steel and construction industries (Pereira

Table 1. Profile description.

S1	Sampling site information: Moeche (A Coruña, Spain). 43° 31.769′ N 8° 00.506′ W. Altitude: 149 m. Quarry spoil. Undulating, slope: class 2. Vegetation: *Festuca rubra* L. and *Juncus* sp. L. Soil information: Spolic Technosol. Quarry tailing. Wet, drainage: class 0. Stony (class 4) with serpentinite stones and gravel on the surface. No signs of erosion. Descriptive: AC: 0–15 cm. Gley 2′4/10 BG. Sandy texture. Unstructured. Without consistency in wet and dry, slightly sticky and without plasticity. Few roots. C: +15 cm. Mixture of spolic materials (serpentinite fragments).
S2	Sampling site information: Moeche (A Coruña, Spain). 43° 31.718′ N 8° 00.571′ W. Altitude: 149 m. Quarry spoil. Located on a flat or nearly flat area, slope: class 1. Vegetation: *Festuca rubra* L, *Salix atrocinerea* Brot and *Juncus* sp. L. Soil information: Spolic Technosol. Quarry tailing. Wet, drainage: class 1. Stony (class 4) with serpentinite stones and gravel on the surface. No signs of erosion. Descriptive: AC: 0–15 cm. 10 YR 4/3. Sandy loam texture. Moderate crumbly structure. Slightly hard, very friable, slightly sticky and plastic. Abundant roots. C: +15 cm. Mixture of spolic materials (serpentinite fragments).
S3	Sampling site information: Moeche (A Coruña, Spain). 43° 31.700′ N 8° 00.598′ W. Altitude: 140 m. Quarry spoil. Hilly, slope: class 5. No vegetation. Soil information: Spolic Technosol. Quarry tailing. Wet, drainage: class 2. Stony (class 4) with gravel, stones and boulders of serpentinite with chrysotile abundance. Evidence of water erosion. Descriptive: AC: 0–25 cm. Gley 1/5 GY. Sandy loam texture. Unstructured. Without consistency in wet and dry, sticky and without plasticity. C: +25 cm. Mixture of spolic materials (serpentinite fragments, with chrysotile abundance).
S4	Sampling site information: Moeche (A Coruña, Spain). 43° 31.674′ N 8°00.608′ W. Altitude: 148 m. Cut zone. Hilly, slope: class 5. No vegetation (coverage < 1 %). Soil information: Lithic Leptosol. Wet, drainage: class 2. Stony (class 4) with stones and boulders of serpentinite with chrysotile abundance. Evidence of water erosion. Descriptive: A: 0–10 cm. Gley 5/5 GY. Loamy sand texture. Unstructured. Without consistency in wet and dry, sticky and without plasticity. R: +10 cm. Serpentinite (mainly composed of serpentine and chrysotile)
CS	Sampling site information: Moeche (A Coruña, Spain). 43° 31.690′ N 8° 00.687′ W. Altitude: 132 m. Reforested area outside the quarry. Sloping: class 2. Vegetation: *Pinus pinaster* Ait, *Festuca rubra* L. (the most abundant) and very few of *Ulex europaeus* L. and *Rubus ulmifolius* Schot. Soil information: Mollic Leptosol (Control soil). Wet, drainage: class 2. Stoniness: class 1 (with gravels and stones of serpentinite). Slight evidence of water erosion. Descriptive: A: 0–30 cm. 7′5 YR 3/3. Loam texture. Moderate crumb structure. Slightly hard, highly friable, slightly sticky and slightly plastic. Abundance of roots of various sizes. R: (+30 cm). Serpentinite.

et al., 2007). The quarry is now abandoned, and hardly any rehabilitation work has been carried out.

Four zones were selected (Fig. 1): three in different quarry spoils (S1, S2 and S3) and one (S4) in the cut zone (natural soils, whose parent material is the living rock: serpentinite). The control soil (CS) was sampled outside the quarry, in an area which has been reforested and treated with fertilizer and animal manure (Fig. 1, Table 1).

In each selected area, three sub-areas were selected with different degrees of plant cover and diversity, as well as different degrees of slope (Table 1). *Festuca rubra* L. and *Juncus* sp. L. were chosen for this study because they are the most abundant species.

In each of the sub-areas, three surface soil (20 cm, and less than 1 m distance among them) samples were collected using an Eijkelkamp sampler and then stored in polyethylene bags. The soil samples from each sub-area were pooled, air dried, sieved (2 mm), and homogenized in a Fritsch Laborette rotary sample divider, thus obtaining a composite sample of each sub-area. Each one of these composite samples was divided into three sub-samples to perform different analyses.

The soil profiles were described according to the FAO (2006) guidelines and the descriptions are shown in Table 1. The soil colours were determined using revised standard soil colour charts (Munsell Soil Color Charts, 2000).

In each zone, several specimens of *Festuca rubra* and *Juncus* sp. were sampled. In the laboratory, the specimens were

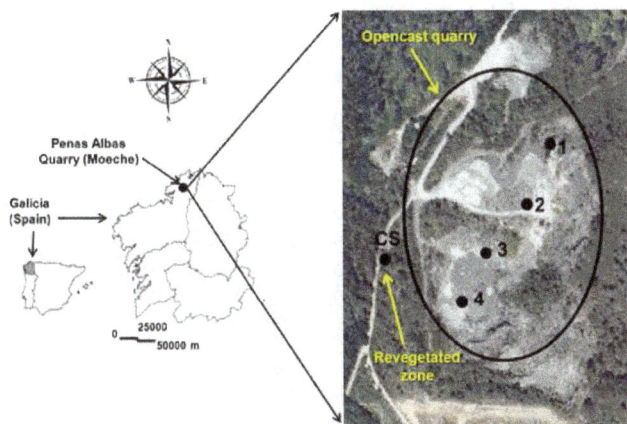

Figure 1. Study area.

washed several times with bidistilled water to remove the remaining soil particles adhered to their surface. Subsequently, the roots and shoots were separated and dried in an oven at 60 °C until reaching constant weight. Afterwards they were crushed and stored in hermetically sealed polythene bags, ready for use.

Three sub-samples per soil and plant were finally used for all of the analytical measurements, meaning that all of the analyses were performed in triplicate.

2.2 Methods

Soil pH, Kjeldahl-N, and organic C (OC) were determined, respectively, with a pH electrode in 2 : 1 water / soil extracts, according to Bremner and Mulvaney (1982) and following the Walkley and Black (1934) procedure. The iron (Fe), manganese (Mn) and aluminium (Al) oxide contents were determined using the dithionite–citrate method (Sherdrick and McKeague, 1975; Soil Conservation Service US, 1972). The concentration in the extract was determined by ICP-OES in a Perkin Elmer Optima 4300 DV apparatus. The effective cation exchange capacity (ECEC) and exchangeable cation content were determined according to Hendershot and Duquette (1986). Al, Ca, K, Mg and Na were extracted with 0.1 M $BaCl_2$, and their concentrations were determined by ICP-OES as above.

Particle size distribution was determined after oxidizing the organic matter with H_2O_2, separating the upper fraction (50 mm) by sieving and using the lower fraction in the internationally endorsed procedure (Day, 1965).

The total metal content was analysed by the fusion method with $Li_2B_4O_7$–$LiBO_2$, mixing 0.5 g of the sample with 3.5 g of $Li_2B_4O_7$–$LiBO_2$ flux (50/50 w/w) and 0.1 g of LiI in a platinum crucible (Hill, 2008). The mixture was fused in a propane-Perl induction heated machine (Claisse) for 20 min. The content of the crucibles was hot-poured into Teflon precipitate flasks containing 100 mL of HNO_3 and then magnetically shaken to help dissolve the fused mixture, which was

then transferred to a 500 mL flask and made up to volume with 5 % HCl. The final solution was analysed by ICP-OES and the control was a standard aqueous multi-element dissolution.

In order to determine the bioavailable Co, Cr and Ni content in soils, five extractants were selected – specifically, the most widely used by numerous authors. In accordance with Houba et al. (2000), soil samples were extracted with 0.01 M $CaCl_2$; EDTA (0.01 M Na_2-EDTA + 1 M CH_3COONH_4) was used following AFNOR (1994). Extractions with DTPA (0.005 M DTPA + 0.1 M TEA + 0.01 M $CaCl_2$) were done in accordance with Lindsay and Norvell (1978). Bidistilled water (BDW) was used following Pueyo et al. (2004) and a 10 mM mixture of five different low molecular weight organic acids (LMWOA) was used according to the instructions of Feng et al. (2005). The LMWOA composition was acetic, lactic, citric, malic and formic acids with molar concentration ratio of 4 : 2 : 1 : 1 : 1, respectively. In all the extractions the concentration of Co, Cr and Ni was determined by ICP-OES.

The selected plants (shoot and root) were also analysed for total Co, Cr and Ni contents after being extracted with H_2O_2 and HNO_3 in a microwave oven (Bell et al., 2000; Lago-Vila et al., 2014). As above, the Co, Cr and Ni concentrations in the supernatant were analysed by ICP-OES.

The extraction efficiency (EF) of each of the extractants accounts for the amount of metal released from the soil with each extractant compared to the total in the soil. It was estimated by the proportion of the total content extracted by each one, as given by

$$EF = 100C_e/C_t, \tag{1}$$

where C_e and C_t are the metal extracted and total metal content respectively (mg kg^{-1}).

It is a useful parameter to better understand the ability of the extractants used to release the metals studied. With EF data it is possible to compare the proportion of the metal released as it is related to the total content.

The translocation factor (TF) was estimated as the ratio between the trace metal content (mg kg^{-1}) in shoot (C_s) and the one in the roots (Cr):

$$TF = C_s/C_r. \tag{2}$$

TF > 1 indicates that the plant translocates metals effectively from roots to shoot (Baker and Brooks, 1989).

The ratio of metal concentration in the plant to soil was used to determine the bioconcentration factor (BF):

$$BF = C_p/C_{so}, \tag{3}$$

where C_p and C_{so} are metal concentrations in the plant (shoot and root) and in the soil, respectively. Hyperaccumulator plants are used to show BF values greater than 1 and sometimes even ranging from 50 to 100 (McGrath, 1998).

Table 2. Some physicochemical characteristics of the soils (mean values and standard deviation).

	Unit	S1	S2	S3	S4	CS
pH$_{(H_2O)}$		7.98 ± 0.06 ab	7.81 ± 0.06 b	8.05 ± 0.03 a	7.87 ± 0.08 b	5.99 ± 0.16 c
pH$_{(KCl)}$		7.11 ± 0.01 c	6.96 ± 0.03 d	7.92 ± 0.02 a	7.72 ± 0.03 b	4.76 ± 0.03 e
Total N	g kg^{-1}	ul	0.42 ± 0.00 b	ul	ul	2.75 ± 0.05 a
OC		0.30 ± 0.07 c	3.58 ± 0.45 b	ul	0.38 ± 0.071 b	11.29 ± 0.97 a
Fe oxides		2.24 ± 0.07 c	15.65 ± 0.15 b	1.61 ± 0.02 d	1.30 ± 0.03 e	23.98 ± 0.16 a
Mn oxides		0.12 ± 0.00 d	0.63 ± 0.01 b	0.16 ± 0.00 c	0.11 ± 0.00 d	0.89 ± 0.01 a
Al oxides		0.09 ± 0.01 e	1.71 ± 0.00 b	0.21 ± 0.00 c	0.18 ± 0.01 d	4.31 ± 0.06 a
				Exchangeable cation and ECEC		
ECEC	cmol$_{(+)}$ kg^{-1}	5.21 ± 0.15 d	19.59 ± 0.89 a	6.15 ± 0.41 c	5.66 ± 0.25 cd	16.76 ± 0.46 b
Ca^{2+}		1.54 ± 0.07 c	7.20 ± 0.36 a	3.84 ± 0.25 b	3.86 ± 0.18 b	3.83 ± 0.11 b
K^{+}		0.36 ± 0.09 b	0.50 ± 0.03 a	0.10 ± 0.01 d	0.07 ± 0.01 d	0.28 ± 0.02 c
Mg^{2+}		3.11 ± 0.13 c	11.50 ± 0.51 b	2.05 ± 0.15 d	1.61 ± 0.05 d	12.13 ± 0.31 a
Na^{2+}		0.19 ± 0.06 b	0.38 ± 0.02 a	0.16 ± 0.03 bc	0.12 ± 0.01 c	0.40 ± 0.02 a
Al^{3+}		0.01 ± 0.01 b	0.01 ± 0.01 b	ul	ul	0.12 ± 0.00 a
				Particle size distribution		
Sand	%	89.39 ± 0.75 a	59.83 ± 0.95 d	74.68 ± 0.80 c	82.16 ± 0.71 b	26.73 ± 0.56 e
Silt		6.56 ± 0.54 e	26.06 ± 0.72 b	13.75 ± 0.38 c	10.76 ± 0.50 d	46.56 ± 1.05 a
Clay		4.05 ± 0.28 e	14.11 ± 0.24 b	11.56 ± 0.73 c	7.08 ± 0.24 d	26.71 ± 0.51 a

ul: undetectable level (detection limit 0.1 mg kg^{-1}). OC: organic carbon. ECEC: effective cationic exchange capacity. For each parameter, values followed by different letters differ significantly with $p < 0.05$.

2.3 Statistical analysis

The data obtained in the analytical determinations were analysed with the statistical program IBM-SPSS Statistics 19 (SPSS, Inc., Chicago, IL). The results obtained in all the determinations were the average of the standard deviation of three analyses and were expressed on a dry material basis. One-factor ANOVA was carried out, together with homogeneity of variance tests for the variables found. In the case of homogeneity of variance, the minimum significant distance test among soil properties was carried out as a post-hoc test ($p < 0.05$), or otherwise Dunnett's T3 test. A bivariate correlation analysis between extracted Co, Cr and Ni and their content in plants was also carried out ($p < 0.05$ and $p < 0.01$), calculating Pearson's correlation coefficient.

3 Results and discussion

3.1 Characterization of the soils

The pH$_{H2O}$ (Table 2) was between 8 (S3) and 6 (CS), and the pH$_{KCl}$ varied between 7.9 (S3) and 4.8 (CS). The soils are basic, which affects the retention of trace elements positively. However, CS was slightly acid, probably because of the vegetation and its organic matter content, as discussed below.

The soil samples with the lowest OC contents (S3 < S1 < S4; Table 2) are those with the lowest vege-

tation coverage. In S3 there is no vegetation and in S1 and S4 only herbs like *Festuca* or *Juncus* grow and the percentage of coverage is small (Table 1). The opposite happens in the places (S1 and SC) where trees like *Pinus* or *Salix* and bushes like *Rubus* are present (Table 1). They are the samples with the highest OC contents (Table 2). Summarizing, the OC content is very low in the quarry soils (Table 2), and it is directly related to the presence of vegetation (Table 1). This is because the serpentinite barely contributes to the C and N pools of the soils (Corti et al., 2002). In fact, the N content in the soils is very low: it was not detected in S1, S3 and S4 and in S2 it is moderately low. The N content in CS is slightly high (Table 2). These results agree with the levels found in soils developed over tailings from extractive activities (Mendez and Maier, 2008). The higher content in CS is due to the received treatments and to the plants growing there.

The oxide contents are low in all of the soils except for CS (Table 2). The highest levels are of iron oxides. The lack of Mn and Al oxides, especially in S3 and S4, is directly related to the parent material, where the mineral chrysotile is in high proportion and lacks Fe, Mn and Al. In the rest of the soils, serpentine $(Mg,Al,Fe,Mn,Ni,Zn)_{2-3}(Si,Al,Fe)_2O_5(OH)_4$ (Neuendorf et al., 2005) predominates in the parent material and their oxide content is higher, especially of iron oxides.

The effective cation exchange capacity is high in S2 and SC (Table 2) and low in the rest of the soils. All of the soils

Table 3. Total content of metals ($mg\,kg^{-1}$) – mean values and standard deviation.

C	S1	S2	S3	S4	CS
Al	$6623 \pm 112\,e$	$22\,675 \pm 374\,b$	$13\,775 \pm 273\,d$	$18\,728 \pm 354\,c$	$42\,890 \pm 678\,a$
Ba	$8 \pm 2\,c$	$41 \pm 5\,b$	$8 \pm 2\,c$	$6 \pm 2\,c$	$87 \pm 7\,a$
Ca	$6589 \pm 167\,e$	$10\,044 \pm 184\,d$	$28\,579 \pm 476\,a$	$23\,416 \pm 345\,b$	$22\,008 \pm 253\,c$
Co	$110 \pm 11\,b$	$147 \pm 10\,a$	$80 \pm 6\,cd$	$76 \pm 2\,d$	$97 \pm 14\,bc$
Cr	$1672 \pm 110\,b$	$2605 \pm 37\,a$	$1366 \pm 49\,c$	$1472 \pm 116\,c$	$2689 \pm 82\,a$
Cu	$145 \pm 7\,d$	$150 \pm 8\,d$	$327 \pm 11\,a$	$209 \pm 12\,c$	$291 \pm 12\,b$
Fe	$52\,808 \pm 235\,c$	$77\,775 \pm 346\,a$	$39\,747 \pm 232\,e$	$43\,449 \pm 227\,d$	$74\,310 \pm 354\,b$
K	$2722 \pm 97\,b$	$4074 \pm 110\,a$	$661 \pm 39\,c$	$733 \pm 44\,c$	$4156 \pm 87\,a$
Mg	$303\,045 \pm 890\,a$	$205\,696 \pm 742\,c$	$186\,262 \pm 635\,d$	$207\,809 \pm 958\,b$	$102\,653 \pm 386\,e$
Mn	$900 \pm 85\,c$	$1602 \pm 88\,b$	$751 \pm 30\,d$	$802 \pm 52\,cd$	$1850 \pm 122\,a$
Na	$283 \pm 25\,c$	$3264 \pm 287\,b$	$569 \pm 96\,c$	$358 \pm 71\,c$	$10\,579 \pm 312\,a$
Ni	$2039 \pm 107\,a$	$1861 \pm 62\,b$	$1342 \pm 32\,d$	$1499 \pm 89\,c$	$1470 \pm 82\,cd$
Sr	$19 \pm 2\,b$	$21 \pm 3\,b$	$15 \pm 1\,c$	$13 \pm 1\,c$	$41 \pm 3\,a$
V	$1 \pm 0\,e$	$52 \pm 5\,b$	$16 \pm 4\,d$	$39 \pm 6\,c$	$68 \pm 5\,a$
Zn	$34 \pm 2\,c$	$63 \pm 5\,b$	$58 \pm 5\,b$	$32 \pm 2\,c$	$115 \pm 8\,a$

For each element, values followed by different letters differ significantly with $p < 0.05$.

are saturated in bases and Ca^{2+} and Mg^{2+} predominate. The latter is the highest in S1, S2 and CS (Table 2). Ca^{2+} predominates in S3 and S4 (Table 2). In S1, S2 and CS there is an imbalance with Ca and Mg and S1 and CS are considered magnesic or hypermagnesic (Chardot et al., 2007). They high Mg contents also causes problems with K in S3, S4 and CS. These imbalances strongly limit plant production, as a deficit of Ca may occur, despite the fact that there is a high content in the soils. The exchangeable K^+ varies between $0.5\,cmol_{(+)}kg^{-1}$ (S2) and $0.07\,cmol_{(+)}kg^{-1}$ (S4), and Na^+ between $0.4\,cmol_{(+)}kg^{-1}$ (CS) and $0.1\,cmol_{(+)}kg^{-1}$ (S4). The soils have virtually no exchangeable Al^{3+}, except CS (Table 2), which contributes to the moderately acidic character of this soil.

3.2 Total metal content

The levels of Co, Cr and Ni are high compared to the contents in soils developed over other materials in the region (Macías et al., 1993). Ni and Cr are potentially toxic elements and the content in the soils is very high (Table 3). The levels of Co are also high but in this case there are few toxicity data for higher plants (Li et al., 2009). Studies have been carried out into how Co toxicity affects soil microbes and invertebrates (Chatterjee and Chatterjee, 2000; Lock et al., 2006) and they have revealed that Co is relatively toxic to plants when given in high doses, but there is still little information regarding the toxicity of Co to higher plants.

Chromium is the most abundant heavy metal in all the studied soils, followed by Ni and Co, except in S1 where the most abundant is Ni. S2 has the highest amount of all three metals. The total Co values are between $147\,mg\,kg^{-1}$ (S2) and $76\,mg\,kg^{-1}$ (S3), Cr contents range from $2689\,mg\,kg^{-1}$ (CS) to $1366\,mg\,kg^{-1}$ (S3), and the Ni contents between

$2039\,mg\,kg^{-1}$ (S1) and $1342\,mg\,kg^{-1}$ (S3). Most of these contents exceed the intervention limits stipulated in different guides (DEFRA and Environmental Agency, 2006; RIVM, 2001). Strict adherence to the limits stipulated in the guides would imply huge investments for the governments or companies involved, to decrease those highest values in order to minimize related environmental and health problems.

3.3 Soil extractions and extraction efficiency

The soil extractions were carried out using different reagents. The results (Table 4) show they pose different extraction capacities for each of the three metals studied.

In all cases the reagent that extracts the most Co, Ni and Cr is $0.01\,M\,CaCl_2$, but this fact does not mean that it is ideal for estimating the availability, as this depends on different aspects of the soil and metal in question. Nevertheless, it is one of the most common reagent used as it has more or less the same ionic strength as the average salt concentration in many soil solutions and it is simple, easy to perform, cheap and in routine daily use in laboratories (Houba et al., 2000).

BDW is the reagent that extracts the least amount of Co, Cr and Ni in most of the cases – the extraction efficiency is always less than 0.07% (Fig. 2).

Soil 1 (S1) has the highest amount of available Co when comparing the results obtained with all the reagents (except S2 when extracted with DTPA, Table 4). The control soil (CS) is the one with the least amount of available Co (Table 4).

The highest efficiency with $CaCl_2$ is for Co, reaching 27.6% of the total content in S1 but it does not reach 4% of the total content of the other studied metals. Besides, all the other extractants are more effective for Co than for Cr and Ni.

Table 4. Mean values and standard deviation of metal extracted from soils, and content in plants $(\text{mg}\,\text{kg}^{-1})$.

Extractant		Metal extracted from the soils				
		CaCl$_2$	EDTA	DTPA	LMWOA	BDW
		Co				
	S1	25.59 ± 0.81 A,a	3.47 ± 0.07 B,a	0.98 ± 0.06 D,b	1.71 ± 0.19 C,a	0.07 ± 0.01 E,a
	S2	25.27 ± 0.40 A,a	2.49 ± 0.25 B,b	1.16 ± 0.05 D,a	1.65 ± 0.17 C,a	0.07 ± 0.02 E,a
	S3	17.67 ± 0.27 A,b	0.89 ± 0.05 B,d	0.39 ± 0.02 C,d	0.51 ± 0.26 BC,b	0.02 ± 0.01 D,b
	S4	16.81 ± 0.44 A,b	1.17 ± 0.09 B,c	0.47 ± 0.03 C,c	0.46 ± 0.01 C,b	0.01 ± 0.01 D,b
	CS	3.29 ± 0.18 A,c	0.84 ± 0.06 B,d	0.17 ± 0.03 D,e	0.38 ± 0.01 C,b	0.02 ± 0.01 E,b
		Cr				
	S1	4.02 ± 0.05 A,c	0.28 ± 0.01 C,b	0.07 ± 0.00 D,b	0.68 ± 0.05 B,c	0.11 ± 0.02 D,bc
	S2	7.65 ± 0.17 A,a	0.44 ± 0.04 C,b	0.10 ± 0.01 D,b	0.84 ± 0.02 B,ab	0.38 ± 0.11 C,a
	S3	5.74 ± 0.27 A,b	0.33 ± 0.02 C,b	0.09 ± 0.02 C,b	0.72 ± 0.17 B,bc	0.09 ± 0.02 C,c
	S4	7.13 ± 0.24 A,a	0.36 ± 0.02 C,b	0.07 ± 0.01 D,b	0.88 ± 0.01 B,a	0.04 ± 0.01 D,c
	CS	4.83 ± 0.24 A,bc	1.11 ± 0.07 B,a	0.42 ± 0.04 C,a	0.58 ± 0.02 C,c	0.20 ± 0.04 D,b
		Ni				
	S1	273.86 ± 6.71 A,a	51.25 ± 0.80 B,b	24.83 ± 1.04 D,b	42.58 ± 2.41 C,a	1.14 ± 0.12 E,a
	S2	153.00 ± 3.23 A,b	32.22 ± 1.11 B,c	19.99 ± 0.13 C,c	12.15 ± 0.57 D,b	1.31 ± 0.21 E,a
	S3	99.05 ± 2.66 A,d	12.03 ± 0.60 B,e	4.98 ± 0.26 C,e	12.50 ± 1.09 B,b	0.31 ± 0.12 D,b
	S4	111.35 ± 2.21 A,c	15.56 ± 0.93 B,d	6.37 ± 0.10 D,d	12.69 ± 0.22 C,b	0.14 ± 0.03 E,b
	CS	96.01 ± 2.69 A,d	67.34 ± 1.14 B,a	66.08 ± 1.40 B,a	6.68 ± 0.07 C,c	0.44 ± 0.14 D,b

Soil	Plant	Plant metal content					
		Co	Cr	Ni	Co	Cr	Ni
			Shoot			Root	
S1	*Festuca*	5.17 ± 0.64 b	15.63 ± 2.03 b	71.19 ± 2.94 b	15.37 ± 1.60 b	19.63 ± 1.79 cd	212.59 ± 18.97 b
S1	*Juncus*	1.10 ± 0.00 c	0.73 ± 0.32 c	21.01 ± 2.81 c	46.83 ± 5.91 a	25.53 ± 2.02 bc	290.38 ± 19.13 a
S2	*Festuca*	8.17 ± 1.70 a	82.84 ± 6.09 a	109.22 ± 9.61 a	13.74 ± 0.63 b	77.04 ± 8.00 a	170.72 ± 8.73 c
S2	*Juncus*	0.92 ± 0.32 c	1.29 ± 0.32 c	9.40 ± 3.14 d	11.21 ± 1.02 b	29.49 ± 3.40 b	84.34 ± 11.07 d
CS	*Festuca*	0.73 ± 0.32 c	15.51 ± 2.28 b	26.46 ± 1.14 c	1.14 ± 0.00 c	16.85 ± 1.43 d	66.64 ± 5.15 d

In each row, for metal concentration in the extracts, values followed by the different capital letter differ significantly ($p < 0.05$); in each column, values followed by the different lowercase letters differ significantly ($p < 0.05$). LMWOA: low-molecular-weight organic acids. BDW: bidistilled water. In each column (for shoot or root) values followed by different letters differ significantly ($p < 0.05$).

The amount of Cr extracted in all cases is low (Table 4) and always less than 1 % of the total content (Fig. 2), therefore there is no evidence of available Cr and it is probably in strongly retained forms. In the case of Cr, CaCl$_2$ is also the most efficient, followed by LMWOA, EDTA, BDW and DTPA, although there are hardly any significant differences between the last three.

According to CaCl$_2$ and LMWOA, S1 is the soil with the highest amount of available Ni (Table 4) while, according to EDTA and DTPA, it is CS. CaCl2 extractions also indicate that CS is the soil with the least amount of Ni in available form (Table 4). Focusing on the proportion of the total content that is extracted, Ni is also more efficiently extracted with CaCl$_2$, followed by EDTA, LMWOA, DTPA and BDW. In S2 and CS, DTPA shows higher extraction efficiency than LMWOA (Fig. 2). Focusing on the soil proper-

ties (Table 1) S2 and CS have lower pH, higher OM content, higher ECEC, and higher exchangeable Ca and Mg contents, and these are soil properties that influence the retention of the metals, therefore they influence the availability of the studied metals.

In general, the lowest extraction efficiency for all of the metals in the study was detected in the soils with the highest pH (S1, S3 and S4), and the more basic the soil, the stronger the retention of the metal cations.

In general, the sequences of greater to lesser extraction capacity differ depending on the metal. In the case of Ni the sequence is different for different soils and it can be related

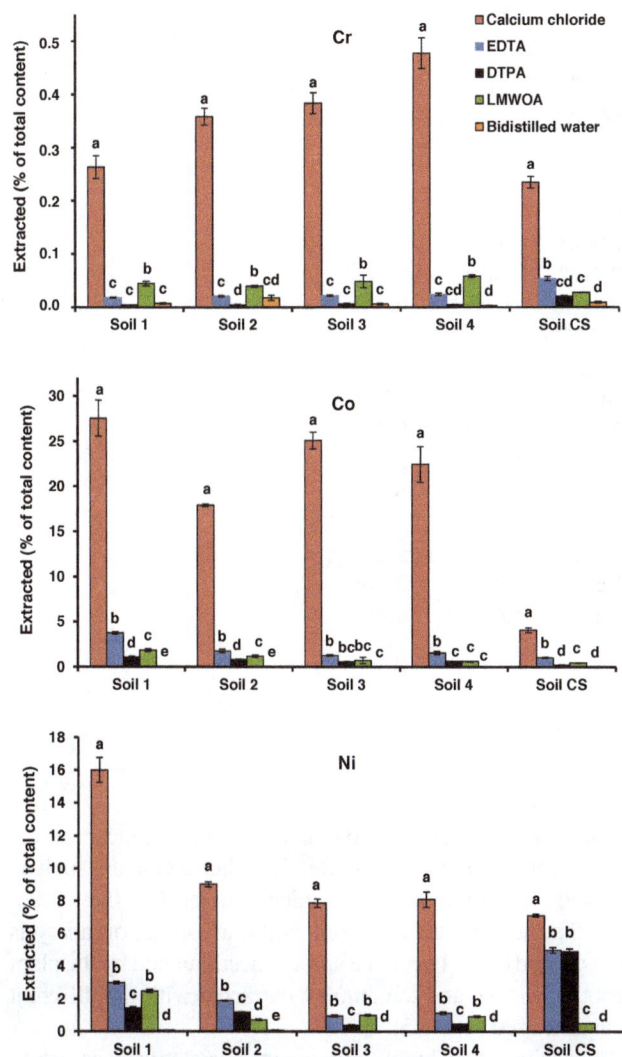

Figure 2. Extraction efficiency. In each soil, bars with different letters indicate significantly different EF values ($p < 0.05$) for each metal. Hanging bars are the standard deviation.

to the organic matter content of the soils:

$$Co : CaCl_2 > EDTA > LMWOA > DTPA > BDW \quad (4)$$

$$Cr : CaCl_2 > LMWOA > EDTA \approx DTPA \approx BDW \quad (5)$$

$$Ni(OM) : CaCl_2 > EDTA > DTPA$$
$$> LMWOA > BDW \quad (6)$$

$$Ni(noOM) : CaCl_2 > EDTA > LMWOA$$
$$> DTPA > BDW. \quad (7)$$

Although a kind of trend was found in the CaCl$_2$ and BDW extractions, the extraction efficiency (Fig. 2) does not only depend on the reagent used, but also on the characteristics of both the soil and the metal. Therefore the available content must be related more to the content in plant than to the total content of the soil.

3.4   Metal content in the plants

The content in both the shoot and roots of *Festuca rubra* and *Juncus* sp. are shown in Table 4. In general, *Festuca* absorbs the largest amounts of the three metals, both in its shoot and roots, except in soil S1, where the roots of *Juncus* sp. accumulated a larger amount of Co and Ni. The Ni is accumulated in the highest amounts by both plants, followed by Cr and Co.

Both species accumulate more Ni, Cr and Co in the roots than in the shoot, except in S2, where *Festuca* accumulated more Cr in the shoot than in its roots.

Li et al. (2009) indicated that plants can accumulate small amounts of Co, and that their absorption and distribution depends on the species being controlled by different mechanisms. The absorption of Co^{2+} by the roots involves active transportation through the cell membranes, although the molecular mechanisms involved are still unknown (Li et al., 2009). Its distribution may involve organic complexes, although the low mobility of Co^{2+} in the plants restricts its transportation from the roots to the shoot, as seen in this study (Table 4).

In turn, soil properties also influence heavy metal availability for plants (Li et al., 2009). There is very little useful information available to quantify the effect of soil properties on the toxicity of Co in different plant species. On the whole, the baseline information is insufficient to evaluate the risks posed by Co in order to support the adoption of new guidelines in the European Union (European Commission, 2003). It has been suggested that threshold toxicity levels should be standardized using the exchangeable Ca content of the soil, as this content is correlated with the CECE; this means that it is indicative of the sorption capacity of the soil, which influences the solubility of Co.

Calcium can reduce the toxicity of Co for plants by competing for binding sites in the root cells and Li et al. (2009) suggested that the Ca^{2+} ion competes with different metallic ions for the binding sites, thus reducing their toxicity. The exchangeable Ca content in the studied soils can influence the low Co content found in the plants; nevertheless, there is a good relationship between the amount of Co extracted (with the reagent that best represents the availability) and its concentration in the plants.

Chromium is accumulated in higher amounts in the roots than in the shoot. These results agree with those of other authors (Adki et al., 2013; Rafati et al., 2011). They indicated that the lowest amounts are always in the vegetative and reproductive organs. They found that Cr distribution is crops is stable and does not depend on soil properties and concentration of the element.

As mentioned above, Ni was absorbed in greater amounts (except in plants from CS, Table 4) and this is probably because of the high pH of the soils from the quarry area. In general, the uptake of Ni usually declines at high soil solution pH values due to the formation of less soluble complexes (Yusuf et al., 2011). These complexes can remain on the soil

Table 5. Pearson's correlation between extracted Co, Cr and Ni and its content in plants.

Plant	Extractant				
	$CaCl_2$	EDTA	DTPA	LMWOA	BDW
Co					
Festuca shoot	0.876[b]	Nc	0.925[b]	0.943[b]	0.875[b]
Festuca root	0.957[b]	0.944[b]	0.962[b]	0.991[b]	0.987[b]
Juncus shoot	0.904[b]	0.913[b]	0.916[b]	0.949[b]	0.943[b]
Juncus root	0.688[a]	0.890[b]	Nc	0.694[a]	0.678[a]
Cr					
Festuca shoot	0.862[b]	Nc	Nc	0.896[b]	0.794[a]
Festuca root	0.858[b]	Nc	Nc	0.903[b]	0.751[a]
Juncus shoot	Nc	Nc	Nc	0.906[b]	Nc
Juncus root	Nc	Nc	Nc	0.803[b]	Nc
Ni					
Festuca shoot	Nc	Nc	Nc	0.692[a]	0.899[b]
Festuca root	0.875[b]	Nc	Nc	0.895[a]	0.837[b]
Juncus shoot	0.909[b]	Nc	Nc	0.969[b]	Nc
Juncus root	0.906[b]	Nc	Nc	0.996[b]	Nc

[a] Correlation is significant at $p < 0.01$ (bilateral). [b] Correlation is significant at $p < 0.05$ (bilateral). Nc: No correlation. $N = 9$.

surfaces in available forms. CS is the soil with the highest content of exchangeable Ca, which affects the decrease of Ni absorption, as demonstrated by Yusuf et al. (2011).

A plant growing in a soil containing heavy metals can be considered a hyperaccumulator if it concentrates in its shoot without suffering from toxicity problems, up to 1 % of Mn or Zn, 0.1 % of As, Co, Cr, Cu, Ni, Pb, Sb, Se and Tl or 0.01 % of Cd (Verbruggen et al., 2009). Also, according to Mongkhonsin et al. (2011), Reeves and Baker (2000) and Tappero et al. (2007), considering a plant a hyperaccumulator of Cr is based on three criteria: that the Cr concentration in the shoot $> 50\,mg\,kg^{-1}$, that the concentration of Cr in the aerial biomass is 10–500 times greater than in the non-metallophytes ($0.2–5\,mg\,kg^{-1}$ of Cr), and that the Cr concentration in the shoot is greater than in the roots.

Thus, none of the plant species we evaluated behave like hyperaccumulators, as the amounts of metals absorbed by the plants are less than the criteria indicated. Only the *Festuca* growing in S2 contains more Cr in the shoot than in the root ($> 50\,mg\,kg^{-1}$); therefore it could be considered as having a certain hyperaccumulator capacity (Reeves and Baker, 2000).

It is well known that the total content of heavy metals in soils is not suitable for establishing the mobility, availability and therefore the possible toxicity of trace elements (Pueyo et al., 2004).

Some authors, like Roy and McDonald (2013), have suggested that the combined soluble and exchangeable fractions from the Tessier (1979) method are correlated with Cd and

Zn uptake in plants more so than the total soil concentration, but did not find any correlation for other elements, such as Pb and Cu. In this paper, in order to determine the extractant that best predicts bioavailability, a correlation analysis was carried out between the amount accumulated by the plant (root or shoot), and the amount extracted with the different reagents used (Table 5).

A positive and highly significant correlation was established ($p < 0.01$) between the amount of Co extracted by practically all of the extractants used and the amount accumulated by *Festuca* and *Juncus*, except between the Co extracted with DTPA and the content in the roots of *Juncus*.

A positive and highly significant correlation was also found between the amount of Cr accumulated in both the shoot and roots of *Festuca* and the amount extracted by $CaCl_2$, LMWOA and BDW. The Cr content in *Juncus* is only correlated with the content extracted with LMWOA.

In the case of Ni, the correlation is between the content in the shoot of *Festuca* and the amount extracted by BDW, as well as between the concentration in the root of the plant and the amount extracted by $CaCl_2$, LMWOA and BDW. In the case of *Juncus* the correlation is between the content in both the shoot and the root and the amount extracted by $CaCl_2$ and LMWOA.

It can therefore be deduced from these results that above all, LMWOA is the extractant that best predicts the bioavailability of Cr, Ni and Co for these plants in the soils from the Moeche quarry. This is a rhizosphere-based extraction method that simulates the rhizosphere conditions and takes

Table 6. Translocation and bioconcentration factors (mean values and standard deviation).

Translocation factor Soil		Festuca	Juncus
		Co	
S1		0.34 ± 0.04 a	0.02 ± 0.00 b
S2		0.59 ± 0.12 a	0.08 ± 0.02 a
CS		0.64 ± 0.28 a	NP
		Cr	
S1		0.80 ± 0.11 b	0.03 ± 0.01 a
S2		1.08 ± 0.09 a	0.04 ± 0.01 a
CS		0.92 ± 0.07 ab	NP
		Ni	
S1		0.33 ± 0.04 b	0.07 ± 0.01 a
S2		0.64 ± 0.03 a	0.11 ± 0.02 a
CS		0.40 ± 0.01 b	NP

Bioconcentration factor Soil	Plant		Co	Cr	Ni
S1	Festuca	Shoot	3.02 ± 0.38 b	22.99 ± 2.99 b	1.67 ± 0.07 c
S1	Juncus		0.64 ± 0.00 c	1.07 ± 0.46 c	0.49 ± 0.07 d
S2	Festuca		4.95 ± 1.03 a	98.62 ± 9.64 a	8.99 ± 0.79 a
S2	Juncus		0.56 ± 0.19 c	1.54 ± 0.38 c	0.77 ± 0.26 d
CS	Festuca		1.92 ± 0.83 b	26.74 ± 3.93 b	3.96 ± 0.17 b
S1	Festuca	Root	8.99 ± 0.94 b	28.87 ± 2.62 c	4.99 ± 0.44 d
S1	Juncus		27.39 ± 1.02 a	37.54 ± 2.98 b	6.82 ± 0.45 c
S2	Festuca		8.33 ± 0.38 b	91.71 ± 2.05 a	14.05 ± 0.72 a
S2	Juncus		6.79 ± 0.62 c	35.11 ± 4.06 b	6.94 ± 0.91 c
CS	Festuca		3.00 ± 0.00 d	29.05 ± 2.47 c	9.98 ± 0.77 b

In each column (for TF) values followed by different letters differ significantly ($p < 0.05$). NP: No plant. In each column (for BF, and for shoot or root) values followed by different letters differ significantly ($p < 0.05$).

into account the effect of soil–root interactions as a whole, at least to some extent (Feng et al., 2005).

3.5 Translocation and bioconcentration factors

The highest TF values correspond to *Festuca* (Table 6), although they are not >1 except for Cr (1.08) in the plants growing in S2 (Table 6). Of the three metals in this study, Cr is translocated the best in *Festuca*, followed by Co and Ni.

The TF values in *Juncus* are very low, no higher than 0.11, and Ni is the metal that is translocated the best. Based on these results, *Juncus* acts as a phytostabilizer, because it fixes the metals in its roots, reducing their mobility within the plant and soil. There is also a small amount of metal that accumulates in its shoot. Both the accumulation in roots and shoot reduces the risk of metals transferring to other compartments of the ecosystem.

The bioconcentration factor (Table 6) links the available content in the soils with the amount absorbed by the plants. The bioconcentration factor (BF) in the studied plants was determined by calculating the ratio of metal concentration in the plant (C_p; root and shoot) to soil (C_{LMWOA}).

The BF (Table 6) in the shoot of *Juncus* is generally very low and <1, except for Cr, due to the low or inexistent translocation of Ni and Co from the root. The BF in its root is >1, with the highest value corresponding to Cr, while the values for Co and Ni confirm that, as previously indicated, the plant accumulates both metals in the root and does not transfer them to the shoot, behaving as a phytostabilizer.

Moreover, these results indicate that *Festuca* is a phytostabilizer of Co and Ni, which is consistent with Simon (2005) for Cd, Cu, Pb and Zn; however we have confirmed that it is an accumulator of Cr, as that is transferred to the shoot.

In addition, the BF for the shoot and root of *Festuca* is >1 in all of the cases, which means that *Festuca* behaves as an accumulator, especially of Cr.

4 Conclusions

The levels of Co, Cr and Ni in the studied soils exceed the intervention values indicated in different reference guides. Although $CaCl_2$ is the reagent with the highest extraction efficiency for all of the soils and metals studied, the extractant that best predicts the bioavailability of the metals is the mixture of low molecular weight organic acids.

Festuca generally accumulates the largest amount of Co, Cr and Ni and also has the highest translocation factor, although it is only >1 in the case of Cr. Furthermore, the bioconcentration factor is >1 for Cr in the shoot and root of *Festuca* and *Juncus*, and is also >1 for Co and Ni in *Festuca*. It is <1 for these two elements in the shoot of *Juncus*.

Juncus seems to be a suitable plant for phytostabilization, while *Festuca* is a phytostabilizer of Co and Ni, and an accumulator of Cr.

Acknowledgements. This research was supported by the Xunta de Galicia (project EM2013/018). F. A. Vega and D. Arenas-Lago would like to thank the Ministry of Science and Innovation and the University of Vigo for the Ramón y Cajal and FPI-MICINN grants, respectively.

Edited by: A. Cerdà

References

Adamo, P., Dudka, S., Wilson, M. J., and McHardy, W. J.: Distribution of trace elements in soils from the Sudbury smelting area (Ontario, Canada), Water Air Soil Pollut., 137, 95–116, 2002.

Adki, V. S., Jadhav, J. P., and Bapat, V. A.: *Nopalea cochenillifera*, a potential chromium (VI) hyperaccumulator plant, Environ. Sci. Pollut. Res., 20, 1173-1180, 2013.

AFNOR: Qualité des sols, Méthodes d'analyses-Recueil de normes françaises, Association française de normalization, Paris, 1994.

Agrawal, A., Kumar, V., and Pandey, B. D.: Remediation options for the treatment of electroplating and leather tanning effluent containing chromium-a review, Miner. Process. Extract. Metall. Rev., 27, 99–130, 2006.

Ali, H., Khan, E., and Sajad., M. A.: Phytoremediation of heavy metals "concepts and applications", Chemosphere, 91, 869–881, 2013.

Asensio, V., Vega, F. A., Andrade, M. L., and Covelo, E. F.: Tree vegetation and waste amendments to improve the physical condition of copper mine soils, Chemosphere, 90, 603–610, 2013.

Baker, A. J. M. and Brooks, R. R.: Terrestrial higher plants which hyperaccumulate metallic elements-a review of their distribution, ecology and phytochemistry, Biorecovery, 1, 81-126, 1989.

Bell, P. F., Xie, B., Higby, J. R., and Aminha, N.: Digestion of NIST peach leaves using sealed vessels and inexpensive microwave ovens, Commun. Soil Sci. Plan., 31, 1897–1903, 2000.

Bidar, G., Pruvot, C., Garcon, G., Verdin, A., Shirali, P., and Douay, F.: Seasonal and annual variations of metal uptake, bioaccumulation, and toxicity in *Trifolium repens* and *Lolium perenne* grow-

ing in a heavy metal-contaminated field, Environ. Sci. Pollut. R., 16, 42–53, 2009.

Biro, K., Pradhan, B., Buchroithner, M., and Makeschin, F.: Land use/land cover change analysis an its impact on soil properties in the Northern part of Gadarif region, Sudan, Land Degrad. Dev., 24, 90–102, 2013.

Bremner, J. M. and Mulvaney, C. S.: Nitrogen-total, in: Method of Soil Analysis, Part 2, Chemical and Microbiological Properties, Agronomy Monographs No 9, edited by: Page, A. L., Miller, R. H., and Keeney, R. S., American Society of Agronomy and Soil Science Society of America, Madison, WI, USA, 595–624, 1982.

Castroviejo, R., Armstrong, E., Lago, A., Martínez Simón, J. M., and Argüelles, A.: Geología de las mineralizaciones de sulfuros masivos en los cloritoesquistos de Moeche (complejo de Cabo Ortegal, A Coruña), Boletín Geológico y Minero, 115, 3–34, 2004.

Chatterjee, J., and Chatterjee, C.: Phytotoxicity of cobalt, chromium and copper in cauliflower, Environ. Pollut., 109, 69–74, 2000.

Chardot, V., Echevarria, G., Gury, M., Massoura, S., and Morel, J.L.: Nickel bioavailability in an ultramafic toposequence in the Vosges Mountains (France), Plant soil, 293, 7–21, 2007.

COMG, Cámara Oficial Minera de Galicia, www.camaraminera.org, 2013.

Corti, G., Ugolini, F. C., Agnelli, A., Certini, G., Cuniglio, R., Berna, F., and Fernández Sanjurjo, M. J.: The soil skeleton, a forgotten pool of carbon and nitrogen in soil, Eur. J. Soil Sci., 53, 283–298, 2002.

Day, P. R.: Particle Fractionation and Particle Size Analysis, in: Methods of Soil Analysis, edited by: Black, C. A., American Society of Agronomy, Madison, USA, 545–566, 1965.

De Souza Braz, A. M., Fernandes, A. R., and Alleoni, L. R. F.: Soil attributes after the conversion from forest to pasture in Amazon, Land Degrad. Dev., 24, 33–38, 2013.

DEFRA and Environmental Agency: Assessing Risks from Land Contamination - a Proportionate Approach. Soil Guideline Values: The Way Forward, Department for Environment, Food and Rural Affairs, London, UK, 32 pp., 2006.

Deng, H., Ye, Z. H., and Wong, M. H.: Lead and zinc accumulation and tolerance in populations of six wetland plants, Environ. Pollut., 141, 69–80, 2006.

Doubková, P., Suda, J., and Sudová, R.: The symbiosis with arbuscular mycorrhizal fungi contributes to plant tolerance to serpentine edaphic stress, Soil Biol. Biochem., 44, 56–64, 2012.

European Commission: Technical Guidance Document on Risk Assessment, European Commission, European Commission Joint Research Centre, Ispra, Italy, 2003.

FAO: Guidelines for soil description, Food and Agriculture Organization of the United Nations, Rome, 2006.

Feng, M. H., Shan, X. Q., Zhang, S. Z., and Wen, B.: Comparison of a rhizosphere-based method with other one-step extraction methods for assessing the bioavailability of soil metals to wheat, Chemosphere, 59, 939–949, 2005.

Fernández-Calviño, D., Garrido-Rodríguez, B., López-Periago, J. E., Paradelo, M., and Arias-Estévez, M.: Spatial distribution of copper fractions in a vineyard soil, Land Degrad. Dev., 24, 556–563, 2013.

Garbisu, C. and Alkorta, I.: Phytoextraction a cost-effective plant-based technology for the removal of metals from the environment, Bioresour. Technol., 71, 229–236, 2001.

Hendershot, W. H., and Duquette, M.: A simple barium chloride method for determining cation exchange capacity and exchangeable cations, Soil Sci. Soc. Am. J., 50, 605-608, 1986.

Hill, S. J.: Inductively Coupled Plasma Spectrometry and its applications, in: School of Earth, Ocean and Environmental Sciences, Blackwell Publishing Ltd, University of Plymouth, UK, 2008

Houba, V. J. G., Temminghoff, E. J. M., Gaikhorst, G. A., and Van Vark, W.: Soil analysis procedures using 0.01 M calcium chloride as extraction reagent, Commun. Soil Sci. Plant Anal., 31, 1299–1396, 2000.

Knight, B. P., Zhao, F. J., McGrath, S. P., and Shen, Z. G.: Zinc and cadmium uptake by the hyperaccumulator *Thlaspi caerulescens* in contaminated soils and its effect on the concentration and chemical speciation of metals in soil solution, Plant Soil, 197, 71–78, 1997.

Lago-Vila, M., Arenas Lago, D., Vega, F. A., and Andrade, M. L.: Phytoavailable content of metals in soils from copper mine tailings (Touro mine, Galicia, Spain), J. Geochem. Explor., 147, 159–166, 2014.

Li, H. F., Gray, C., Mico, C., Zhao, F. J., and McGrath, S. P.: Phytotoxicity and bioavailability of cobalt to plants in a range of soils, Chemosphere, 75, 979–986, 2009.

Liu, Y., Su, C., Zhang, H., Li, X., and Pei, J.:Interaction of soil heavy metal pollution with industrialisation and the landscape pattern in Taiyuan city, China, PLoS ONE, 9, e105798, doi:10.1371/journal.pone.0105798, 2014.

Lindsay, W. L. and Norwell, W. A.: Development of a DTPA soil test for zinc, iron, manganese and copper, Soil Sci. Soc. Am. J., 42, 421–428, 1978.

Lock, K., De Schamphelaere, K. A. C., Becaus, S., Criel, P., Van Eeckhout, H., and Janssen, C. R.: Development and validation of an acute biotic ligand model (BLM) predicting cobalt toxicity in soil to the potworm *Enchytraeus albidus*, Soil Biol. Biochem., 38, 1924–1932, 2006.

Macías, F., Veiga, A., and Calvo, R.: Influencia del material geológico y detección de anomalías en el contenido de metales en horizontes superficiales de suelos de la provincia de A Coruña, Cad. Lab. Xeolóxico de Laxe, 18, 317–323, 1993.

Mandal, D. and Sharda, V. N.: Appraisal of soil erosion risk in the Eastern Himalayan region of India for soil conservation planning, Land Degrad. Dev., 24, 430–437, 2013.

Martin, T. A. and Ruby, M. V.: Review of in situ Remediation Technologies for Lead, Zinc and Cadmium in Soil, Remed. J., 14, p. 35, 2004.

McGrath, S. P.: Phytoextraction for soil remediation, in: Plants that hyperaccumulate heavy metals, edited by: Brooks, R. R., CAB, International, Wallingford, UK, 261–287 pp., 1998.

Mendez, M. O. and Maier, R. M.: Phytoremediation of mine tailings in temperate and arid environments, Rev. Environ. Sci. Biotechnol., 7, 47–59, 2008.

Milder, A. I., Fernández-Santos, B., and Martínez-Ruiz, C.: Colonization patterns of woody species on lands mined for coal in Spain: Preliminary insights for forest expansion, Land Degrad. Dev., 24, 39–46, 2013.

Mongkhonsin, B., Nakbanpote, W., Nakai, I., Hokura, A., and Jearanaikoon, N.: Distribution and speciation of chromium accumulated in *Gynura pseudochina* (L.) DC, Environ. Exp. Bot., 74, 56–64, 2011.

Munsell Soil Color Charts, Revised Washable Edition, New Windsor, NY, 2000.

Naidu, R., Bolan, N. S., Megharaj, M., Juhasz, A. L., Gupta, S. K., Clothier, B. E., and Schulin, R.: Chemical bioavailability in terrestrial environments, in: Developments in Soil Science, 32, Chemical bioavailability in terrestrial environments, edited by: Hartemink, A. E., McBratney, A. B., and Naidu, R., Elsevier, Oxford, UK, 1–6, 2008.

Neuendorf, K. K. E., Mehl, J. P., and Jackson, J. A., (Eds.): Glossary of Geology (5th Edn.), American Geological Institute, Alexandria, Virginia, 2005.

Paz-Ferreiro, J., Lu, H., Fu, S., Méndez, A., and Gascó, G.: Use of phytoremediation and biochar to remediate heavy metal polluted soils: a review, Solid Earth, 5, 65–75, doi:10.5194/se-5-65-2014, 2014.

Pereira, D., Yenes, M., Blanco, J. A., and Peinado, M.: Characterization of serpentinites to define their appropriate use as dimension stone, Geol. Soc. Spec. Pub., 271, 55–62, 2007.

Proctor, J. and Woodell, S. R. J.: The ecology of serpentine soils, Adv. Ecol. Res., 9, 255–365, 1975.

Pueyo, M., López-Sánchez, J. F., and Rauret, G.: Assessment of $CaCl_2$, $NaNO_3$ and NH_4NO_3 extraction procedures for the study of Cd, Cu, Pb and Zn extractability in contaminated soils, Anal. Chim. Acta, 504, 217–226, 2004.

Rafati, M., Khorasani, N., Moattar, F., Shirvany, A., Moraghebi, F., and Hosseinzadeh, S.: Phytoremediation Potential of *Populus alba* and *Morus alba* for Cadmium, Chromium and Nickel Absorption from Polluted Soil, Int. J. Environ. Res., 5, 961–970, 2011.

Reeves, R. D. and Baker, A. J. M.: Phytoremediation of Toxic Metals: Using Plants to Clean up the Environment, John Wiley and Sons Inc, New York, 2000.

RIVM: Technical Evaluation of the Intervention Values for Soil/Sediment and Groundwater, RIVM Report 71701023, National institute of Public Health and the Environment, Bilthoven, The Netherlands, 2001.

Roy, M., and McDonald, L.M.: Metal uptake in plants and health risk assessments in metal-contaminated smelter soils, Land Degrad. Dev., doi:10.1002/ldr.2237, in press, 2013.

Sherdrick, B. H. and McKeague, J. A.: A comparison of extractable Fe and Al data using methods followed in the USA and Canada, Can. J. Soil Sci., 55, 77–78, 1975.

Simon, L.: Stabilization of metals in acidic mine spoil with amendments and red fescue (*Festuca rubra* L.) growth, Environ. Geochem. Health, 27, 289–300, 2005.

Soil Conservation Service US:Dithionite Citrate Method,in: Soil, Survey Laboratory Methods and Procedures for Collecting Soil Samples, Soil Survey Investigation, Department of Agriculture, Washington DC, USA, 1972.

Tappero, R., Peltier, E., Gräfe, M., Heidel, K., Ginder-Vogel, M., Livi, K., Rivers, M., Marcus, M., Chaney, R., and Sparks, D.: Hyperaccumulator *Alyssum murale* relies on a different metal storage mechanism for cobalt than for nickel, New Phytol., 175, 641–654, 2007.

Tessier, A., Campbell, P. G. C., and Bisson, M.: Sequential extraction procedure for the speciation of particulate trace metals, Anal. Chem., 51, 844–851, 1979.

Turitzin, S. N.: Nutrient limitations to plant growth in a California serpentine grassland, Am. Mid. Nat., 107, 95–99, 1991.

Vacca, A., Bianco, M. R., Murolo, M., and Violante, P.: Heavy metals in contaminated soils of the rio sitzerry floodplain (sardinia, Italy): Characterization and impact on pedodiversity, Land Degrad. Dev., 23, 350–364, 2012.

Verbruggen, N., Hermans, C., and Schat, H.: Molecular mechanisms of metal hyper accumulation in plants, New Phytol., 181, 759–776, 2009.

Walker, R. B.: Factors affecting plant growth on serpentine soils, in: The Ecology of Serpentine Soils, edited by: Whittaker, R. H., A Symposium Ecology, 35, 259–266, 1954.

Walkey, A. and Black, I. A.: An Examination of Degtjareff method for determining soil organic matter and a proposed modification of the cromic titration method, Soil Sci., 34, 29–38, 1934.

Yang, D., Zeng, D. H., Zhang, J., Li, L. J., and Mao, R.: Chemical and microbial properties in contaminated soils around a magnesite, in northeast China, Land Degrad. Dev., 23, 256–262, 2012.

Yusuf, M., Fariduddin, Q., Hayat, S., and Ahmad, A.: Nickel: An Overview of Uptake, Essentiality and Toxicity in Plants, Bull. Environ. Contam. Toxicol., 86, 1–17, 2011.

Zhao, G., Mu, X., Wen, Z., Wang, F., and Gao, P.: Soil erosion, conservation, and Eco-environment changes in the Loess Plateau of China, Land Degrad. Dev., 24, 499–510, 2013.

11

Adsorption, desorption and fractionation of As(V) on untreated and mussel shell-treated granitic material

N. Seco-Reigosa[1], L. Cutillas-Barreiro[2], J. C. Nóvoa-Muñoz[2], M. Arias-Estévez[2], E. Álvarez-Rodríguez[1], M. J. Fernández-Sanjurjo[1], and A. Núñez-Delgado[1]

[1]Department of Soil Sciences and Agricultural Chemistry, Higher Polytechnic School, University Santiago de Compostela, 27002 Lugo, Spain
[2]Department of Plant Biology and Soil Sciences, Faculty of Sciences, University Vigo, 32004 Ourense, Spain

Correspondence to: A. Núñez-Delgado (avelino.nunez@usc.es)

Abstract. As(V) adsorption and desorption were studied on granitic material, coarse and fine mussel shell and granitic material amended with 12 and 24 t ha^{-1} fine shell, investigating the effect of different As(V) concentrations and different pH as well as the fractions where the adsorbed As(V) was retained. As(V) adsorption was higher on fine than on coarse shell. Mussel shell amendment increased As(V) adsorption on granitic material. Adsorption data corresponding to the unamended and shell-amended granitic material were satisfactory fitted to the Langmuir and Freundlich models. Desorption was always < 19 % when the highest As(V) concentration (100 mg L^{-1}) was added. Regarding the effect of pH, the granitic material showed its highest adsorption (66 %) at pH < 6, and it was lower as pH increased. Fine shell presented notable adsorption in the whole pH range between 6 and 12, with a maximum of 83 %. The shell-amended granitic material showed high As(V) adsorption, with a maximum (99 %) at pH near 8, but decreased as pH increased. Desorption varying pH was always < 26 %. In the granitic material, desorption increased progressively when pH increased from 4 to 6, contrary to what happened to mussel shell. Regarding the fractionation of the adsorbed As(V), most of it was in the soluble fraction (weakly bound). The granitic material did not show high As(V) retention capacity, which could facilitate As(V) transfer to water courses and to the food chain in case of As(V) compounds being applied on this material; however, the mussel shell amendment increased As(V) retention, making this practice recommendable.

1 Introduction

Igneous rocks, as granite, have low As concentrations (< 5 mg kg^{-1}), and background levels in soils are between 5 and 10 mg kg^{-1} (Smedley and Kinniburgh, 2002), although As levels are much higher in certain polluted soils. As pollution can be very relevant in mine sites where oxidation of sulfides such as pyrite takes place and in areas treated with certain biocides and fertilizers (Matschullat, 2000). As is an element that can accumulate in living beings and may cause severe affectations, especially when it is in inorganic form (Smith et al., 2000; Ghimire et al., 2003), with the potential to provoke environmental and public health issues. In fact, the recommended threshold level for As in drinking water is 10 μg L^{-1} (WHO, 2011).

When As-based products are spread on soils or spoils with the aim of fertilizing, controlling plagues or promoting revegetation, risks of soil and water pollution, and subsequent transfer to the food chain, must be taken into account. As indicated in previous works, the use of wood preservative compounds including arsenic or of As-based herbicides could cause arsenic pollution episodes in forest areas (Smith et al., 1998) and cultivation soils (Gur et al., 1979), in both cases increasing risks of soil and water pollution (Clothier et al., 2006). In this way, it is interesting to determine As retention capacity corresponding to solid substrates receiving the spreading of the pollutant, both individually or treated with complementary materials that can affect As retention/release potential. In this regard, some previous works have investigated the effectiveness of mussel shell waste amendment to increase As retention on diverse solid materials (Seco-Reigosa et al., 2013a, b; Osorio-López et al., 2014), and this amendment could also be useful to increase As retention on granitic substrates (such as mine spoils or exposed C horizons), which has not been studied up to now.

As concentration in natural waters is mainly controlled by interactions between solids and solution, as adsorption/desorption, which are affected by pH and other environmental parameters. Clays, organic matter and Fe, Al and Mn oxyhydroxides can protonate or deprotonate as a function of pH, facilitating retention of anions such as arsenate when they are positively charged and promoting progressive anions release when pH rises and surface charge becomes increasingly negative (Smith et al., 1999; Fitz and Wenzel, 2002); however, at high pH values and in the presence of sulfate and carbonate, co-precipitation of As with oxyhydroxides and sulfates, or even as calcium arsenate, may occur (García et al., 2009). This could explain that certain soils show maximum As adsorption at pH near 10.5 (Goldberg and Glaubig, 1988). In this way, Zhang and Selim (2008) indicate that carbonate can play an important role in arsenate retention in solid substrates having high pH value. In fact, calcite has been related to As retention in calcareous soils and carbonate-rich environments due to adsorption/precipitation of $CaCO_3$ and As forming inner sphere complexes (Alexandratos et al., 2007; Mehmood et al., 2009; Yolcubal and Akyol, 2008; Zhang and Selim, 2008), which could be relevant in granitic materials that were amended with mussel shell to promote As retention.

The study of risks of soil and water As pollution, and the investigation of potential means to diminish it are just a part of global concerns affecting soil (and, subsequently, other environmental compartments). In the last years, numerous studies have indicated that restoration needs to recover soil functionality, and this call is taking place all over the world (Ahmad et al., 2013; Johnston et al., 2013; Mao et al., 2014; Moreno et al., 2014; Novara et al., 2014; Roy and McDonald, 2015; Sacristán et al., 2015; Sadeghi et al., 2015; Srivastava et al., 2014). Some authors indicate that this task should be accomplished with a broad view (Brevik et al., 2015) by considering how soils can interfere with human health (Brevik and Sauer, 2015).

In view of that, the objectives of this work are (a) to determine As(V) retention/release capacity corresponding to a granitic material, fine mussel shell and coarse mussel shell, as well as to the granitic material amended with 12 or 24 t ha^{-1} fine mussel shell, for different As(V) concentrations and pH values; (b) to examine fitting of adsorption data to the Langmuir and Freundlich models; and (c) to determine the fractions where the adsorbed As(V) was retained, which is in relation with stability of retention. As far as we know, no equivalent studies were made previously with the combination of materials here used.

2 Materials and methods

2.1 Materials

We used different solid materials: (a) granitic material from Santa Cristina (Ribadavia, Ourense Province, Spain) (latitude 42°17′33.81″ N; longitude 8°7′21.75″ W; altitude 162 m a.s.l.) similar to a C horizon derived from the evolution of a rocky substrate, nowadays exposed to the atmosphere after the elimination of the upper horizons, then needing organic matter and nutrients to be restored, as granitic mine spoils do; (b) finely (< 1 mm) and coarsely (0.5–3 mm) crushed mussel shell from the factory Abonomar S.L. (A Illa de Arousa, Pontevedra province, Spain) that had been previously studied by Seco-Reigosa et al. (2013b); (c) mixtures of the granitic material +12 and 24 t ha^{-1} fine mussel shell (which showed higher adsorption potential than coarse shell in preliminary trials); concretely, considering an effective soil depth of 20 cm and a soil bulk density of 1 g cm^{-3}, samples of 400 g of the granitic material were mixed with 6 or 12 g of fine mussel shell per kg of granitic material and then shaken for 48 h in 2 L polypropylene bottles to achieve homogenization. The granitic material was sampled in a zigzag manner (20 cm depth), with 10 subsamples taken to perform the final one. These samples were transported to the laboratory to be air dried and sieved through 2 mm. Finally, chemical determinations and trials were carried out on the < 2 mm fraction.

2.2 Methods

2.2.1 Characterization of the solid materials

The Robinson pipette procedure was used according to Gee and Bauder (1986) to characterize the particle-size distribution of the materials studied. For each particle-size determination 20 g of sample were used. A pH meter (model 2001, Crison, Spain) was used to measure pH in water (10 g of solid sample, with solid : liquid relationship 1 : 2.5) (McLean, 1982). C and N were measured on 5 g samples using an elemental TruSpec CHNS auto-analyzer (LECO, USA) (Chatterjee et al., 2009). Available P was determined as per Olsen and Sommers (1982) using 5 g samples. A NH$_4$Cl 1 M solution was used on 5 g samples to displace the exchangeable cations, and then Ca, Mg and Al were quantified by atomic absorption spectroscopy and Na and K by atomic emission spectroscopy (AAnalyst 200, Perkin Elmer, USA) (Sumner and Miller, 1996); the effective cationic exchange capacity (eCEC) was calculated as the sum of all these cations (Kamprath, 1970). Total concentrations of Na, K, Ca, Mg, Al, Fe and Mn, as well as As, Cd, Co, Cr, Cu, Ni and Zn, were determined using ICP-MS (ICP mass spectrometry) (820-NS, Varian, USA) after nitric acid (65 %) microwave-assisted digestion on 1 g samples (Nóbrega et al., 2012). Different selective solutions were used to obtain Al and Fe fractions (Álvarez et al., 2013) from 1 g samples: total non-crystalline Al and Fe (Al_o, Fe_o), total Al and Fe bound to organic matter (Al_p, Fe_p), non-crystalline inorganic Al and Fe (Al_{op}, Fe_{op}), Al bound to organic matter in medium and low-stability complexes (Al_{cu}), Al bound to organic matter in high-stability complexes (Al_{pcu}), Al bound to organic matter in medium-stability complexes (Al_{cula}) and Al bound to organic matter in low-stability complexes (Al_{la}).

2.2.2 Adsorption/desorption as a function of added As(V) concentration

The methodology of Arnesen and Krogstrad (1998) was used to study As(V) adsorption/desorption as a function of the added concentration of the element.

The materials used were triplicate samples of the granitic material, coarse and fine mussel shell and granitic material amended with 12 and 24 t ha^{-1} fine mussel shell.

In the adsorption experiment, 3 g of each solid sample were added with 30 mL NaNO$_3$ 0.01 M dissolutions containing 0, 0.5, 5, 10, 25, 50 or 100 mg L^{-1} of As(V) prepared from analytical grade Na$_2$HAsO$_4$.7H$_2$O (Panreac, Spain). The resulting suspensions were shaken for 24 h, centrifuged at 4000 rpm for 15 min and finally filtered using acid-washed paper. In the equilibrium dissolutions, pH was measured using a glass electrode (Crison, Spain) and dissolved organic carbon (DOC) was determined by means of UV-visible spectroscopy (UV-1201, Shimadzu, Japan) and As(V) using ICP-mass (Varian 800-NS, USA). Adsorbed As was calculated as the difference between added As(V) and As(V) remaining in the equilibrium solution.

Desorption studies were carried out at the end of the adsorption trials, adding 30 mL of a NaNO$_3$ 0.01 M solution to each sample, shaking for 24 h, centrifuging at 4000 rpm for 15 min and filtering through acid-washed paper. Desorbed As(V), DOC and pH were determined by triplicate in all samples.

Adsorption data were fitted to the Freundlich (Eq. 1) and Langmuir (Eq. 2) models.

The Freundlich equation can be formulated as follows:

$$q_e = K_F \ C_e^n, \tag{1}$$

where q_e is the As(V) adsorption per unit of mass of the adsorbent, C_e is the equilibrium concentration of the dissolved As, K_F is a constant related to the adsorption capacity and n is a constant related to the adsorption intensity.

The Langmuir equation formulation is formulated as follows:

$$q_e = X_m \ K_L \ C_e/(1 + K_L \ C_e), \tag{2}$$

where X_m is the maximum adsorption capacity and K_L is a constant related to the adsorption energy.

The statistical package SPSS 19.0 (IBM, USA) was used to perform the fitting of the adsorption experimental data to Freundlich and Langmuir models.

2.2.3 As(V) adsorption/desorption as a function of pH

Adsorption trials were performed using triplicate samples (1 g each) of fine mussel shell and granitic material, as well as granitic material +12 t ha^{-1} fine mussel shell, that were added with 10 mL of solutions containing 5 mg L^{-1} As(V) and different concentrations of HNO$_3$ (0.0025, 0.0038, 0.005, 0.0075 M) or NaOH (0.0025, 0.0038, 0.005, 0.0075 M), including NaNO$_3$ 0.01 M as background electrolyte. To elaborate control samples, each of the solid materials were added with 10 mL of solutions containing NaNO$_3$ 0.01 M and 5 mg L^{-1} As(V) but without HNO$_3$ or NaOH. After 24 h of shaking, all samples were centrifuged for 15 min at 4000 rpm and then filtered through acid-washed paper. The resulting liquid phase was analyzed for pH, DOC and As(V); finally, adsorbed As(V) was calculated as the difference between added As(V) concentration and that remaining in the equilibrium solution.

Desorption trials consisted of triplicate samples (1 g each) of fine mussel shell and granitic material that were added with 10 mL of solutions containing 100 mg L^{-1} As(V), including NaNO$_3$ 0.01 M as background electrolyte. After a shaking period of 24 h, all samples were centrifuged for 15 min at 4000 rpm and then filtered through acid-washed paper, this time discarding the liquid phase. The remaining solid phase was added with 30 mL of solutions containing NaNO$_3$ 0.01 M and diverse HNO$_3$ or NaOH concentrations, aiming to provide a wide pH range in order to achieve desorption for different pH values. After shaking for 24 h, all samples were centrifuged for 15 min at 4000 rpm and filtered through acid-washed paper. The resulting liquid was analyzed for pH, DOC and As(V); finally, desorbed As(V) was calculated as the difference between the amount retained in the adsorption phase and that released to the equilibrium solution in this desorption phase, and it was expressed as percentage of the total amount adsorbed.

2.2.4 Fractionation of the As(V) adsorbed at three different incubation times

Granitic material, fine mussel shell and granitic material +12 t ha^{-1} fine mussel shell samples were added with a NaNO$_3$ 0.01 M solution containing 100 mg L^{-1} As(V) (1 : 10 solid : solution ratio), shaken for 24 h and filtered through acid-washed paper. The resulting liquid phase was analyzed for pH, DOC and As(V). Finally, the adsorbed As(V) was fractionated using the BCR (Bureau of Reference) procedure modified by Rauret et al. (1999), using the four steps indicated by Nóvoa-Muñoz et al. (2007), finally obtaining an acid soluble fraction, a reducible fraction, an oxidizable fraction and a residual fraction. The fractionation was performed for three different incubation times: 24 h, 1 week and 1 month.

Table 1. General characteristics of the solid materials (average values for three replicates, with coefficients of variation always $< 5\,\%$).

		Coarse mussel shell	Fine mussel shell	Granitic material
C	%	12.67 ± 0.07	11.43 ± 0.11	0.11 ± 0.00
N	%	0.36 ± 0.01	0.21 ± 0.02	0.04 ± 0.00
C / N		35.00 ± 0.94	55.65 ± 4.13	2.80 ± 0.00
pH_{H_2O}		9.11 ± 0.13	9.39 ± 0.01	5.72 ± 0.04
Ca_e	$cmol\,kg^{-1}$	12.64 ± 0.52	24.75 ± 0.22	0.18 ± 0.00
Mg_e	$cmol\,kg^{-1}$	0.58 ± 0.02	0.72 ± 0.04	0.13 ± 0.00
Na_e	$cmol\,kg^{-1}$	5.24 ± 0.08	4.37 ± 0.02	0.27 ± 0.01
K_e	$cmol\,kg^{-1}$	0.31 ± 0.00	0.38 ± 0.00	0.31 ± 0.01
Al_e	$cmol\,kg^{-1}$	0.04 ± 0.00	0.03 ± 0.00	1.63 ± 0.08
eCEC	$cmol\,kg^{-1}$	18.82 ± 0.43	30.25 ± 0.21	2.53 ± 0.12
Al saturation	%	0.21 ± 0.01	0.11 ± 0.00	64.55 ± 1.73
P_{Olsen}	$mg\,kg^{-1}$	23.21 ± 0.64	54.17 ± 1.25	2.56 ± 0.12
Ca_T	$mg\,kg^{-1}$	298085 ± 6290	280168 ± 2193	$< 0.01 \pm 0.00$
Mg_T	$mg\,kg^{-1}$	1020 ± 22	980.6 ± 44.9	355.2 ± 17.3
Na_T	$mg\,kg^{-1}$	5508 ± 114	5173 ± 95	102.4 ± 4.2
K_T	$mg\,kg^{-1}$	80.57 ± 1.75	202.1 ± 2.6	1434 ± 49
Al_T	$mg\,kg^{-1}$	93.89 ± 3.02	433.2 ± 13.9	5980 ± 154
Fe_T	$mg\,kg^{-1}$	3534 ± 22	1855 ± 92	3505 ± 125
Mn_T	$mg\,kg^{-1}$	5.70 ± 0.22	33.75 ± 1.35	23.96 ± 0.51
Cu_T	$mg\,kg^{-1}$	3.20 ± 0.13	6.72 ± 0.33	7.15 ± 0.34
Zn_T	$mg\,kg^{-1}$	7.71 ± 0.19	7.66 ± 0.45	18.10 ± 0.28
Cd_T	$mg\,kg^{-1}$	0.02 ± 0.00	0.07 ± 0.01	$< 0.01 \pm 0.00$
Ni_T	$mg\,kg^{-1}$	5.64 ± 0.21	8.16 ± 0.24	0.97 ± 0.04
Cr_T	$mg\,kg^{-1}$	1.32 ± 0.05	4.51 ± 0.17	2.71 ± 0.12
Co_T	$mg\,kg^{-1}$	0.68 ± 0.03	1.02 ± 0.04	0.41 ± 0.01
As_T	$mg\,kg^{-1}$	0.48 ± 0.07	1.12 ± 0.06	2.94 ± 0.07
Al_o	$mg\,kg^{-1}$	85.00 ± 1.97	178.3 ± 2.82	1425 ± 38
Al_p	$mg\,kg^{-1}$	62.67 ± 1.25	78.67 ± 1.14	462.7 ± 9.6
Al_{cu}	$mg\,kg^{-1}$	7.57 ± 0.21	22.87 ± 0.57	150.2 ± 6.5
Al_{la}	$mg\,kg^{-1}$	2.47 ± 0.09	2.60 ± 0.02	137.4 ± 3.4
Al_{op}	$mg\,kg^{-1}$	22.33 ± 1.16	99.67 ± 1.37	962.3 ± 12.6
Al_{pcu}	$mg\,kg^{-1}$	55.10 ± 2.03	55.80 ± 1.16	312.5 ± 5.7
Al_{cula}	$mg\,kg^{-1}$	5.10 ± 0.12	20.27 ± 0.71	12.75 ± 0.57
Fe_o	$mg\,kg^{-1}$	42.67 ± 1.18	171.0 ± 2.23	224.3 ± 2.56
Fe_p	$mg\,kg^{-1}$	7.67 ± 0.18	37.67 ± 0.89	54.33 ± 1.17
Fe_{op}	$mg\,kg^{-1}$	35.00 ± 1.21	133.3 ± 1.88	170.0 ± 2.14

X_e: exchangeable concentration of the element; X_T: total concentration of the element; Al_o, Fe_o: Al and Fe extracted with ammonium oxalate; Al_p, Fe_p: Al and Fe extracted with sodium pyrophosphate; Al_{cu}: Al extracted with copper chloride; Al_{la}: Al extracted with lanthanum chloride; Al_{op}: Al_o-Al_p; Al_{pcu}: Al_p-Al_{cu}; Al_{cula}: Al_{cu}-Al_{la}; Fe_{op}: Fe_o-Fe_p.

2.2.5 Statistical analysis

Tests for normality, correlation and analysis of variance were performed using the statistical package SPSS 19.0 (IBM, USA).

3 Results and discussion

3.1 Characterization of the solid materials

Table 1 shows that the granitic material had low C and N percentages (indicating low organic matter content) and acid pH (5.7), whereas pH was alkaline for fine and coarse mussel shell (9.4 and 9.1, respectively). Total Ca and Na contents were higher for fine and coarse mussel shell, whereas the granitic material presented the lowest effective eCEC

Table 2. Desorption results (average $\pm$ standard deviation, in mg kg^{-1}, with percentage values between brackets) corresponding to fine and coarse mussel shell and to the unamended and shell-amended (12 and 24 t ha^{-1}) granitic material.

Added As (mg L^{-1})	Fine shell	Coarse shell	GM	GM+12 t ha^{-1}	GM+24 t ha^{-1}
0	$0.02 \pm 0.00(0.0)$	$0.04 \pm 0.00(0.0)$	$0.01 \pm 0.00(0.0)$	$0.02 \pm 0.00(0.0)$	$0.07 \pm 0.00(0.0)$
0.5	$0.25 \pm 0.01(6.9)$	$0.22 \pm 0.01(7.6)$	$0.10 \pm 0.00(2.3)$	$0.38 \pm 0.01(9.9)$	$0.51 \pm 0.02(10.7)$
5	$2.68 \pm 0.08(7.5)$	$2.22 \pm 0.10(7.9)$	$0.90 \pm 0.03(2.0)$	$3.24 \pm 0.12(6.6)$	$5.72 \pm 0.16(12.3)$
10	$6.18 \pm 0.19(9.0)$	$3.49 \pm 0.14(6.2)$	$2.98 \pm 0.11(3.8)$	$9.85 \pm 0.21(10.2)$	$12.6 \pm 0.2(14.2)$
25	$13.0 \pm 0.3(8.2)$	$17.7 \pm 0.6(49.4)$	$10.1 \pm 0.4(6.4)$	$34.8 \pm 1.2(16.6)$	$29.1 \pm 0.6(15.0)$
50	$25.8 \pm 0.6(9.9)$	$37.2 \pm 1.2(46.4)$	$25.8 \pm 1.1(9.5)$	$65.4 \pm 2.1(25.1)$	$33.6 \pm 0.7(10.1)$
100	$45.6 \pm 1.3(8.4)$	$39.0 \pm 1.4(7.0)$	$54.7 \pm 1.7(10.7)$	$98.2 \pm 2.3(18.9)$	$72.7 \pm 1.9(12.3)$

GM: granitic material.

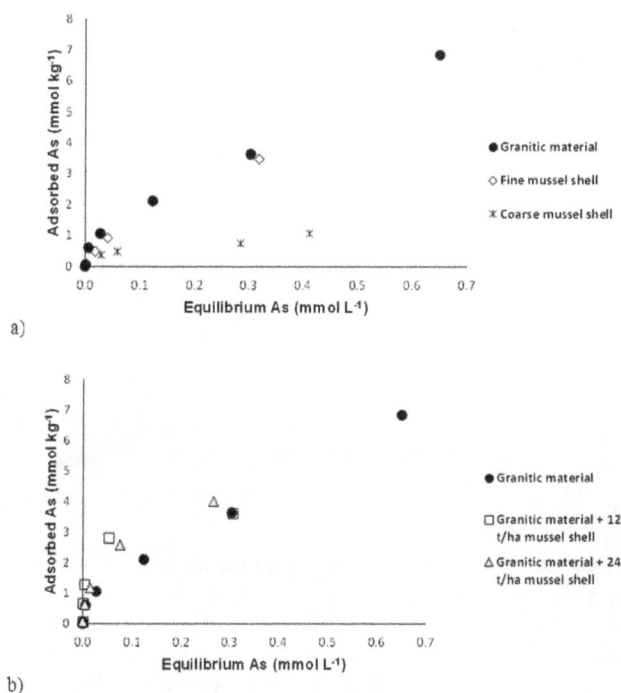

Figure 1. Adsorption curves for the individual materials (**a**) and for the unamended and shell-amended (12 or 24 t ha^{-1}) granitic material (**b**). Average values of three replicates, with coefficients of variation always < 5 %.

(eCEC < 4 cmol kg^{-1}) as well as high Al saturation (64.5 %) and total Al concentrations. Regarding Al forms, amorphous Al$_o$ compounds were clearly more abundant in the granitic material, whereas those bound to organic matter (Al$_p$) had low presence in all of the studied materials, with most of the amorphous Al being in inorganic form (Al$_{op}$). Similarly, the low organic-C content of the granitic material and coarse and fine mussel shells justified that most Fe was bound to inorganic forms (Fe$_{op}$). Additionally to that shown in Table 1, the particle size distribution of the granitic material was 60 % sand, 23 % clay and 17 % silt.

3.2 Adsorption/desorption as a function of added As(V) concentration

Figure 1a shows that As(V) adsorption was equivalent on granitic material and fine mussel shell and higher than on coarse mussel shell. The different behavior for both mussel shell materials (higher As adsorption on fine than on coarse mussel shell) can be in relation with the higher surface area of fine shell (1.4 m^2 g^{-1}) than that of coarse shell (1 m^2 g^{-1}), as previously stated by Peña-Rodríguez et al. (2013). Figure 1b indicates that As(V) adsorption increased when granitic material was amended with mussel shell. Adsorption curves in Fig. 1 show type C layout (Giles et al., 1960) for granitic material and fine and coarse mussel shell (Fig. 1a), exhibiting a rather constant slope when the added arsenic concentration was increased. This kind of adsorption curve is generally associated with the existence of a constant partition between the adsorbent surface and the equilibrium solution in the contacting layer or to a proportional increase of the adsorbent surface taking place when the amount of adsorbed arsenic increases, as indicated by Seco-Reigosa et al. (2013b), who found the same type of adsorption curve studying arsenic retention on pine sawdust and on fine mussel shell. The granitic material treated with mussel shell shows adsorption curves that are near C type (Fig. 1b).

Figure 2 shows that percentage adsorption progressively decreased on granitic material when the As(V) concentration added was > 10 mg L^{-1}. The 24 t ha^{-1} mussel shell amendment caused slightly increase in percentage adsorption, whereas the 12 t ha^{-1} amendment did not result in systematic increased percentage adsorption.

Regarding desorption, Table 2 shows released As(V) concentrations and percentages (referred to the amounts previously adsorbed). The highest desorption percentage (49 %) corresponded to coarse mussel shell when 25 mg L^{-1} As(V) were added. When 100 mg L^{-1} As(V) were added, percentage desorption was always < 19 %. Mussel shell amendment (12 and 24 t ha^{-1}) increased As(V) desorption, which could be in relation with the fact that arsenate bind strongly to

Table 3. Fitting of the adsorption results to the Freundlich and Langmuir models.

	Freundlich			Langmuir		
	K_F $(L^n \, kg^{-1} \, mmol^{(1-n)})$	n (dimensionless)	R^2	K_L $(L \, mmol^{-1})$	X_m $(mmol \, kg^{-1})$	R^2
Fine shell	10.8 ± 0.8	0.86 ± 0.08	0.987	–	–	
Coarse shell	38.7 ± 11.4	3.14 ± 0.55	0.991	–	–	
GM	9.0 ± 0.5	0.68 ± 0.06	0.991	1.0 ± 0.6	16.7 ± 6.0	0.978
GM+12 t ha^{-1}	7.7 ± 0.9	0.41 ± 0.09	0.938	9.2 ± 8.0	6.9 ± 1.6	0.866
GM+24 t ha^{-1}	10.8 ± 1.0	0.61 ± 0.08	0.977	1.6 ± 1.3	16.1 ± 7.5	0.951

GM: granitic material; 12 and 24 t ha^{-1}: doses of the fine mussel shell amendments; - fitting was not possible due to estimation errors being too high.

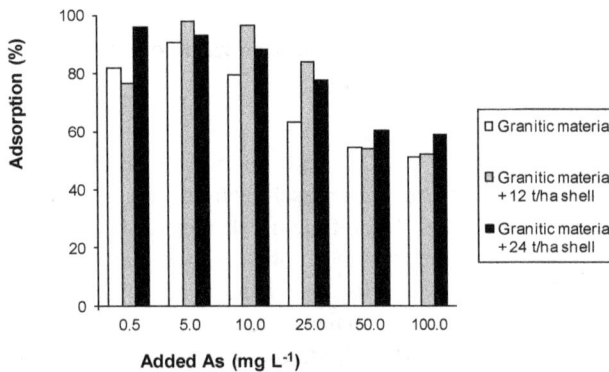

Figure 2. Relationship between added As(V) (mg L^{-1}) and As(V) percentage adsorption for the unamended and shell-amended (12 or 24 t ha^{-1}) granitic material. Average values for three replicates, with coefficients of variation always < 5 %.

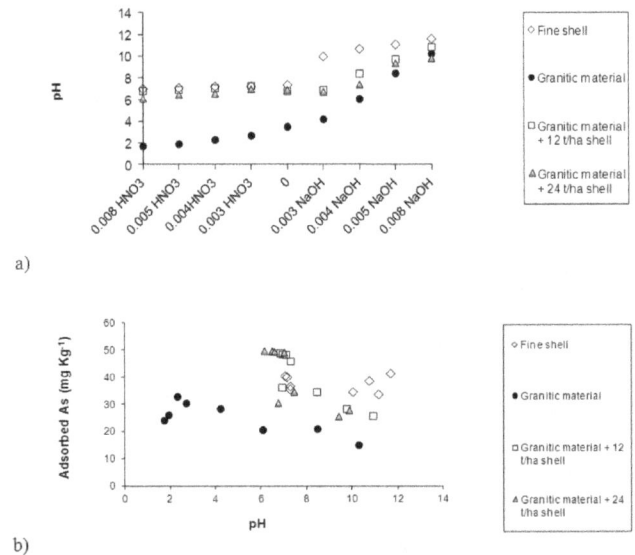

Figure 3. (a) Time-course evolution of pH for the solid materials as a function of the various molar concentrations of added HNO$_3$ and NaOH; **(b)** relationship between adsorption (mg kg^{-1}) and pH value for fine shell and the unamended and shell-amended granitic material. Average values for three replicates, with coefficients of variation always < 5 %.

the surface of oxides and hydroxides in clearly acid environments (pH between 3.5 and 5.5; Silva et al., 2010), whereas increased pH values (from above 5 for clay minerals to above 12 for calcite) favor desorption (Golberg and Glaubig 1988). Any case, most of the adsorbed As(V) did not desorb, indicating notable irreversibility of the process.

Adsorption data were adjusted to the Freundlich and Langmuir models (Table 3), finding that the unamended and shell-amended granitic material fitted well to both models, whereas fine and coarse mussel shell can be fitted only to the Freundlich model. Maji et al. (2007) found satisfactory adjustment to both Freundlich and Langmuir models studying As(V) adsorption on lateritic substrates, while Yolcubal and Akyol (2008) obtained better fitting to the Freundlich model using carbonate-rich solid substrates.

3.3 As(V) adsorption/desorption as a function of pH

3.3.1 Adsorption

Figure 3 shows the repercussion on As(V) adsorption of adding different HNO$_3$ and NaOH molar concentrations to fine mussel shell and to the unamended and shell-amended

granitic material. The acid concentrations added to fine shell were not permitted to reach pH < 7 (Fig. 3a), whereas the addition of alkaline solutions was allowed to achieve pH values near 12 for this material. The granitic material exhibited the lowest buffer potential (possibly related to its low colloids content), presenting pH values between 2 and 10. Mussel shell amendment increased the buffer potential of this granitic material, especially when the 24 t ha^{-1} dose was used.

Figure 3b shows that As(V) adsorption on the granitic material (expressed in mg kg^{-1}) progressively decreased from pH 4 as a function of increasing pH value, whereas the mussel shell amendment increased As(V) adsorption. The granitic material contains variable charge compounds (such as Fe and Al oxyhydroxides, kaolinite-type clays and organic matter), positively charged at acid pH, facilitating retention of $H_2AsO_4^-$ and $HAsO_4^{2-}$ (Smedley and Kinniburgh, 2002; Xu et al., 2002; Yan et al., 2000) but suffering progressive deprotonation and increase of negative charge as pH increases, which can lower As(V) adsorption (Fitz and Wenzel, 2002). However, the effect of lowering As(V) adsorption due to pH increase did not occur when granitic material was amended with mussel shell, which must be related to the additional As(V) adsorption capacity associated with calcium carbonate present in mussel shell, establishing cationic bridges when pH values are higher (Alexandratos et al., 2007). Salameh et al. (2015) found that arsenic was completely removed by charred dolomite samples (another alkaline material) over a wide range of pH (2–11). Our granitic material suffered just slight changes in As(V) adsorption in the pH range 3.5 to 6.9, which can be related to the effective adsorption that As(V) experience in a wide range (4–11) (Stanic et al., 2009).

Expressing As(V) adsorption as percentage with respect to the amount added, the maximum for the unamended granitic material (66 %) took place at pH < 6, progressively decreasing from that point as a function of increasing pH value. Fine mussel shell adsorbed As(V) notably on the pH range 6–12, with maximum value of 83 %. When the granitic material was amended with fine mussel shell, As(V) adsorption reached 99 % at pH near 8 and then progressively decreased as pH increased.

In the case of the shell-amended granitic material, significant ($p < 0.005$) statistical correlations existed between adsorbed As(V) and pH ($r = 0.926$ and $r = 0.880$ for the 12 and $24\,t\,ha^{-1}$ mussel shell doses, respectively), whereas no correlation was found between both parameters in the case of mussel shell by itself. The latter can be due to the absence of anionic exchange with OH- groups when As(V) anions adsorb on mussel shell, contrary to that happening to other anions on different adsorbent materials (Arnesen and Krogstad, 1998; Bower and Hatcher, 1967; Gago et al., 2012; Huang and Jackson, 1965). However, anions other than OH^- can be released, as is the case for SO_4^{2-}, PO_4^{3-} or organic anions, which is in concordance with the correlations found between adsorbed As(V) and DOC ($r = 0.810$, for fine shell, and $r = 0.919$ and $r = 0.913$, for the granitic material amended with 12 and $24\,t\,ha^{-1}$ mussel shell, respectively, $p < 0.005$). Moreover, other mechanisms that can be responsible for anion retention (such as retention on calcite or H and van der Waals bindings) do not implicate OH^- release (Boddu et al., 2003). Different authors remark on the influence of pH on As(V) adsorption (Maji et al., 2007; Partey et al., 2008; Stanic et al., 2009), but in the case of our granitic material, Al, Fe, Al_o, Fe_o, organic matter and organoaluminum complexes, contents must also be relevant.

Fine and coarse mussel shell presented alkaline pH (9.39 and 9.11, respectively, Table 1), making the dominant As species $HAsO_4^{2-}$ (Yan et al., 2000), which can bind to the surface of carbonates such as calcite by means of inner sphere complexes with octahedral Ca (Alexandratos et al., 2007).

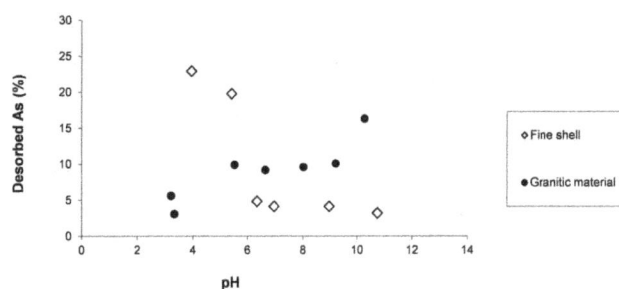

Figure 4. Relationship between As(V) desorption (%) and pH value for fine shell and for the granitic material (average values for three replicates, with coefficients of variation always < 5 %) when $100\,mg\,L^{-1}$ As(V) were added to the adsorbents.

3.3.2 Desorption

Figure 4 shows that, when a concentration of $100\,mg\,L^{-1}$ As(V) was added, As(V) desorption from fine shell and granitic material was always < 26 % of the amount previously adsorbed, considering the whole pH range studied (2–12). Two different behaviors took place: (a) As(V) desorption from granitic material clearly increased as pH increased between 4 and 6, and (b) As(V) desorption from mussel shell clearly decreased as pH increased between 4 and 6. Moreover, As(V) desorption from mussel shell continued to be low at pH > 6, slowly decreasing, whereas release from the granitic material further increased when pH > 6.

As(V) desorption from mussel shell clearly increased at pH < 6 in accordance with that detected by Goldberg and Glaubig (1988), who found that As adsorption on calcite increased from pH 6 to 10 (then decreasing release), attaining maximum adsorption at pH between 10 and 12 and then decreasing at higher pH values. Di Benedetto et al. (2006) indicated that As(V) can be incorporated to calcite in alkaline conditions by preventing its mobilization even in situations where oxyhydroxides do not exhibit adsorption potential. Alexandratos et al. (2007) found that arsenate anions have great affinity for calcite at pH around 8, establishing strong bindings due to inner sphere complexes with AsO_4^{3-} binding to the mineral surface through Ca cationic bridges. All these facts are in accordance with the low As(V) release suffered by our mussel shell samples at pH > 6 (Fig. 4).

3.4 Fractionation of the As(V) adsorbed at three different incubation times

Figure 5 shows that the As(V) soluble fraction (exchangeable and bound to carbonates) is quantitatively the most important in all samples (especially in the unamended and shell-amended granitic material), representing at 24 h of incubation contents that ranged between a minimum of 69 % in fine mussel shell and a maximum of 88 % in the $12\,t\,ha^{-1}$ shell-amended granitic material. The soluble fraction corresponds to the most mobile As(V), which is weakly retained mainly

24 h

a)

1 week

b)

1 month

c)

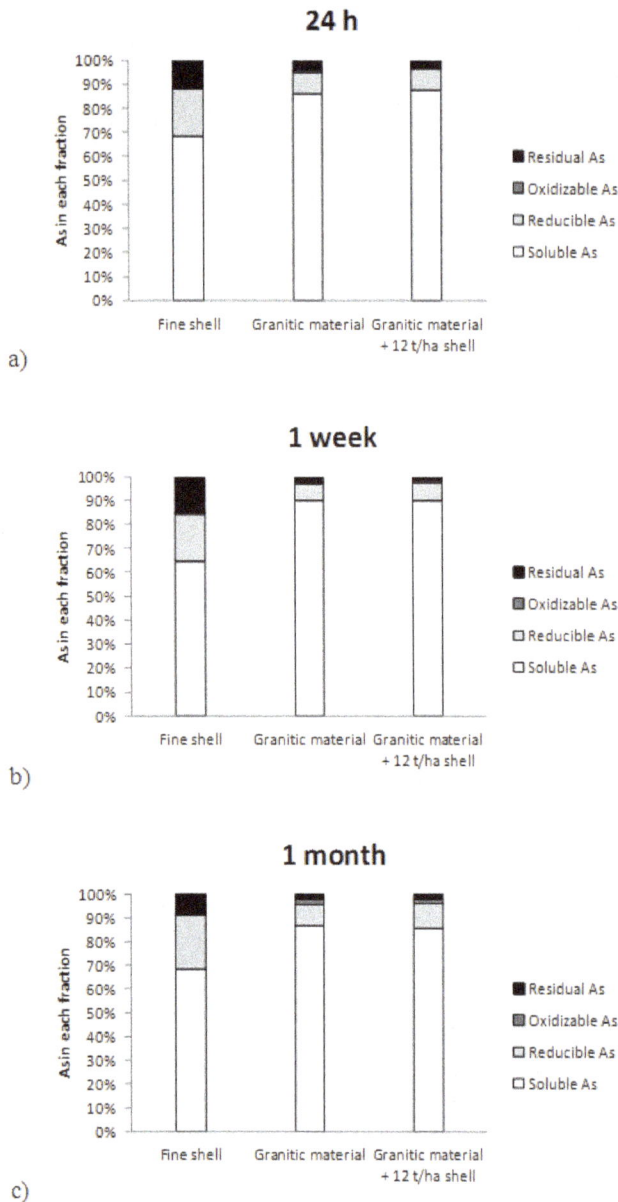

Figure 5. Percentages of the various fractions of As(V) adsorbed after 24 h (**a**), 1 week (**b**) and 1 month (**c**) of incubation. Average values for three replicates, with coefficients of variation always < 5 %.

due to anionic exchange mechanisms (Keon et al., 2001) and which is associated to high risks of toxicity. Moreover, Taggart et al. (2004) indicate that As(V) derived from anthropogenic pollution incorporates to the most mobile fractions of solid substrates in great percentage. In our materials, the As(V) reducible fraction (associated to Al and Fe oxides and oxyhydroxides) represented between 9 and 19 % of the As(V) adsorbed at 24 h of incubation (Fig. 5), whereas the As(V) residual fraction (that incorporated to the structure of minerals) always constituted < 16 % of the amount adsorbed. Finally, the As(V) oxidizable fraction (associated to

organic matter and as sulfides) was always < 2.6 % (Fig. 5), attributable to the low organic content of the solid materials here studied. The increase of incubation time from 24 h to 1 week and to 1 month, as well as the $12\,\mathrm{t\,ha^{-1}}$ shell amendment of the granitic material, did not cause statistically significant modifications in the percentage content of each fraction of the adsorbed As(V) (Fig. 5).

The As(V) reducible fraction (bound to Al and Fe oxides and oxyhydroxides) correlated positively with DOC ($r = 0.957$ at 24 h, and $r = 0.954$ at 1 week incubation time, $p < 0.005$), suggesting that arsenate compete with organic groups to bind on oxides and oxyhydroxides. Additionally, the As(V) residual fraction correlated with total Fe ($r = 0.980$ at 24 h, and $r = 0.973$ at 1 month incubation time, $p < 0.005$), suggesting the existence of re-adsorption and co-precipitation processes with Fe minerals.

4   Conclusions

The granitic material studied here presented lower As(V) adsorption capacity than the fine and coarse mussel shells used. Furthermore, As(V) retention on the granitic material was weak, implying scarce capacity to attenuate acute toxic effects of an eventual As(V) pollution episode. Fine shell showed moderate As(V) retention potential (higher than that of coarse shell). The amendment of 12 and $24\,\mathrm{t\,ha^{-1}}$ fine mussel shell on the granitic material increased As(V) retention, thus justifying this management practice. Most of the adsorbed As(V) did not desorb in a wide range of pH, with higher risk corresponding to the granitic material when pH increased from pH value 6. The adsorbed As(V) was retained mainly on the soluble fraction, with weak bindings, also facilitating release.

Acknowledgements. This study was funded by the *Ministerio de Economía y Competitividad* (government of Spain), grant numbers CGL2012-36805-C02-01 and -02.

Edited by: A. Cerdà

References

Ahmad, S., Ghafoor, A., Akhtar, M. E., and Khan, M. Z.: Ionic displacement and reclamation of saline-sodic soils using chemical amendments and crop rotation, Land Degrad. Dev., 24, 170–178, 2013.

Alexandratos, V. G., Elzinga, E. J., and Reeder, R. J.: Arsenate uptake by calcite: Macroscopic and spectroscopic characterization of adsorption and incorporation mechanisms, Geochim. Cosmochim. Acta, 71, 4172–4187, 2007.

Álvarez, E., Fernández-Sanjurjo M. J., Núñez, A., Seco, N., and Corti, G.: Aluminium fractionation and speciation in bulk and rhizosphere of a grass soil amended with mussel shells or lime, Geoderma, 173/174, 322–329, 2013.

Arnesen, A. K. M. and Krogstad, T.: Sorption and desorption of fluoride in soil polluted from the aluminium smelter at Ardal in Western Norway, Water Air Soil Poll., 103, 357–373, 1998.

Boddu, V. M., Abburi, K., Talbott, J. L., and Smith, E. D.: Removal of hexavalent chromium from wastewater using a new composite chitosan biosorbent, Environ. Sci. Technol., 37, 4449–4456, 2003.

Bower, C. A. and Hatcher, J. T.: Adsorption of fluoride by soils and minerals, Soil Sci., 103, 151–154, 1967.

Brevik E. C. and Sauer T. J.: The past, present, and future of soils and human health studies, Soil, 1, 35–46, 2015.

Brevik, E. C., Cerdà, A., Mataix-Solera, J., Pereg, L., Quinton, J. N., Six, J., and Van Oost, K.: The interdisciplinary nature of Soil, Soil, 1, 117–129, 2015.

Chatterjee, A., Lal, R., Wielopolski, L., Martin, M. Z., and Ebinger, M. H.: Evaluation of Different Soil Carbon Determination Methods, Cr. Rev. Plant Sci., 28, 164–178, 2009.

Clothier, B. E., Green, S. R., Vogeler, I., Greven, M. M., Agnew, R., van den Dijssel, C. W., Neal, S., Robinson, B. H., and Davidson, P.: CCA transport in soil from treated-timber posts: pattern dynamics from the local to regional scale, Hydrol. Earth Syst. Sci. Discuss., 3, 2037–2061, doi:10.5194/hessd-3-2037-2006, 2006.

Di Benedetto, F., Costagliola, P., Benvenuti, M., Lattanzi, P., Romanelli, M., and Tanelli, G.: Arsenic incorporation in natural calcite lattice. Evidence from electron spin echo spectroscopy, Earth Planet. Sci. Lett., 246, 458–465, 2006.

Fitz, W. J. and Wenzel, W. W.: Arsenic transformations in the soil-rhizosphere-plant system: fundamentals and potential application to phytoremediation, J. Biotechnol., 99, 259–278, 2002.

Gago, C., Romar, A., Fernández-Marcos, M. L., and Álvarez, E.: Fluorine sorption by soils developed from various parent materials in Galicia (NW Spain), J. Colloid Interf. Sci., 374, 232–236, 2012.

García, I., Diez, M., Martín, F., Simón, M., and Dorronsoro, C.: Mobility of Arsenic and Heavy Metals in a Sandy-Loam Textured and Carbonated Soil, Pedosphere, 19, 166–175, 2009.

Gee , G. W. and Bauder, J. W.: Particle-size analysis, in: Methods of Soil Analysis, Part 1, Physical and mineralogical methods, ASA, Madison, USA, 383–409, 1986.

Ghimire, K. N., Inoue, K., Yamagchi, H., Makino, K., and Miyajima, T.: Adsortive separation of arsenate and arsenite anions from aqueous medium by using orange waste, Water Res., 37, 4945–4953, 2003.

Giles, C. H., MacEwan, T. H., Nakhwa, S. N., and Smith, D.: A system of classification of solution adsorption isotherms, and its use in diagnosis of adsorption mechanisms and in measurement of specific surface area of solids, J. Chem. Soc., 111, 3973–3993, 1960.

Goldberg, S. and Glaubig, R. A.: Anion sorption on a calcareous, montmorillonitic soil-arsenic, Soil Sci. Soc. Am. J., 52, 1297–1300, 1988.

Gur, A., Gil, Y., and Bravd, B.: The efficacy of several herbicides in the vineyard and their toxicity to grapevines, Weed Res., 19, 109–116, 1979.

Huang, P. M. and Jackson, M. L.: Mechanism of reaction of neutral fluoride solution with layer silicates and oxides in soils, Soil Sci. Soc. Am. P., 29, 661–665, 1965.

Johnston, C. R., Vance, G. F., and Ganjegunte, G. K.: Soil property changes following irrigation with coalbed natural gas water: role

of water treatments, soil amendments and land suitability, Land Degrad. Dev., 24, 350–362, 2013.

Kamprath, E. J.: Exchangeable aluminium as a criterion for liming leached mineral soils, Soil Sci. Soc. Am. P., 34, 252–54, 1970.

Keon, N. E., Swartz, C. H., Brabander, D. J., Harvey, C., and Hemond, H. F.: Validation of an arsenic sequential extraction method for evaluating mobility in sediments, Environ. Sci. Technol., 35, 2778–2784, 2001.

Mao, W., Kang, S., Wan, Y., Sun, Y., Li, X., and Wang, Y.: Yellow river sediment as a soil amendment for amelioration of saline land in the Yellow river delta, Land Degrad. Dev., doi:10.1002/ldr.2323, 2014.

Maji, S. K., Pal, A., Pal, T., and Adak, A.: Adsorption thermodynamics of arsenic on laterite soil, Surface Sci. Technol., 22, 161–176, 2007.

Matschullat, J.: Arsenic in the geosphere – a review, Sci. Total Environ., 249, 297–312, 2000.

McLean, E. O.: Soil pH and Lime Requirement, in: Methods of Soil Analysis, Part 2, Chemical and Microbiological Properties, ASA, Madison, USA, 199–223, 1982.

Mehmood, A., Hayat, R., Wasim, M., and Akhtar, M. S.: Mechanisms of Arsenic Adsorption in Calcareous Soils, J. Agric. Biol. Sci., 11, 59–65, 2009.

Moreno-Ramón, H., Quizembe, S. J., and Ibáñez-Asensio, S.: Coffee husk mulch on soil erosion and runoff: experiences under rainfall simulation experiment, Solid Earth, 5, 851–862, doi:10.5194/se-5-851-2014, 2014.

Nóbrega, J. A., Pirola, C., Fialho, L. L., Rota, G., de Campos C. E. K. M. A. J., and Pollo, F.: Microwave-assisted digestion of organic samples: How simple can it become?, Talanta, 98, 272–276, 2012.

Novara, A., Gristina, L., Guaitoli, F., Santoro, A., and Cerdà, A.: Managing soil nitrate with cover crops and buffer strips in Sicilian vineyards, Solid Earth, 4, 255–262, doi:10.5194/se-4-255-2013, 2013.

Nóvoa-Muñoz, J. C., Queijeiro, J. M., Blanco-Ward, D., Álvarez-Olleros, C., García-Rodeja, E., and Martínez-Cortizas, A.: Arsenic fractionation in agricultural acid soils from NW Spain using a sequential extraction procedure, Sci. Total Environ., 378, 18–22, 2007.

Olsen, S. R. and Sommers, L. E.: Phosphorus, in: Methods of Soil Analysis, Part 2, Chemical and Microbiological Properties, ASA, Madison, USA, 403–430, 1982.

Osorio-López, C., Seco-Reigosa, N., Garrido-Rodríguez, B., Cutillas-Barreiro, L., Arias-Estévez, M., Fernández-Sanjurjo, M.J., Álvarez-Rodríguez, E., and Núñez-Delgado, A.: As(V) adsorption on forest and vineyard soils and pyritic material with or without mussel shell: Kinetics and fractionation, J. Taiwan Inst. Chem. Eng., 45, 1007–1014, 2014.

Partey, F., Norman, D., Ndur, S., and Nartey, R.: Arsenic sorption onto laterite iron concretions: Temperature effect, J. Colloid Interf. Sci., 321, 493–500, 2008.

Peña-Rodríguez, S., Bermúdez-Couso, A., Nóvoa-Muñoz, J.C., Arias-Estévez, M., Fernández-Sanjurjo, M.J., Álvarez-Rodríguez, E., and Núñez-Delgado, A.: Mercury removal using ground and calcined mussel shell, J. Environ. Sci., 2512, 2476–2486, 2013.

Rauret, G., López-Sánchez, J. F., Sahuquillo, A., Rubio, R., Davidson, C. M., Ure, A. M., and Quevauviller, J.: Improvement of the

BCR three step sequential extraction procedure prior to the certification of new sediment and soil reference materials, J. Environ. Monitor., 1, 57–61, 1999.

Roy, M. and McDonald, L. M.: Metal uptake in plants and health risk assessments in metal-contaminated smelter soils, Land Degrad. Dev., doi:10.1002/ldr.2237, 2015.

Sacristán, D., Peñarroya, B., and Recatalá, L.: Increasing the knowledge on the management of Cu-contaminated agricultural soils by cropping tomato (*Solanum lycopersicum* L.), Land Degrad. Dev., doi:10.1002/ldr.2319, 2015.

Sadeghi, S. H. R., Gholami, L., Sharifi, E., Khaledi Darvishan, A., and Homaee, M.: Scale effect on runoff and soil loss control using rice straw mulch under laboratory conditions, Solid Earth, 6, 1–8, doi:10.5194/se-6-1-2015, 2015.

Salameh, Y., Albadarin, A. B., Allen, S., Walker, G., and Ahmad, M. N. M.: Arsenic(III,V) adsorption onto charred dolomite: Charring optimization and batch studies, Chem. Eng. J., 259, 663–671, 2015.

Seco-Reigosa, S., Peña-Rodríguez, S., Nóvoa-Muñoz, J. C., Arias-Estévez, M., Fernández-Sanjurjo, M. J., Álvarez-Rodríguez, E., and Núñez-Delgado, A.: Arsenic, chromium and mercury removal using mussel shell ash or a sludge/ashes waste mixture, Environ. Sci. Pollut. Res., 20, 2670–2678, 2013a.

Seco-Reigosa, N, Bermúdez-Couso, A., Garrido-Rodríguez, B., Arias-Estévez, M., Fernández-Sanjurjo, M.J., Álvarez-Rodríguez, E., and Núñez-Delgado, A. As(V) retention on soils and forest by-products and other waste materials, Environ. Sci. Pollut. Res., 20, 6574–6583, 2013b.

Silva, J., Mello, J. W. V., Gasparon, M., Abrahão, W. A. P., Ciminellic, V. S. T., and Jong, T.: The role of Al-Goethites on arsenate mobility, Water Res., 4419, 5684–5692, 2010.

Smedley, P. L. and Kinninburgh, D. G.: A review of the source, behaviour and distribution of arsenic in natural waters, Applied Geochem., 17, 517–568, 2002.

Smith, E., Naidu, R., and Alston, A. M.: Arsenic in the soil environment, a review, Adv. Agron., 64, 149–195, 1998.

Smith, E., Naidu, R., and Alston, A. M.: Chemistry of arsenic in soils: I. Sorption of arsenate and arsenite by four Australian soils, J. Environ. Qual., 28, 1719–1726, 1999.

Smith, A. H., Arroyo, A.P., Mazumder, D. N. G., Kosnett, M. J., Hernandez, A. L., Beeris, M., Smith, M. M., and Moore, L. E.: Arsenic-induced skin lesions among Atacameño people in Northern Chile despite good nutrition and centuries of exposure, Environ. Health Persp., 108, 617–620, 2000.

Srivastava, P. K., Gupta, M., Singh, N., and Tewari, S. K.: Amelioration of sodic soil for wheat cultivation using bioaugmented organic soil amendment, Land Degrad. Dev., doi:10.1002/ldr.2292, 2014.

Stanic, T., Dakovic, A., Zivanovic, A., Tomasevic-Canovic, M., Dondur, V., and Milicevic, S.: Adsorption of arsenic (V by iron (III – modified natural) zeolitic tuff), Environ. Chem. Lett., 7, 161–166, 2009.

Sumner, M. E. and Miller, W. P.: Cation exchange capacity and exchange coefficients, in: Methods of Soil Analysis, Part 3, Chemical Methods, ASA, Madison, USA, 437–474, 1996.

Taggart, M. A., Carlisle, M., Pain, D. J., Williams, R., Osborn, D., Joyson, A., and Meharg, A. A.: The distribution of arsenic in soil affected by the Aznalcóllar mine spill, S.W. Spain, Sci. Total Environ., 323, 137–52, 2004.

WHO (World Health Organization): Guideline for Drinking Water Quality (4th Edn.), WHO Press, Geneva, Switzerland, 2011.

Xu, Y.H., Nakajima, T., and Ohki, A.: Adsorption and removal of arsenic (V) from drinking water by aluminum-loaded Shirasuzeolite, J. Hazard. Mat., 92, 275–287, 2002.

Yan, X., P., Kerrich, R. and Hendry, M. J.: Distribution of arsenic(III), arsenic(V) and total inorganic arsenic in pore-waters from a thick till and clay-rich aquitard sequence, Saskatchewan, Canada, Geochim. Cosmochim. Acta, 64, 2637–2648, 2000.

Yolcubal, I. and Akyol, N. H.: Adsorption and transport of arsenate in carbonate-rich soils: Coupled effects of nonlinear and rate-limited sorption, Chemosphere, 73, 1300–1307, 2008.

Zhang, H. and Selim, H. M.: Reaction and transport of arsenic in soils: equilibrium and kinetic modeling, Adv. Agron., 98, 45–115, 2008.

Identifying areas susceptible to desertification in the Brazilian northeast

R. M. S. P. Vieira[1], **J. Tomasella**[1,2], **R. C. S. Alvalá**[2], **M. F. Sestini**[1], **A. G. Affonso**[1], **D. A. Rodriguez**[1], **A. A. Barbosa**[2], **A. P. M. A. Cunha**[2], **G. F. Valles**[1], **E. Crepani**[1], **S. B. P. de Oliveira**[3], **M. S. B. de Souza**[3], **P. M. Calil**[4], **M. A. de Carvalho**[2], **D. M. Valeriano**[1], **F. C. B. Campello**[5], and **M. O. Santana**[5]

[1]Instituto Nacional de Pesquisas Espaciais, São José dos Campos, Brazil
[2]Centro Nacional de Monitoramento e Alertas de Desastres Naturais, Cachoeira Paulista, Brazil
[3]Fundação Cearense de Meteorologia e Recursos Hídricos, Fortaleza, Brazil
[4]Secretaria de Agricultura Agropecuária e Abastecimento de Goiás, Goiânia, Brazil
[5]Secretaria de Extrativismo e Desenvolvimento Rural Sustentável, Brasília, Brazil

Correspondence to: R. M. S. P. Vieira (rita.marcia@inpe.br)

Abstract. Approximately 57 % of the Brazilian northeast region is recognized as semi-arid land and has been undergoing intense land use processes in the last decades, which have resulted in severe degradation of its natural assets. Therefore, the objective of this study is to identify the areas that are susceptible to desertification in this region based on the 11 influencing factors of desertification (pedology, geology, geomorphology, topography data, land use and land cover change, aridity index, livestock density, rural population density, fire hot spot density, human development index, conservation units) which were simulated for two different periods: 2000 and 2010. Each indicator were assigned weights ranging from 1 to 2 (representing the best and the worst conditions), representing classes indicating low, moderate and high susceptibility to desertification. The results indicate that 94 % of the Brazilian northeast region is under moderate to high susceptibility to desertification. The areas that were susceptible to soil desertification increased by approximately 4.6 % (83.4 km^2) from 2000 to 2010. The implementation of the methodology provides the technical basis for decision-making that involves mitigating actions and the first comprehensive national assessment within the United Nations Convention to Combat Desertification framework.

1 Introduction

Drylands (arid, semi-arid and dry sub-humid areas) cover approximately 41 % of the Earth's surface and approximately 10 to 20 % of these regions are experiencing degradation processes (Deichmann and Eklundh, 1991; Reynolds et al., 2007), resulting in a decline in agricultural productivity, loss of biodiversity and the breakdown of ecosystems. According to the United Nations Conference to Combat Desertification (UNCCD), when land degradation happens in the world's drylands it often creates desert-like conditions. Land degradation occurs everywhere but is defined as desertification when it occurs in the drylands, resulting from various factors, including climatic variations and human activities (UN, 1979; UNCCD, 2012). The vegetation is composed of scrublands patches (high plant cover) interspersed with herbaceous patches (low plant cover)(Aguiar and Sala, 1999). This heterogeneity is induced by overgrazing, one of the main causes of the increase of bare soil that facilitates water and wind erosion and accelerates the desertification process (Cerdà and Lavee, 1999; Kröpfl et al., 2013; Pulido-Fernández et al., 2013; Ziadat and Taimeh, 2013).

Forty-four percent of global agricultural areas and almost 2 billion people are located over the drylands, and the majority (90 %) are in developing countries (D'Odorico et al., 2013). Overexploitation of natural resources in extremely vulnerable regions can accelerate land degradation and desertification process, affecting ecosystem functions and de-

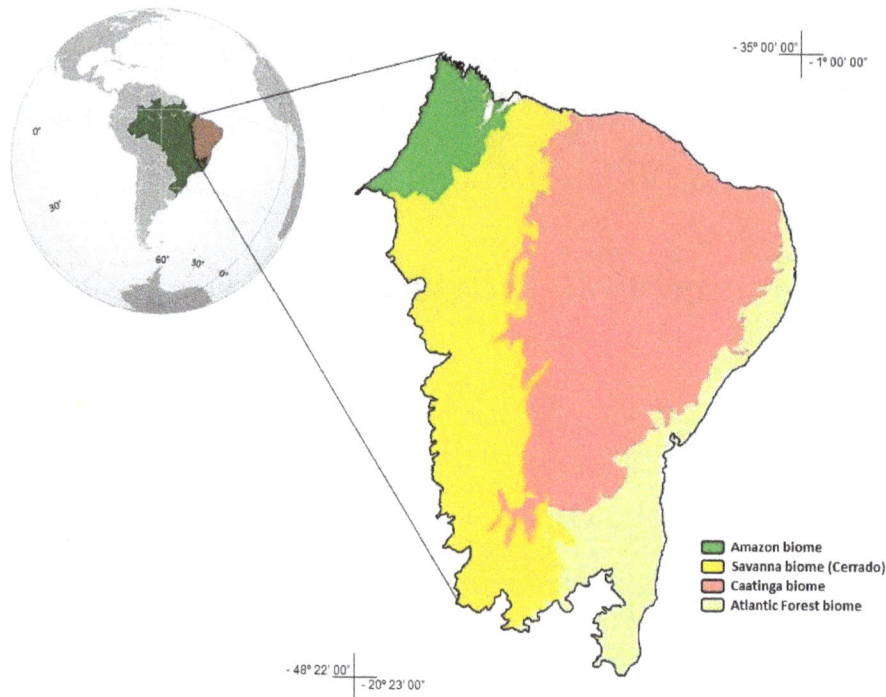

Figure 1. Study area location and its main biomes.

creasing productivity, biodiversity and landscape heterogeneity, and represents a major threat to the environment and human welfare (Mainguet, 1994; Reynolds and Stafford Smith, 2002; Montanarella, 2007; Salvati and Zitti, 2008; Cerdà et al., 2010; Santini et al., 2010; Kashaigili and Majaliwa, 2013; Pulido-Fernández et al., 2013; Bisaro et al., 2014).

In South America, the United Nations Convention to Combat Desertification report (ONU, 1997) concluded that, until 2025, one-fifth of the productive land could be affected by the desertification process. The most susceptible areas are located in Argentina, Bolivia, Chile, Mexico, Peru and Brazil (Arellano-Sota et al., 1996). In Brazil, the most critical desertification hot spots are located in the semi-arid northeast. In this region the climate is one of the factors that control the desertification process. Soil type, geology, landscape, vegetation, socioeconomic factors and land management also are considered important aspects of this process (IBGE, 2004). The main causes of desertification in this region are (i) deforestation to produce fuel wood and explore clay deposits; (ii) intensive land use employing poor agricultural methods, such as slash and burn, harvesting and land clearing; (iii) salinization; and (iv) extensive herding and overgrazing (Nimer, 1988).

Considering that the Brazilian semi-arid region is the world's most populous dry land region (Marengo, 2008), with more than 53 million inhabitants and a human population density of approximately 34 inhabitants per km^2 (IBGE, 2010), and that global climate change scenarios indicate that the region will be affected by increased aridity in the next

century, this area is seen as one of the world's most vulnerable regions to climatic change (IPCC, 2007).

The UNCCD recognizes desertification as an environmental problem with huge human, social and economic costs (Hulme and Kelly, 1993).

The most accepted definition currently states that desertification is land degradation in arid, semi-arid and dry subhumid areas resulting from various factors, including climatic variations and human activities (UN, 1979). Due to the complex social interactions and the biophysical processes, the identification and assessment of the desertification areas have been addressed through a multidisciplinary framework across different spatial and temporal scales (e.g., Prince et al., 1998; Diouf and Lambin, 2001; Thornes, 2004; Santini et al., 2010).

Several methods have been successfully applied for desertification analysis based on indicators and indices (Kepner et al., 2006; Sommer et al., 2011). For instance, the MEDALUS methodology, developed for the European Mediterranean environment, is widely used because of its simplicity and flexibility. The MEDALUS methodology is based on the environmentally sensitive area index (ESAI; Parvari et al., 2011; Salvati et al., 2011; Izzo et al., 2013; Jafari and Bakhshandehmeh, 2013). In order to identify areas potentially affected by land degradation, the method analyzes four main variables: climate, soil, vegetation and land management (Kosmas et al., 1999, 2006; Lavado Contador et al., 2009). It has been validated on regional and local scales (Basso et al., 2000; Brandt et al., 2003; Salvati and Bajocco, 2011) and was

Table 1. Indicators of land degradation/desertification.

Indicators	Scale/Spatial resolution	Period	Source
Geology	1 : 500 000/90 m	2010	INPE/MMA
Geomorphology	1 : 500 000/90 m	2010	INPE/MMA
Pedology	1 : 500 000/90 m	2010	INPE/MMA
Land use and land cover	1 : 500 000/90 m	2000 and 2010	INPE/MMA
Aridity index	1 : 500 000/5 km	1970–2000	INMET/CPTEC
Slope angle	1 : 500 000/90 m	2010	INPE
Rural population density	Per municipality	2000 and 2010	IBGE
Livestock density	Per municipality	2000 and 2010	IBGE
Fire hot spot density	1 : 500 000/1 km	1999–2003 and 2008–2012	CPTEC
Human development	Per municipality	2000 and 2010	FJP
Conservation units	1 : 500 000/90 m	2010	MMA

CPTEC – Center for Weather Forecasting and Climate Research; INMET – National Institute of Meteorology; FJP – João Pinheiro Foundation, INPE – National Institute For Space Research; MMA– Ministry of the Environment; IBGE – Brazilian Institute of Geography and Statistics.

applied to quantify the impact of mitigation policies against desertification (Basso et al., 2012).

Symeonakis et al. (2014) estimated the environmental sensitivity areas on the island of Lesvos (Greece) through a modified ESAI, which included 10 additional parameters related to soil erosion, groundwater quality, demographic and grazing pressure, for two dates (1990 and 2000). This study identified areas that are critically sensitive on the eastern side of the island mainly due to human-related factors that were not previously identified.

Although several studies have been conducted to detect desertification or to identify the drivers (indicators) of the process in critical hot spots in the Brazilian northeast (Matallo Júnior, 2001; Lemos, 2001; Sampaio et al., 2003; Aquino and Oliveira, 2012), there have been no studies addressing the entire region.

Crepani et al. (1996) developed a methodology based on the concept of the eco-dynamic principles, proposed by Tricart (1977), and on the relationship between morphogenesis and pedogenesis to identify areas that are susceptible to soil erosion. The author provided an integrated view of the physical environment and the conceptual basis for developing human–nature relationships. However, this study did not include socioeconomic and management indicators as parameters that can influence soil loss.

Therefore, this paper presents a novel approach which integrates the MEDALUS project and the methodology developed by Crepani et al., 1996 to identify areas that are susceptible to desertification in the northeastern region of Brazil and the northern regions of the states of Minas Gerais and Espírito Santo by combining social, economic and environmental indices. This study was conducted considering two reference periods: early 2000s and 2010. The results will be useful for providing basic information for the diagnosis and prognosis of desertification in the region and providing subsidies for the technical support for mitigation and adaptation actions.

2 Study area

The study area is located in the equatorial zone (1–21° S, 32–49° W), totaling an area of 1 797 123 km^2, which corresponds to 20 % of the Brazilian territory (Fig. 1).

The climatology of the northeast of Brazil includes three different rainfall regimes: (i) in the south-southwest area, the rainy season occurs from October through February, which is associated with the displacement of cold fronts coming from the south; (ii) in the north of the region, rainfall occurs from February to May, which is associated with the southward movement of the Intertropical Convergence Zone; and finally, (iii) in a narrow area that is close to the coast at the east, the rainy season occurs from April through August, triggered by temperature differences between the oceans and the sea shore (Kousky, 1979; Marengo, 2008). The evaporation rate in the region is very high and can reach 1000 mm yr^{-1} in the coastal region and up to 2000 mm yr^{-1} in the interior (IICA, 2001), based on 11 stations distributed in the semi-arid region and on historical series (Molle, 1989). Annual evaporation average is 2700 to 3300 mm, with the highest values occurs from October to December and the lowest from April to June.

Because of the high evaporation rates and the short duration of the wet season, most of the rivers are temporary, and flash floods occur only during the rainy season (MMA-IBAMA, 2010).

In the northeast region of Brazil, natural vegetation includes rainforests, riparian forests, savannas and montane forests, among others (Foury, 1972). However, the natural vegetation that dominates 62 % of Brazilian semi-arid region is *caatinga* (MMA, 2007). *Caatinga* vegetation is composed of shrubs and small trees, usually thorny and decidu-

Table 2. Land use and land cover classes.

Land use and land cover classes	Description
Evergreen forest	Evergreen broadleaf closed/open
Water body	Rivers, streams, canals, lakes, ponds or puddles
Beach	Beach area
Seasonal forest	Type of forest characterized by trees that seasonally shed their leaves
Restinga	Herbaceous and arbustive vegetation, distributed along the coastal zone
Urban area	Cities and towns
Savanna (Cerrado)	Grasslands, shrublands and woodlands
Fluviomarine	Mangrove
Alluvial	Similar characteristics to the evergreen forest but differs because of its physiographical position (alluvial plain)
Campo Maior complex	Prevailingly herbaceous vegetation; presence of carnaubais (coconut type) in flood plains
Steppe Savanna (*caatinga*)	Vegetation typical of the Brazilian semi-arid region characterized by xeric shrubland and thorn forest that primarily consists of small, thorny trees that shed their leaves seasonally
Shrimp farming	Producing shrimp
Pasture	Pasture area (both natural and planted)
Agriculture	Cultivated areas (temporally and permanent crops)
Baixada Maranhense	Low plain area that is flooded in the rainy season, creating large lagoons
Bare soil	Bare soil areas without natural covering
Dunes	Sand dunes along the coast
Rock outcrops	Exposed rock areas
Salt fields	Areas where sea salt is produced

ous, that lose their leaves in the early dry season. *Caatinga* is a highly dynamic ecosystem that responds quickly to climatic conditions. The dominant factor that controls the structure and distribution of vegetation is the precipitation, with an annual mean of 500–800 mm and high spatial and temporal variability (Hastenrath and Heller, 1977; Oliveira et al., 2006). *Caatinga*, in comparison with other xeric areas in South America, presents climatic distinctiveness that resulted in numerous important morphological and physiological adaptations to aridity by many species of plants (Mares et al., 1985). Nowadays, more than 10 % of the semi-arid area has already undergone a very high degree of environmental degradation, being susceptible to desertification (Oyama and Nobre, 2004).

3   Methods

To identify areas susceptible to desertification, we evaluated 11 indicators of susceptibility to desertification (Table 1) based on previous studies of the area (Vasconcelos Sobrinho, 1978; Ferreira et al., 1994; Matallo Júnior, 2001; Lemos, 2001). From Table 1, each indicator was sub-divided into various uniform classes. Each class received a weight factor, related to the potential influence on desertification process, that ranged between 1 (low susceptibility) and 2 (high susceptibility), producing 11 susceptibility maps (SM). The weight factors were assigned based on previous analyses of the literature (Crepani et al., 1996, Torres et al., 2003; Alves,

2006; Santini et al., 2010; Symeonakis et al., 2013). These indicators were grouped into two groups as described below.

3.1   Physical indicators

3.1.1   Slope data, geology, geomorphology and pedology maps

The basic topographic data set used was a 30 m spatial resolution digital elevation model (DEM), derived from TOPA-DATA, which was developed based on Shuttle Radar Topography Mission data (Farr and Kobrick, 2000; Van Genderen et al., 1987). The DEM was processed to derive elevation and slope angle and used to identify breakline surface discontinuities where changes occurred in the vertical curvature which are linked to lithological, pedological, geomorphological and vegetation characteristics. Therefore, breaklines often indicate the boundary between adjacent units on a map.

Geomorphology and geology maps were extracted from RADAMBRASIL Project (Projeto RADAMBRASIL 1973–1981) and from the Geological Survey of Brazil (CPRM – Companhia de Pesquisa de Recursos Minerais), both with a spatial scale of 1 : 1 000 000. These basic maps were digitized and then rescaled to the scale of 1 : 500 000 using the processed DEM, following the procedure suggested by Valeriano and Rossetti (2012).

Soil maps (EMBRAPA, 1999) were rescaled from 1 : 5 000 000 to 1 : 500 000 based on the topographic map information. The Brazilian System of Soil Classification is based on soil pedogenetic characteristics, and also uses morphological, physical, chemical and mineralogical criteria (Camargo et al., 1987). The system is hierarchical and "opened" which allows the inclusion of new classes and enables the classification of all soil types that occur in Brazil.

3.1.2 Aridity index (AI)

The aridity index is considered to be one of the most important indicators of areas that are susceptible to desertification (UNESCO, 1979; Sampaio et al., 2003). In this study, the AI was obtained by the following formula:

$$AI = P/PET, \tag{1}$$

where P is the precipitation and PET is the potential evapotranspiration calculated using the Penman–Monteith equation (Monteith, 1965).

3.2 Socioeconomic indicators

3.2.1 Land use and land cover maps

Between 2000 and 2010, northeast Brazil was the fastest-growing economic (IBGE, 2010) region of the country and has been undergone severe land use and land cover changes. Therefore, it is crucial to asses if the combination of both effects – fast growth and severe land use changes – have impacted the susceptibility to desertification/degradation of the region. Thus, 90 Landsat-TM images (30 m resolution) of the dry period (July to September) of 2010 and 2011 were selected and geocoded based on the orthorectified Landsat images from the Global Land Cover Facility (NASA). These images were used to update the land use and land cover map derived by the ProVeg Project (Vieira et al., 2013), which was based on Landsat images from 2000. Additionally, land use and land cover maps from the PROBIO (Project for Conservation and Sustainable Use of Biological Diversity) (MMA, 2007) project, with a spatial scale of 1 : 500 000, and high-resolution images from Google Earth were used as auxiliary data. The land use and land cover classes mapped in this study are presented on Table 2.

3.2.2 Rural population density

These data were extracted from IBGE census data (available at http://downloads.ibge.gov.br/downloads_estatisticas.htm). The rural area boundaries and the number of inhabitants were defined considering information for both 2000 and 2010.

3.2.3 Livestock density

Livestock density data, based on the total number of cattle and goat herds per municipality in 2000 and 2010, were extracted from IBGE agricultural census.

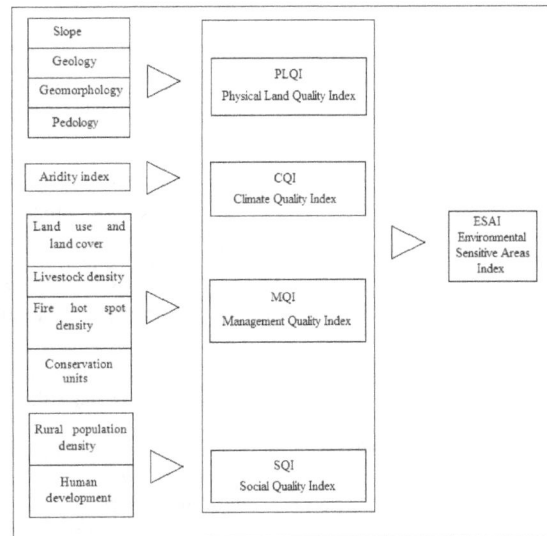

Figure 2. Combination of indicators for the determination of the ESAI; adapted from Benabderrahmane and Chenchouni (2010).

3.2.4 Fire hot spot density

Fire hot spot data were obtained from INPE's Fire Monitoring Project (INPE, 2012). Fire hot spot density maps were derived for two periods: (i) the average number of satellite hot spots from 1999 to 2003, which was used to represent the year 2000, and (ii) the average for the period 2008 to 2012, which was used as an indicator for the year 2010. To convert point data to continuous smooth surfaces, Kernel density estimation was applied to fire hot spots point using a 50 km radius (Koutsias et al., 2004; de la Riva et al., 2004). This estimator improves visualization and enables comparison with continuous environmental variables (Silverman, 1986).

3.2.5 Conservation units

Conservation unit data were obtained from the Ministry of the Environment. In the present study, the number of conservation units for 2000 and 2010 did not change. There are two basic categories of conservation units: integral protection units and the conservation units for sustainable use (Rocco, 2002). The former forbids the use of natural resources and includes national parks, ecological stations, biological reserves and wildlife sanctuaries. The latter includes national forests, extractive reserves and sustainable development reserves where the sustainable use and the management of natural resources are allowed under certain regulations.

3.2.6 Human development index (HDI)

The HDI indicators for the years 2000 and 2010 were obtained from the João Pinheiro Foundation (http://atlasbrasil.org.br/2013/). Population data, as well as HDI, are essential to understand the territorial dynamics. The calculation

Table 3. Classes and weights of parameters used for environment quality assessment.

Susceptibility class	Geomorphological types and features	Susceptibility weight
Low	Terrace formations structural and flat tops landforms; the roughness of the topographic relief is characterized by being very slightly dissected; flat relief and planation surface without intense erosive action.	1.00
	Flat and convex tops landforms; the roughness of the topographic relief is characterized by being lightly to moderately dissected; flat relief and planation surface with significant erosive action; slightly undulating relief with gentle slopes.	1.25
Moderate	Convex tops landforms; the roughness of the topographic relief is characterized by being moderately dissected; undulating relief with steep slopes.	1.50
High	Convex and sharp tops; the roughness of the topographic relief is characterized by being highly dissected; strong undulating relief with very steep slopes; karstic relief.	1.75
Geology type		
Low	Quartzite, metaquartzite, banded iron formation, metagranodiorite, metatonalite	1.00
	Rhyolite, granite, dacite, metasyenogranite, monzogranite, syenogranite, magnetite, metadiorite, metagabbro	1.05
	Granodiorite, quartz-diorite, granulite	1.10
	Migmatite, gneiss, orthogneiss	1.15
	Nepheline syenite, trachyte, quartz-monzonite, quartz-syenite	1.20
	Andesite, basalt	1.25
	Gabbro, anorthosite	1.30
Moderate	Biotite, quartz-muscovite, itabirite, metabasite, mica schist	1.35
	Amphibolite, kimberlite	1.40
	Hornblende, tremolite	1.45
	Schists	1.50
High	Phyllite, metasiltite	1.55
	Slate rock, metargillite	1.60
	Marble	1.65
	Quartz arenites (sandstones), ortoquartizites	1.70
	Conglomerates	1.75
	Arkoses	1.80
	Siltstones, Argillite	1.85
	Shale	1.90
	Limestone, dolostone	1.95
	Unconsolidated sediments (colluvial and alluvial deposits, sandy deposits, etc.)	2.00

Table 3. Continued.

Susceptibility class	Geomorphological types and features	Susceptibility weight
	Soil type (EMBRAPA, 1999)	
Low	Latosols, organic soils, hydromorphic soils, humic soils	1.00
Moderate	Podzolic soils, brunizem, planosol, brunizem, structured dusky red earth	1.33
High	Cambisol	1.66
	Non-cohesive soils, immature soils, laterites, rocky outcrop	2.00
	Slope (%)	
Low	2–6	1.00
Moderate	6–18	1.50
High	>18	2.00

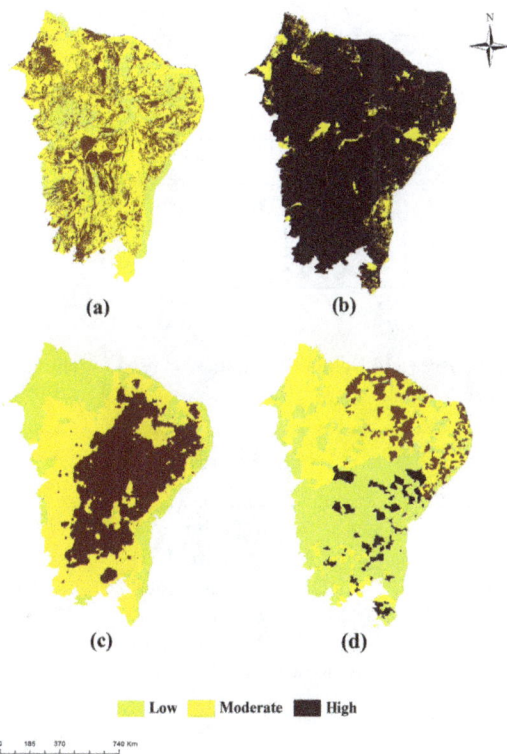

Figure 3. (a) Physical land quality index; (b) management quality index; (c) climate quality index; (d) social quality index.

of the HDI includes three kinds of data: longevity, education and economic income. HDI scale ranges from 0 to 1, where values from 0 to 0.49 represent low HDI, 0.5 to 0.59 medium HDI, 0.60 to 0.79 high HDI, and 0.8 to 1.0 very high HDI. According to the Atlas of Human Development of Brazil 2013, developed by a partnership between United Nations Development Program (UNDP, 2010), the Institute of

Applied Economic Research and the João Pinheiros Foundation the Brazil have reduced the inequalities between its sub-indices of education, income and longevity in 2010.

3.3 Environmentally sensitive area index

The methodology used to map susceptible areas to desertification was based on the MEDALUS methodology (Mediterranean Desertification and Land Use, by Kosmas et al., 1999), which uses geometric means of environment-state and response indicators. Each index is estimated from a combination of indicators of desertification, which depends on geology, pedology, land management, human occupation and conservation policies (Fig. 2).

These maps were then grouped according to four quality indexes (Kosmas et al., 1999).

– Physical land quality index (PLQI):

$$\text{PLQI} = (I_s \cdot I_g \cdot I_{gm} \cdot I_d)^{1/4}, \tag{2}$$

where I_s is the soil SM, I_g is the geology SM, I_{gm} is the geomorphology SM and I_d is the slope SM.

– Management quality index (MQI):

$$\text{MQI} = (I_{uc} \cdot I_p \cdot I_{fq} \cdot I_{ucob})^{1/4}, \tag{3}$$

where I_{uc} is conservation units SM, I_p is the livestock density SM, I_{fq} is the fire density SM and I_{ucob} is the land use and land cover SM.

– Climate quality index (CQI):

$$\text{CQI} = I_a, \tag{4}$$

where I_a is the aridity index SM.

Table 4. Classes and weights of parameters used for management quality assessment.

Susceptibility class	Land use/land cover change classes	Susceptibility weight
Low	Evergreen forest, water body, beach, urban area	1.00
	Deciduous forest	1.40
	Restinga	1.45
Moderate	Savanna (Cerrado), fluviomarine pioneer, alluvial pioneer	1.50
	Complex of Campo Maior, Baixada Maranhense	1.55
	Caatinga	1.60
	Shrimp farming, pasture	1.80
	Agriculture	1.90
High	Bare soil, dunes, rocky outcrop	2.00
	Livestock density data	
Low	0 to 30	1.00
Moderate	30 to 75	1.50
High	above 75	2.00
	Fire density data	
Low	0 to 1000	1.00
Moderate	1000 to 2000	1.50
High	above 2000	2.00
	UC data	
Low	Integral protection units	1.00
Moderate	Conservation units for sustainable use	1.50
High	Without conservation unit	2.00

– Social quality index (SQI):

$$\text{SQI} = (I_{\text{HDI}} \cdot I_{\text{Pop}})^{1/2}, \qquad (5)$$

where I_{HDI} is the human development index SM and I_{pop} is rural population density SM.

The geo-database was developed using SPRING (Câmara, et al., 1996).

Finally, to obtain an ESAI, the geometric mean is calculated among the variables inside each factor through the following equation:

$$\text{ESAI} = (\text{PLQI} \cdot \text{MQI} \cdot \text{CQI} \cdot \text{SQI})^{1/4}. \qquad (6)$$

Based on these calculations, three types of ESAs were assigned: (a) low-susceptibility areas (ESAI $1.00 \geq 1.25$), (b) moderate-susceptibility areas (ESAI $1.25 \geq 1.50$) and (c) high-susceptibility areas (ESAI > 1.50).

3.4 Validation

In this study, the 2010 susceptibility map was validated using the method proposed by Van Genderen et al. (1978). This method assumes that the probability of making f interpretation errors when taking x samples from a remote-sensing-based classification map follows a binomial probability distribution function. The method allows the determination of

Table 5. Classes and weights of parameters used for climate quality assessment.

Susceptibility class	Climate types	Susceptibility weight
Low	Wet sub-humid (AI above 0.65)	1.00
Moderate	Dry sub-humid (AI between 0.51 to 0.65)	1.50
High	Semi-arid (AI between 0.21 to 0.50)	2.00

the minimum sample size required for validating the map, avoiding the risk of accepting a map with low accuracy.

Based on this methodology, 110 random samples were selected from the low-, medium- and high-susceptibility classes and compared with high-resolution images from Google Earth (Ginevan, 1979; Congalton and Green, 1999) and in situ images. Thus, the points from high-susceptibility classes were compared to their corresponding images to observe the degraded areas of exposed soil.

4 Results and discussion

This work presents the first effort to identify the areas that are most susceptible to desertification in the semi-arid region of Brazil through a system that enables continuous and integrated analysis of the factors that provide the best explanation of the desertification processes.

The weight factors assigned to each indicator are described in Tables 3, 4, 5 and 6.

Analyses from 11 indicators stress that areas with predominantly humid and sub-humid climate are potentially susceptible to desertification due to inadequate soil management, which is a key factor for adaptation and mitigation of climate change (IPCC, 2007).

On the MEDALUS methodology, variables like HDI and conservation units were not included. However, these two indicators were considered important in the semi-arid region Brazil based on the fact that the region has relatively low development indexes and several inadequate land uses practices, and previous studies in other regions of Brazil (Trancoso et al., 2010) have shown that conservation enforcement in protected areas is crucial for avoiding degradation.

4.1 Physical land quality index

In terms of soil types, the northeast and southern portions of the region are largely covered by Podzolic soils (23 %) that are more prone to erosion due to the low permeability of the B clayey horizon. Lithosols (21 % of the area) occur in the semi-arid region, associated with rock outcrops. Lastly, the Latosols (18 %) dominate the northwest region, associated with Savanna vegetation, where the relief is plain and favors the mechanized agriculture increasing soil compaction (Cavaliere et al., 2006; Araújo et al., 2007).

The eastern part of the study area is dominated by crystalline rocks. However, there is a predominance of sedimentary basins located in coastal regions and in the western part of the study area. To the south of the region, extensive karst formations can be found. Most of the study area consists of flat and undulating relief, but the occurrence of steep formations and the presence of inselbergs have also been noted.

According to the spatial distribution of the physical land quality index (Fig. 3a), 52 % of the study area has a moderate susceptibility. The areas with high susceptibility are on soil types that are more vulnerable to erosion processes, such as podzols (23 %) and lithosols (21 %).

4.2 Management quality index

The analyses showed an increase of 3 % of the area with high susceptibility for a period of 11 years between 2000 and 2010 (Table 7). Areas with high susceptibility reached 87 % (1 571 033 km²) of the studied area in 2000, while in 2010 the percentage increased to 90 % (1 622 716 km²). Among the factors that might be contributing to the increase in area

Table 6. Classes and weights of the parameters used for social quality assessment.

Susceptibility class	Human development index Per municipality	Susceptibility weight
Low	0.70 to 1.00	1.00
Moderate	0.60 to 0.70	1.50
High	0 to 0.60	2.00
	Rural population density	
Low	0 to 25	1.00
Moderate	25 to 50	1.50
High	above 50	2.00

are shrimp farming, agriculture, livestock and fire hot spots. Analyzing the results of use land and land cover, it is possible to observe that the natural vegetation is being replaced by pastures and agriculture. According to the land use/cover map developed by Vieira et al. (2013), the typical vegetation of the semi-arid of Brazil, known as *caatinga*, has been replaced by pasture and agricultural activities. Approximately 40 % of the *caatinga* has been converted to these uses, and the remaining area is being transformed at a rate of 0.3 % per year (IBAMA/MMA, 2010).

In recent years, agribusiness has become one of the most dynamic segments in the northeastern states with the production of fruits, such as papayas, melons, grapes, watermelons, pineapples and mangos. The activities related to shrimp farming covered an area of 69.7 km² in 2000, which increased to 136.7 km² in 2010. Northeastern Brazil is responsible for 94 % of all shrimp production in Brazil (Ferreira, 2008).

Even though areas located in sub-humid and humid areas are less vulnerable from a climatic point of view, they are susceptible to land degradation and desertification due to inadequate land use and management. In the northwestern portion of study area, for example, the deforestation is one of main causes to land degradation. The natural vegetation is being replaced by pasture and agriculture, increasing from 106 568 in 2000 to 143 323 km² in 2010 and from 10 425 in 2000 to 20 100 km² in year 2010. In livestock areas of the region, fire is routinely used as a method for clearing land from bushes and for the re-establishment of pasture (Miranda, 2010). In the present work, the number of fire hot spot increased from 26 181 in 2000 to 73 429 in 2010.

4.3 Climate quality index

According to the climate quality index (Fig. 3c, Table 7), 42 % of the area is a highly susceptible semi-arid climate, while 38 % is classified as moderate susceptible dry sub-humid. Finally, 20 % of the area, where the climate is sub-humid to humid, is considered as having a low susceptibil-

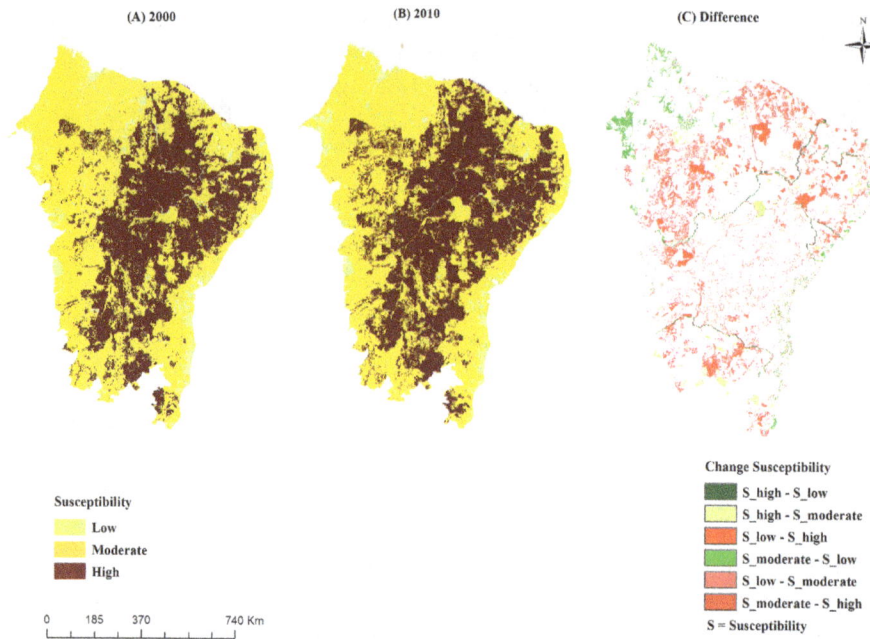

Figure 4. Environmental susceptibility area for **(a)** 2000 and **(b)** 2010. **(c)** Difference between 2000 and 2010.

Table 7. Percentage of the land area covered by each susceptibility class of the four quality indices in 2000 and 2010.

Index	Susceptibility class	2000 (%)	2010 (%)
Physical land quality index (PLQI)	Low	24.5	24.5
	Moderate	52.7	52.7
	High	22.9	22.9
Management quality index (MQI)	Low	1.0	0.8
	Moderate	11.6	8.9
	High	87.4	90.3
Climate quality index (CQI)	Low	19.5	19.5
	Moderate	38.2	38.2
	High	42.3	42.3
Social quality index (SQI)	Low	42.4	48.1
	Moderate	34.8	32.9
	High	22.8	19.0

ity. From a climatic point of view, rainfall exceeds 1250 mm in the coastal region annual. To the west, annual rainfall is around 1500 mm, while in the semi-arid interior annual rainfall is less than 1000 mm, ranging from 350 to 750 mm (IBGE, 1996).

4.4 Social quality index

The social quality index showed that 42 % of the region had low susceptibility in 2000, while the value increased to 48 % in 2010 (Table 7). According to IBGE (2010), the HDI improved in this period in response to the country's economic growth. The region is marked by socioeconomic inequality;

the highest HDI is in the northern (0.682) and eastern (0.684) regions and the lowest is in the northeast (0.631).

4.5 Susceptibility areas to desertification

The areas susceptible to desertification in the Brazilian semi-arid region for both 2000 and 2010, as well as the changes that occurred between these periods, are presented in Fig. 4. The results showed that 94 % of the semi-arid region is moderately (59.4 %) or highly (35 %) environmentally sensitive for both periods: 2000 (94.4 %) and 2010 (94 %). High-sensitivity areas increased from 35 to 39.6 %, which corresponds to 83 348 km². Moderate regions decreased almost 5 % (89 856 km²), while low-sensitivity areas increased from

5.6 % (2000) to 6 % (2010). The most susceptible areas were mapped, both in 2000 and 2010, in the central-eastern regions that include the four desertification hot spots officially recognized by the Brazilian Ministry of the Environment: Gilbués (PI), Irauçuba (CE), Cabrobó (PE) and Seridó (RN) (MMA, 2007).

The results also showed several areas with high susceptibility, specifically in the south of the study area. According to the field survey, desertification in this area is increasing due to inadequate soil management and indiscriminate deforestation (MMA, 2005). The human activities are the dominant factor for desertification expansion. However, in the northwest of the study area, several spots showed low susceptibility. Government incentives in the last decades have turned this region into a tropical fruit producer (Araujo and Silva, 2013).

From these results, it is clear that the management quality index is the main driver of desertification in the study region (Fig. 3b). Therefore, mitigation actions for reducing the susceptibility to degradation in the region depend heavily on changes in management practices towards more sustainable land use.

Finally, it is important to note that the validation results indicated that the environment susceptibility map has an accuracy of 85 %, which is considered acceptable due to the extent and complexity of the study area.

5 Final considerations

The environmentally sensitive area index calculated in the present study allowed a better understanding of the degradation/desertification process in the Brazilian semi-arid region. The study showed that desertification susceptibility ranges from moderate to high in the Brazilian semi-arid region.

From a climatic point of view, the humid and sub-humid areas have low vulnerability. However, when management issues associated with land use are taken into consideration, these areas become potentially susceptible to degradation.

The northwestern part of the study area is highly susceptible to land degradation due to inadequate soil management associated with intensive agricultural land expansion. In the last 50 years, the area received millions of migrants looking for better opportunities created by agriculture expansion.

This study is the first effort to produce a comprehensive diagnosis of the desertification processes for the entire region and combines the existing experience from previous studies in the region with a consolidated methodology. Additionally, new indicators were included in the methodology of this study, such as HDI (social indicator) and conservation units (management indicator), because previous knowledge indicated that they would be relevant in the study area.

In addition, it was possible to obtain a database with biophysical and social information on the same scale and resolution, which allowed the integrated analysis of the desertification indicators.

One of the major issues facing humanity today is the development of knowledge in regards to the occupation of land in regions affected by desertification in a sustainable way. Then it becomes critical to define adaptation alternatives for living in semi-arid regions. Furthermore, it can be applied in multi-scale studies, showing the magnitude of the risk in different areas and the factors that may contribute to triggering the process. The approach was based on the use of indicators that are routinely surveyed in the area, allowing for continuous monitoring of the desertification processes. The proposed methodology proved to be a useful, timely and cost-effective tool to identify areas that are susceptible to degradation/desertification.

Acknowledgements. The authors are grateful to the Brazilian Ministry of the Environment and Inter-American Institute for Cooperation on Agriculture (IICA) for providing logistical and financial support, to Soil EMBRAPA, from Recife, for supplying the soil data and to the National Council for Scientific and Technological Development.

Edited by: A. Cerdà

References

Aguiar, M. R. and Sala, O. E.: Patch structure, dynamics and implications for the functioning of arid ecosystems, Trends Ecol. Evol., 14, 273–277, 1999.

Aquino, C. M. S. and Oliveira, J. B. V.: Avaliação de indicadores biofísicos de degradação/desertificação no núcleo de São Raimundo Nonato, Revista Equador (UFPI), 1, 44–59, 2012.

Alves, H. P. F.: Vulnerabilidade socioambiental na metrópole paulistana: uma análise sociodemográfica das situações de sobreposição espacial de problemas e riscos sociais e ambientais, Revista Brasileira de Estudos de População, 23, 43–59, 2006.

Araujo, G. J. F. de and Silva, M. M da.: Crescimento econômico no semiárido brasileiro: o caso do polo frutícola Petrolina/Juazeiro, Caminhos de Geografia, Uberlândia, 14, 246–264, 2013.

Araujo, R., Goedert, W. J., and Lacerda, M. P. C. Qualidade do solo sob diferentes usos e sob Cerrado nativo, Revista Brasileira de Ciência do Solo, 31, 1099–1108, 2007.

Arellano-Sota, C., Frisk, T., Izquierdo, J., Prieto-Celi, M., Thelen, K. D., and Vita, A.: FAO/UNEP – Program on desertification control in Latin America and the Caribbean, Desertification Control Bulletin, Nairobi, 29, 56–62, 1996.

Benabderrahmane, M. C. and Chenchouni, H.: Assessing environmental sensitivity areas to desertification in eastern Algeria using Mediterranean Desertification and Land Use "MEDALUS" model, Int. J. Sustain. Water Environ. Sys., 1, 5–10, 2010.

Basso, B., De Simone, L., Cammarano, D., Martin, E. C, Margiotta, S., Grace, P. R., Yeh, M. L., and Chou, T. Y.: Evaluating Responses to Land Degradation Mitigation Measures in Southern Italy, Int. J. Environ. Res., 6, 367–380, 2012.

Basso, F., Bove, E., Dumontet, S., Ferrara, A., Pisante, M., Quaranta, G., and Taberner, M.: Evaluating environmental sensitivity at the basin scale through the use of geographic information systems and remotely sensed data: an example covering the Agri basin (Southern Italy), Catena, 40, 19–35, 2000.

Bisaro, A., Kirk, M., Zdruli, M., Zdruli, P., and Zimmermann, W.: Global drivers setting desertification research priorities: insights from a stakeholder consultation forum, Land Degrad. Dev., 25, 5–16, doi:10.1002/ldr.2220, 2014.

Brandt, J., Geeson, N., and Imeson, A.: A desertification indicator system for Mediterranean Europe, DESERTLINKS Project, UK, 2003.

Câmara, G., Souza, R. C. M., Freitas, U. M., and Garrido, J.: SPRING: Integrating Remote Sensing and GIS by Object-Oriented Data Modelling, Computer & Graphics, 20, 395–403, 1996.

Camargo, M. N., Klamt, E., and Kauffman, J. H.: Soil classification as used in Brazilian soil surveys, ISRIC, Wageningen, Annual Report, 1987.

Cavalieri, K. M. V., Tormena, C. A., Vidigal Filho, P. S., Gonçalves, A. C. A., and Costa, A. C. S.: Efeitos de sistemas de preparo nas propriedades físicas de um Latossolo Vermelho distrófico, Revista Brasileira de Ciência do Solo, Viçosa, 30, 137–147, 2006.

Cerdà, A. and Lavée, H.: The effect of grazing on soil and water losses under arid and mediterranean climates, Implications for desertification, Pirineos, 153–154, 159–174, 1999.

Cerdà, A., Lavee, H., Romero-Diaz, A., Hooke, J., and Montanarella, L.: Soil erosion and degradation in Mediterranean-type ecosystems preface, Land Degrad. Dev., 21, 71–74, doi:10.1002/ldr.968, 2010.

Congalton, R. G. and Mead, R. A.: A quantitative method to test for consistency and correctness, in: Photointerpretaion, Photogrametric Engineering and Remote Sensing, 49, 69–74, 1983.

Congalton, R. G. and Green, K.: Assessing the accuracy of remotely sensed data: principles and practices, Boca Raton: CRC Lewis Press, 137 pp., 1999.

Crepani, E., Medeiros, J. S., Azevedo, L. G, Duarte V., Hernandez, P., and Florenzano, T.: Curso de Sensoriamento Remoto Aplicado ao Zoneamento Ecológico-Econômico, INPE, São José dos Campos-SP, 1996.

De la Riva, J., Perez-Cabello, F., Lana Renault, N., and Koutsias, N.: Mapping forest fire occurrence at a regional scale, Remote Sens. Environ., 92, 363–369, 2004.

Deichmann, U. and Eklundh, L.: Global Digital Datasets for Land Degradation Studies: A GIS Approach, GRID Case Study Series, 4, 1991.

Departamento Nacional da Produção Mineral – DNPM: Projeto RADAMBRASIL – Levantamento dos Recursos Naturais, Vol. 1, 2, 3, 4 e 21, Geomorfologia, 1973–1981.

Diouf, A. and Lambin, E. F.: Monitoring land-cover changes in semi-arid regions: remote sensing and field observations in the Ferlo, Senegal, J. Arid Environ., 48, 129–148, doi:10.1006/jare.2000.0744, 2001.

D'Odorico, P., Carr, J. A., Laio, F., Ridolfi, L., and Vandoni, S.: Feeding humanity through global food trade, 2, 458–469, doi:10.1002/2014EF000250, 2013.

EMBRAPA: Centro Nacional de Pesquisa de Solos, Sistema brasileiro de classificação de solos, Brasília, Embrapa Produção de Informação, Rio de Janeiro, 412 pp., 1999.

Farr, T. and Kobrick, M.: Shuttle radar topography mission produces a wealthy of data, American Geophysical Union Eos, 81, 583–585, doi:10.1029/EO081i048p00583, 2000.

Ferreira, D. G, Melo, H. P., Neto, F. R. R, Nascimento, P. J. S., and Rodrigues, V.: Avaliação do quadro de desertificação no Nordeste do Brasil: diagnósticos e perspectivas, in: Conferência Nacional de Desertificação, Fundação Esquel Brasil, Brasília, 1994, 7–55, 1994.

Ferreira, D. G., Melo, J. V., and Costa Neto L. X.: Influência da Carcinicultura sobre a salinização do solo em áreas do município de Guamaré/RN, Holos, 24, 72–80, 2008.

Foury, A. P.: As matas do nordeste brasileiro e sua importância econômica, Boletim de Geografia, 31, 14–131, 1972.

Ginevan, M.: Testing land use map accuracy: another look, Photogrammetric Engineering and Remote Sensing, 45, 1371–1377, 1979.

Hastenrath, S. and Heller, L.: Dynamics of climatic hazards in northeast Brazil, Quart. J. Roy. Meteor. Soc, 103, 77–92, 1977.

Hulme, M. and Kelly, M.: Exploring the links between: Desertification and Climate Change, Environment, 35, 5–11, doi:10.1080/00139157.1993.9929106, 1993.

IBAMA/MMA: Monitoramento do desmatamento nos biomas brasileiros por satélites, Monitoramento do Bioma Caatinga 2002 a 2008, Centro de Sensoriamento Remoto – CSR/IBAMA: http://www.ambiente.gov.br/estruturas/sbf_chm_rbbio/_arquivos/relatrio_tcnico_caatinga_72.pdf (last access: 10 March 2012), 2010.

IBGE: Recursos Naturais e Meio Ambiente: Uma visão do Brasil, Departamento de Recursos Naturais e Estudos Ambientais, 2, Rio de Janeiro, 208 pp., 1996.

IBGE: Estudos de pesquisas e informações geográficas, Coordenação de Recursos Naturais e Estudos Ambientais e Coordenação de Geografia, 4, 389 pp., 2004.

IBGE: Estados: População, Instituto Brasileiro de Geografia, Rio de Janeiro: http://censo2010.ibge.gov.br/apps/atlas/ (last access: 10 April 2013), 2010.

IPCC: Climate change 2007: the physical science basis, Contribution of Working Group I to the Fourth Assessment Report of the Intergovernmental Panel on Climate Change, edited by: Solomon, S., Qin, D., Manning, M., Chen, Z., Marquis, M., Averyt, K. B., Tignor, M., and Miller, H. L., Cambridge University Press, Cambridge, United Kingdom and New York, NY, USA, 2007.

INPE: Instituto Nacional de Pesquisas Espaciais, Portal do Monitoramento de Queimadas e Incêndios: http://www.inpe.br/queimadas, last access: 2 August 2012.

IICA: Instituto Interamericano de Cooperação para Agricultura, Projeto Áridas: https://books.google.com.br/books?id=NO8qAAAAYAAJ, last access: 27 March 2011.

Izzo, M., Araujo, N., Aucelli, P. P. C., Maratea, A., and Sánchez, A.: Land sensitivity to desertification in the Dominican Republic: an adaptation of the ESA methodology, Land Degrad. Dev., 24, 486–498, 2013.

Jafari, R. and Bakhshandehmehr, L.: Quantitative mapping and assessment of environmentally sensitive areas to desertification in central Iran, Land Degrad. Dev., 2013.

Kashaigili, J. J. and Majaliwa, A. M.: Implications of land use and land cover changes on hydrological regimes of the Malagarasi

River, Tanzania, Journal of Agricultural Sciences and Application (JASA), 2, 45–50, doi:10.1002/ldr.2241, 2013.

Kepner, W. G., Rubio, J. L., Mouat, D. A., and Pedrazzini, F. (Eds.): Desertification in the Mediterranean Region: A Security Issue, NATO Security through Science Series-C, Environmental Security, Springer: Dordrecht, the Netherlands, 2006.

Kosmas, C., Kirkby, M., and Geeson, N.: Manual on key indicators of desertification and mapping environmentally sensitive areas to desertification, European Commission: Brussels, 1999.

Kosmas, C., Tsara, M., Moustakas, N., Kosma, D., and Yassoglou, N.: Environmental sensitive areas and indicators of desertification In Desertification in the Mediterranean region, A security issue, NATO Security Through Science Series, 3, 2006.

Koutsias, N., Kalabokidis, K. D, and Allgower, B.: Fire occurrence patterns at landscape level: beyond positional accuracy of ignition points with kernel density estimation methods, Nat. Resour. Modell., 17, 359–376, doi:10.1111/j.1939-7445.2004.tb00141.x, 2004.

Kousky, V. E.: Frontal influences on Northeast Brazil, Mon. Weather Rev., 107, 1140–1153, 1979.

Kröpfl, A. I., Cecchi, G. A., Villasuso, N. M., and Distel, R. A.: Degradation and recovery processes in Semi-Arid patchy rangelands of northern Patagonia, Argentina, Land Degrad. Dev., 24, 393–399, doi:10.1002/ldr.1145, 2013.

Lavado Contador, J. F., Schnabel, S., Gómez Gutiérrez, A., and Pulido Fernández, M.: Mapping sensitivity to land degradation in Extremadura, SW Spain, Land Degrad. Dev., 20, 129–144, doi:10.1002/ldr.884, 2009.

Lemos, J. J. S.: Níveis de degradação no Nordeste do brasileiro, Revista Econômica do Nordeste, 32, 406–429, 2001.

Mainguet, M.: What is Desertification?, Definitions and Evolution of the Concept, Desertification Natural Background and Human Mismanagement, Springer, Berlin, 1–16, 1994.

Mares, M. A., Willig, M. R., and Lacher, T.: The Brazilian caatinga in South American zoogeography: Tropical mammals in a dry region, J. Biogeogr., 12, 57–69, 1985.

Marengo, J. A.: Vulnerabilidade, impactos e adaptação à mudança do clima no semi-árido do Brasil, Parcerias Estratégicas, 27, 149–75, 2008.

Matallo Júnior, H.: Indicadores de desertificação: histórico e perspectivas, UNESCO, Brasília, 126 pp., 2001.

MMA – Ministério do Meio Ambiente: Programa de ação nacional de combate à desertificação e mitigação dos efeitos da seca – PAN-BRASIL, Ministério do Meio Ambiente, Secretaria de Recursos Hídricos, 213 pp., 2005.

MMA – Ministério do Meio Ambiente: Atlas das áreas suscetíveis à desertificação do Brasil, Universidade Federal da Paraíba, Secretaria de Recursos Hídricos – SRH, 131 pp., 2007.

MMA-IBAMA – Ministério do Meio Ambiente: Monitoramento dos desmatamentos nos biomas brasileiros por satélite, Acordo de cooperação técnica MMA/IBAMA, Monitoramento do bioma Caatinga: 2002 a 2008, available at: http://siscom.ibama.gov.br/monitorabiomas/caatinga/relatrio_tecnico_caatinga_72.pdf (last cccess: 10 July 2013), 2010.

Mirando, H. S.: Efeitos do regime do fogo sobre a estrutura de comunidades de cerrado: Resultados do projeto Fogo, Brasília-DF, IBAMA, 144 pp., 2010.

Molle, F.: Perdas por evaporação e infiltração em pequenos açudes, Série Brasil: SUDENE, Hidrologia, 25, 11–70, 1989.

Montanarella, L.: Trends in land degradation in Europe, edited by: Sivakumar, M. V. and N'diangui, N., Climate and land degradation, Berlin: Springer, 83–104, 2007.

Monteith, J. L.: Evaporation and Environment, in: The state and movement of water in living organism, 19th Symp., Soc. Exptl. Biol., 205–234, 1965.

Nimer E.: Desertificação: realidade ou mito?, Revista Brasileira de Geografia, 50, 7–39, 1988.

Oliveira, M. B. L., Santos, A. J. B., Manzi, A. O., Alvala, R. C. S., Correia, M. F., and Moura, M. S. B.: Exchanges of energy and carbon flux between caatinga vegetation and the atmosphere in northeastern Brazil, Rev. Bras. Meteor., 21, 166–174, 2006.

ONU: Convenção das Nações Unidas de Combate à Desertificação nos países afetados por seca grave e/ou desertificação, particularmente na África, Brasília, Ministério do Meio Ambiente, dos Recursos Hídricos e da Amazônia Legal, 89 pp., 1997.

Oyama, M. D. and Nobre, C. A.: Climatic consequences of a large-scale desertification in northeast Brazil: A GCM simulation study, J. Climate, 17, 3203–3213, 2004.

Parvari, S. H, Pahlavanravi, A., Nia, A. R. M., Dehvari, A., and Parvari, D.: Application of methodology for mapping environmentally sensitive areas (ESAs) to desertification in dry bed of Hamoun wetland (Iran), Int. J. Nat. Resour. Mar. Sci., 1, 65–80, 2011.

Pulido-Fernández, M., Schnabel, S., Lavado-Contador, J. F., Miralles Mellado, I., and Ortega-Pérez, R.: Soil organic matter of Iberian open woodland rangelands as influenced by vegetation cover and land management, Catena, 109, 13–24, doi:10.1016/j.catena.2013.05.002, 2013.

Prince, S. D., Colstoun, E. B., and Kravitz, L. L.: Evidence from rain-use efficiencies does not indicate extensive Sahelian desertification, Glob. Change Biol., 4, 359–374, doi:10.1046/j.1365-2486.1998.00158.x, 1998.

Reynolds, J. F. and Stafford, S. D. M.: Global Desertification: Do Humans Cause Deserts?, 88, Dahlem University Press, Berlin, 2002.

Reynolds, J. F., Stafford, S. D. M., Lambin, E. F., Turner, I. B. L., Mortimore, M., Batterbury, S. P. J., Downing, T. E, Dowlatabadi, H., Fernández, R. J, Herrick, J. E., Huber-Sannwald, E., Jiang, H., Leemans, R., Lynam, T., Maestre, F. T., Ayarza, M., and Walker, B.: Global desertification: building a science for dryland development, Science, 316, 847–851, doi:10.1126/science.1131634, 2007.

Rocco, R.: Legislação Brasileira do Meio Ambiente, DP&A Editora, Rio de Janeiro, 238 pp., 2002.

Salvati, L. and Zitti, M.: Long term demographic dynamics along an urban-rural gradient: implications for land degradation, Biota, 8, 61–69, 2008.

Salvati, L., Bajocco, S., Ceccarelli, T., Zitti, M., and Perini, L.: Towards a process based evaluation of land susceptibility to soil degradation in Italy, Ecol. Indic., 11, 1216–1227, 2011.

Santini, M., Caccamo, G., Laurenti, A., Noce, S., and Valentini, R.: A multicomponent GIS framework for desertification risk assessment by an integrated index, Appl. Geogr., 30, 394–415, doi:10.1016/j.apgeog.2009.11.003, 2010.

Sampaio, E. V. S. B, Araújo, M. S. B., and Sampaio, Y. S. B.: Propensão à desertificação no semi-árido brasileiro, Revista de Geografia, 22, 59–76, 2003.

Silverman, B. W.: Density estimation for statistics and data analysis, Chapman and Hall, 1986.

Sommer, S., Zucca, C., Grainger, A., Cherlet, M., Zougmore, R., Sokona, Y., Hill, J., Della Peruta, R., Roehrig, J., and Wang, G.: Application of indicator systems for monitoring and assessment of desertification from national to global scales, Land Degrad. Dev., 22, 184–197, doi:10.1002/ldr.1084, 2011.

Symeonakis, E., Karathanasi, N., Koukoulas, S., and Panagopoulos, G.: Monitoring Sensitivity to land degradation and desertification with the environmentally sensitive area index: the case of Lesvos Island, Land Degrad. Dev., 22, 184–197, doi:10.1002/ldr.2285, 2014.

Torres, H. G., Marques, E., Ferreira, M. P., and Bitar, S.: Pobreza e espaço: padrões de segregação em São Paulo, Estudos Avançados, IEA, 17, 97–128, 2003.

Thornes, J. B.: Stability and instability in the management of Mediterranean desertification, in: Environmental modelling: Finding simplicity in complexity, edited by: Wainwright, J. and Mulligan, M., Chichester, UK, Wiley, 303–315, 2004.

Trancoso, R., Filho, A. C., Tomasella, J., Schietti, J., Forsberg, B. R., and Miller, R. P.: Deforestation and conservation in major watersheds of the Brazilian Amazon, Environ. Conserv., 36, 277–288, doi:10.1017/S0376892909990373, 2010.

Tricart, J.: Ecodinâmica, IBGE-SUPREN (Recursos Naturais e Meio Ambiente), 91 pp., 1977.

UNESCO – United Nations Educational: Scientific and Cultural Organization, Map of the world distribution of arid regions: Map at scale 1 : 25 000 000 with explanatory note, MAB Technical Notes 7, UNESCO, 1979.

United Nations Development Programme (UNDP): Institute of Applied Economic Research (IPEA), Brazilian Institute of Geography and Statistics (IBGE) & João Pinheiro Foundation (FJP), 2010, Atlas of Human Development in Brazil, Brasília, DF: UNDP, 2010.

United Nations Convention to Combat Desertification (UNCCD) Regions: Africa, Bonn: UNCCD, available at: http://www.unccd.int/en/regional-access/Pages/countries.aspx?place=_31 (last access: 05 May 2013), 2012.

Valeriano, M. M. and Rossetti, D. F.: TOPODATA: Brazilian full coverage refinement of SRTM data, Appl. Geogr., 32, 300–309, 2012.

Van Genderen, J. L., Lock, B. F., and Vass, P. A.: Remote Sensing: Statistical testing of thematic map accuracy, Remote Sens. Environ., 7, 3–14, 1978.

Vasconcelos Sobrinho, J. O.: Metodologia para a identificação de processos de desertificação, Manual de indicadores, SUDENE-DDL, 20 pp., 1978.

Vieira, R. M. S. P, Cunha, A. P. M. A., Alvalá, R. C. S., Carvalho, V. C., Ferraz Neto, S., and Sestini, M. F.: Land use and land cover map of a semi-arid Region of Brazil for meteorological and climatic models, Revista Brasileira de Meteorologia, 28, 129–138, 2013.

Ziadat, F. M. and Taimeh, A. Y.: Effect of rainfall intensity, slope and land use and antecedent soil moisture on soil erosion in an arid environment, Land Degrad. Dev., 24, 582–590, doi:10.1002/ldr.2239, 2013.

13

Use of phytoremediation and biochar to remediate heavy metal polluted soils: a review

J. Paz-Ferreiro[1,2]**, H. Lu**[1,3]**, S. Fu**[1]**, A. Méndez**[2]**, and G. Gascó**[2]

[1]Key Laboratory of Vegetation Restoration and Management of Degraded Ecosystems, South China Botanical Garden, Chinese Academy of Sciences, Guangzhou 510650, China
[2]Departamento de Edafología, ETSI Agrónomos, Universidad Politécnica de Madrid, Avenida Complutense 3, Madrid 28050, Spain
[3]University of Chinese Academy of Sciences, Beijing 100049, China

Correspondence to: J. Paz-Ferreiro (jorge.paz@upm.es)

Abstract. Anthropogenic activities are resulting in an increase of the use and extraction of heavy metals. Heavy metals cannot be degraded and hence accumulate in the environment, having the potential to contaminate the food chain. This pollution threatens soil quality, plant survival and human health. The remediation of heavy metals deserves attention, but it is impaired by the cost of these processes. Phytoremediation and biochar are two sound environmental technologies which could be at the forefront to mitigate soil pollution. This review provides an overview of the state of the art of the scientific research on phytoremediation and biochar application to remediate heavy-metal-contaminated soils. Research to date has attempted only in a limited number of occasions to combine both techniques, however we discuss the potential advantages of combining both, and the potential mechanisms involved in the interaction between phytoremediators and biochar. We identified specific research needs to ensure a sustainable use of phytoremediation and biochar as remediation tools.

1 Introduction

Industrialisation and technical advances have led to an increase in the use of heavy metals and heavy metal pollution. Contrary to organic substances, heavy metals are non-degradable and accumulate in the environment. While some soils can have a high background level of heavy metals due to volcanic activity or weathering of parent materials, in other soils anthropogenic activities, including smelting, mining, use of pesticides, fertilisers and sludges are responsible for these high levels of heavy metals.

Soil heavy metal pollution has a pernicious effect on soil microbial properties (Yang et al., 2012) and on the taxonomic and functional diversity of soils (Vacca et al., 2012). Soil heavy metal pollution poses a risk to the environment and to human health (Roy and McDonald, 2014) due to biomagnification (increases in metal concentration as the element passes from lower to higher trophic levels). Some of these elements can be essential for living organisms while some others are non-essential. Even concentrations of essential elements beyond a certain threshold will have pernicious health effects, as they interfere with the normal metabolism of living systems. It is not the purpose of this article to review the adverse effects of heavy metals on human or plant health. Kabata-Pendias and Pendias (2001) provide a list of toxic effects of heavy metals on plants and the mechanism involved, while a summary of adverse effects of heavy metals on human health was provided by Ali et al. (2013). We would like to remind the reader that studies on heavy metal pollution are focused on As, Cd, Cr, Hg and Pb as they are toxic, non-essential heavy metals, and on Cu, Ni and Zn which, although essential, can cause health problems in humans or can result in phytotoxicity at high concentrations.

With an increasing amount of literature on heavy metal remediation, we aim to summarise the state of art of two of these techniques situated at the forefront of remediation practices: phytoremediation, with a focus on phytoextraction, and biochar soil amendment, and to discuss their mechanism and how we could combine them to improve remediation efforts.

2   Biochar

Biochar is a porous, carbonaceous product obtained from the pyrolysis of organic materials. Numerous materials can be used as feedstocks, including sludges, plant materials and manures. Although the use of charcoal (wood biochar) has been common since preterit times, the idea of using other feedstocks for biochar production is new and relatively unexplored. Typically biochars have high cation exchange capacity and are alkaline. Biochar has many potential benefits on soil properties as an increase in soil biological activity (Lehmann et al., 2011; Paz-Ferreiro et al., 2014), diminishing soil greenhouse gas emissions from agricultural sources and thus enhancing soil carbon sequestration due to its elevated content of recalcitrant forms of carbon (Gascó et al., 2012). The changes brought about by biochar addition to the soil will cause alterations in soil quality (Paz-Ferreiro and Fu, 2014) with the potential to increase agricultural yields (Jeffery et al., 2011; Liu et al., 2013). The multiple benefits of biochar for soil have been compiled recently in the book by Lehmann and Joseph (2009). However, little information was available in this book about the effect of biochar on soil heavy metals. Although there is a recent review on the role of biochar to remediate polluted soils, with a particular interest in the metalloid arsenic (Beesley et al., 2011), our article has a more focused scope on the combination of phytoremediation and biochar with respect to heavy metal remediation. Moreover, in the last years there have been an increasing number of articles devoted to understanding the interaction between heavy metals, vegetation and biochar.

3   Mechanism of interaction between biochar and heavy metals

Biochar characteristics are a function of several factors, including the type of feedstock, the particle size of the feedstock and temperature and conditions of pyrolysis. The wide range of characteristics that biochar might posses makes some particular materials more suitable than others to remediate different heavy metals. Therefore, when selecting a biochar for remediation purposes, scientists should be aware not only of soil type and characteristics but also on biochar properties. Moreover, it should also be considered that key biochar properties such as surface area, pH, ash and carbon contents can be affected by post-treatments and thus enhance biochars' ability to immobilise heavy metals (Lima et al., 2014).

Before reviewing the mechanisms implied in the interaction between biochar and heavy metal it is necessary to note that biochars act on the bioavailable fraction of soil heavy metals and that they can reduce also their leachability.

One of the characteristics of biochars is possessing large surface areas, which implies a high capacity for complex heavy metals on their surface. Surface sorption of heavy metals on biochar has been demonstrated on multiple occasions using scanning electron microscopy (Beesley and Marmiroli, 2011; Lu et al., 2012). This sorption can be due to complexation of the heavy metals with different functional groups present in the biochar, due to the exchange of heavy metals with cations associated with biochar, such as Ca^{+2} and Mg^{+2} (Lu et al., 2012), K^+, Na^+ and S (Uchimiya et al., 2011c), or due to physical adsorption (Lu et al., 2012). Also oxygen functional groups are known to stabilise heavy metals in the biochar surface, particularly (Uchimiya et al., 2011c) for softer acids like Pb^{+2} and Cu^{+2}. In addition, Méndez et al. (2009) observed that Cu^{+2} sorption was related to the elevated oxygenated surface groups and also with high average pore diameter, elevated superficial charge density and Ca^{+2} and Mg^{+2} exchange content of biochar. Possibly, sorption mechanisms are highly dependent on soil type and the cations present in both biochar and soil. Some other compounds present in the ash, such as carbonates, phosphates or sulphates (Cao et al., 2009; Karimi et al., 2011; Park et al., 2013) can also help to stabilise heavy metals by precipitation of these compounds with the pollutants.

Alkalinity of biochar can also be partially responsible for the lower concentrations of available heavy metals found in biochar-amended soils. Higher pH values after biochar addition can result in heavy metal precipitation in soils. Biochar pH value increases with pyrolysis temperature (Wu et al., 2012), which has been associated with a higher proportion of ash content (Cantrell et al., 2012).

Biochar can also reduce the mobility of heavy metals, altering their redox state of those (Choppala et al., 2012). As an example, biochar addition could lead to the transformation of Cr^{+6} to the less mobile Cr^{+3} (Choppala et al., 2012).

The relative contribution of the different mechanisms to heavy metal immobilisation by different biochar remains unknown, although some authors like Houben et al. (2013a) postulate that it is mostly a pH effect.

4   Studies on the effect of biochar on soil heavy metals

Table 1 shows a brief summary of the latest papers about the effect of biochar on soil heavy metals. Fellet et al. (2011) tried to use biochar to remediate a multicontaminated mine soil. Biochar addition did not result in the decrease of the total heavy metal content of the soil, however, biochar addition reduced the bioavailability of Cd, Pb and Zn and the mobility (measured using a leaching experiment) of Cd, Cr and Pb.

Park et al. (2011) studied the effect of two biochars in a heavy-metal-spiked soil and a naturally strongly polluted soil. They performed a sequential extraction of some heavy metals. They found chicken manure biochar effective reducing extractable concentrations of Cd and Pb, but not Cu

Table 1. Studies considering the effect of biochar application on soil heavy metals. Blank indicates not specified in the article.

Feedstock (temperature)	Soil type	Heavy metals	Reference
Sewage sludge	Haplic Cambisol	Cu, Ni, Zn, Cd, Pb	Méndez et al. (2012)
Rice husk, rice straw and rice bran (400 °C)	Technosol	As, Cd, Pb, Zn	Zheng et al. (2012)
Wastewater sludge (550 °C)	Chromosol (Australian system)	As, Cd, Cr, Cu, Pb, Ni, Se, Zn, Sb, B, Ag, Ba, Be, Co, Sn, Sr	Hossain et al. (2010)
		Cu, Pb, Zn	Sizmur et al. (2011)
Broiler litter (350 and 700 °C), pecan shells (450 °C)	Abruptic Durixeralfs	Cu, Cd, Ni	Uchimiya et al. (2010)
Pecan shell (450 °C), broiler litter samples (700 °C)	Typic Kandiudult and Abruptic Durixeralfs	Cu	Uchimiya et al. (2011a)
Chicken manure (550 °C), green waste (550 °C)		Cd, Cu, Pb	Park et al. (2011)
Forest green waste (600–800 °C)	Peat	Cu	Buss et al. (2012)
Dairy manure (350 and 700 °C), paved feedlot manure (350 and 700 °C), poultry litter (350 and 700 °C), turkey litter (350 and 700 °C), separated swine solids (350 and 700 °C)	Typic Kandiudult	Pb, Cu, Ni, Cd	Uchimiya et al. (2012a)
Mix of hardwoods (400 °C)		As, Cd, Zn	Beesley and Marmiroli (2011)
Mix of hardwoods (400 °C)	Technosol	Pb, Cu	Karami et al. (2011)
Orchard prune residue (500 °C)	Technosol	Cd, Cr, Cu, Ni, Pb, Tl, Zn	Fellet et al. (2011)
Eucalyptus		As, Cd, Cu, Pb, Zn	Namgay et al. (2010)
Wheat straw (350–550 °C)	Technosol	Cd	Cui et al. (2011)
Wheat straw (350–550 °C)	Technosol	Cd	Cui et al. (2012)
Rice straw	Ultisol	Cu, Cd, Pb	Jiang et al. (2012)
Orchard prune residues (500 °C)	Technosol	As	Beesley et al. (2013)
Miscanthus (600 °C)		Cd, Zn, Pb	Houben et al. (2013a)
Chicken manure (550 °C), green waste (550 °C)		Cd, Pb	Park et al. (2013)
De-inking paper sludge (300 and 500 °C)	Vertisol	Ni	Méndez et al. (2014)

concentration, while green waste biochar was more effective in diminishing all of the heavy metals studied. Heavy metal fractions that were bonded to organic matter increased after biochar addition. Both biochars also decreased Cd and Pb presence in soil pore water.

Uchimiya et al. (2012a) analysed the effects on soil heavy meals concentrations of 10 biochars prepared from 5 feedstocks at 2 different temperatures. They observed that manures with a high or low proportion of ash or P were less effective to immobilise heavy metals. In contrast, biochars prepared at 700 °C were more effective, which could be attributed to transformations in the material, including the removal of nitrogen containing heteroaromatic and leachable aliphatic functional groups. They found Cu and Pb relatively easy to stabilise in soil, while Cd and Ni response depended strongly on the type of biochar added to the soil.

Beesley and Marmiroli (2011) detected a retention of As, Cd and Zn on biochar surfaces. These authors proved that sorption of the metal was produced at the biochar surface and that this process was not immediately reversible. Leachate concentrations of Cd and Zn were reduced 300- and 45-fold, respectively. However, leachate concentrations of As did not diminish.

Namgay et al. (2010) reported that the concentrations of Cd, As and Pb in maize shoots decreased after biochar application. Beesley et al. (2013) reported interesting results, finding that As can increase in soil pore water after biochar addition, but transfer to the plant be reduced. This would imply that at least some biochars could pose no risk of increasing heavy metals in plants and hence are safe in terms of food chain transfer, but leaching of As to nearby waters must be considered. Karami et al. (2011) added biochar to a mine soil polluted with Pb and Cu. They found that biochar addition reduced pore water Pb concentrations to half their values in the mine soil. When biochar was combined with greenwaste compost the levels of Pb concentrations in the pore water were 20 times lower than in the control. Jiang et al. (2012) found that the acid-soluble fractions of Pb^{+2} and Cu^{+2} diminished by 18.8–77.0 % and 19.7–100.0 %, respectively, depending on biochar concentration. However, only 5.6–14.1 % of acid-soluble Cd^{+2} was immobilised. Park et al. (2013) compared the sorption capacity of two biochars, made from chicken manure and from green waste. They found chicken manure biochar more effective to immobilise Cd and Pb compared to green waste biochar. Both biochars presented a higher sorption capacity for Pb, possibly as a consequence of precipitation and complexation of Pb with carbonate, sulphate and phosphate present in the biochar.

Hydrochars could also be used for soil heavy metal immobilisation, however there is a lack of studies on the topic. Hydrochars are produced after pyrolysis of organic-matter-rich materials in the presence of subcritical liquid water. This technique can be applied to obtain pyrolysed products from wet feedstocks. In principle, the adsorption capacity of hydrochars seems to be reduced compared to biochars or other adsorbents due to the fewer functional groups containing oxygen present on hydrochar surfaces. However, Xue et al. (2012) have demonstrated experiments that the use of activated hydrochars could overcome these problems. They performed a series of batch and columns experiments to show how this type of hydrochar could reduce Pb on water. The potential applicability of hydrochar to address soil heavy metal pollution remains untested. However, hydrochars tend to be acidic and could possess phytotoxic or genotoxic risks (Busch et al., 2013), which would deem them unsuitable in restoration projects.

There is a lack of studies concerning how pyrolysis conditions affect biochar properties as heavy metal sorbent. To fill this gap, Uchimiya et al. (2011b) performed an experiment using wood and grass biochars prepared at five different temperatures and another one (Uchimiya et al., 2012b) used poultry litter prepared at four different temperatures to study lead retention. From the first experiment they suggested using biochars prepared at high temperature (650 °C to 800 °C) for remediation purposes. In addition they recommended performing acid or other oxidant post-treatment in order to increase oxygen-containing surface functional groups (carboxyl, carbonyl and hydroxyl) which have a great importance in relation to heavy metal sorption into biochar. In the case of the chicken litter biochar, they found that lower production temperatures were more suitable than higher ones due to the stabilising effect. Higher rates of amendment were necessary in their experiments for chicken manure biochar to get the same remediation effect as plant-derived biochars.

It is expected that as biochar is in contact with soil for a prolonged period of time, oxidation, both biotic and abiotic, would result in the alteration of biochar, a process known as aging. This process, which would result in the formation of carboxylic, phenolic, carbonyl, quinones and hydroxyl functional groups and which can be emulated under laboratory conditions was studied by Uchimiya et al. (2010). These authors found that the immobilisation of heavy metals by biochar was related to the metal lability, this means that heavy metal immobilisation followed the order $Cu^{+2} > Cd^{+2} > Ni^{+2}$. Heavy metal immobilisation was not affected by biochar aging, except for a small increase in Ni observed in soils with aged biochar.

Earthworms can be added to soil at some stages of ecological restoration due to their well-established positive effects on soil properties as organic matter content, soil formation, soil aeration and nutrient cycling. Sizmur et al. (2011) tested a polluted soil collected in the vicinity of a Cu mine using biochar in combination with compost and earthworms (*Lumbricus terrestris*). They found all treatments (biochar alone, biochar + compost, and biochar + compost + earthworms) to reduce the amount of heavy metals compared to the control soils. A limiting aspect when using earthworms with remediation purposes is that their addition to soil could lead to the mobilisation of heavy metals and hence to an increase of plant heavy metal concentrations. Interestingly, Sizmur et al. (2011) found that the treatments containing biochar and earthworms did not result in higher heavy metal mobility or plant availability.

As a consequence of heavy metal immobilisation, biochars can reduce the phytotoxicity of polluted soils, resulting in increases in the percentage of germinated seeds and root length (Ahmad et al., 2012).

All of the above experiments have been conducted under laboratory conditions. We would urge scientists to design experiments to help to demonstrate the benefits of biochar against heavy metal pollution under field conditions, as done by Zheng et al. (2012) and Cui et al. (2011, 2012). Zheng et al. (2012) studied the effect of three biochars on different heavy metals (see Table 2) using a multi-polluted soil planted with rice. They found Cd, Pb and Zn to be reduced on rice shoots, in particular when using straw-derived biochar. However, as in rice shoots was increased by biochar addition. More importantly, we believe that this is one of the first studies considering the effects of biochar particle size on plant heavy metals. The authors found that decreases in particle size resulted in less Cd, Zn and Pb accumulating in the rice plants. Similarly, Cui et al. (2011) and Cui et al. (2012) found reduced Cd uptake in paddy fields and in a soil cropped with wheat, respectively. Both studies consisted of two annual measurements, so the need to reapply biochar after more extended periods of time remains to be explored.

5 Phytoremediation

Phytoremediation is an umbrella term for a series of techniques that combine the disciplines of soil microbiology and chemistry and plant physiology (Cunningham and Ow, 1996). Currently the most extended practice for soil heavy metal remediation does not address the problem of contamination, as it consists of encapsulation or digging and dumping. Immobilisation or extraction can be expensive and, as a consequence, phytoremediation can be considered comparatively attractive as it can be used at a relatively low cost to restore or partially decontaminate a site compared to other options, as the cost is 5 % that of other alternative methods (Prasad, 2003). Other advantages would include its good perception as a remediation technique among the general public and being more environmentally friendly than other options, as the introduction of vegetation in the polluted area can also help to prevent erosion or contaminant leaching. Phytoremediation consists in the use of plants to remove contaminants from the environment or to transform them into less harmful forms (see Table 3 for a summary of phytoremediation techniques). Phytoremediation is a relatively new technology, as research studies have been mostly conducted from 1990 onwards.

Phytoextraction is the main and most promising technique to remove soil heavy metals. It is based on the use of hyperaccumulators which uptake heavy metals and then translocate them to aboveground tissues (Table 1). One common way of defining a hyperaccumulator is as a plant that can store heavy metals at a level 100-fold greater than common plants without yield reduction (Chaney et al., 2007). On other occasions, these types of plants are defined on their basis to accumulate more than $100\,mg\,kg^{-1}$ dry weight of Cd, more than $1000\,mg\,kg^{-1}$ of Cu, Co, Cr, Ni or Pb, or more than $10\,000\,mg\,kg^{-1}$ of Mn or Zn (Baker and Brooks, 1989). Some other authors have mentioned that these values are conservative and propose these criteria to be lowered (van der Ent et al., 2013). Species used for phytoextraction must not only accumulate high amounts of the target element but also have a high growth rate, tolerate the toxic effects of the heavy metals, be adapted to local environment and climate, be resistant to pathogen and pests, be easy to cultivate and repulse herbivores to avoid food chain contamination (Ali et al., 2013).

To date, more than 400 species have been identified as hyperaccumulators, including more than 300 Ni hyperaccumulators (Li et al., 2003). In contrast with Ni, only a few plant species have demonstrated the potential to accumulate Cd, Cu, Pb, and Zn (Brooks, 1998). Many phytoremediators belong to the taxonomical order of Brassicales and phytoremediators are also abundant in Asterales, Solanales, Poales, Malpighiales, Fabales, Caryophyllales and Rosales (Shao et al., 2011). The amount of metal extracted from the soil depends not only on the plant species utilised but also on the type of soil and climate of the region (Shao et al., 2011).

The mechanism and reasons of phytoaccumulation remain unknown. Metal concentrations are higher in the shoots compared to the roots, suggesting that there could be an ecological role, leading to protection against insect, herbivore or fungal attack, by making the leaves toxic or unpalatable.

Phytoextraction has three main purposes: firstly, to remove the contaminant from the soil or contain it, secondly phytoextraction of elements that have market value and finally gradually improving soil quality to cultivate crops with higher market value (Vangrosveld et al., 2009).

There are a number of problems associated with the effectiveness of this remediation technique. Phytoremediation might not be suitable in areas were the heavy metal concentration is too elevated, as plants could show symptoms of phytotoxicity. In addition, most of the phytoaccumulators have slow growth rate or produce few biomass, limiting the amount of metal uptaken.

Manipulation of soil pH, soil nutrient content or soil organic matter can also be undertaken to improve metal hyperaccumulation. In this sense, these additional agronomic practices can be carried out when heavy metal concentrations in the soil are too elevated to reduce plant stress (Adriano et al., 2004; Gabos et al., 2011; de Abreu et al., 2012). Thus, liming can allow the decrease of the heavy metal available fraction, therefore enabling vegetative growth, while fertilisers can improve phytoextractor growth. On the other hand, both liming and fertiliser addition can alter the mobility

Table 2. Studies considering the effect of biochar application on soil heavy metals in combination with phytoremediators. Blank indicates not specified in the article.

Feedstock (temperature)	Soil type	Pollutants	Plant species	Reference
Mix of hardwoods (400 °C)	3 soils	As	*Miscanthus x giganteus*	Hartley et al. (2009)
Miscanthus (600 °C)		Cd, Pb, Zn	*Brassica napus* L.	Houben et al. (2013b)
Pruning residues from orchards (550 °C), fir tree pellets (350–400 °C) and manure pellets mixed with fir tree pellets (350–400 °C)	Technosol	Cd, Cr, Cu, Fe, Ni, Pb, Tl and Zn	*Anthyllis vulneraria* subsp. *polyphylla* (Dc.) Nyman, *Noccaea rotundifolium* (L.) Moench subsp. *cepaeifolium* and *Poa alpina* L. subsp. *alpina*	Fellet et al. (2014)

Table 3. Summary of the different techniques of phytoremediation.

Technique	Description
Phytoextraction	Plants accumulate contaminants in harvestable biomass i.e., shoots
Phytofiltration	Sequestration of pollutants from contaminated waters by plants
Phytostabilisation	Limiting the mobility and bioavailability of polluting substances by prevention of migration or immobilisation
Phytovolatilisation	Conversion of pollutants to volatile form followed by their release to the atmosphere
Phytodegradation	Degradation of organic xenobiotics by plant enzymes within plant tissues
Rhizodegradation	Degradation of organic xenobiotics in the rhizosphere by rhizospheric microorganisms
Phytodesalination	Removal of excess salts from saline soils by halophytes

and speciation of soil heavy metals. As an example, Li et al. (2012) found that Cd removal from soil was enhanced by the phytoaccumulator *Amaranthus hypocondriacus* after NPK or NP fertilisation due to an increase on plant biomass. However, they found that N alone did not increase plant biomass and led to a limited increment in phytoextraction. Other studies (Huang et al., 2013) have found that P fertilisers can decrease soil pH, enhancing the mobility of Cd and leading to increased phytoextraction by *Sedum alfredii*. When adding a phosphate fertiliser to promote phytoremediation, the choice of amendment should be carefully chosen as cations (K^+, Na^+, Ca^{+2} or NH_4^+) associated with the phosphate could affect the mobility of heavy metals (Bolan et al., 2003; Huang et al., 2013). Indeed, plant growth (Oo et al., 2014) and the mobility of different elements in the soil (Ahmad et al., 2013) can be related to soil salinity. For example, Stevens et al. (2003) observed that Zn^{+2} and Pb^{+2} mobility increased with the increment of electrical conductivity. Differences in soil pH caused by the addition of different phosphate fertilisers can also lead to differences in phytoextraction (Mandal et al., 2012). Urea has also been used to alleviate plant stress and improve B phytoextraction by the plant species *Brassica juncea* (Giansoldati et al., 2012).

Organic amendments such as chicken manure have also been shown to increase growth of the species *Rorippa globosa* (Wei et al., 2011). Chicken manure addition resulted in a decrease in soil extractable Cd and thus, the concentration of Cd in the shoots was lower in soils amended with chicken manure than in soils amended with urea or in the controls (soil + phytoremediator). However, the total concentration of metal extracted in the shoots was in both cases higher than in the control. Other materials such as pig manure vermicompost can also be used to improve plant yield and assist phytoremediation, as demonstrated by Wang et al. (2012) in an experiment using Cd as target heavy metal and *Sedum alfredii* as phytoremediator. Indeed, the use of organic amendments has numerous applications, for example, Siebielec and Chaney (2012) have demonstrated the effectiveness of biosolids compost in the rapid stabilisation of Pb and Zn and revegetation of military range contaminated soils increasing tall fescue growth by more than 200 %, while Clemente et al. (2012) recovered land contaminated by mining activity with Cd, Cu, Pb, and Zn by a combination of the halophytic shrub *Atriplex halimus* L. with pig slurry.

The use of chelators such as citric acid or EDTA has also been sometimes advised to assist phytoremediation, with the aim of increasing the mobility of soil heavy metals and thus plant extraction (Zhou et al., 2007; Freitas et al., 2013). However, we should bear in mind that the use of chelators can originate other environmental problems, including toxicity for plants and metal leaching (Zhou et al., 2007).

In addition, experiments should be done to account for the potential impact of climate change on the capability of phytoextractors to accumulate heavy metals, which at the moment is uncertain (Rajkumar et al., 2013).

Finally, we would like to remark that pot experiments are a good first approach to evaluate the potential of a phytoextractor, but they cannot substitute field experiments as the uptake of heavy metals is higher in pots than for the same soil in the field (see for example, Marschner, 1986). This can be due to differences in soil moisture or microclimate and to the fact that field-grown plants can reach down to less polluted soil.

Phytostabilisation is another phytoremediation technique and has been used mostly in relation with the stabilisation and containment of mine tailings (Conesa et al., 2007; Méndez et al., 2007). Thus, the vegetative cover diminishes eolian dispersion while roots prevent water erosion and leaching and contributes to the immobilisation of heavy metals. Mechanisms involved in phytostabilisation include precipitation, root sorption, complexation or metal valence reduction. Phytostabilisation, contrary to phytoextraction, primarily focuses on heavy metal sequestration within the rhizosphere but not in plant tissues.

6 Combining biochar and phytoremediation

There is an abundance of reports in the literature about amendments, such as lime and compost being used to reduce the bioavailability of heavy metals (Komárek et al., 2013) and thus having the potential to be combined with phytoremediators (de Abreu et al., 2012). Biochar, as reviewed before, can also stabilise heavy metals in soils and thus reduce plant uptake. However, until recently there was a lack of experiments trying to combine both approaches to soil remediation.

Biochar is commonly reported in the literature to increase plant growth, hence there is a potential of biochar to increase the yield of phytoremediators. This increase in plant productivity is highly heterogeneous and has overall been quantified as 10 % (Jeffery et al., 2011; Liu et al., 2013). However, there are several factors that limit the accuracy of the figure provided by Jeffery et al. (2011) and Liu et al. (2013) and that could skew the data. To date, most of the field experiments have been conducted in the short term, being limited to a period of 1–2 yr and there are a high relative number of laboratory mesocosm incubations (with a duration of 1–2 months) included in the data set. Also, the data set in this review comprises a higher number of experiments in tropical latitudes compared to temperate ones. Finally, we should bear in mind that a high heterogeneity in the response was detected, depending on the type of soil and plant utilised.

Improvements in plant yield after biochar addition are often attributed to increased water and nutrient retention, improved biological properties and CEC, effects on nutrient cycling and turnover and improvements in soil pH. Many of these effects are interrelated and potentially they could act synergistically. In general, acid soils with a coarse texture or a medium texture are more prone to produce increases in crop productivity (Jeffery et al., 2011; Liu et al., 2013). In the last years the scientific community has also raised awareness over the improvement of plant responses to disease as an additional benefit of biochar soil amendment (Graber et al., 2010). As said before, biochar can alter

soil microbial community, possibly including an increase in beneficial organisms that produce antibiotics and can protect plants against pathogens. Another mechanism could be compounds included in biochar such as 2-phenoxyethanol, benzoic acid, hydroxy-propionic and butyric acids, ethylene glycol and quinones suppressing some of the pathogens present in the microbiota (Graber et al., 2010; Elad et al., 2011).

In principle, biochar prepared from any material would have the potential to increase plant yield and thus be used in combination with phytoremediation. However, the use of sewage sludge biochar would be unadvised due to its generally negative effect on crop performance (Jeffery et al., 2011). Caution should also be taken with the presence of heavy metals in sewage sludge biochars, although some studies (Méndez et al., 2012; Hossain et al., 2010) show that the metals present in the biochar are not in mobile forms.

It is also worth mentioning that for a long time, phytoextractors were considered to be non-mycorrhizal. However, in the last year it has been demonstrated that hyperaccumulators can form symbiosis with arbuscular mycorrhizal fungi (AMF) and these enhance plant growth and lead to higher contents of metal extracted (Al Agely et al., 2005; Orlowska et al., 2011). Positive effects of biochar have usually been found in arbuscular mycorrhizal fungi, although exceptions can be found in nutrient-rich soils (Lehmann et al., 2011).

There has been a recent interest about the possibility of combining phytoremediation with other potential plant uses, such as using plants that can be used to obtain bioenergy (de Abreu et al., 2012). While heavy-metal-contaminated areas are not suitable for food production, planting biocrops could promote soil organic matter stocks and reduce soil pollutants (Hartley et al., 2009). Willow and poplar have been commonly used as biocrops and they can be utilised for phytoremediation purposes due to their high uptake of heavy metals and fast growing rates (Baum et al., 2009). Recently, Hartley et al. (2009) observed no increase on As transfer to plants in three soils planted with Miscanthus and amended with hardwood biochar. They warned, however, that alkalyne biochars could mobilise As. It is a well-known fact that As behaves differently to other metals with respect to pH, as As mobility is reduced in acid soils due to adsorption on iron oxide surfaces. The results of Hartley et al. (2009) show that biochar can be used in combination with Miscanthus for phytostabilisation. More recent research has proved that biochar can have an added environmental benefit, improving the greenhouse gas balance of other bioenergy crops such as Miscanthus (Case et al., 2014).

Biochar and phytoremediation techniques have been used recently (see Table 3) to target at Cd-polluted soils (Houben et al., 2013b) using Brassica napus L. as Cd and Zn phytoextractor in combination with Miscanthus biochar and for the case of multicontaminated soils using different biochars and plant species (Fellet et al., 2014). Houben et al. (2013b) observed that phytoextraction of Cd and Zn by Brassica napus was impeded by biochar and, due to the lower BCF

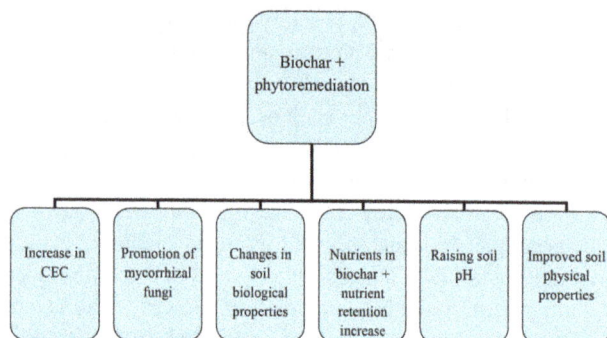

Fig. 1. An overview of the potential positive effects attained by combining phytoremediation and biochar in heavy metal pollution remediation.

achieved in pots with biochar, suggested using biochar and *Brassica napus* as a phytostabilisation alternative. While Fellet et al. (2014) used three biochars, produced from pruning residues from orchards, fir tree pellets and fir tree pellets mixed with manure at two different doses. Fellet et al. (2014) observed higher concentrations of Pb in plants grown with the fir tree pellets biochar. However, no increase in yield was obtained with this treatment, and the value of the translocation index, although significantly higher than in the control, was insufficient for the purposes of phytoextraction. Overall, they found the manure biochar to immobilise more heavy metals and also to produce the most noticeable increase in plant biomass, thus, making manure biochar more suitable for phytostabilisation purposes.

It seems plausible that one of the best approaches to combine biochar and phytoextractors would be in multicontaminated soils, where both can target at different elements. Biochar could also be used as a soil conditioner prior to plant colonisation in acidic, polluted mine tailings. However, these two approaches in relation with phytoextraction remains to be tested.

7   Conclusions and research needs

Biochar and phytoremediation techniques have the potential to be combined in the remediation on heavy metal polluted soils (see Fig. 1). Biochar can reduce the bioavailability and leachability of heavy metals in the soil. On the other hand phytoextractors can reduce the amount of soil heavy metals in polluted areas.

We anticipate that in the next years there will be a growing interest to study the interaction between phytoremediators and biochars and we identify the next areas as the ones warranting research.

Biochars have highly heterogeneous properties, which should be understood as maximising the efficacy of soil remediation. We should comprehend, firstly, how these properties are relevant for heavy metal adsorption and how they

contribute to the different mechanism of heavy metal immobilisation, and secondly how to optimise the choice of pyrolysis conditions and feedstocks in order to produce the desired products.

Most experiments utilising biochar or phytoremediators alone and not in combination have been carried out under laboratory conditions. In the case of phytoremediators this can result in an overestimation of heavy metal extraction.

For biochar most of the experiments (both in field and under laboratory conditions) have been conducted in the short term, which poses an interrogation on the long-term fate of these heavy metals. In fact it could be expected that, due to aging processes, the ability of biochar to sequester heavy metals decreases with time. More research will be needed to understand the aging process in biochar.

Thus, well-designed, large-scale and long-term field trials will be essential to evaluate the feasibility on the approach proposed in this article. The economics of these new remediation processes should be assessed against other options.

Acknowledgements. J. Paz-Ferreiro thanks the Chinese Academy of Sciences for financial support (fellowship for young international scientists number 2012Y1SA0002).

Special Issue: "Environmental benefits of biochar"
Edited by: G. Gascó, A. Méndez, A. M. Tarquis, J. Paz-Ferreiro, and A. Cerdà

References

Adriano, D. C., Wenzel, W. W., Vangrosveld, J., and Nolam, N. S.: Role of assisted natural remediation in environmental clean-up, Geoderma, 122, 121–142, 2004.

Ahmad, M., Lee, S. S., Yang, J. E., Ro, H. M., Lee, Y. H., and Ok, Y. S.: Effects of soil dilution and amendments (mussel shell, cow bone and biochar) on Pb availability and phytotoxicity in military shooting range soil, Ecotox. Environ. Safe., 79, 225–231, 2012.

Ahmad, S., Ghafoor, A., Akhtar, M. E., and Khan, M. Z.: Ionic displacement and reclamation of saline-sodic soils using chemical amendments and crop rotation, Land Degrad. Dev., 24, 170–178, 2013.

Al Agely, A., Sylvia, D. M., and Ma, L. Q.: Mycorrhizae increase arsenic uptake by the hyperaccumulator Chinese brake fern (Pteris vittata L.), J. Environ. Qual., 6, 2181–2186, 2005.

Ali, H., Khan, E., and Sajad, M. A.: Phytoremediation of heavy metals – Concepts and applications, Chemosphere, 91, 869–881, 2013.

Baker, A. J. M. and Brooks, R. R.: Terrestrial higher plants which hyperaccumulate metallic elements. A review of their distribution, ecology and phytochemistry, Biorecovery, 1, 81–126, 1989.

Baum, C., Leinweber, P., Weih, M., Lamersdorf, N., and Dimitriou, I.: Effects of short rotation coppice with willows and poplar on soil ecology, Agric. Forestry Res., 3, 183–196, 2009.

Beesley, L. and Marmiroli, M.: The immobilisation and retention of soluble arsenic, cadmium and zinc by biochar, Environ. Pollut., 159, 474–480, 2011.

Beesley, L., Moreno-Jiménez, E., Gómez-Eyles, J.L., Harris, E., Robinson, B., and Sizmur, T.: A review of biochars' potential role in the remediation, revegetation and restoration of contaminated soils, Environ. Pollut., 159, 3269–3282, 2011.

Beesley, L., Marmiroli, M., Pagano, L., Pigoni, V., Fellet, G., Fresno, T., Vamerali, T., Bandiera, M., and Marmiroli, N.: Biochar addition to an arsenic contaminated soil increases arsenic concentrations in the pore water but reduces uptake to tomato plants (Solanum lycopersicum L.), Sci. Total Environ., 454–455, 598–603, 2013.

Bolan, N. S., Adriano, D. C., and Naidu, R.: Role of phosphorus in (im)mobilization and bioavailability of heavy metals in the soil–plant system, Rev. Environ. Contam. T., 177, 1–44, 2003.

Brooks, R. R.: Geobotany and hyperaccumulator, in: Plants that hyperaccumulate heavy metals, their role in phytoremediation, microbiology, archaeology, mineral exploration and phytomining, edited by: Brooks, R. R., Wallingford, UK: CAB, International, 55–94, 1998.

Busch, D., Stark, A., Kammann, C. I., and Glaser, B.: Genotoxic and phytotoxic risk assessment of fresh and treated hydrochar from hydrothermal carbonization compared to biochar from pyrolysis, Ecotox. Environ. Safe., 97, 59–66, 2013.

Buss, W., Kammann, C., and Koyro, H. W.: Biochar reduces copper toxicity in Chenopodium quinoa Willd. in a sandy soil, J. Environ. Qual., 41, 1157–1165, 2012.

Cantrell, K. B., Hunt, P. G., Uchimiya, M., Novak, J. M., and Ro, K. S.: Impact of pyrolysis temperature and manure source on physicochemical characteristics of biochar, Bioresource Technol., 107, 419–428, 2012.

Cao, X. D., Ma, L. N., Gao, B., and Harris, W.: Dairy-manure derived biochar effectively sorbs lead and atrazine, Environ. Sci. Technol., 43, 3285–3291, 2009.

Case, S. D. C., McNamara, N. P., Reay, D. S., and Whitaker, J.: Can biochar reduce soil greenhouse gas emissions from a Miscanthus bioenergy crop?, GCB Bioenergy, 6, 76–89, 2014.

Chaney, R. L., Angle, J. S., Broadhurst, C. L., Peters, C. A., Tappero, R. V., and Sparks, D. L.: Improved understanding of hyperaccumulation yields commercial phytoextraction and phytomining technologies, J. Environ. Qual., 36, 1429–1443, 2007.

Choppala, G. K., Bolan, N. S., Megharaj, M., Chen, Z., and Naidu, R.: The influence of biochar and black carbon on reduction and bioavailability of chromate in soils, J. Environ. Qual., 41, 1175–1184, 2012.

Clemente, R., Walker, D. J., Pardo, T., Martínez-Fernández, D., and Bernal, M. P.: The use of a halophytic plant species and organic amendments for the remediation of a trace elements-contaminated soil under semi-arid conditions, J. Hazard. Mater., 223–224, 63–71, 2012.

Conesa, H. M., Faz, A., and Arnaldos, R.: Initial studies for the phytostabilization of a mine tailing from the Cartagena-La Unión mining district (SE Spain), Chemosphere, 66, 38–44, 2007.

Cui, L., Li, L., Zhang, A., Pan, G., Bao, D., and Chang, A.: Biochar amendment greatly reduces rice Cd uptake in a contaminated paddy soil: A two-year field experiment, Bioresources, 6, 2605–2618, 2011.

Cui, L., Pan, G., Li, L., Yan, J., Zhang, A., Bian, R., and Chang, A.: The reduction of wheat Cd uptake in contaminated soil via biochar amendment: A two-year field experiment, Bioresources, 7, 5666–5676, 2012.

Cunningham, S. D. and Ow, D. W.: Promises and prospects of root zone of crops phytoremediation, Plant Physiol., 110, 715–719, 1996.

de Abreu, C. A., Coscione, A. R., Pires, A. M., and Paz-Ferreiro, J.: Phytoremediation of a soil contaminated by heavy metals and boron using castor oil plants and organic matter amendments. J. Geochem. Explor., 123, 3–7, 2012.

Elad, Y., Cytryn, E., Meller Harel, Y., Lew, B., and Graber, E. R.: The Biochar effect: Plant resistance to biotic stresses, Phytopathol. Mediterr., 50, 335–349, 2011.

Fellet, G., Marchiol, L., Delle Vedove, G., and Peressotti, A.: Application of biochar on mine tailings: effects and perspectives for land reclamation, Chemosphere, 83, 1262–1297, 2011.

Fellet, G., Marmiroli, M., and Marchiol, L.: Elements uptake by metal accumulator species grown on mine tailings amended with three types of biochar, Sci. Total Environ., 468–469, 598–608, 2014.

Freitas, E. V., Nascimento, C. W., Souza, A., and Silva, F. B.: Citric acid-assisted phytoextraction of lead: A field experiment, Chemosphere, 92, 213–217, 2013.

Gabos, M. B., Casagrande, G., Abreu, C. A., and Paz-Ferreiro, J.: Use of organic matter to mitigate multicontaminated soil and sunflower plants as phytoextractor, R. Bras. Eng. Agric. Ambiental, 15, 1298–1306, 2011.

Gascó, G., Paz-Ferreiro, J. and Méndez, A.: Thermal analysis of soil amended with sewage sludge and biochar from sewage sludge pyrolysis, J. Therm. Anal. Calorim., 108, 769–775, 2012.

Giansoldati, V., Tassi, E., Morelli, E., Gabellieri, E., Pedron, F., and Barbafieri, M.: Nitrogen fertilizer improves boron phytoextraction by Brassica juncea grown in contaminated sediments and alleviates plant stress, Chemosphere, 87, 1119–1125, 2012.

Graber E. R., Meller-Harel, Y., Kolton, M., Cytryn, E., Silber, A., Rav David, D., Tschansky, L., Borenshtein M., and Elad, Y.: Biochar impact on development and productivity of pepper and tomato grown in fertigated soilless media, Plant Soil, 337, 481–496, 2010.

Hartley, W., Dickinson, N. M., Riby, P., and Lepp, N. W.: Arsenic mobility in brownfield soils amended with green waste compost or biochar and planted with Miscanthus, Environ. Pollut., 157, 2654–2662, 2009.

Hossain, M. K., Strezov, V., Chan, K. Y., and Nelson, P. F.: Agronomic properties of wastewater sludge biochar and bioavailability of metals in production of cherry tomato (Lycopersicon esculentum), Chemosphere, 78, 1167–1171, 2010.

Houben, D., Evrard, L., and Sonnet, P.: Mobility, bioavailability and pH-dependent leaching of cadmium,zinc and lead in a contaminated soil amended with biochar, Chemosphere, 92, 1450–1457, 2013a.

Houben, D., Evrard, L., and Sonnet, P.: Beneficial effects of biochar application to contaminated soils on the bioavailability of Cd, Pb and Zn and the biomass production of rapeseed (Brassica napus L.), Biomass Bioenerg., 57, 196–204, 2013b.

Huang, H., Wang, K., Zhu, Z., Li, Y., He, Z., Yang, X. E., and Gupta, D. K.: Moderate phosphorus application enhances Zn mobility and uptake in hyperaccumulator Sedum alfredii, Environ. Sci. Pollut. R., 20, 2844–2853, 2013.

Jeffery, S., Verheijen, F. G. A., van der Velde, M., and Bastos, A. C.: A quantitative review of the effects of biochar application to soils on crop productivity using meta-analysis, Agr. Ecosyst. Environ., 144, 175–187, 2011.

Jiang, J., Xu, R. K., Jiang, T. Y., and Li, Z.: Immobilization of Cu(II), Pb(II) and Cd(II) by the addition of rice straw derived biochar to a simulated polluted Ultisol, J. Hazard. Mater., 229–230, 145–150, 2012.

Kabata-Pendias, A. and Pendias, H.: Trace elements in soil and plants, 3rd Edn., CRC press, 403 pp., 2001.

Karami, N., Clemente, R., Moreno-Jiménez, E., Lepp, N., and Beesley, L.: Efficiency of green waste compost and biochar soil amendments for reducing lead and copper mobility and uptake to ryegrass (Lolium perenne), J. Hazard. Mater., 191, 41–48, 2011.

Komárek, M., Vaněk, A., and Ettler, V.: Chemical stabilization of metals and arsenic in contaminated soils using oxides-a review, Environ. Pollut., 172, 9–22, 2013.

Lehmann, J. and Joseph, S.: Biochar for environmental management: science and technology. Earthscan, London and Sterling, VA USA, 2009.

Lehmann, J., Rillig, M. C., Thies, J., Masiello, C. A., Hockaday, W. C., and Crowley, D.: Biochar effects on soil biota. A review, Soil Biol. Biochem., 43, 1812–1836, 2011.

Li, N. Y., Fu, Q. L., Zhuang, P., Guo, B., Zou, B., and Li, Z. A.: Effect of fertilizers on Cd uptake of Amaranthus Hypochondriacus, a high biomass, fast growing and easily cultivated potential Cd hyperaccumulator, Int. J. Phytoremediat., 14, 162–173, 2012.

Li, Y. M., Chaney, R., Brewer, E., Roseberg, R., Angle, J. S., Baker, A., Reeves, R., and Nelkin, J.: Development of a technology for commercial phytoextraction of nickel: economic and technical considerations, Plant Soil, 249, 107–115, 2003.

Lima, I. M., Boykin, D. L., Klasson, K. T., and Uchimiya, M.: Influence of post-treatment strategies on the properties of activated chars from broiler manure, Chemosphere, 95, 96–104, 2014.

Liu, X., Zhang, A., Ji, C., Joseph, S., Bian, R., Li, L., Pan, G., and Paz-Ferreiro, J.: Biochar's effect on crop productivity and the dependence on experimental conditions- a meta-analysis of literature data, Plant Soil, 373, 583–594, 2013.

Lu, H., Zhang, Y. Y., Huang, X., Wang, S., and Qiu, R.: Relative distribution of Pb^{2+} sorption mechanisms by sludge-derived biochar, Water Res., 46, 854–862, 2012.

Mandal, A., Purakayastha, T. J., Patra, A. K., and Sanyal, S. K.: Phytoremediation of arsenic contaminated soils by Pteris Vittata L. I. Influence of phosphatic fertilizers and repeated harvests, Int. J. Phytoremediat., 14, 978–995, 2012.

Marschner, H.: Mineral nutrition in higher plants, Academic, London, 1986.

Méndez, A., Barriga, S., Fidalgo, J. M., and Gascó, G.: Adsorbent materials from paper industry waste materials and their use in Cu(II) removal from water, J. Hazard. Mater., 165, 736–743, 2009.

Méndez, A., Gómez, A., Paz-Ferreiro, J., and Gascó, G.: Effects of biochar from sewage sludge pyrolysis on Mediterranean agricultural soils, Chemosphere, 89, 1354–1359, 2012.

Méndez, A., Paz-Ferreiro, J., Araujo, F., and Gasco, G.: Biochar from pyrolysis of de-inking paper sludge and its use in the treatment of a nickel polluted soil, J. Anal. Appl. Pyrol., doi:10.1016/j.jaap.2014.02.001, in press, 2014.

Méndez, M. O., Glenn, E. P., and Maier, R. M.: Phytostablization potential of quailbush for mine tailings, J. Environ. Qual., 36, 245–253, 2007.

Namgay, T., Singh, B., and Singh, B. P.: Influence of biochar application to soil on the availability of As, Cd, Cu, Pb, and Zn to maize (Zea mays L.), J. Aust. Soil Res., 48, 638–647, 2010.

Oo, A. N., Iwai, C. B., and Saenjan, P.: Soil properties and maize growth in saline and nonsaline soils using cassava-industrial waste compost and vermicompost with or without earthworms, Land Degrad. Dev., doi:10.1002/ldr.2208, in press, 2014.

Orlowska, E., Przybylowicz, W., Orlowski, D., Turnau, K., and Mesjasz-Przybylowicz, J.: The effect of mycorrhiza on the growth and elemental composition of Ni hyperaccumulating plant Berkheya coddii Roessler, Environ. Pollut., 159, 3730–3738, 2011.

Park, J. H., Choppala, G. H., Bolan, N. S., Chung, J. W., and Chuasavathi, T.: Biochar reduces the bioavailability and phytotoxicity of heavy metals, Plant Soil, 348, 439–451, 2011.

Park, J. H., Choppala, G. H., Lee, S. J., Bolan, N., Chung, J. W., and Edraki, M.: Comparative sorption of Pb and Cd by biochars and its implication for metal immobilization in soil, Water Air Soil Poll., 224, 1711, doi:10.1007/s11270-013-1711-1, 2013.

Paz-Ferreiro, J. and Fu, S.: Biological indices for soil quality evaluation: perspectives and limitations, Land Degrad. Dev., doi:10.1002/ldr.2262, in press, 2014.

Paz-Ferreiro, J., Fu, S., Méndez, A., and Gasco, G.: Interactive effects of biochar and the earhworm Pontoscolex corethrurus on plant productivity and soil enzymes activities, J. Soils Sediments, doi:10.1007/s11368-013-0806-z, in press, 2014.

Prasad, M. N. V.: Phytoremediation of metal-polluted ecosystems: hype for commercialization, Russ. J. Plant Physiol., 50, 686–700, 2003.

Rajkumar, M., Prasad, M. N. V., Swaminathan, S., and Freitas, H.: Climate change driven plant–metal–microbe interactions, Environ. Int., 53, 74–86, 2013.

Roy, M. and McDonald, L. M.: Metal uptake in plants and health risk assessments in metal-contaminated smelter soils, Land Degrad. Dev., doi:10.1002/ldr.2237, in press, 2014.

Shao, H., Chu, L. Y., Xu, G., Yan, K., Zhang, L. H., and Sun, J. N.: Progress in phytoremediating heavy-metal contaminated soils, in: Detoxification of Heavy Metals, edited by: Sherameti, I. and Varma, A., Springer, 73–90, 2011.

Siebielic, G. and Chaney, R. L.: Testing amendments for remediation of military range contaminated soil, J. Environ. Manage., 108, 8–13, 2012.

Sizmur, T., Wingate, J., Hutchings, T., and Hodson, M. E.: Lumbricus terrestris L. does not impact on the remediation efficiency of compost and biochar amendments, Pedobiologia, 54, S211–S216, 2011.

Stevens, D. P., McLaughlin, M. J., and Heinrich, T.: Determining toxicity of lead and zinc runoff in soils: salinity effects on metal partitioning and on phytotoxicity, Environ. Toxicol. Chem., 22, 3017–3024, 2003.

Uchimiya, M., Lima, I. M., Klasson, K. T., and Wartelle, L. H.: Contaminant immobilization and nutrient release by biochar soil amendment: Roles of natural organic matter, Chemosphere, 80, 935–940, 2010.

Uchimiya, M., Klasson, K. T., Wartelle, L. H., and Lima, I. M.: Influence of soil properties on heavy metal sequestration by biochar amendment: 1. Copper sorption isotherms and the release of cations, Chemosphere, 82, 1431–1437, 2011a.

Uchimiya, M., Wartelle, L. H., Klasson, K. T., Fortier, C. A., and Lima, I. M.: Influence of pyrolysis temperature on biochar property and function as a heavy metal sorbent in soil, J. Agr. Food Chem., 59, 2501–2510, 2011b.

Uchimiya, M., Chang, S. C., and Klasson, K. T.: Screening biochars for heavy metal retention in soil: Role of oxygen functional groups, J. Hazard. Mater., 190, 432–444, 2011c.

Uchimiya, M., Cantrell, K. B., Hunt, P. G., Novak, J. M., and Chang, S. C.: Retention of heavy metals in a Typic Kandiudult amended with different manure-based biochars, J. Environ. Qual., 41, 1138–1149, 2012a.

Uchimiya, M., Bannon, D. I., Wartelle, L. H., Lima, I. M., and Klasson, K. T.: Lead retention by broiler litter biochars in small arms range soil: Impact of pyrolysis temperature, J. Agr. Food Chem., 60, 5035–5044, 2012b.

Vacca, A., Bianco, M. R., Murolo, M., and Violante, P.: Heavy metals in contaminated soils of the Rio Sitzerri floodplain (Sardinia, Italy): Characterization and impact on pedodiversity, Land Degrad. Dev., 23, 250–364, 2012.

van der Ent, A., Baker, A. J. M., Reeves, R. D., Pollard, A. J., and Schat, H.: Hyperaccumulators of metal and metalloid trace elements: facts and fiction, Plant Soil, 362, 319–334, 2013.

Vangronsveld, J., Herzig, R., Weyens, N., Boulet, J., Adriaensen, K., Ruttens, A., Thewys, T., Vassilev, A., Meers, E., Nehnevajova, E., Van der Lelie, D., and Mench, M.: Phytoremediation of contaminated soils and groundwater: lessons from the field, Environ. Sci. Pollut. R., 16, 765–794, 2009.

Wang, K., Zhang, J., Zhu, Z., Huang, H., Li, T., He, Z., Yang, X., and Alva, A.: Pig manure vermicompost (PMVC) can improve phytoremediation of Cd and PAHs co-contaminated soil by Sedum alfredii, J. Soils Sediments, 12, 1089–1099, 2012.

Wei, S., Zhu, J., Zhou, Q.X., and Zhan, J.: Fertilizer amendment for improving the phytoextraction of cadmium by a hyperaccumulator Rorippa globosa (Turcz.) Thell. J. Soils Sediments, 11, 915-922, 2011.

Wu, W., Yang, M., Feng, Q., McGrouther, K., Wang, H., Lu, H., and Chen, Y.: Chemical characterization of rice straw-derived biochar for soil amendment, Biomass Bioenerg., 47, 268–276, 2012.

Xue, Y. W., Gao, B., Yao, Y., Inyang, M., Zhang, M., Zimmerman, A. R., and Ro, K. S.: Hydrogen peroxide modification enhances the ability of biochar (hydrochar) produced from hydrothermal carbonization of peanut hull to remove aqueous heavy metals: Batch and column tests, Chem. Eng. J., 200–202, 673–680, 2012.

Yang, D., Zeng, D. H., Li, L. J., and Mao, R.: Chemical and microbial properties in contaminated soils around a magnesite mine in Northeast China, Land Degrad. Dev., 23, 256–262, 2012.

Zheng, R. L., Cai, C., Liang, J. H., Huang, Q., Chen, Z., Huang, Y. Z., Arp, H. P. H, and Sun, G. X.: The effects of biochars from rice residue on the formation of iron plaque and the accumulation of Cd, Zn, Pb, As in rice (Oryza sativa L.) seedlings, Chemosphere, 89, 856–863, 2012.

Zhou, Q. X., Cui, S., Wei, S. H., Zhang, W., Cao, L., and Ren, L. P.: Effects of exogenous chelators on phytoavailability and toxicity of Pb in Zinnia elegans Jacq, J. Hazard. Mater., 146, 341–346, 2007.

Thermal shock and splash effects on burned gypseous soils from the Ebro Basin (NE Spain)

J. León[1], **M. Seeger**[2,4], **D. Badía**[3], **P. Peters**[2], **and M. T. Echeverría**[1]

[1]Dept. of Geography and Land Management, University of Zaragoza, Spain
[2]Soil Physics and Land Management, Wageningen University, the Netherlands
[3]Dept. of Agricultural Science and Environment, University of Zaragoza, Spain
[4]Physical Geography, Trier University, Germany

Correspondence to: J. León (fcojleon@unizar.es)

Abstract. Fire is a natural factor of landscape evolution in Mediterranean ecosystems. The middle Ebro Valley has extreme aridity, which results in a low plant cover and high soil erodibility, especially on gypseous substrates. The aim of this research is to analyze the effects of moderate heating on physical and chemical soil properties, mineralogical composition and susceptibility to splash erosion. Topsoil samples (15 cm depth) were taken in the Remolinos mountain slopes (Ebro Valley, NE Spain) from two soil types: Leptic Gypsisol (LP) in a convex slope and Haplic Gypsisol (GY) in a concave slope. To assess the heating effects on the mineralogy we burned the soils at 105 and 205 °C in an oven and to assess the splash effects we used a rainfall simulator under laboratory conditions using undisturbed topsoil subsamples (0–5 cm depth of Ah horizon). LP soil has lower soil organic matter (SOM) and soil aggregate stability (SAS) and higher gypsum content than GY soil. Gypsum and dolomite are the main minerals (> 80 %) in the LP soil, while gypsum, dolomite, calcite and quartz have similar proportions in GY soil. Clay minerals (kaolinite and illite) are scarce in both soils. Heating at 105 °C has no effect on soil mineralogy. However, heating to 205 °C transforms gypsum to bassanite, increases significantly the soil salinity (EC) in both soil units (LP and GY) and decreases pH only in GY soil. Despite differences in the content of organic matter and structural stability, both soils show no significant differences ($P < 0.01$) in the splash erosion rates. The size of pores is reduced by heating, as derived from variations in soil water retention capacity.

1 Introduction

Fire is a natural factor of landscape evolution in Mediterranean ecosystems. Forest fires change the vegetation cover, the soil properties and trigger higher erosion rates that can contribute to rejuvenate the gullies (Hyde et al., 2007). The important socioeconomic changes that occurred in the last decades have contributed to an increase in forest fires (Shakesby, 2011), altering the fire regimes in terms of frequency, size, seasonality and recurrence as well as fire intensity and severity (Keeley, 2009; Doerr and Cerdà, 2005) causing severe effects on soils, water and vegetation (Bento-Gonçalves et al., 2012; Guénon et al., 2013).

Fire affects soil properties directly by heat impact and ash incorporation and reduction or elimination of plant cover (Bodí et al., 2014). Raindrop impact on burnt soil can lead to the structural degradation of the soil surface (Bresson and Boiffin, 1990; Poesen and Nearing, 1993; Ramos et al., 2003). Aggregate breakdown liberates small soil particles forming a surface crust with low permeability to air and water (Llovet et al., 2008; Mataix Solera et al., 2011). Fire severity affects the susceptibility of soils to degradation (Neary et al., 1999; Shakesby, 2011). The effects of heat on soil organic matter content (Mataix-Solera et al., 2002; González-Pérez et al., 2004), on structural stability (Mataix-Solera et al., 2011), on hydrophobic response (Bodí, 2012; Giovannini, 2012), and on infiltration capacity (Cerdà, 1998) have been also investigated. These characteristics represent factors in soil erodibility and soil degradation risk (Shakesby, 2011; Giovannini, 2012). This is why the vegetation cover and the litter are key factors on soil erosion after forest fires (Prats et al., 2013), which determines the debris flow formation (Riley et al., 2013). Besides, ash plays an important role in soil protection after the forest fire and after the first storms and winds (Cerdà and Doerr, 2008; León et al., 2013; Pereira et al., 2013).

In the central Ebro Valley (NE Spain), the tectonic history contributed to developing evaporative rock and then saline and gypsum soils (Dominguez et al., 2013). Moreover, the climate, lithology and relief promote the development of soils whose main constituent is gypsum ($CaSO_4 \cdot 2\,H_2O$), named gypseous soils (Herrero and Porta, 2000), occupying 7.2 % of the total area (Aznar et al., 2013a). The global distribution of gypseous soils is associated with regions of arid and semiarid climate (FAO, 1990; Verheye and Boyadgiev, 1997) represents edaphic modifications that exert strong effects in this important agricultural area of the central Ebro Valley. Moreover, the gypseous are common through out the badlands within Mediterranean environments and represent the dominant sediment source within these areas (Nadal et al., 2013).

Temperature controls some of the changes that occur in the soil as a result of the fire: protein degradation and biological tissue death at 40–70 °C; dehydration of roots or death at 48–54 °C; death of seeds at 70–90 °C; death of edaphic microorganisms at 50–121 °C; and destructive distillation and combustion of about 85 % of the organic horizon at 180–300 °C (Neary et al., 1999). In terms of temperatures attained during actual fires, Pérez-Cabello et al. (2012) measured maximum values between 400–800 °C during a prescribed fire, and heat transfer values of up to 110 °C during controlled burn plots in semiarid shrubland. The lower temperatures are associated with low vegetation cover characterized by small shrub patches on gypseous soils. Although the temperatures reached in burned semiarid woodlands may not be very high, they may be enough to cause some edaphic changes. Research on gypseous soils has focused on genesis and classification (Herrero and Porta, 2000; Badía et al., 2013), plant recovery (Badía and Martí, 2000), erosion processes (Gutiérrez and Gutiérrez, 1998) and mineralogy (Herrero and Porta, 2000; Herrero et al., 2009), but few studies address postfire hydrological response (León et al., 2011) and erodibility (León et al., 2012).

Rainfall simulations are a remarkably useful tool to assess changes on soil properties and erosion by raindrop impact, especially in semiarid areas, where the precipitation regime is irregular, having intense and short-duration events (Seeger, 2007; Cerdà et al., 2009; León et al., 2012).

The aim of this research is to analyze the heating effects of a moderate fire on physical, chemical soil properties, mineralogical composition and susceptibility to splash erosion by simulated rainfall.

2 Materials and methods

2.1 Study area

The gypseous soils were sampled in the Zuera Mountains in the central sector of the Ebro Basin (NE Spain) near the town of Remolinos. This area has been regularly affected by wildfires that promoted the development of shrub communities (*Retama sphaerocarpa* L., *Rosmarinus officinalis* L., *Lygeum spartum*, *Gypsophila struthium* subsp. *hispanica* and *Ononis tridentata*) and small patches of forest (*Pinus halepensis* Mill. with an understory of *Quercus coccifera* L.), covering the north slopes (Ruiz, 1990).

The research sites were selected on irregular relief (200–748 m), where gypseous soils predominate on the low elevation slopes (Badía et al., 2013). The climate is continental–Mediterranean, with a mean annual precipitation up to 450 mm, with maxima autumn, spring and extreme temperatures that can vary between −7.1 and 36.5 °C. The mean annual evapotranspiration reaches 1200 mm (using FAO56 by the Penman–Monteith method) and it is enhanced by strong winds, which makes the water deficit to be one of the highest in Europe (Herrero and Synder, 1997).

2.2 Soil sampling and preparation

Topsoil blocks (20 cm × 20 cm × 15 cm) were sampled in two different geomorphic position at the head of the slope with unburned gypseous soils, classified as Leptic Gypsisol (skeletic) by IUSS (2007), with a sequum Ahy-R (which we will call henceforth LP), and at the foot of the slope, with burned gypseous soils, Haplic Gypsisol (humic) with a sequum Ah-By-Cy (which we will call henceforth GY). The soils are described in more detail in Badía et al. (2013). The samples had different textures: GY was sandy loam, and LP was loamy. The GY topsoil has a different distribution of the main mineralogical components; gypsum, dolomite, calcite and quartz were around 18–25 % each. Gypsum and dolomite are the main minerals (> 80 %) in the LP topsoil. Clay minerals (kaolinite and illite) are scarce in both soils.

Fifteen replicates of each soil block were taken in clean cylinders (5 cm × 6 cm) and closed at the bottom with a metallic plate (Fig. 1). The samples were dried for one month at 25 °C, under controlled conditions, and heated to 35 °C (all samples), 105 °C (ten samples of each soil) and 205 °C (five samples of both soils) for 30 min to observe the soil changes after the heating.

2.3 Rainfall simulation

We used a portable rainfall simulator with a Lechler nozzle (Ref. 460.608.30) at 2 m height (Iserloh et al., 2011), in the Kraaijenhofvan de Leur Laboratory for water and sediment dynamics at Wageningen University. Small rainfall collectors were used for the detailed determination of the rainfall intensity on the simulation area (1.16 m^2). Drop size

Fig. 1. Sample preparation: **(A)** block extraction, **(B)** subsamples of the blocks; and **(C)** sub-sample preparation.

(d_{50} 1–1.5 mm) and the kinetic energy (5.81 J m^{-2} mm^{-1}) were measured with a Thies laser disdrometer (Fister et al., 2011). The simulation time was 20 min with rain intensity of 53 mm h^{-1} and demineralized water (with pH 7.1 and EC (soil salinity) of 36.7 μS cm^{-1}) was used. The simulation plot was divided into four subplots that previously were calibrated with the laser disdrometer (Fig. 2).

2.4 Soil analysis

Eight parameters were measured on the gypseous soils: pH, EC, soil organic matter (SOM), gypsum content (GC), soil aggregates stability (SAS), matric potential (pF or Ψ_p^m), soil texture and mineralogy of fine fraction (< 2 mm).

Soil samples were sieved to 2 mm. The pH was measured in a 1 : 5 dilution with distilled water with a pH meter, while the EC was measured in a 1 : 10 dilution (25 °C), with a conductivimeter after filtering the diluted sample (pore size filter 0.45 μm). For measuring SOM by weight difference, about 30 g of material was crushed and sieved to 2 mm, dried at 105 °C (24 h) and heated to 550 °C (3 h). SOM was measured by gravimetry. Gypsum content was measured by thermo-gravimetry (Vieillefon, 1979; Herrero and Porta, 2000; Lebron et al., 2009). SAS was measured using an Eijkelkamp wet sieving device (Schinner et al., 1996). To measure the matric potential (pF) we used sand box method of Eijkelkamp at pF 0, 1, 1.5 and 2. The volumetric pressure plate extractor (Richards, 1947) was used for measuring the pF 3 and 4.2 by breaking up and sieving the sample to 2 mm. The soil texture was measured with a Malvern Mastersizer 2000, correcting the clay value according to the Taubner et al. (2009) equation ($y = 3089x - 2899$). Mineralogical composition of the fine fraction (< 2 mm) of each soil type and after each heat treatment was determined separately. The analysis was performed using a Siemens D500 diffractometer, in the laboratory of Geology at Trier University. The diffractograms were evaluated by Diffrac Plus Release 2000 EVA 6.0.0.1. The splash effect was measured by differences of weight, before and after the rainfall simulation and two weeks after the experiment, after air drying at 35 °C. To avoid loss of material by percolation, samples were closed at the bottom with a metal plate.

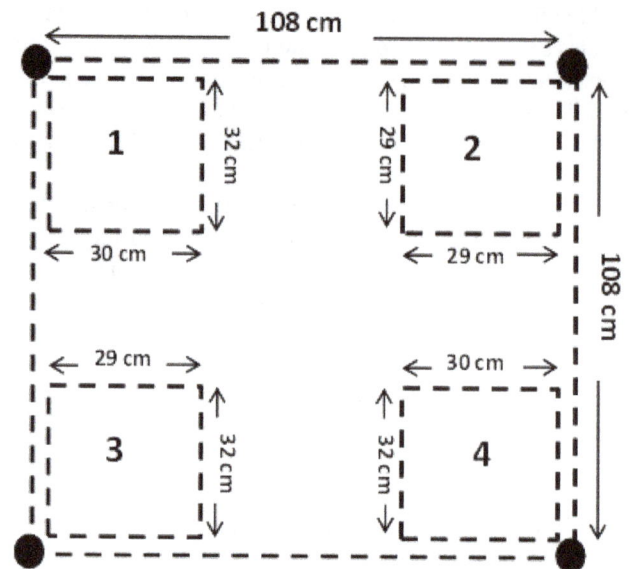

Fig. 2. Distribution of the subplots within the rainfall simulator wetting area.

2.5 Data analysis

A two-factor analysis of variance (Tukey's HSD $p < 0.05$) was used to test for differences between the response by soil type and temperature (independent variable), for splash, pH, EC, SOM, SAS and gypsum content (dependent variable), for each set of treatments. The separation of means was made according to Tukey's honestly significant difference test at an alpha level of 0.05 for all the parameters analyzed. Previous to ANOVA (analysis of variance), all variables were tested for normality using the Kolmogorov–Smirnov test. Percentage data (SOM, SAS and gypsum content) were transformed using arcsine of the square root and the others were ln-transformed to improve the normality of the data. The analyses were performed using the RStudio version 3 statistical software (Robinson and Hamann, 2011).

3 Results and discussion

3.1 Thermal shock

The values of SOM, SAS and GC are significantly different for both soil types (see Table 1). The LP soil sample, placed on head slope, has lower SOM, SAS and higher GC than the GY soil sample, a more developed soil on foot slope with higher plant cover.

Heating significantly decreases pH and increases EC and soil loss by splash effect for both soil types ($p < 0.05$). The decrease in pH is highest at treatment of 205 °C, with GY being more susceptible to a decrease of pH than LP (7.8–7.9, respectively; Table 1). The EC increased significantly with temperature, especially at 205 °C, doubling its value (from

Table 1. Mean value, standard deviation, mean comparison and their probability (ANOVA) according to the soils (S), temperature (T) and both ($S \times T$), of the studied variable. (***) $p < 0.001$; (**) $p < 0.01$; (*) $p < 0.1$. For each soil property (line), different letters (a, b, c) show significant differences ($P < 0.05$) among means, by post hoc Tukey test.

	Soil unit								
Soil	Leptic Gypsisol			Haplic Gypsisol			Probability		
Temp (°C)	25	105	205	25	105	205	Soil (S)	Temperature (T)	$S \times T$
pH	7.9 ± 0.07 bc	7.9 ± 0.00 c	7.8 ± 0.04 b	7.8 ± 0.00 b	7.8 ± 0.06 bc	7.4 ± 0.01 a	0.041*	0.039 *	< 0.001***
EC (mS cm^{-1})	2.1 ± 0.01 a	2.1 ± 0.00 a	4.2 ± 0.18 b	2.1 ± 0.01 a	2.1 ± 0.01 a	4.1 ± 0.02 b	0.974	< 0.001***	< 0.001***
SOM (%)	3.4 ± 0.11 ac	3.3 ± 0.03 ab	2.3 ± 0.01 a	3.9 ± 0.09 bc	4.8 ± 0.14 c	4.5 ± 0.11 bc	< 0.001***	0.506	0.003**
SAS (%)	33.8 ± 2.34 a	38.9 ± 10.12 ab	35.7 ± 1.61 ab	55.5 ± 11.56 ab	60.7 ± 1.44 b	52.4 ± 1.69 ab	< 0.001***	0.868	0.025*
Gypsum (%)	39.9 ± 2.13 c	46.2 ± 0.87 c	45.0 ± 4.60 c	15.5 ± 1.52 ab	22.3 ± 3.94 b	11.1 ± 0.34 a	< 0.001***	0.983	< 0.001***
Splash (kg m^{-2})	0.5 ± 0.3 a	0.7 ± 0.2 a	0.5 ± 0.5 a	0.8 ± 0.2 a	0.6 ± 0.2 a	1.4 ± 0.2 a	0.384	0.014*	0.240

2.1 dS m^{-1} for 35 °C to 4.2 dS m^{-1} for 205 °C, see Fig. 3). The SOM decreased significantly in the LP at a temperature of 205 °C but not in GY topsoil, which can be related to SOM quality. The GY topsoil, more organic than the LP topsoil, is not significantly affected by the heat (Table 1). SAS was lower on LP than on GY, independent of the treatment (33.8 and 55.5 %, respectively), and the gypsum content in GY soil (15.5 ± 1.5 %) is lower than in the LP soil (39.9 ± 2.1 %). After heating, gypsum content undergoes a considerable decrease. Another remarkable difference between soils is that the water retention capacity available to plants is greater in the GY than LP topsoil (Fig. 4).

The statistical significance of all the relationships between the dependent variables (splash, pH, EC, SOM, SAS and gypsum content) are given in Table 1. The differences are not significantly different between the parameters with the same lowercase letter (i.e., the splash for both soil types). The statistical significance is shown by soil types (LP and GY) and temperature effect (35, 105 and 205 °C), and it was significant ($p < 0.05$), using the ANOVA. Notably, the temperature effect shows statistical significance for the pH, EC and splash ($p < 0.05$).

In this study a small, but significant decrease was observed in pH at 205 °C in both soils. This reduction could be a result of the oxidation, the exposure of new surfaces, the dehydration of colloids and the consequent decrease of the soil buffer action (Giovannini et al., 1990). Similar pH decrease was obtained by Badía and Martí (2003) after heating a gypseous soil at 250 °C.

Mineralogical components of these gypseous soils can be modified by heating above 50 °C (Vieillefon, 1979; Herrero and Porta, 2000; Lebron et al., 2009) to transform the gypsum ($CaSO_4 \cdot 2 H_2O$) into bassanite ($CaSO_4 \cdot 1/2 H_2O$). The intensity of the changes in the soil depends on the temperature reached at different depths, on the duration of the heat pulse and the temperature peaks, and the stability of the various components of the soil (Gónzalez-Pérez et al., 2004; Terefe et al., 2008; Granged et al., 2011). Therefore the study was concentrated on the upper 6 cm of the soil samples. The range of heating 105–205 °C was employed because during

a fire only a small part of the heat is transmitted to the first few centimeters of soil (Badía et al., 2014). Pérez-Cabello et al. (2012) have recorded, maximum temperatures ranging from 400 to 800 °C on the soil surface and 29–110 °C in the upper soil centimeters (6.5, 2.5 and 1 cm of soil depth). Other authors recorded 50 °C at 2.5 cm of depth under shrub and 90 °C at the same depth under a less dense shrub (Luchessi et al., 1994). Aznar et al. (2013b) recorded 250 °C at 1 cm depth and down to 150 °C at 2 cm depth in a Gypsic Haploxerept soil (IUSS, 2007) in an experimental fire.

The increase of EC at 205 °C may be related to solubilization and incorporation of cations from the ashes (Badía and Martí, 2003) or to the increased amount of soluble inorganic ions resulting from the combustion of soil organic matter (Certini, 2005). The trend of reduced SOM and SAS by heating in LP soil was explained by some authors (Cerdà, 1998; Giovannini, 2012) as caused by organic matter decline and the consequent destruction of aggregates and increased soil erodibility (Llovet et al., 2009; Mataix-Solera et al., 2011).

Increases in organic matter after low intensity fires (Mataix-Solera et al., 2011) may explain the increase in soil aggregate stability, even with the passage of time (Bento-Gonçalves et al., 2012; Martín et al., 2012). The reduction of soil aggregate stability after heating may be related to a decrease of soil organic matter (DeBano et al., 1998; Cerdà, 1998; Badía and Martí, 2003), and changes in the mineral composition of the soil (Varela et al., 2002). Giovannini (2012) observed that soil aggregate stability increased at about 150° and again about 500 °C, from the combustion of the soil organic matter, as transformations of iron oxides cemented soil aggregates. Badía and Martí (2003) observed that organic matter in gypseous soils drops significantly from 2.8 to 2.2 % when heated to 250 °C. This decrease is accompanied by a significant reduction of the structural stability from 70 to 50 %. Novara et al. (2011) observed a redistribution of the OM by water erosion and degradation at the upper part of the hillslopes. Despite the variety of results and explanations provided by different authors, similar values of EC, pH, OM and SAS have been found in other works in gypseous soils

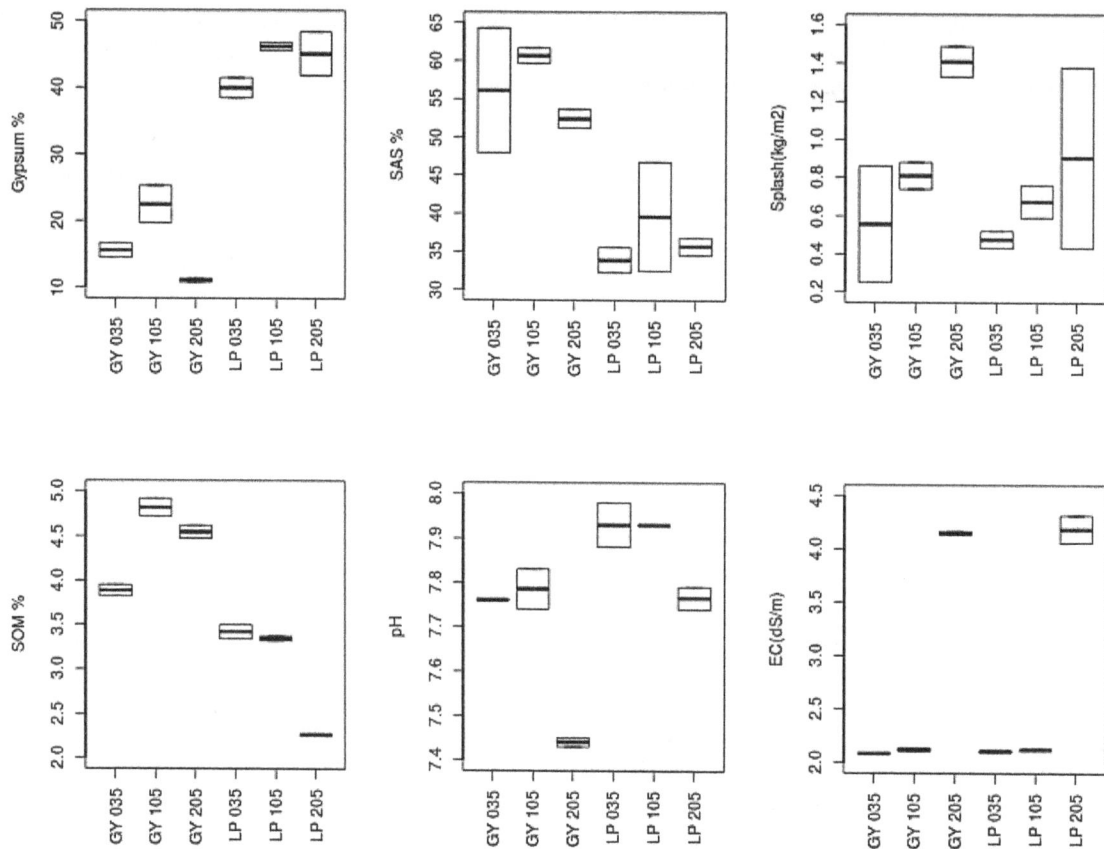

Fig. 3. Box plot of the soil properties.

(Badía and Martí, 2000, 2003; Cantón et al., 2001; Badía et al., 2008, 2013; Ries and Hirt, 2008; Lebron et al., 2009).

The water potential increases after heating at 105 °C can relate to experimental fires of low intensity; the water retention values decrease both at the surface and at depth, and the bulk density increases in the outermost layer due to the reduction of micropores (Ralston and Hatchell, 1971; Boyer et al., 1994). Wahlenberg et al. (1939), observed in annual fires that bulk density increased and soil porosity was reduced due to aggregate dispersion by impact of rain, which can clog pores (Moehring et al., 1966; Cerdà and Doerr, 2005). Giovannini (2012) found a decrease in porosity in sandy soil when the temperature increased, most notably at 170–220 °C, and increases in bulk density and decreases of soil porosity related to the decrease of soil organic matter by soil heating. However, Mallik and Fitzpatrick (1996) concluded that soil porosity increased directly after fire.

3.2 Mineralogical changes

The main mineralogical component that is affected by heat is the gypsum. The gypsum content in the topsoil sample was higher in LP, than in GY (Tables 1 and 3). The gypsum content was higher in the Ahy horizon of LP than in the GY, as the gypsum content did not vary with heating at 105 °C

in both soils, but it was significantly reduced at 205 °C in GY. The thermal increase transforms the gypsum into bassanite ($CaSO_4 \cdot 1/2\,H_2O$). In contrast, the LP soil with 50 % gypsum in its fine earth fraction is partly transformed into bassanite when it was heated at 205 °C.

Bassanite could not be found in the samples heated at 105 °C, but it was in those heated at 205 °C. When the gypsum is heated up to 105 °C, only a small fraction (13–19 %) of gypsum is removed; moreover, subsequent partial rehydration in the laboratory results in the formation of bassanite at relative moisture below saturation (Lebron et al., 2009). Some data show that, at 105 °C, the gypsum crystal loses 13–19 % of its mass (Artieda et al., 2006), corresponding to the two water molecules. Consequently Lebron et al. (2009) concluded that the temperature at which total water disappears is around 163 °C, since fast heating does not allow enough time for the water to diffuse through the crystal. For this reason in the mineralogical analysis bassanite did not show up until temperatures reached 205 °C.

3.3 Splash effects

The soil loss due to splash effect was higher in GY ($0.8 \pm 0.2\,kg\,m^{-2}$) than in LP soils ($0.5 \pm 0.3\,kg\,m^{-2}$). The first soil experienced a significant soil loss during the rainfall

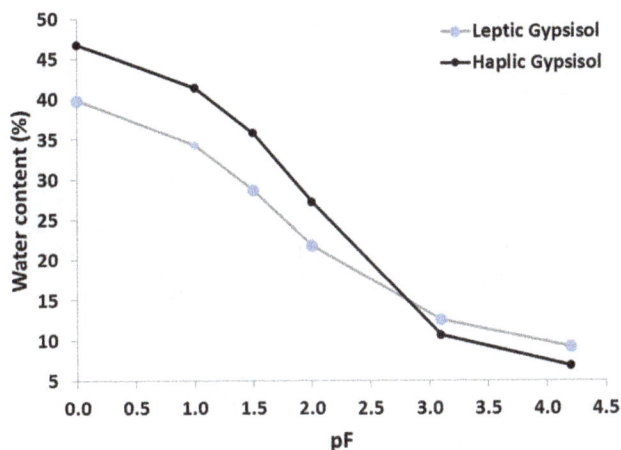

Fig. 4. Matric potential (pF) values at collected samples from the field.

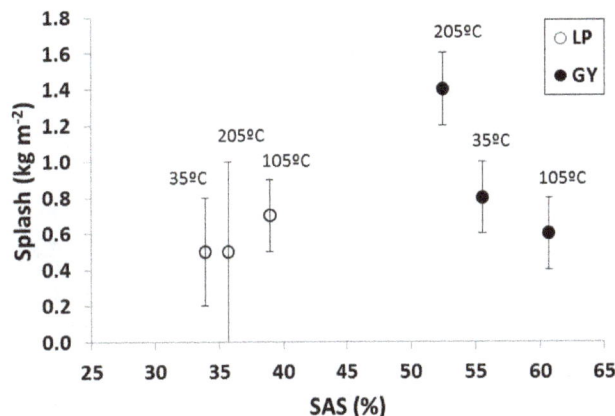

Fig. 5. Relationship between soil aggregate stability and splash erosion in both soils (means and standard deviations).

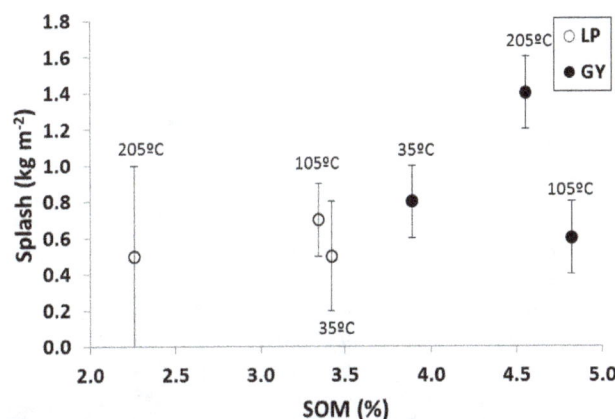

Fig. 6. Relationship between soil organic matter and splash erosion in both soils (means and standard deviations).

simulation experiment after heating $(1.4 \pm 0.2 \, \mathrm{kg \, m^{-2}})$, whilst the latter one was not sensitive to heating (Table 2).

The pF curve was similar for both soils, decreasing as the soil dries. The pF was higher in GY (46.7 ± 2.5) than in LP (39.9 ± 1.0) to values of 3 and 4.2 pF, and from here the trend is reversed. The LP soil $(9.2 \pm 2.4 \,\%)$ had higher pF than GY (6.8 ± 0.3) (Fig. 4).

The splash rates increased when soil aggregate stability (SAS) increased in LP $(0.5–0.7 \, \mathrm{kg \, m^{-2}})$. In GY the splash rates decreased when SAS increased $(0.6–1.4 \, \mathrm{kg \, m^{-2}})$ (Fig. 5).

There is no trend in the relationship between splash rate and soil organic matter. The splash rates increase when SOM decreases in LP $(0.5 \pm 0.5 \, \mathrm{kg \, m^{-2}})$. In GY the splash rates decrease when the SOM increases $(0.6 \pm 0.2 \, \mathrm{kg \, m^{-2}})$ (Fig. 6).

Splash erosion increases significantly by heating in GY but not in LP topsoil. The SAS and the splash rate are correlated positively in LP. While SAS is correlated negatively with the splash rate in GY (see Fig. 5), the soil aggregate stability is significantly correlated with soil type. In dry soil samples, crusting and cracking are detected, particularly in the LP. Herrero et al. (2009) explained that upon drying the salt migration is an important weathering process related to the formation and widening of crack and fissures, which can transmit additional water for further salt solution and crystallization. Splash production was higher in LP than in GY, in the samples heated at 105 °C; a possible explanation is that the SAS and SOM were lower in LP. The impact of raindrops on bare soil aggregates destroys the aggregates affected by fire (Moore and Singer, 1990; Fernández-Raga, 2013). Lower infiltration rates and higher surface runoff are explained by clogging pores and favoring entrainment of particles and nutrients, which may result in the formation of physical crusts (Mc Intyre, 1958; Mataix-Solera et al., 2011). Nevertheless, Bresson and Boiffin (1990) and Ries and Hirt (2008) did not

find any clear separation in inter-aggregates because of the short transport path or the transport via splash.

The difference between pF 3 and pF 4.2, indicating the amount of water capillary absorbable, has been reduced after wetting by rain and subsequent heating (see Table 2), but more experiments should be conducted to confirm these changes. The burned samples (GY) show a lower porosity (see Fig. 5), and the splash production is higher than in unburned samples (LP). However, the opposite occurs with the heated samples, because the porosity in GY is higher than in LP. Gypseous soil had a significantly low infiltration value, because the growth of gypsum crystals in pre-existing pores decreases water flow (Poch and Verplancke, 1997). The SOM and the splash rate are positively correlated in LP. While the soil aggregate stability with the splash rate is correlated negatively in GY to 105 °C, changing to positively at 205 °C (Fig. 7). Moreover, gypseous soils have low organic matter content and aggregate stability that elicits a slow micropore flow (Martí et al., 2001). Lasanta et al. (2000) and Desir et al. (1995) observed the same results in similar soils

Table 2. Mean value and standard deviation of the values of the pF (the base 10 logarithm of the water potential in centimeters). LP and GY before and after the rainfall simulation. (**) Difference between pF 4.2 and pF 3.

Treatments		pF							
Rainfall	Temp	0	1	1.5	2	3	4.2	Dif.**
	(°C)	0 kPa	1 kPa	5 kPa	10 kPa	100 kPa	1500 kPa	Dif.**
LP before	Field sample	39.9 ± 1.05	34.3 ± 2.12	28.7 ± 1.37	21.8 ± 0.78	12.6 ± 2.65	9.2 ± 2.48	3.4
LP after	35	37.9	32.8	29.3	23.4	27.6 ± 2.93	19.6 ± 0.30	8.0
	105	43.9	40.7	35.2	27.4	18.8 ± 0.06	15.0 ± 0.02	3.7
	205	42.0	40.5	35.2	27.8	21.2 ± 0.10	17.4 ± 0.13	3.9
GY before	Field sample	46.7 ± 2.51	41.4 ± 1.69	35.8 ± 0.80	27.2 ± 0.31	10.6 ± 0.74	6.8 ± 0.38	3.8
GY after	35	41.2	41.0	36.4	28.6	27.6 ± 0.56	15.0 ± 0.24	12.6
	105	50.4	44.7	38.7	30.0	15.9 ± 0.56	11.9 ± 0.13	3.9
	205	44.6	40.0	35.1	28.3	13.7 ± 0.71	10.0 ± 0.24	3.7

Table 3. Mineralogy of the soil (2 mm mesh).

	Soil mineralogy (%)					
Soil	Leptic Gypsisol			Haplic Gypsisol		
Temp (°C)	25	105	205	25	105	205
Gypsum	50	49	15	26	7	0
Bassanite	0	0	33	0	0	10
Calcite	4	0	2	18	24	25
Dolomite	33	39	20	23	28	27
Quartz	6	6	14	19	24	20
Feldspar	0	0	0	4	5	4
Illite	7	3	10	10	8	10
Kaolinite	0	3	6	0	4	4

where a microcrust was developed. The precipitation as microcrystalline gypsum could relate to root channels, where the moisture conditions were different from those within the groundmass (Aznar et al., 2013a). The progressive loss of weigh could be due by dissolution of some soluble mineral within the sample in the wetting phase or by the enhanced dehydration of some mineral as a consequence of oven heating. When water dissolves gypsum, some new pore space increases the water intake in the following saturation phase and this progressive enlargement of pre-existing pore volume may enhance further weathering through solution (Cantón et al., 2001).

4   Conclusions

Heating decreases significantly the pH and increases EC and the splash erosion. SOM and SAS are not modified by moderate heating. The SOM, positively correlated with the SAS, is higher in the GY than in the LP topsoil according to their slope position and plant cover. Nonetheless soil loss by splash is three times higher in GY than in LP topsoil. Soil loss by splash increased significantly ($P < 0.05$) at 205 °C only in GY topsoil. Heating at 205 °C caused a partial dehydration of gypsum to bassanite in both gypseous soils (LP and GY).

Acknowledgements. This research was supported by the Ministry of Science and Innovation BES-2008-003056, the CETSUS project (CGL2007-66644-C04-04/HIDCLI) and the Geomorphology and Global Change Research Group (D.G.A., 2011). The Spanish Army has supported this work at the San Gregorio CENAF. We acknowledge the support by Catheline Stoff, for providing ideas, material and instruments. We thank Kevin Hyde for the final review of the manuscript.

Edited by: A. Cerdà

References

Artieda, O., Herrero, J., and Drohan, P. J.: A refinement of the differential water loss method for gypsum determination in soils, Soil Sci. Soc. Am. J., 70, 1932–1935, 2006.

Aznar, J. M., Poch, R. M., and Badía, D.: Soil catena along gypseous woodland in the middle Ebro Basin: soil properties and micromorphology relationships, Span. J. Soil Sci., 3, 28–44, 2013a.

Aznar, J. M., González-Pérez, J. A., Badía, D., and Martí, C.: At what depth are the properties of a gypseous forest topsoil affected by burning?, Land Degrad. Dev., doi:10.1002/ldr.2258, 2013b.

Badía, D. and Martí, C.: Seeding and mulching treatments as conservation measures of two burned soils in Central Ebro Valley, NE Spain, Arid Soil Res. Rehab., 14, 219–232, 2000.

Badía, D. and Martí, C.: Plant ash and heat intensity effects on chemical and physical properties of two contrasting soils, Arid Land Res. Manag., 17, 23–41, 2003.

Badía, D., Martí, C., Aguirre, J., Echeverría, M. T., and Ibarra, P.: Erodibility and hydrology of Arid Burned Soils: soil type and revegetation effects, Arid Land Res. Manag., 22, 286–295, 2008.

Badía, D., Martí, C., Aznar, J. M., and León, J.: Influence of slope and parent rock on soil genesis and classification in semi-arid mountainous environments, Geoderma 193–194, 13–21, doi:10.1016/j.geoderma.2012.10.020, 2013.

Badía, D., Martí, C., Aguirre, J., Aznar, J. M., González-Pérez, J. A., De la Rosa, J., León, J., M., Echeverría, M. T., and Ibarra, P.: Wildfire effects on nutrients and organic carbon of a Rendzic Phaeozem in NE Spain: changes at cm-scale topsoil, Catena, 113, 267–275, 2014.

Bento-Gonçalves, A., Vieira, A., Úbeda, X., and Martin, D.: Fire and soils: key concepts and recent avances, Geoderma, 191, 3–13, doi:10.1016/j.geoderma.2012.01.004, 2012.

Bodí, M. B., Doerr, S. H., Cerdà, A., and Mataix-Solera, J.: Hydrological effects of a layer of vegetation ash on underlying wettable and water repellent soil, Geoderma, 191, 14–23, 2012.

Bodí, M. B., Martin, D. A., Balfour, V. N., Santín, C., Doerr, S. H., Pereira, P., Cerdà, A., and Mataix-Solera, J.: Wildland fire ash: Production, composition and eco-hydro-geomorphic effects, Earth-Sci. Rev., 130, 103–127, 2014.

Boyer, W. D. and Miller, J. H.: Effects of burning and brush treatments on nutrient and soil physical properties in young long-leaf pine stands, Forest Ecol. Manag., 70, 311–318, 1994.

Bresson, L. M. and Boiffin, J.: Morphological characterization of soil crust development stages on an experimental field, Geoderma, 47, 301–325, 1990.

Cantón, Y., Solé-Benet, A., Queralt, I., and Pini, R.: Weathering of a gypsum-calcareous mudstone under semi-arid environment at Tabernas, SE Spain: laboratory and field-based experimental approaches, Catena, 44, 111–132, 2001.

Cerdà, A.: Changes in overland flow and infiltration after a rangeland fire in a Mediterranean scrubland, Hydrol. Process., 12, 1031–1042, 1998.

Cerdà, A. and Doerr, S. H.: The influence of vegetation recovery on soil hydrology and erodibility following fire: an eleven-year investigation, Int. J. Wildland Fire, 14, 423–437, 2005.

Cerdà, A. and Doerr, S. H.: The effect of ash and needle cover on surface runoff and erosion in the immediate post-fire period, Catena, 74, 256–263, doi:10.1016/S0341-8162(02)00027-9, 2008.

Cerdà, A. Giménez-Morera, A., and Bodí, M. B.: Soil and water losses from new citrus orchards growing on sloped soils in the western Mediterranean basin, Earth Surf. Proc. Land., 34, 1822–1830, doi:10.1002/esp.1889, 2009.

Certini, G.: Effects of fire on properties of forest soils: a review, Oecologia, 143, 1–10, 2005.

DeBano, L. F., Neary, D. G., and Ffolliot, P. F.: Fire's effects on ecosystems, John Wiley & Sons, Inc., 1998.

Desir, G., Sirvent, J., Gutierrez, M., and Sancho, C.: Sediment yield from gypsiferous degraded áreas in the middle Ebro Basin, Phys. Chem. Earth, 20, 385–393, 1995.

Doerr, S. and Cerdà, A.: Fire effects on soil system functioning: new insights and future challenges, Int. J. Wildland Fire, 14 339–342, 2005.

Domínguez-Beisiegel, M., Herrero, J., and Castañeda, C.: Saline wetlands' fate in inland deserts: an example of 80 years' decline in Monegros, Spain, Land Degrad. Dev., 24, 250–265, doi:10.1002/ldr.1122, 2013.

Ellison, W. D.: Studies of raindrop erosion, Agr. Eng., 25, 131–136, 181–182, 1944.

FAO: Management of gypsiferous soils, Soils Bull. 62, FAO, Rome, 1990.

Fernández-Raga, M.: Splash erosion in recently-burnt area in North-West Spain, Geophys. Res. Abstracts, 15, EGU2013-216, 2013.

Fister, W., Iserloh, T., Ries, J. B., and Schmidt, R. G.: Comparison of rainfall characteristics of a small portable rainfall simulator and a portable wind and rainfall simulator, Z. Geomorphol., 55, 109–126, 2011.

Giovannini, G.: Fire in agricultural and forestal ecosystems: the effects on soil. Edizioni ETS, Pisa, Italy, 86 pp., 2012.

Giovannini, G., Lucchesi, S., and Giachetti, M.: Beneficial and detrimental effects of heating on soil quality, in: Fire in Ecosystem Dynamics – Mediterranean and Northern Perspectives. Third International Symposium on Fire Ecology, edited by: Goldammer, J. G. and Jenkins, M. J., Freiburg University, 95–102, 1990.

González-Pérez, J. A., González-Vila, F. J., Almendros, G., and Knicker, H.: The effect of fire on soil organic matter – a review, Environ. Int., 30, 855–870, 2004.

Granged, A. J. P., Jordán, A., Zavala, L. M., and Bárcenas, G.: Fire-induced changes in soil water repellency increased fingered slow and runoff rates following the 2004 Huelva wildfire, Hydrol. Process., 25, 1614–1629, 2011.

Guénon, R., Vennetier, M., Dupuy, N., Roussos, S., Pailler, A., and Gros, R.: Trends in recovery of Mediterranean soil chemical properties and microbial activities after infrequent and frequent wildfires, Land Degrad. Dev., 24, 115–128, doi:10.1002/ldr.1109, 2013.

Gutiérrez, M. and Gutiérrez, F.: Geomorphology of the Tertiary gypsum formations in the Ebro Depression (Spain), Geoderma, 87, 1–29, 1998.

Herrero, J. and Porta, J.: The terminology and the concepts of gypsum-rich soils, Geoderma, 96, 47–61, 2000.

Herrero, J. and Snyder, R. L.: Aridity and irrigation in Aragon, Spain, J. Arid Environ., 35, 535–547, 1997.

Herrero, J., Artieda, O., and Hudnall, W. H.: Gypsum, a tricky material, Soil Sci. Soc. Am., 73, 1757–1763, 2009.

Hyde, K., Woods, S. W., and Donahue, J.: Predicting gully rejuvenation after wildfire using remotely sensed burn severity data, Geomorphology, 86, 496–511, 2007.

Iserloh, T., Ries, J. B., Cerdà, A., Echeverría, M. T., Fister, W., Geißler, C., Kuhn, N. J., León, F. J., Peters, P., Schindewolf, M., Schmidt, J., Scholten, T., and Seeger, M.: Comparative measurements with seven rainfall s simulators on uniform bare fallow land, Z. Geomorphol., 57, 11–26, 2012.

Keeley, J. E.: Fire intensity, fire severity and burn severity: a brief review and suggests usage, Int. J. Wildland Fire, 18, 116–126, 2009.

Lasanta, T., García-Ruiz, J. M., Pérez-Rontomé, C., and Sancho-Marcén, C.: Runoff and sediment yield in a semi-arid environment: the effect of land management after farmland abandonment, Catena, 38, 265–278, 2000.

Lebron, I., Herrero, J., and Robinson, D. A.: Determination of gypsum content in dryland soils exploiting the gypsum-bassanite phase change, Soil. Sci. Soc. Am. J., 73, 403–411, 2009.

León, F. J., Mataix-Solera, J., Echeverría, M. T., and Badía, D.: Wildfire effects and temporal changes of water repellency in a gypsiferous soils, Geophys. Res. Abstracts, 13, EGU2011-3978, 2011.

León, J., Echeverría, M. T., Badía, D., Martí, C., and Álvarez, C.: Effectiveness of Wood chips cover at reducing erosion in two contrasted burnt soils, Z. Geomorphol., 57, 27–37, 2013.

Llovet, J., Josa, R., and Vallejo, V. R.: Thermal shock and rain effects on soil surface characteristics: a laboratory approach, Catena, 74, 227–234, 2008.

Llovet, J., Ruiz-Valera, M., Josa, R., and Vallejo, V. R.: Soil responses to fire in Mediterranean forest landscapes in relation to the previous stage of land abandonment, Int. J. Wildland Fire, 18, 222–232, 2009.

Mallik, A. U. and Fitzpatrick, E. A.: Thin section studies of *Calluna* heathland soils subject to prescribed burning, Soil Use Manag., 12, 143–149, 1996.

Martí, C. Badía, D., and Buesa, M. A.: Aggregate stability in Altoaragón soils: comparison of wet steving and lab rainfall simulation methods, Edafología, 8, 21–30, 2001.

Martín, A., Díaz-Raviña, M., and Carballas, T.: Short- and medium-term evolution of soil properties in Atlantic forest ecosystems affected by wildfires, Land Degrad. Dev., 23, 427–439, 2012.

Mataix-Solera, J. and Doerr, S. H.: Hydrophobicity and aggregate stability in calcareous topsoils from fire-affected pine forests in south-eastern Spain, Geoderma, 118, 77–88, 2004.

Mataix-Solera, J., Gómez, I., Navarro-Pedreño, J., Guerrero, C., and Moral, R.: Soil organic matter and aggregates affected by wildfire in a *Pinus halepensis* forest in Mediterranean environment, Int. J. Wildland Fire, 11, 107–114, 2002.

Mataix-Solera, J., Cerdà, A., Arcenegui, V., Jordán, A., and Zavala, L. M.: Fire effects on soil aggregation: A review, Earth-Sci. Rev., 109, 44–60, 2011.

Mc Intyre, D. S.: Soil splash and the formation of surface crust by raindrop impact, Soil Sci., 85, 261–266, 1958.

Moehring, D. M., Grano, C. X., and Basset, J. R.: Properties of forested loess soils after repeated prescribed burns, US For. Serv. Res. Note, SO-40, 4 pp., 1966.

Moore, D. C. and Singer, M. J.: Crust formation effects on soil erosion processes, Soil Soc. Am. J., 54, 1117–1123, 1990.

Nadal-Romero, E., Torri, D., and Yair, A.: Updating the badlands experience, Catena, 106, 1–3, 2013.

Neary, D. G., Klopatek, C. C., DeBano, L. F., and Ffolliot, P.: Fire effects on belowground sustainability: a review and synthesis, Forest Ecol. Manag., 122, 51–71, 1999.

Novara, A., Gristina, L., Bodí, M. B., and Cerdà, A.: The impact of fire on redistribution of soil organic matter on a mediterranean hillslopes under maquia vegetation type, Land Degrad. Dev., 22, 530–536, 2011.

Pereira, P., Cerdà, A., Úbeda, X., Mataix-Solera, J., Martin, D., Jordán, A., and Burguet, M.: Spatial models for monitoring the spatio-temporal evolution of ashes after fire – a case study of a burnt grassland in Lithuania, Solid Earth, 4, 153–165, 2013.

Pérez-Cabello, F., Cerdà, A., de la Riva, J., Echeverría, M. T., García-Martín, A., Ibarra, P., Lasanta, T., Montorio, R., and Palacios, V.: Micro-scale post-fire surface cover changes monitored using high spatial resolution photography in a semiarid environment: A useful tool in the study of post-fire soil erosion processes, J. Arid Environ., 76, 88–96, 2012.

Poch, R. M. and Verplancke, H.: Penetration resistance of gypsiferous horizons, Eur. J. Soil Sci., 48, 535–543, 1997.

Poesen, J. W. A. and Nearing, M. A.: Soil surface sealing and crusting, Catena Supplement 24, Catena Verlag, Cremlinge-Destedt, Germany, 1993.

Prats, S. A., Malvar, M. C., Simões-Vieira, D. C., MacDonald, L., and Keizer, J. J.: Effectiveness of hydromulching to reduce runoff and erosion in a recently burnt pine plantation in central Portugal, Land Degrad. Dev., doi:10.1002/ldr.2236, 2013.

Ralston, C. W. and Hatchell, G. E.: Effect of prescribed burning on physical properties of soil, in: Proceedings, Prescribed Burning Symposium, 14–16 April 1971, Charleston SC. US. For. Serv. Southeastern For. Exp. Stn., Asheville, NC, 68–85, 1971.

Ramos, M. C., Nacci, S., and Pla, I.: Effects of raindrop impact and its relationship with aggregate stability to different disaggregation forces, Catena, 53, 365–376, 2003.

Ries, J. B. and Hirt, U.: Permanence of soil surface on abandoned farmland in the Central Ebro Basin/Spain, Catena, 72, 282–296, 2008.

Riley, K. L., Bendick, R., Hyde, K. D., and Gabet, E. J.: Frequency–magnitude distribution of debris flows compiled from global data, and comparison with post-fire debris flows in the western U.S., Geomorphology, 191, 118–128, 2013.

Robinson A. P. and Hamann J. D.: Forest Analytics with R. Use R! Springer, ISBN 978-1-4419-7761-8, 2011.

Ruiz, J.: Mapa forestal de España. Hoja 7-4: Zaragoza. E. 1:200.000. Ed. Icona-MAPA, Madrid, 1990.

Schinner, F., Ohlinger, R., Kandeler, E., and Margesin, R.: Methods in soil biology, Springer-Verlag, Berllin, 1996.

Seeger, M.: Uncertainty of factors determining runoff and erosion processes as quantified by rainfall simulations, Catena, 71, 56–67, 2007.

Shakesby, R. A.: Post-wildfire soil erosion in the Mediterranean: Review and future research directions, Earth Sci. Rev., 105, 71–100, 2011.

Taubner, H., Roth, B., and Tippkotter, R.: Determination of soil texture: comparison of the sedimentation method and the laser-diffraction analysis, J. Plant Nutr. Soil Sci., 172, 161–171, 2009.

Terefe, T., Mariscal-Sancho, I., Peregrina, F., and Espejo, R.: Influence of heating on various properties of six Mediterranean soils. A laboratory study, Geoderma, 143, 273–280, 2008.

Ulery, A. L., Graham, R. C., and Amrhein, C.: Wood-ash composition and soil pH following intense burning, Soil Sci., 156, 358–364, 1993.

Varela, M. E., de Blas, E., Benito, E., and López, I.: Changes induced by forest fires in the aggregate stability and water repellency of soils in NW Spain, in: Forest Fire Research & Wildland Fire Safety, edited by: Viegas, D. X., Ed. Millpress, Rotterdam, 2002.

Verheye, W. H. and Boyadgiev, T. G.: Evaluating the land use potential of gypsiferous soils from field pedogenic characteristics, Soil Use Management, 13, 97–103, 1997.

Vieillefon, J.: Contribution a l'amélioration de l'étude analytique des sols gypseux, Cahiers ORSTOM, Pédologie, 17, 195–223, 1979.

Wahlenberg, W. G., Greene, S. W., and Reed, H. R.: Effects of fire and cattle grazing on longleaf pine lands as studied at McNeill, Mississippi, US Dep. Agr. Tech. Bull., 683, 52 pp., 1939.

Short-term changes in soil Munsell colour value, organic matter content and soil water repellency after a spring grassland fire in Lithuania

P. Pereira[1], X. Úbeda[2], J. Mataix-Solera[3], M. Oliva[4], and A. Novara[5]

[1]Environmental Management Center, Mykolas Romeris University, Ateities g. 20, 08303 Vilnius, Lithuania
[2]GRAM (Mediterranean Environmental Research Group), Department of Physical Geography and Regional Geographic Analysis, University of Barcelona, Montalegre, 6, 08001 Barcelona, Spain
[3]Environmental Soil Science Group, Department of Agrochemistry and Environment, Miguel Hernández University, Avda. de la Universidad s/n, Elche, Alicante, Spain
[4]Institute of Geography and Territorial Planning, University of Lisbon Alameda da Universidade, 1600-214, Lisbon, Portugal
[5]Dipartimento di Scienze agrarie e forestali, University of Palermo, 90128 Palermo, Italy

Correspondence to: P. Pereira (paulo@mruni.eu)

Abstract. Fire is a natural phenomenon with important implications on soil properties. The degree of this impact depends upon fire severity, the ecosystem affected, topography of the burned area and post-fire meteorological conditions. The study of fire effects on soil properties is fundamental to understand the impacts of this disturbance on ecosystems. The aim of this work was to study the short-term effects immediately after the fire (IAF), 2, 5, 7 and 9 months after a low-severity spring boreal grassland fire on soil colour value (assessed with the Munsell colour chart), soil organic matter content (SOM) and soil water repellency (SWR) in Lithuania. Four days after the fire a $400\,m^2$ plot was delineated in an unburned and burned area with the same topographical characteristics. Soil samples were collected at 0–5 cm depth in a $20\,m \times 20\,m$ grid, with 5 m space between sampling points. In each plot 25 samples were collected (50 each sampling date) for a total of 250 samples for the whole study. SWR was assessed in fine earth ($< 2\,mm$) and sieve fractions of 2–1, 1–0.5, 0.5–0.25 and $< 0.25\,mm$ from the 250 soil samples using the water drop penetration time (WDPT) method. The results showed that significant differences were only identified in the burned area. Fire darkened the soil significantly during the entire study period due to the incorporation of ash/charcoal into the topsoil (significant differences were found among plots for all sampling dates).

SOM was only significantly different among samples from the unburned area. The comparison between plots revealed that SOM was significantly higher in the first 2 months after the fire in the burned plot, compared to the unburned plot. SWR of the fine earth was significantly different in the burned and unburned plot among all sampling dates. SWR was significantly more severe only IAF and 2 months after the fire. In the unburned area SWR was significantly higher IAF, 2, 5 and 7 months later after than 9 months later. The comparison between plots showed that SWR was more severe in the burned plot during the first 2 months after the fire in relation to the unburned plot. Considering the different sieve fractions studied, in the burned plot SWR was significantly more severe in the first 7 months after the fire in the coarser fractions (2–1 and 1–0.5 mm) and 9 months after in the finer fractions (0.5–0.25 and $< 0.25\,mm$). In relation to the unburned plot, SWR was significantly more severe in the size fractions 2–1 and $< 0.25\,mm$, IAF, 5 and 7 months after the fire than 2 and 9 months later. In the 1–0.5- and 0.5–0.25 mm-size fractions, SWR was significantly higher IAF, 2, 5 and 7 months after the fire than in the last sampling date. Significant differences in SWR were observed among the different sieve fractions in each plot, with exception of 2 and 9 months after the fire in the unburned plot. In most cases the finer fraction ($< 0.25\,mm$) was more water repellent than the others. The comparison between plots for each sieve fraction showed significant differences in all cases IAF, 2 and 5 months after the fire. Seven months after the fire significant differences were only observed in the finer fractions (0.5–0.25 and < 0.25 mm) and after 9 months no significant differences were identified. The correlations between soil Munsell colour value and SOM were negatively significant in the burned and unburned areas. The correlations between Munsell colour value and SWR were only significant in the burned plot IAF, 2 and 7 months after the fire. In the case of the correlations between SOM and SWR, significant differences were only identified IAF and 2 months after the fire. The partial correlations (controlling for the effect of SOM) revealed that SOM had an important influence on the correlation between soil Munsell colour value and SWR in the burned plot IAF, 2 and 7 months after the fire.

1 Introduction

Fire is a natural phenomenon important to many ecosystems worldwide. It is accepted that fire plays an important role in plant adaptations and ecosystem development and distribution (Pausas and Kelley, 2009). It is well known that fire is a common occurrence and important disturbance in boreal ecosystems and a factor in the forest ecology of the region (Vanha-Majamaa et al., 2007). These ecosystems are strongly adapted to fire disturbance (Granstrom, 2001; Hylander, 2011; Pereira et al., 2013a, b). However, climate change, recent land-use change and fire suppression policies, may have important implications on the fire regime, fire severity and the role of fire in boreal environments (De Groot et al., 2013; Kouki et al., 2012; Van Bellen et al., 2010).

Fire has been recognized to be a soil-forming factor (Certini, 2014). Despite this, little research has been carried out on soil properties from boreal grassland ecosystems (Pereira et al., 2013a, c). The majority of studies on fire impacts on grassland soils have been carried out in tropical (Coetsee et al., 2010; Michelsen et al., 2004), subhumid (Knapp et al., 1998), desert (Ravi et al., 2009a; Whitford and Steinberger, 2012), arid (Vargas et al., 2012), semiarid (Dangi et al., 2010; Ravi et al., 2009b; Xu and Wan, 2008), temperate (Harris et al., 2007) and Mediterranean environments (Marti-Roura et al., 2013; Novara et al., 2013; Úbeda et al., 2005).

After a fire, the degree of direct and indirect impacts on soils (e.g. ash and soil erosion, water balance, organic matter, hydrophobicity, ash nutrient input, and microbiological changes) has consequences for the complex spatio-temporal distribution and availability of nutrients (Kinner and Moody, 2010; Malkinson and Wittenberg, 2011; Moody et al., 2013; Pereira et al., 2011; Sankey et al., 2012; Shakesby, 2011). The spatio-temporal extent of fire impacts depends on the fire severity, topography of the burned area and the post-fire meteorological conditions.

Fire can change soil colour. In fires of high severity the temperatures increase soil redness, especially at temperatures of 300–500 °C (Terefe et al., 2008) or > 600 °C (Ketterings and Bigham, 2000; Ulery and Graham, 1993), which is attributed to the destruction of the organic matter and increase in iron oxides such as hematite (Terefe et al., 2005). In contrast, low-severity fires darken the soil as a result of the incorporation of ash/charcoal into the soil surface and matrix (Eckmeier et al., 2007). These authors observed that soil lightness of colour had a significant negative correlation with charcoal carbon. Despite this knowledge, little is known about the soil lightness changes in the immediate period after the fire, when the major changes in soil properties and ash transport happen (Pereira et al., 2013a; Scharenbroch et al., 2012).

Few studies have been carried out about fire effects on soil colour lightness in comparison to unburned soils. Eckmeier et al. (2007) studied the effects of a slash-and-burn fire on soil lightness compared to soil in an unburned plot. However, the study was carried out immediately after the fire and 1 year after the fire. Major changes were not observed in detail in the year after the fire. The changes in soil lightness after fire can have implications for temperature (albedo increase or decrease) and microbiological activity (Certini, 2005; Gomez-Heras et al., 2006). Thus it is important to have high-resolution studies of fire effects on soil lightness.

Fire affects also the soil organic matter (SOM) chemical composition and quantity. Fire can increase or decrease SOM depending on the type of fire and severity, a parameter which considers the effects of biophysical variables such as topography, soil type, vegetation species and ecosystem affected (Certini et al., 2011; González-Peréz et al., 2004; Knicker, 2007). Low-severity fires can increase SOM in the immediate period after, due to the incorporation of charred material (De Marco et al., 2005), and high-severity fires tend to consume the major part of SOM due to the high temperatures (Neff et al., 2005). Depending on the rainfall and topography, important amounts of SOM can be also lost by erosion some months after a fire (Novara et al., 2011).

The soil Munsell colour value, chroma and hue are useful methods to estimate SOM content (Spielvogel et al., 2004; Viscarra Rossell et al., 2006). The Munsell colour value is used to describe the lightness of the soil, chroma measures the colour intensity and hue the shade of the soil (Thwaites, 2002). Usually, SOM content is negatively correlated with soil hue, value and chroma (Ibañez-Ascencio et al., 2013; Viscarra Rossell et al., 2006). However, this relationship depends on the SOM composition. In soils with high organic carbon, soil darkening is attributed to the composition and quantity of black humic substances (Schulze et al., 1993). Soil colour estimation has been carried out using visual observation in the field (Post et al., 1983), in a laboratory environment (Torrent et al., 1980; Scharenbroch et al., 2012), using diffuse reflectance spectrophotometers (Spielvogel et al.,

2004; Torrent and Barron, 1983) and more recently, smartphone applications (Gomez-Robledo et al., 2013).

It is widely known that fire can induce soil water repellency (SWR), with implications for soil infiltration, water and nutrient availability and an increase of runoff and erosion (DeBano et al., 2000; Mataix-Solera et al., 2013; Varela et al., 2005). The fire impacts on SWR depend on type of soil affected, temperature reached, fire severity, fire recurrence, time of residence, type and amount of vegetation combusted, ash produced and pre- and post-fire soil moisture content (Bodí et al., 2011; Doerr et al., 2000; Jordán et al., 2011; MacDonald and Huffman, 2004; Mataix-Solera and Doerr, 2004; Tessler at al., 2012; Vogelmann et al., 2012). Previous studies observed that after a fire, SWR is especially changed in soils that were wettable before the fire compared to those that are hydrophobic (Gimeno-Garcia et al., 2011). In wettable soils, fire usually increases SWR (Granged et al., 2011; Mataix-Solera and Doerr, 2004), meanwhile in hydrophobic soils, fire can slightly reduce or have no impact on SWR (Doerr et al., 1998; Jordán et al., 2011; Neris et al., 2013). However, this effect depends on fire severity. Rodriguez-Alleres et al. (2012) reported that moderate-to-high severity fires can increase SWR in naturally repellent soils. Soil heating increases SWR due the volatilization of organic compounds in the litter and topsoil. The heating of the soil surface layer develops a pressure gradient in the heated layer, causing the upward movement into the atmosphere of these compounds, while others move downwards. The decrease of soil temperature with depth forces SOM compounds to condense onto soil particles at or below the soil surface. Soil heating can redistribute and concentrate the natural substances in soil and litter, facilitate the bonding of these substances to soil particles, and increase their hydrophobicity as a result of conformational changes in their structural arrangement (Doerr et al., 2009). Heat changes the SOM composition through thermal alteration and chemical transformation. Heating also induces an increase in the content of aromatic compounds, the formation of complex high-molecular-weight compounds and low-molecular-weight oxo- and hydroxyacids (Atanassova and Doerr, 2011). Soil moisture controls SWR. Doerr and Thomas (2000) observed in coarse-textured burned and unburned soils that SWR disappeared when soil moisture exceeded 28 %. MacDonald and Huffman (2004) noted soil moisture thresholds where soils became hydrophilic were 10 % for unburned sites, 13 % for areas burned with low severity and 26 % for sites burned at moderate and high severity. Post-fire changes in SWR are not well understood. Doerr et al. (2009) stated that more detailed studies are needed to determine (i) the duration of fire-induced SWR in different vegetation types and (ii) the relative roles of physical, chemical, and biological factors in breaking down post-fire SWR.

Spring grassland fires are frequent in Lithuania. After the winter, farmers burn the dead grass in order to improve fields for spring and summer crops (Pereira et al., 2012a). Thus,

it is important to know the effects of these fires on soil properties in order to understand the impacts of this practice and their persistence in time, especially in this environment where few studies have been carried out. This study contributes to a better understanding of fire effects and short-term changes in soil properties in boreal grasslands. At this time, the use of fire for landscape management is forbidden in Lithuania but, frequently, farmers set fires and leave the area until the fires are extinguished, leading on many occasions to loss of infrastructure and impacts on natural resources (Mierauskas, 2012; Pereira et al., 2012a).

The aim of this work was to study the short-term temporal effects of a low-severity spring grassland fire on some surface soil properties (0–5 cm) such as soil colour value (assessed with the Munsell colour chart), SOM content and SWR, in order to observe if this grassland fire induced relevant short-term impacts on these soil properties. The study focused on the upper soil layer because previous studies have shown that fire effects on soil are especially limited to the first 5 cm (Marion et al., 1991; Blank et al., 2003), and especially in low-severity fires, where soil temperatures rarely exceed 100 °C at the surface and 50 °C at 5 cm (Agee, 1973).

2 Materials and methods

2.1 Study site and design

On 15 April 2011 an area of 20–25 ha near Vilnius (Lithuania) was affected by a wildfire. The burned area is located at coordinates 54°42' N, 25°08' E with an elevation of 158 m a.s.l. (above sea level). According to the local farmers, the fire was attributed to human causes resulting from the burning of grass and wood residues (Pereira et al., 2012a). The characteristics of the study area are described in Table 1. Fire severity was considered low based on the predominance of black ash and unburned patches (Pereira et al., 2013a). Four days after the fire, a plot of 400 m² was delineated (20 m × 20 m, with a grid with 5 m spacing between sampling points) in an unburned and burned area with the same topographical characteristics (flat area). In total, 25 samples (topsoil, 0–5 cm) were collected in each plot, immediately after burning (IAF) and 2, 5, 7 and 9 months later. Samples were stored in plastic bags, taken to the laboratory and air-dried for 24 h to constant weight. Subsequently, the samples were carefully sieved through a 2 mm mesh.

2.2 Laboratory analysis

The soil colour value was assessed using the Munsell colour chart (Viscarra Rossel et al., 2006) in the 2 mm sieved fraction. The Munsell value gives information about soil darkness/lightness. Low values correspond to dark soils and high values to light soils (Eckmeier et al., 2007). All the soil value analyses were carried out by the same person under the same light conditions. SOM content was

Table 1. Main characteristics of the study area.

Geological substrate (Kadunas et al., 1999)	Glacio-lacustrine deposits
Soil type (WRB, 2006)	Albeluvisols
[a, b]Texture (% sand, silt and clay) (USDA, 2004)	9.4 ($\pm$3.07), 63.5 ($\pm$8.14), 27.1 ($\pm$5.21) (Silt loam)
[a]pH	7.2 ($\pm$0.15)
[a]Organic matter content (%)	6.5 ($\pm$1.16)
Mean annual rainfall (mm) (Bukantis, 1994)	735
Mean annual temperature (°C) (Bukantis, 1994)	8.8
Dominant vegetation	Fall dandelion (*Leontodon autumnalis* L.) and sweet vernal grass (*Anthoxanthum odaratum* L.)

[a] Values based on unburned soil samples ($N = 25$).
[b] Sand: 2–0.05 mm, silt: 0.05–0.002 mm, clay: < 0.002 mm.

estimated by the loss-on-ignition (LOI) method using approximately 1 g of soil heated to 900 °C for 4 h (Avery and Bascomb, 1974) after drying at 105 °C for 24 h to remove the moisture. LOI was calculated according to the formula LOI = (Weight$_{105}$ − Weight$_{900}$) / Weight$_{105}$ × 100.

Soil texture of unburned samples was analysed using the Bouyoucos method (Bouyoucos, 1936) and pH with 1 : 2.5 deionized water (Table 1). Soil water repellency was assessed in the samples sieved through the 2 mm mesh (fine earth) and in the subsamples of all of the 250 samples divided into different soil sieve fractions of 2–1, 1–0.5, 0.5–0.25 and < 0.25 mm, as used in previous studies (Jordán et al., 2011; Mataix-Solera and Doerr, 2004). Soil sieving was done on the dried samples and the separation of fractions was carried out carefully, in order to not destroy the aggregates (Mataix-Solera and Doerr, 2004). In total 1250 SWR subsamples were analysed. Between 5 and 7 g of soil of each sample and subsample were placed in 60 mm diameter plastic dishes and exposed to a controlled laboratory environment (temperature of 20 °C and 50 % of air relative humidity) for 1 week in order to avoid potential effects of atmospheric conditions on SWR (Doerr, 1998; Doerr et al., 2005). The persistence of SWR was measured with the water drop penetration time (WDPT) method that involves placing three drops of distilled water onto the soil surface and registering the time required for the complete penetration of the drops (Wessel, 1988). The average time of the three drops was used to assess the WDPT of each sample and subsample. WDPT classes were assessed according to Doerr (1998) (Table 2).

Table 2. WDPT classes used in this work. Water drop penetration time measured in seconds (s) (according to Doerr, 1998).

WDPT classes	Wettable	Low	Strong	Severe
WDPT interval (s)	< 5	6–60	61–600	601–3600

2.3 Statistical analysis

Data normality and homogeneity of the variances were tested with the Shapiro–Wilk test (Shapiro and Wilk, 1965) and Levene test, respectively. Data were considered normal and homogeneous at a $p > 0.05$. In this study, data did not follow the normal distribution and displayed heteroscedasticity. Thus the alternative non-parametric Kruskal–Wallis ANOVA (analysis of variance) test (K–W) was used to analyse differences among sampling dates and SWR according to the aggregate sieve fractions in each plot. The comparison between plots was carried out with the Mann–Whitney U test (MU). If significant differences at a $p < 0.05$ were observed after the K–W test, a Tukey HSD (honestly significant difference) post-hoc test was applied.

Correlations between the variables were carried out with the Pearson coefficient of correlation after variables SQR transformation, in order for the data to meet normality requirements. In the case of SWR, the coefficient of correlation just considered the fine-earth samples. A partial correlation was carried out between Munsell colour value and SWR, using SOM content as a control variable in order to observe if SOM influenced the correlation between Munsell colour value and SWR. Significant correlations were considered at a $p < 0.05$. Statistical analyses were carried out with STATISTICA 6.0 (Statsoft Inc., 2006).

3 Results

3.1 Soil Munsell colour value

The soil colour in the burned and unburned plots was in the soil Munsell 10YR hue for all the samples. The Munsell colour value was significantly different among sampling dates in the burned plot (K–W = 35.37, $p < 0.001$), but not in the unburned area (K–W = 9.20, $p > 0.05$) (Fig. 1). Soil was significantly darker in the burned than in the unburned plot for all sampling dates, IAF (MU = 1, $p < 0.001$), 2 months (MU = 69, $p < 0.001$), 5 months (MU = 46, $p < 0.001$), 7 months (MU = 56, $p < 0.001$) and 9 months later (MU = 84, $p < 0.001$).

3.2 Soil organic matter

SOM content was not significantly different among sampling dates in the burned plot (K–W = 6.60, $p > 0.05$), but it was in the unburned area (K–W = 20.96 $p < 0.001$) (Fig. 2). SOM

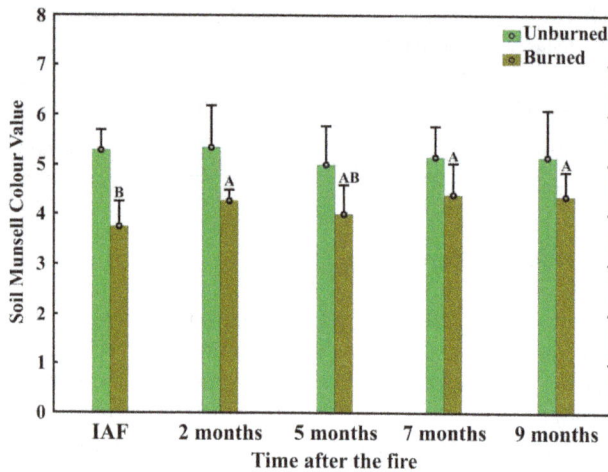

Fig. 1. Evolution of soil Munsell value in the unburned and burned plots in the post-fire sampling dates (bars represent ± standard deviation). Different letters indicate significant differences ($p < 0.05$) among times. IAF (Immediately After the Fire).

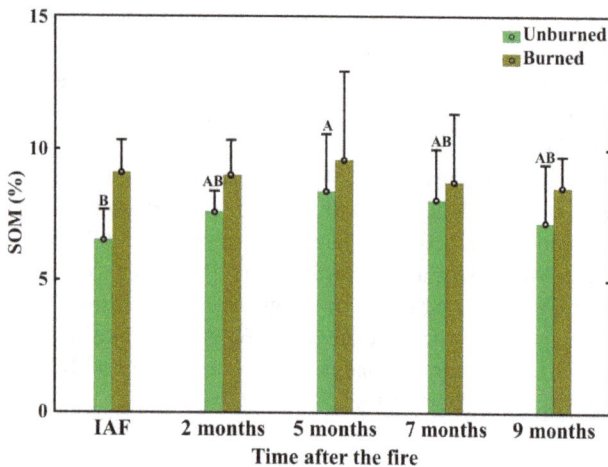

Fig. 2. Evolution of SOM content in the unburned and burned plots in the post-fire sampling dates (bars represent ± standard deviation). Different letters indicate significant differences ($p < 0.05$) among times.

content was significantly higher in the burned plot than in the unburned plot IAF (MU = 31, $p < 0.001$) and 2 months after the fire (MU = 116, $p < 0.001$). Five (MU = 266, $p > 0.05$), 7 (MU = 299, $p > 0.05$) and 9 months (MU = 254, $p > 0.05$) after the fire no significant differences were observed between plots.

3.3 Soil water repellency

The SWR of the fine earth was significantly different among sampling dates in the burned (K–W = 94.18, $p < 0.001$) and unburned plots (K–W = 45.65, $p < 0.001$) (Fig. 3). With time a decrease of SWR was observed in the burned area. In the unburned area SWR was significantly more severe IAF, 2,

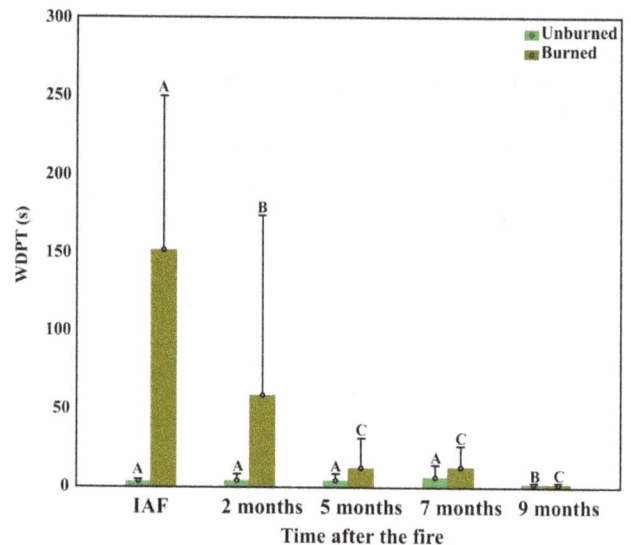

Fig. 3. Evolution of SWR (composite sample) in the unburned and burned plots in the post-fire sampling dates (bars represent ± standard deviation). Different letters indicate significant differences ($p < 0.05$) among times.

5 and 7 months after the fire than 9 months later. SWR was significantly high in the burned soil in the first two sampling dates, IAF (MU = 0, $p < 0.001$) and 2 months after the fire (MU = 26, $p < 0.001$). No significant differences were observed between plots 5 (MU = 249, $p > 0.05$), 7 (MU = 238, $p > 0.05$) and 9 months (MU = 267, $p > 0.05$) after the fire.

In relation to the analysed sieved soil fractions, significant differences were observed in SWR among all sieve fractions in the burned and unburned areas (Table 3a). In the burned area significant differences were observed in the coarser sieve fractions (2–1 and 1–0.5 mm) in the first 7 months after the fire, whereas in the finer fractions (0.5–0.25 and < 0.25 mm), significant differences among fractions were not identified until 9 months later (Table 4). In the unburned area's aggregate-size fractions of 2–1 and < 0.25 mm SWR was more severe IAF, 5 and 7 months after the fire than 2 and 9 months after the fire. In the size fractions 1–0.5 and 0.5–0.25 mm, SWR was significantly more persistent IAF, 2, 5 and 7 months after the fire than 9 months after the fire (Table 4).

The SWR was higher in the finer fractions (0.5–0.25 and < 0.25 mm) than in the coarser fractions (2–1 and 1–0.5 mm) (Table 4). Significant differences were observed in the studied sieve fractions in SWR in each plot during the experimental period, with the exception of 2 and 9 months after the fire in the unburned plot (Table 3b). In the unburned and burned plots for all sampling dates, the SWR in the finer fraction (< 0.25 mm) was significantly more severe than in the other sieve fractions, except for IAF and 5 months after the fire in the unburned plot, where no significant differences were observed between 0.5–0.25 mm and < 0.25 mm sieve fractions

Table 3. Results of Kruskal–Wallis ANOVA and Mann–Whitney tests for SWR according to the analysed sieved fractions, (a) time, (b) soil sieved fractions in the same plot and (c) between plots in each soil sieved fraction.

(a)	Sieved fraction mm	Plots	K–W	p
	2–1	Unburned	43.07	***
		Burned	75.25	***
	1–0.5	Unburned	35.39	***
		Burned	78.17	***
	0.5–0.25	Unburned	41.17	***
		Burned	87.28	***
	< 0.25	Unburned	62.89	***
		Burned	89.44	***
(b)	Sampling date	Plot	K–W	p
	IAF	Unburned	25.14	***
		Burned	33.29	***
	2 months	Unburned	4.06	n.s.
		Burned	24.35	***
	5 months	Unburned	41.30	***
		Burned	9.07	*
	7 months	Unburned	36.21	***
		Burned	27.07	***
	9 months	Unburned	4.25	n.s.
		Burned	8.60	*
(c)	Sampling date	Sieve fractions mm	MU	p
	IAF	2–1	30	***
		1–0.5	30	***
		0.5–0.25	13.50	***
		< 0.25	15	***
	2 months	2–1	39	***
		1–0.5	10.50	***
		0.5–0.25	22.50	***
		< 0.25	13.00	***
	5 months	2–1	30.50	***
		1–0.5	29.50	***
		0.5–0.25	67	***
		< 0.25	164	*
	7 months	2–1	255	n.s.
		1–0.5	265	n.s.
		0.5–0.25	193	*
		< 0.25	196	*
	9 months	2–1	298.5	n.s.
		1–0.5	297.5	n.s.
		0.5–0.25	299	n.s.
		< 0.25	225	n.s.

n.s.: non-significant at a $p < 0.05$. < 0.05*, and < 0.001***.
IAF (immediately after the fire).

(Table 4). Significant differences were also found in SWR between both plots IAF, 2 and 5 months after the fire. Seven months after the fire significant differences were only observed in the finer fractions (0.5–0.25 and < 0.25 mm) and 9 months later no significant differences were identified between plots in any of the sieve fractions (Table 3c).

In the unburned plot, for all the sampling dates and aggregate sieve fractions analysed, samples were predominantly wettable (Fig. 4a, c, e, i), with the exception of 7 months after the fire where the finer fraction (< 0.25 mm) samples were classified as "low". In the burned plot the SWR was classified mainly as "low" (Fig. 4b, d, f, h, j). However, SWR was classified as strong and severe IAF in the finer fraction (< 0.25 mm). With time SWR persistence was reduced in all the fractions and 9 months after the fire the samples were all wettable, with SWR < 5 s (Fig. 4i, j).

3.4 Correlation between variables

In the unburned area the correlations between soil Munsell colour value and SOM were always negatively significant ($p < 0.05$). The correlations between soil Munsell colour value and SWR and between SOM and SWR were not significant in any case (Table 5). The correlations between Munsell colour value and SOM in the burned area were negatively significant for all sampling dates. However, the correlations between Munsell colour value and SWR and between SOM and SWR were only significant IAF, 2, and 7 months after the fire (7 months later only in the correlation between Munsell colour value and SWR). The coefficients of correlation decrease with time in all cases (Table 5). The partial correlation results showed that SOM controls the correlation between Munsell colour value and SWR, in the burned plot IAF, 2, and 7 months after the fire. IAF the original correlation was highly significant ($r = -0.81$, $p < 0.001$), being considerably reduced in the partial correlation ($r = 0.41$, $p < 0.01$), 2 months after the fire the original correlation was significant ($r = 0.39$, $p < 0.01$), disappearing in the partial correlation ($r = 0.26$, $p > 0.05$), and 7 months later the original correlation was significant ($r = 0.32$, $p < 0.05$), decreasing in the partial correlation ($r = 0.14$, $p > 0.05$) (Table 5).

4 Discussion

4.1 Soil Munsell colour value

Fire darkened the soil in the immediate period after the fire. Incomplete fuel combustion produces black ash (Úbeda et al., 2009), especially in low-severity fires, as in the present one, where the temperatures do not reach high values (Keeterings and Bigham, 2000). Normally, black ash is incorporated into the soil or can be eroded in the weeks following the fire (Pereira et al., 2013b), contributing to the darkening of the soil following the fire and the reduction of Munsell value as observed in this study and in previous reports (Ulery and

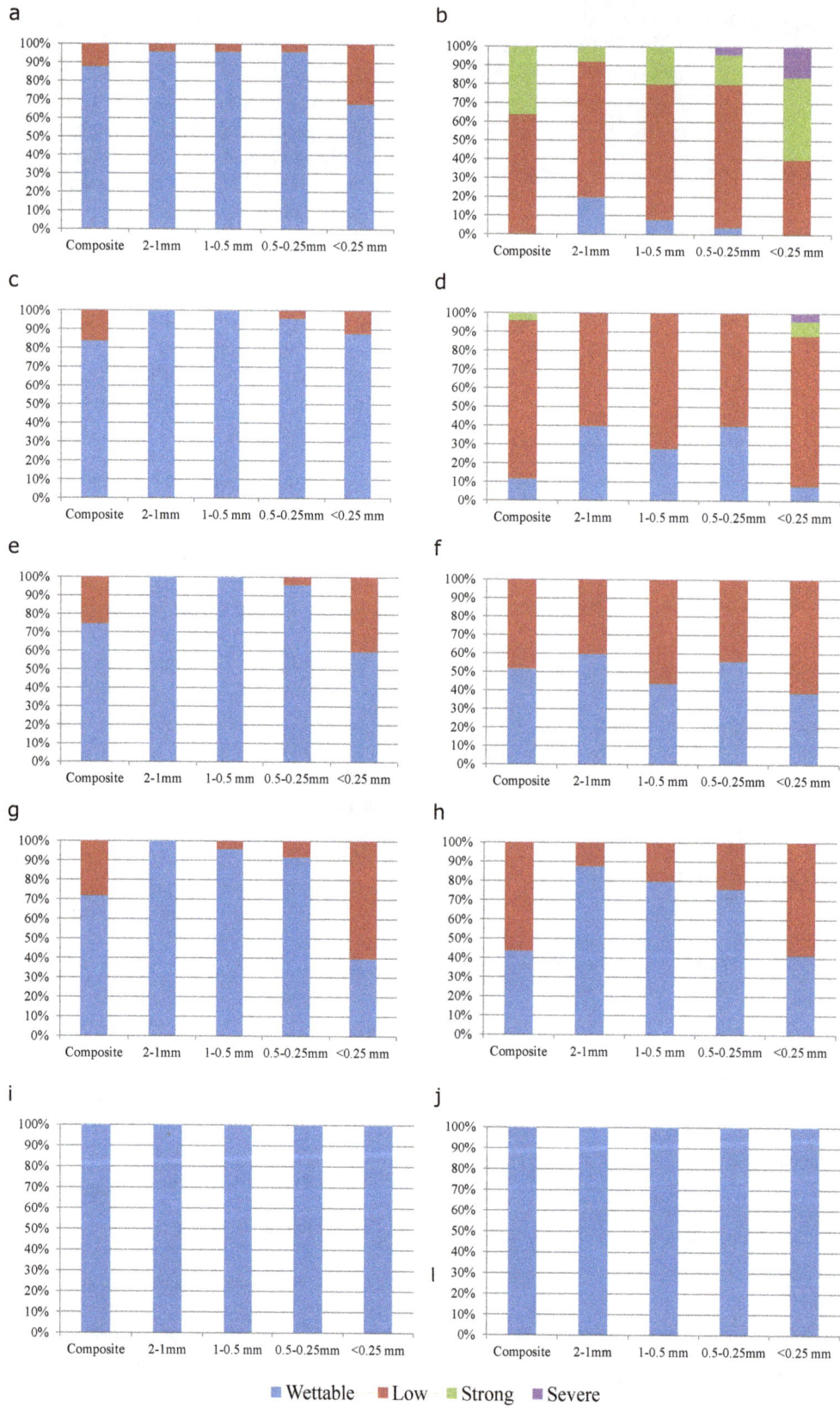

Fig. 4. Relative frequency of SWR for composite and sieved soil fractions, (**a**) unburned, after the fire; (**b**) burned, after the fire; (**c**) unburned, 2 months after the fire; (**d**) burned, 2 months after the fire; (**e**) unburned, 5 months after the fire; (**f**) burned, 5 months after the fire; (**g**) unburned, 7 months after the fire; (**h**) burned, 7 months after the fire; (**i**) unburned, 9 months after the fire; and (**j**) burned, 9 months after the fire.

Table 4. Water drop penetration time (s) in terms of the different size fractions for unburned and burned plots for different sampling dates. Statistical comparisons were carried out between times (upper case) and in each plot (different fractions in the same plot) during the studied sampling dates (lower case). Different letters represent significant differences at $p < 0.05$.

Sampling date	Plots	2–1 mm	1–0.5 mm	0.5–0.25 mm	< 0.25 mm
IAF	Unburned	1.73(0.78)Ab	2.02(1.91)Ab	3.12(7.29)Aab	15.44(37.42)Aa
	Burned	65.74(133.01)Ab	101.13(165.66)Ab	159.65(301.90)Ab	500.44(657.81)Aa
2 months	Unburned	1.57(0.58)B	1.62(0.74)A	1.78(1.36)A	3.21(4.86)B
	Burned	6.60(4.05)Bb	12.24(15.14)Bb	17.88(26.53)Bb	119.13(237.27)Ba
5 months	Unburned	1.72(0.62)Ab	1.73(0.61)Ab	2.69(3.69)Aa	11.66(16.02)Aa
	Burned	6.70(5.02)Bb	8.08(7.32)Bb	9.13(9.86)Cb	39.33(46.50)Ca
7 months	Unburned	2.12(0.79)Ab	2.25(1.92)Ab	2.70(2.42)Ab	11.93(15.56)Aa
	Burned	3.24(1.89)Cb	3.61(2.67)Cb	4.60(4.29)Db	19.04(25.45)Da
9 months	Unburned	1.05(0.15)B	1.08(0.22)B	1.02(0.09)B	1.33(0.25)B
	Burned	1.10(0.30)Cb	1.36(1.20)Cb	1.09(0.34)Eb	1.57 (0.85)Ea

Table 5. Coefficients of correlation between the studied variables in the burned area.

Sampling date		Munsell colour value vs. SOM	Munsell colour value vs. SWR	SOM vs. SWR	Partial correlation (SOM)
IAF	Unburned	-0.63^c	$-0.01^{n.s.}$	$0.01^{n.s.}$	n.c.
	Burned	-0.74^c	-0.81^c	0.75^c	-0.41^b
2 months	Unburned	-0.62^c	$-0.01^{n.s.}$	$0.02^{n.s.}$	– n.c.
	Burned	-0.56^b	-0.39^b	0.34^a	$-0.26^{n.s.}$
5 months	Unburned	-0.47^b	$-0.08^{n.s.}$	$0.17^{n.s.}$	n.c.
	Burned	-0.45^b	$-0.23^{n.s.}$	$0.22^{n.s.}$	n.c.
7 months	Unburned	-0.50^b	$-0.10^{n.s.}$	$0.18^{n.s.}$	n.c.$^{n.s.}$
	Burned	-0.45^b	-0.32^a	$0.17^{n.s.}$	$-0.14^{n.s.}$
9 months	Unburned	-0.41^b	$-0.01^{n.s.}$	$0.01^{n.s.}$	n.c.
	Burned	-0.42^b	$-0.22^{n.s.}$	$-0.07^{n.s.}$	n.c.

Significant at $< 0.05^a$, $< 0.01^b$ and $< 0.001^c$.
n.s.: non-significant at a $p < 0.05$.
n.c.: partial correlation not calculated due to the lack of correlation between Munsell colour value and SWR.

Graham, 1991). With time, despite the significant differences of soil Munsell colour value between plots, the soil became lighter in the burned plot. This may be attributed to the incorporation of burned residues into the first top centimetres of the soil, reducing soil surface darkness (Eckmeier et al. 2007; Pereira et al., 2012b, c; Woods and Balfour, 2011). The black ash cover has implications in the soil environment in the immediate period after the fire (e.g. temperature and water content). The soil blackening decreases the albedo. This leads to an increase of the soil temperature during the day and a more rapid cooling and heat loss at night (Bowman et al., 2009; Hart et al., 2005; Mataix-Solera et al., 2009; Moody et al., 2013; Scharenbroch et al., 2012). These changes in the soil environment may have effects on the soil temperature and consequently on the microbiological activity, since most biological reactions are related to the temperature. Warmer soils after the fire increase the rates of microbiological processes, such as organic matter decomposition and nutrient release, important to plant recovery (Badia and Marti, 2003; Dooley and Treseder, 2012; Hart et al., 2005; Raison and McGarity, 1980). The change in environmental conditions, together with the nutrient availability, rainfall amount after the fire, and warmer temperatures during the spring season, can explain the fact that 2 months after the fire vegetation recovered completely in this burned area. During this period a total of 88 mm of rainfall was registered (Pereira et al., 2012a; 2013a). As a result of this, 2 months after the fire the effects of soil colour on soil temperature may have been reduced. As in other grassland ecosystems, the fast vegetation recovery is an indicator that the ecosystem is resilient to the impacts of this type of fire (Bond and Parr, 2010; Lewis et al., 2009; Morgan, 1999; Wu et al., 2014).

4.2 Soil organic matter

SOM was higher in the burned plot, especially in the first 2 months after the fire. Among sampling dates, a significant difference was only observed in the unburned plot. Previous studies observed that SOM increases in the immediate period after the fire. Vergnoux et al. (2012) identified that in recent fire-affected areas the total organic carbon was significantly higher. In low-severity fires, as in this study, SOM increases temporarily due to the incorporation of ash and charred material into the soil profile (González-Peréz et al., 2004). Short-term increases of SOM in the immediate period after low and medium severity fires were also reported in other studies (De Marco et al., 2005; Gimeno-Garcia et al., 2000; Mataix-Solera et al., 2002; Vogelmann et al., 2012). In this work, during the experimental period significant differences among sampling periods were not observed in the burned plot and this may be related to the fact that the studied plot is located in a flat area and the fast vegetation recovery may have prevented or reduced wind erosion. Soil erosion and SOM transport are accelerated in fire-affected areas due to vegetation removal (Shakesby et al., 2011). Previous studies have shown that losses are high in sloped areas due to water erosion. Gimeno-Garcia et al. (2000) observed that 1 month after an experimental fire carried out in a sloped area, the majority of SOM was washed out due to an extreme rainfall event of more than 30 mm h^{-1}. Also, Novara et al. (2011) identified a redistribution and a major accumulation of SOM on the bottom of the slope after a fire in the Valencia region (Spain). The authors attributed this to transport of burned material by surface wash. In the unburned area significant differences among sampling periods were observed, showing that fire might have changed in the short-term the SOM seasonal variation. The lowest value of SOM was observed IAF (April 2011), increasing in the following months. This reduced SOM content in the beginning of the spring season may be attributed to the lack of fresh litter input and reduced biological activity during the winter due to the low temperatures. In summary, this spring fire of low severity increased SOM which may have contributed to the rapid recovery of the vegetation (Pereira et al., 2013a).

The correlation between soil Munsell colour value and SOM was always significantly negative, but especially high in the immediate sampling dates after the fire in both plots. Darker soils correspond to low Munsell values (Viscarra Rosell et al., 2006; Shields et al., 1968; Conant et al., 2011), independently of the area being affected by fire or not. In burned areas, soil became darker with the increasing content of aromatic carbon, present in high amounts in the charred material produced by fires (Dümig et al., 2009). In soils af-

fected by low-severity fires, the colour is darker due to the incomplete combustion of organic matter (Terefe et al., 2008).

4.3 Soil water repellency

SWR in the fine earth was significantly different among sampling dates in the burned plot until 2 months after the fire, whereas in the unburned plot 9 months after the fire SWR was significantly lower than the previous sampling dates. Fire-induced SWR was reported in previous works in areas affected by low-severity fires (Gleen and Finley, 2010; Granjed et al., 2011; Stoof et al., 2011). Fire changes SWR in previously wettable soils depending on the fuel amount and litter consumed, soil temperature and pre-fire moisture level (Doerr et al., 2000). In this burned plot it is very likely that the direct impacts of fire (e.g. temperature) were minimal since IAF no significant differences were observed in soil moisture between the burned (14.17 % ± 2.83) and unburned (13.59 % ± 2.82) plots (Pereira et al., 2012b). In this case, since the temperature impact on the topsoil was probably minimal, the observed increase of SWR in the burned plot can be attributed to the indirect effect of ash deposition on the topsoil. Miranda et al. (1993) observed that during a prescribed fire in an open grassland, at 2 cm below the soil surface, the temperature ranged from 29 to 38 °C. According to these authors and Heringuer et al. (2002), in grassland fires the soil temperature does not increase importantly and the majority of the heat is lost by convection. Thus, as observed by Vogelman et al. (2012), the increase of soil temperature may not be the responsible for the increase of SWR. The ash produced at low temperature can be hydrophobic (Bodí et al., 2011) and once deposited onto the soil surface can contribute strongly to SWR increases. As in previous works, the ash collected in this burned area (all samples had black colour) was hydrophobic (Pereira et al., 2012a). Ash water repellency is strongly linked to ash chemistry, especially the organic matter content. Dlapa et al. (2013) observed that the wettability of ash decreases with organic matter content. Hydrophobic surfaces are mainly present in organic material, while inorganic material produced at high temperatures is hydrophilic. According to the authors, this explains the different hydrological properties of different types of ash. These results suggested that the incorporation of organic hydrophobic material produced by the fire may have increased temporarily the SWR. In the unburned plot, changes in SWR may be linked with the seasonal variability in this parameter. SWR is a short-term or seasonal phenomenon and depends, among other factors, on climate, the critical soil moisture content above which SWR disappears, texture and organic matter (Doerr et al., 2000; Vogelman et al., 2013). Nine months after the fire (January 2012), the soil was covered by a thick layer of snow and ice. SWR is more severe after dry periods than during wet conditions (Doerr et al., 2000). Buczko et al. (2005) observed in sandy luvisols that SWR was more severe in summer than in autumn/winter. The authors attributed this seasonal variability to the organization of organic amphiphilic compounds that changes during wetting and drying cycles according to the seasonal variations of the soil moisture regime. However, the seasonal variability of organic compounds dissolved into the soil solution may also be relevant. Studies carried out by Arye et al. (2007) observed that SWR decreases with the increase of dissolved organic matter leached out by water. In grassland soils, Farrel et al. (2011) observed that soil-dissolved organic carbon was higher in spring than in autumn and winter due to the reduced microbiological activity and the vegetation's seasonal carbon cycles, which have implications for SOM decomposition. Also, according to Kaiser et al. (2001), the soil samples collected in the summertime are richer in hydrophobic compounds than those collected in winter. Further research is needed in order to understand the dynamics of seasonal variation of SWR in boreal grasslands, especially during the wintertime in snow covered soils.

Two months after the fire, SWR decreased substantially in the burned plot, while SOM maintained the same levels during the whole study period. Vogelmann et al. (2012) also observed after a grassland fire an increase of SWR 2 months after the fire, decreasing thereafter. The preservation of SOM levels may be attributed to the rapid vegetation recuperation in the studied area, which maintained the SOM content levels, but vegetation recovery, rainfall, microbiological and invertebrate activity, may contribute to a decrease in the amount of hydrophobic compounds produced by the fire. The biological activity associated with vegetation recovery has implications on the reduction of SWR (Doerr et al., 2009). Knicker et al. (2013) observed that in fire-affected soils where there is no vegetation cover re-establishment and litter input, the different chemical composition of SOM and pyrogenic organic matter increase the SOM aromaticity with reduced solubility. The inputs of fresh litter from vegetation re-establishment replenish SOM and changes soil chemical composition towards that of an area unaffected by fire (Knicker et al., 2013).

In burned areas, previous reports have shown that after a fire, dissolved organic compounds increased in relation to the unburned plot. Michalzik and Martin (2013) observed that after a low-severity prescribed fire in a pine forest, the amount of dissolved organic carbon was significantly higher in the burned plot than in the unburned area. The authors concluded that the leaching of dissolved organic carbon increased measurably after low-severity fires. Similar findings were registered by Zhao et al. (2012) after a prescribed fire in a wetland located in north-eastern China. The authors identified that the dissolved organic carbon was higher in the burned plot than in the unburned plot, until the second growing season after the fire. The solubility of the dissolved organic fractions depends on pH (Andersson et al., 2000; Impellitteri et al., 2002). Impellitteri et al. (2002) observed that the solubility of humic and fulvic acids in soils increased with increasing pH, while hydrophilic acids remain constant at a pH

range between 3 and 9. The authors found that at a pH between 3 and 6 the hydrophilic acids dominate the dissolved organic fraction, while at a pH between 7 and 9, humic acids were the dominant fraction. Humic and fulvic acids are recognized to be potential sources of SWR (Atanassova and Doerr, 2011; Badía-Villas et al., 2013; DeBano, 2000). Humic acids increase in percentage in the humin fraction after laboratory heating and real fires (González-Peréz et al., 2004). The potentially leached material in the burned area may be primarily composed of humic and fulvic acids, very likely leached in the first 2 months after the fire. The soil pH of the burned plot was in the range of 6.73–7.42 IAF and 7.13–7.66 2 months after the fire (not shown), hence favourable to the leaching of fulvic and especially humic acids. In contrast, pH levels were not the most advantageous to hydrophilic acid leaching. Overall, this may have facilitated the reduction of SWR. Fire induces important changes in pH and increases nutrient availability due to ash deposition, determining the composition of the microbial community. In the short term, the heat impacts on soil induce microbial mortality. Over the long term, there may be changes in soil microbial communities due to the modification of the plant community and soil environment (Hart et al., 2005). In addition to the direct impact of fire, bacterial activity can be increased in the immediate period after the fire due to increases in soil pH and dissolved organic compounds (Bárcenas-Moreno et al., 2011). This increase of soluble carbon in fire-affected soils stimulates the recolonization of some microbes such as heterotrophic bacteria and enhances the basal respiration rates (Mataix-Solera et al., 2009). After the fire, the increase of microbiological activity reduces the SWR, due to the decomposition of waxes and hydrophobic material (Franco et al., 2000; Noordman and Jansen, 2002). This activity contributes to the release of organic nutrients immobilized in aromatic compounds present in charred material and fundamental to plant recovery (Knicker et al., 2013). Microbiological activity stimulates root development, plant growth and vice versa (Cheng and Coleman, 1990; Fu and Cheng, 2002; Vessey, 2003). The plant regrowth protects the soil from raindrop impact (Cerdà and Robichaud, 2009) and root development creates new pathways and preferential water flow, increasing the water infiltration (Lange et al., 2009). The invertebrates' activity may also have contributed to the reduction of soil hydrophobic compounds and changed the hydraulic conductivity in the burned plot studied (Fig. 5). To our knowledge there are no studies about the impact of earthworm activity on SWR in burned soils, however, in contaminated areas, it was reported that earthworms have the capacity to take up hydrophobic compounds (Belfroid and Sijm, 1998; Belfroid et al., 1995). A bibliographic review carried out by Blouin et al. (2013) described that earthworm biomass is positively correlated with water infiltration. Earthworm burrows facilitate root penetration and increase hydraulic conductivity. Soil invertebrates can survive easily after grassland fires, since the severity needed to affect them is normally

Fig. 5. Evidence of earthworm activity (indicated with a red circle) in the burned plot 17 days after the fire.

not achieved (Neary et al., 1999). Previous studies observed that, in the period between 3 and 16 days after a fire in a grassland area, ants constructed their mounds (Pereira et al., 2013b). In other words, the increase of microbiological activity after the fire may have had impacts on the decomposition of hydrophobic material present in the soil particles and aggregates. In the burned area, the decomposition of this material together with the root development and invertebrate activities may have reduced SWR and increased water infiltration, facilitating the transport of the soluble hydrophobic material. These aspects may have had important effects on the SWR decrease 2 months after the fire in the burned area. Also, post-fire wetting and drying cycles (Doerr et al., 2009) and the exceedance of a "critical soil moisture threshold" (Doerr and Thomas, 2000; Huffman and MacDonald, 2004) are related to the SWR decrease. However, Doerr and Thomas (2000) showed that after wetting, SWR is not necessarily re-established when soil becomes dry again. Other factors involved in SWR reduction may be the spatial organization of amphiphilic molecules (Horne and McIntosh, 2000). Differences of SWR among sample times in each sieve fraction of each plot were identified in the burned and unburned plots. In the burned area the coarser-size fractions (2–1 and 1–0.5 mm) demonstrated significant differences in SWR in the first 7 months after the fire, while in the finer-size fractions (0.5–0.25 and < 0.25 mm) significant differences in SWR were observed until 9 months later. This shows that the hydrophobic substances attached to soil fractions disappear faster in the coarser sieve fractions than from the finer ones. This dynamic can be attributed to microbiological activity. Microbes may decompose the organic material at different rates. To our knowledge, no previous works have been conducted on microbial decomposition rates in different size fractions in burned areas. However, Fazle Rabbi et al. (2014) observed in Acrisols collected in a native pasture that the soil organic carbon mineralization was higher in macro-

(250–2000 μm) and microaggregates (53–250 μm) than in the < 53 μm fraction. Fernández et al. (2010) found in non-tilled Entic Haplustoll soils that carbon losses through mineralization were especially observed in intermediate-size fractions (1–4 mm). Wu et al. (2012) identified in grassland soils that microbial biomass and dissolved organic carbon were significantly higher in the > 2000 μm-size fraction, than in the 0–63 μm-size fraction. Also, Jha et al. (2012) observed that water soluble carbon was significantly higher in macroaggregates than in microaggregates. These results may support the hypothesis that the mineralization rates and leaching of hydrophobic organic materials were higher in coarser sieve fractions than in the smaller ones. In relation to the differences observed in the unburned plot, in the coarser (2–1 mm) and the finer fractions (< 0.25 mm) SWR was more persistent IAF, 5 and 7 months after the fire in relation to the other sampling dates, while in the intermediate-size fractions (1–0.5 and 0.5–0.25 mm) SWR was significantly lower 9 months after the fire in comparison to the other sampling dates. The intermediate-size fractions followed the same pattern observed for the fine earth. The main differences were identified 2 months after the fire. It is not clear why this difference occurred in the second sampling date after the fire. In the international literature no previous works were found about the seasonal impacts on SWR according to soil aggregate sizes. Further research is needed to identify the factors responsible for these changes.

In the unburned and burned plots, the SWR was high in the finer fraction (< 0.25 mm). The results obtained in this study are in accordance with previous works in unburned (Arcenegui et al., 2008; Urbanek et al., 2007) and burned soils (Mataix-Solera and Doerr, 2004; Gimeno Garcia et al., 2011; Jordán et al., 2011), which identified that the finer soil fraction was more repellent than the coarser fractions. SWR is mainly attributed to soils with coarser textures that are more susceptible to developing repellent surfaces, due to the smaller specific surface area in relation to fine textured soils (Blas et al., 2010; Doerr et al., 2000). However, it has been observed that when a soil is hydrophobic, the finer fraction is usually more water repellent than the coarser ones (Jordán et al., 2011; Mataix-Solera and Doerr, 2004). In the present study SWR was especially severe in the finer fraction in the immediate sampling dates after the fire in the burned area. This can be attributed to the existence of hydrophobic ash smaller than 0.25 mm and/or the presence of hydrophobic interstitial organic matter that influenced the SWR (Mataix-Solera and Doerr, 2004). In the fine earth significant differences between plots were only identified in the 2 months after the fire. Nevertheless, between each fraction in the different plots, significant differences were observed in the coarser fractions (2–1 and 1–0.5 mm) until 7 months after the fire and in the fine fractions (0.5–0.25 and < 0.25 mm) until 9 months after the fire. The time for the burned plot to return to previous conditions depends also on the soil-size fraction because the rates of mineralization and/or leaching of organic hydrophobic substances may be not equal.

In the burned area the correlations between the Munsell colour value and SOM with SWR were significant only in the first 2 months after the fire (7 months later in the case of Munsell colour value and SWR). In unburned and burned areas SWR can be correlated (Lozano et al., 2013; Martínez-Zavala and Jordán-López, 2009; Mataix-Solera et al., 2002; Mataix-Solera and Doerr, 2004) or not (Blas et al., 2010) with the amount of SOM. The presence of hydrophobic compounds may be related to a certain type of organic material and not to the total SOM content (Doerr et al., 2000). Badía-Villas et al. (2013) observed a significant positive correlation between SWR and pyrolysed carbon, suggesting that SWR is strongly linked with organic materials produced by fire. Also, SWR may be affected by the ionic strength of the soil solution that induces an approximation of charged functional groups in SOM (Hurraß and Shaumann, 2006). These results suggest that the soil became water repellent from the hydrophobic substances produced during the fire, as organic coatings that covered the soil particles and aggregates that with time were decomposed and leached, especially from the coarser fractions. The significant correlations obtained in the first sampling dates after the fire in the burned plot may be the result of the presence of hydrophobic compounds with dark colour. Nevertheless, the partial correlation results showed that SOM controls the correlation of the Munsell colour value and SWR, IAF, 2 and 7 months after the fire, revealing that the original correlations were spurious. This suggests that SOM characteristics may have influenced the SWR. Other studies observed also that SOM has an important influence on SWR correlation with other variables, such as pH and the fungi parameters ergosterol- and glomalin-related soil proteins (Lozano et al., 2013). In fact, SWR must be more controlled by the chemical composition of SOM, than by its amount (DeBano et al., 1970). Horne and McIntosh (2000) observed that SWR was especially determined by amphipathic compounds rather than the organic matter's bulk characteristics. Spielvogel et al. (2004) found that SOM aromatic compounds contribute strongly to the correlation of soil lightness and SOM. The authors observed a strong correlation between soil lightness and aryl C ($r = 0.87$, $p < 0.01$). Also Schmidt et al. (1997) identified that charred material and the presence of aromatic C had important implications in the negative correlation between soil lightness and SOM. These results suggest that SOM characteristics exert significant control on soil Munsell colour values. Also, a soil with the same Munsell value may have different concentrations of aromatic compounds that increase SWR, such as humic and fulvic acids. This shows that the Munsell colour value may not be a good variable to estimate SWR.

5   Conclusions

Fire darkened the soil and increased for a short period the SOM content (first 2 months after the fire). This increase was likely due to the input of partially burned ash into the surface soil that produced an increase in the SWR, due to the characteristics of the burned material. However, this increase was not homogeneous across all aggregate-size fractions. Finer fractions were more water repellent than the coarser ones. In the burned area, the SWR of the finer fractions was more persistent in time (9 months after the fire) than in the coarser fractions (7 months after the fire). The correlations between Munsell colour value and SOM were negatively significant in all cases in the burned and unburned plots. However, the correlations between Munsell colour value and SWR and Munsell colour value and SOM were only significant in the burned area IAF, 2 and 7 months after the fire (in the last sampling date, only between Munsell colour value and SWR). The partial correlations revealed that the correlation between Munsell colour value and SWR IAF, 2 and 7 months after the fire in the burned plot was strongly controlled by SOM, suggesting that organic matter properties may have implications on SWR.

Future research is needed to understand the persistence of the SWR in different sieve fractions, and the factors that control this dynamic, that may be linked with microbiological activity. The different responses of soil-size fractions to SWR after a fire induce considerable temporal variability of fire impacts on SWR and hydrologically related parameters such as infiltration, runoff and soil erosion.

Acknowledgements. The authors would like to acknowledge the Lithuanian Research Council for financing the project LITFIRE, Fire effects on Lithuanian soils and ecosystems (MIP-48/2011), to Comissionat per a Universitats i Recerca del DIUE de la Generalitat de Catalunya, to the Lithuanian Hydrometereological Service for providing meteorological data, to the Spanish Ministry of Science and Innovation for funding through the HYDFIRE project CGL2010-21670-C02-01, FUEGORED (Spanish network of forest fire effects on soils http://grupo.us.es/fuegored/), and to the Cerdocarpa team for the important suggestions to this manuscript. The authors appreciate the paper's English revision by Deborah Martin. We like to acknowledge the important help of Antonio Jordán, Raul Zornoza, and an anonymous reviewer that improved the quality of this manuscript.

Edited by: J. Bockheim

References

Agee, J. K.: Prescribed fire effects on physical and hydrologic properties of mixed-conifer forest floor and soil. Report 143, Univ. California Resources Center, Davis, California, 1973.

Andersson, S., Nilsson, S. I., and Saetre, P.: Leaching of dissolved organic carbon (DOC) and dissolved organic nitrogen in mor humus as affected by temperature and pH, Soil Biol. Biochem., 32, 1–10, 2000.

Arcenegui, V., Mataix-Solera, J., Guerrero,C., Zornoza, R., Mataix-Beneyeto, J., and Garcia-Orenes, F.: Immediate effects of wildfires on water repellency and aggregate stability in Mediterranean soils, Catena, 74, 219–226, 2008.

Arye, G., Nadav, Y., and Chen, Y.: Short-term reestablishment of soil water repellency after wetting: effect on capilarity pressure-saturation relationship, Soil Sci. Am. J., 71, 692–702, 2007.

Atanassova, I. and Doerr, S. H.: Changes in soil organic compound composition associated with heat-induced increases in soil water repellency, Eur. J. Soil Sci., 62, 516–532, 2011.

Avery, B. W. and Bascomb, C. L.: Soil survey laboratory methods, Soil Survey Tech. Monogr. No. 6. Rothamsted Exp. Harpenden, UK, 1974.

Badía, D. and Martí, C.: Effect of simulated fire on organic matter and selected microbiological properties of two contrasting soils, Arid Land Res. Manag., 17, 55–59, 2003.

Badía-Villas, D., González-Peréz, J. A., Aznar, J. M., Arjona-Garcia, B., and Marti-Dalmau, C.: Changes in wáter repellency, aggregation amd organic matter of a mollic horizon burned in laboratory: Soil depth affected by fire, Geoderma, 213, 400–407, 2014.

Bárcenas-Moreno, G., Garcia-Orenes, F., Mataix-Solera, J., Mataix-Beneyeto, J., and Baath, E.: Soil microbiological recolonization after fire in a Mediterranean forest, Biol. Fertil. Soils, 47, 261–272, 2011.

Belfroid, A., van den Berg, M., Seinen, W., Hermens, J., and van Gestel, K.: Uptake, bioavailability and elimination of hydrophobic compounds in earthworms (*Eisenia andrei*) in field-contaminated soil, Environ. Toxicol. Chem., 14, 605–612, 1995.

Belfroid, A. C. and Sijm, D. T. H. M.: Influence of soil organic matter on elimination rates of hydrophobic compounds in the earthworm: Possible causes and consequences, Chemosphere, 37, 1221–1234, 1998.

Blank, R. B., Chambers, J. C., and Zamudio, D.: Restoring riparian corridors with fire: Effects on soil and vegetation, J. Range Manag., 56, 388–396, 2003.

Blas, E., Rodriguez-Alleres, M., and Almendros, G.: Speciation of lipid and humic fractions in soils under pine and eucalyptus forests in northwest of Spain and its effects on water repellency, Geoderma, 155, 242–248, 2010.

Blouin, M., Hodgson, M. E., Delgado, E. A., Baker, G., Brussard, L., Butt, K. R., Dai, J., Dendooven, L., Peres, G., Tondoh, E., Cluzeau, D., and Brun, J. J.: A review of earthworm impact on soil function and ecosystem services. Eur. J. Soil Sci., 64, 161–182, 2013.

Bodí, M. B., Mataix-Solera, J., Doerr, S., and Cerdà, A.: Wettability of ash from burned vegetation and its relationship to Mediterranean plant species type, burn severity and total organic carbon content, Geoderma, 160, 599–607, 2011.

Bond, W. J. and Parr, C. L.: Beyond the forest edge: Ecology, diversity and conservation of the grassy biomes, Biol. Conserv., 143, 2395–2404, 2010

Bouyoucos, G. J.: Directions for making mechanical analysis of soils by the hydrometer method, Soil Sci., 42, 225–230, 1936.

Bowman, D. M. J. S., Balch, J. K., Artaxo, P., Bond, W. J., Carlson, J. M., Cochrane, M. A., D'Antonio, C. M., DeFries, R. S., Doyle, J. C., Harrisson, S. P., Johnston, F. H., Keeley, J. A., Krawchuk, M. A., Kull, C. A., Marston, J. B., Moritz, M. A., Prentice, I. C., Roos, C. I., Scott, A. C., Swetnam, T. W., van der Werf, G. R., and Pyne, S. J.: Fire in the earth system, Science, 324, 481–484, 2009.

Buczko, U., Bens, O., and Huttl, R. F.: Variability of soil water repellency in sandy forest soils with different stand structure under Scots pine (*Pinus sylvestris*) and Beech (*Fagus sylvatica*), Geoderma, 126, 317–336, 2005.

Bukantis, A.: Lietuvos klimatas, Vilniaus Universitetas, leidykla, Vilniaus, 1994.

Cerdà, A. and Robichaud, P.: Fire effects on soil infiltration, edited by: Cerdà, A. and Robichaud, P. R., Science Publishers, 81–103, 2009.

Certini, G.: Effects of fire on properties of forest soils: a review, Oecologia, 143, 1–10, 2005.

Certini, G.: Fire as a soil-forming factor, Ambio, 43, 191–195, doi:10.1007/s13280-013-0418-2, 2014.

Certini, G., Nocentini, C., Knicker, H., Arfaioli, P., and Rumpel, C.: Wildfire effects on soil organic matter quantity and quality in two fire-prone Mediterranean pine forests, Geoderma, 167–168, 148–155, 2011.

Cheng, W. and Coleman, D. C.: Effect of living roots on soil organic matter decomposition, Soil Biol. Biochem., 22, 781–787, 1990.

Coetsee, C., Bond, W. J., and February, E. C.: Frequent fire affects soil nitrogen and carbon in an African savanna by changing woody cover, Oecologia, 162, 1027–1034, 2010.

Conant, R. T., Ogle, S. M., Paul, E. A., and Paustian, K.: Measuring and monitoring soil organic carbon stocks in agricultural lands for climate change, Front. Ecol. Environ., 9, 169–173, 2011.

Dangi, S. R., Stahl, P. D., Pendall, E., Cleary, M. B., and Buyer, J. S.: Recovery of soil microbial community structure after fire in a sagebrush-grassland ecosystem, Land Degrad. Dev., 21, 423–432, 2010.

DeBano, L. F.: The role of fire and soil heating on water repellency in wildland environments, J. Hydrol., 231–232, 195–206, 2000.

DeBano, L. F., Mann, L. D., and Hamilton, L. D.: Translocation of hydrophobic substances into soil by burning organic litter, Soil Sci. Soc. Am. J., 34, 130–133, 1970.

De Groot, W. J., Flannigan, M. D., and Cantin, A. S.: Climate change impacts on future boreal fire regimes, Forest Ecol. Manag., 294, 35–44, 2013.

De Marco, A., Gentile, A. E., Arena, C., and De Santo, A. V.: Organic matter, nutrient content and biological activity in burned and unburned soil of a mediterranean maquis area of southern Italy, Int. J. Wildland Fire, 14, 365–377, 2005.

Dlapa, P., Bodi, M., Mataix-Solera, J., Cerdà, A., and Doerr, S. H.: FT-IR spectroscopy reveals that ash water repellency is highly dependent on ash chemical composition, Catena, 108, 35–43, 2013.

Doerr, S. H.: On standardising the "Water Drop Penetration Time" and the "Molarity of an Ethanol Droplet" techniques to classify soil hydrophobicity: a case study using medium textured soils, Earth Surf. Process. Landf., 23, 663–668, 1998.

Doerr, S. H. and Thomas, A. D.: The role of soil moisture in controling water repellency: New incidences from forest soils in Portugal, J. Hydrol., 231–232, 134–147, 2000.

Doerr, S. H., Shakesby, R., and Walsh, R. P. D.: Spatial variability of soil hydrophobicity in fire-prone Eucalyptus and Pine forests, Portugal, Soil Sci., 163, 313–324, 1998.

Doerr, S. H., Shakesby, R. A., and Walsh, R. P. D.: Soil water repellency: its causes, characteristics and hydro-geomorphological significance, Earth-Sci. Rev., 51, 33–65, 2000.

Doerr, D. H., Douglas, P., Evans, R. C., Morley, C. P., Mullinger, N. J., Bryant, R., and Shakesby, R. A.: Effects of heating and post-heating equilibrium times on soil water repellency, Aust. J. Soil. Res., 43, 261–267, 2005.

Doerr, S. H., Shakesby, R. A., and MacDonald, L. H.: Soil water repellency: A key factor in Post-fire erosion, edited by: Cerdà, A. and Robichaud, P. R., Science Publishers, 197–223, 2009.

Dooley, S. R. and Treseder, K. K.: The effect of fire on microbial biomass: a meta-analysis of field studies, Biogeochemistry, 109, 49–61, 2012.

Dümig, A., Knicker, H., Schad, P., Rumpel, C., Dignac, M. F., and Kogel-Knabner, I.: Changes in soil organic matter composition are associated with forest encroachment into grassland with long-term fire history, Eur. J. Soil Sci., 60, 578–589, 2009.

Eckmeier, E., Gerlach, R., Skjemstad, J. O., Ehrmann, O., and Schmidt, M. W. I.: Minor changes in soil organic carbon and charcoal concentrations detected in a temperate deciduous forest a year after an experimental slash-and-burn, Biogeosciences, 4, 377–383, doi:10.5194/bg-4-377-2007, 2007.

Farrel, M., Hill, P. W., Farrar, J., Bardgett, R. D., and Jones, D. L.: Seasonal variation in soluble soil carbon and nitrogen across a grassland productivity gradient, Soil Biol. Biochem., 43, 835–844, 2011.

Fazle Rabbi, S. M., Wilson, B. R., Lockwood, P. V., Daniel, H., and Young, I. M.: Soil organic carbon mineralization rates in aggregates under contrasting land uses, Geoderma, 216, 10–18, 2014.

Fernández, R., Quiroga, A., Zorati, C., and Noellemeyer, E.: Carbon contents and respiration rates of aggregate size fractions under no-till and conventional tillage, Soil Tillage Res., 109, 103–109, 2010.

Franco, C. M. M., Clarke, P. J., Tate, M. E., and Oades, J. M.: Hydrophobic properties and chemical characterization of natural and water repellent materials in Australian sands, J. Hydrol., 231–232, 47–58, 2000.

Fu, S. and Cheng, W.: Ryzosphere priming effects on the decomposition of soil organic matter in C_4 and C_3 grassland soils, Plant Soil, 238, 289–294, 2002.

Gimeno-Garcia, E., Andreu, V., and Rubio, J. L.: Changes in organic matter, nitrogen, phosphorous and cations in soil as a result of fire and water erosion in a Mediterranean landscape, Eur. J. Soil Sci., 51, 201–210, 2000.

Gimeno-Garcia, E., Pascual, J. A., and Llovet, J.: Water repellency and moisture content spatial variations under *Rosmarinus officinalis* and *Quercus coccifera* in a Mediterranenan burned soil, Catena, 85, 48–57, 2011.

Gleen, N. F. and Finley, C. D.: Fire and vegetation type effects on soil hydrophobicity and infiltration in sagebrush-steppe: I. Field analysis, J. Arid Environ., 74, 653–659, 2010.

Gomez-Heras, M., Smith, B. J., and Fort, R.: Surface temperature differences between minerals in crystalline rocks: implication for granular disaggregation of granites through thermal fatigue, Geomorphology, 78, 236–249, 2006.

Gomez-Robledo, L., Lopez-Ruiz, N., Melgosa, M., Palma, A. J., Capitan-Vallvey, L. F., and Sanchez-Maranon, M.: Using the mobile phone as Munsell soil-colour sensor: an experiment under controlled illumination conditions, Comput. Electron. Agr., 99, 200–208, 2013.

González-Peréz, J. A., Gonzalez-Vila, F. J., Almendros, G., and Knicker, H.: The effect of fire on soil organic matter – review, Environ. Int., 30, 855–870, 2004.

Granged, A. J. P., Zavala, L. M., Jordán, A., and Bárcenas-Moreno, G.: Post-fire evolution of fire properties and vegetation cover in a Mediterranean heathland after a experimental burning: a 3 year study, Geoderma, 164, 85–94, 2011.

Granstrom, A.: Fire management for biodiversity in the European boreal forest, Scand. J. For. Res., 3, 62–69, 2001.

Harris, W. N., Moretto, A. S., Distel, R. A., Boutton, T. W., and Boo, R. M.: Fire and grazing in grasslands of the Argentinian Caldenal: Effects on plant and soil carbon and nitrogen, Acta Oecol., 32, 207–214, 2007.

Hart, S. C., DeLuca, T. H., Newman, G. S., MacKenzie, M. D., and Boyle, S. I.: Post-fire vegetative dynamics as drivers of microbial community structure, For. Ecol. Manage., 220, 166–184, 2005.

Heringuer, A., Jaques, A. V. A., Bissani, C. A., and Tedesco, M.: Caracteristicas de um latossolo vermelho sob pastagem natural sujeita a accao prolongada do fogo e de practicas alternatyvas de manejo, Cienc. Rural, 32, 309–314, 2002.

Horne, D. J. and McIntosh, J. C.: Hydrophobic compounds in sands in New Zealand – extraction, characterization and proposed mechanisms for repellency expression, J. Hydrol., 231–232, 35–46, 2000.

Hurraß, J. and Schaumann, G. E.: Properties of soil organic matter and aqueous extracts of actually water repellent and wettable soil samples, Geoderma, 132, 222–239, 2006.

Hylander, K.: The response of land snail assemblages below aspens to forest fire and clear cutting in Fennoscadian boreal forests, Forest Ecol. Manag., 261, 1811–1819, 2011.

Ibañez-Ascencio, S., Marques-Mateu, A., Moreno-Ramon, H., and Balasch, S.: Statistical relationships between soil colour and soil attributes in semiarid areas, Biosyst. Eng., 116, 120–129, 2013.

Impellitteri, C. A., Lu, Y., Saxe, J. K., Allen, H. E., and Peijnenburg, W. J. G. M.: Correlation of the partitioning of dissolved organic matter fractions with the desorption of Cd, Cu, Ni, Pb, and Zn from 18 Dutch soils, Environ. Int., 28, 401–410, 2002.

Jha, P., Garg, N., Lakaria, B. L., Biswas, A. K., and Rao, A. S.: Soil and residue carbon mineralization as affected by soil aggregate size, Soil Tillage Res., 121, 57–62, 2012.

Jordán, A., Zavala, L. M., Mataix-Solera, J., Nava, A., and Alanis, N.: Effect of fire severity on water repellency and aggregate stability on Mexican volcanic soils, Catena, 84, 136–147, 2011.

Kadunas, V., Budavicius, R., Gregorauskiene, V., Katinas, V., Kliaugiene, E., Radzevicius, A., and Tareskevicius, R.: Geochemical atlas from Lithuania, first ed., Geological Institute, Vilnius, 1999.

Kaiser, K., Guggenberger, G., Haumaier, L., and Zech, W.: Seasonal variations in the chemical composition of dissolved organic matter in organic forest floor layer leachates of old Scots pine (*Pinus Sylvestris* L.) and European beech (*Fagus sylvatica*) stands in northeastern Bavaria, Biogeochemistry, 55, 103–143, 2001.

Ketterings, Q. M. and Bigham, J. M.: Soil colour as an indicator of slash-and-burn fire severity and soil fertility in Sumatra, Indonesia, Soil Sci. Soc. Am. J., 64, 1826–1833, 2000.

Kinner, D. A. and Moody, J.: Spatial variability of steady-state infiltration into a two layer soil system on burned hillslopes, J. Hydrol., 381, 322–332, 2010.

Knapp, A. K., Conrad, S. L., and Blair, J. M.: Determinants of soil CO_2 flux from a sub-humid grassland: Effect of fire and fire history, Ecol. Appl., 8, 760–770, 1998.

Knicker, H.: How does fire affect the nature and stability of soil organic nitrogen and carbon, Biogeochemistry, 85, 91–118, 2007.

Knicker, H., Gonzalez-Vila, F., and Gonzalez-Vazquez, R.: Biodegradability of organic matter in fire affected mineral soils of Southern Spain, Soil Biol. Biochem., 56, 31–39, 2013.

Kouki, J., Hyvarinen, E., Lappalainen, H., Martikainen, P., and Simila, M.: Landscape context affects the success of habitat restoration: Large scale recolonization patterns of saproxylic and fire-associated species in boreal forests, Divers. Distrib., 18, 348–355, 2012.

Lange, B., Lüescher, P., and Germann, P. F.: Significance of tree roots for preferential infiltration in stagnic soils, Hydrol. Earth Syst. Sci., 13, 1809–1821, doi:10.5194/hess-13-1809-2009, 2009.

Lewis, T., Reid, N., Clarke, P. J., and Whalley, R. D. B.: Resilience of a high-conservation-value, semi-arid grassland on fertile clay soils to burning, mowing and ploughing, Austral. Ecol., 35, 461–481, 2009.

Lozano, E., Jimenez-Pinilla, P., Mataix-Solera, J., Arcenegui, V., Bárcenas, G. M., González-Peréz, J. A., Garcia-Orenes, F., Torres, M. P., and Mataix-Benyeto, J.: Biological and chemical factors controlling the patchy distribution of soil water repellency among plant species in a Mediterranean semiarid forest, Geoderma, 207–208, 212–220, 2013.

MacDonald, L. and Huffman, E. L.: Post-fire soil water repellency: Persistence and soil moisture thresholds, Soil Sci. Soc. Am. J., 68, 1729–1734, 2004.

Malkinson, D. and Wittenberg, L.: Post-fire induced soil water repellency – Modelling short and long-term processes, Geomorphology, 125, 186–192, 2011.

Marion, G. M., Moreno, J. M., and Oechel, W. C.: Fire severity, ash deposition, and clipping effects on soil nutrients in Chaparral, Soil Sci. Soc. Am. J., 55, 235–240, 1991.

Martínez-Zavala, L. and Jordán-López, A.: Influence of different plant species on water repellency in Mediterranean heathlands, Catena, 76, 215–223, 2009.

Marti-Roura, M., Casals, P., and Romanya, J.: Long-term retention of post-fire soil mineral nitrogen pools in Mediterranean shrubland and grassland, Plant Soil, 371, 521–531, 2013.

Mataix-Solera, J. and Doerr, S.: Hydrophobicity and aggregate stability in calcareous topsoils from fire-affected pine forests in southeasthern Spain, Geoderma, 118, 77–88, 2004.

Mataix-Solera, J., Gomez, I., Navarro-Pedreno, J., Guerrero, C., and Moral, R.: soil organic matter, Int. J. Wildland Fire, 11, 107–114, 2002.

Mataix-Solera, J., Guerrero, C., Garcia-Orenes, F., Bárcenas, G. M., and Torres, M. P.: Fire effects on soil microbiology, edited by: Cerdà, A. and Robichaud, P. R., Science Publishers, 133–175, 2009.

Mataix-Solera, J., Arcenegui, V., Tessler, N., Zornoza, R., Wittenberg, L., Martínez, C., Caselles, P., Perez-Bejarano, A., Malkinsnon, D., and Jordán, M. M.: Soil properties as key factors controlling water repellency in fire affected areas: Evidences from burned sites in Spain and Israel, Catena, 108, 6–13, 2013.

Michalzik, B. and Martin, S.: Effects of experimental duff fires on C, N and P fluxes into the mineral soil at coniferous and broadleaf forest site. Geoderma, 197–198, 169–176, 2013.

Michelsen, A., Andersson, M., Jensen, M., Jensen, M., Kjoller, A., and Gashew, M.: Carbon stocks, soil respiration and microbial biomass in fire-prone tropical grassland, woodland and forest ecosystems, Soil Biol. Biochem., 36, 1707–1717, 2004.

Mierauskas, P.: Policy and legislative framework overview of fire management in Lithuanian protected areas, Flamma, 3, 1–5, 2012.

Miranda, A. C., Miranda, H. S., Dias, I. F. O., and Dias, B. F. S.: Soil and air temperatures during prescribed cerrado fires, J. Trop. Ecol., 9, 313–320, 1993.

Moody, J. A., Shakesby, R. A., Robichaud, P. R., Cannon, S. H., and Martin, D. A.: Current research issues related to post-wildfire runoff and erosion processes, Earth-Sci. Rev., 122, 10–37, 2013.

Morgan, J. W., Defining grassland fire events and the response of perennial plants to annual fire in temperate grasslands of southeastern Australia, Plant Ecol., 144, 127–144, 1999.

Neary, D. G., Klopatek, C. C., DeBano, L., and Ffolliott, P. F.: Fire effects on bellow ground sustainability: a review and synthesis, Forest Ecol. Manag., 122, 51–71, 1999.

Neff, J. C., Harden, J. W., and Gleixner, G.: Fire effects on soil organic matter content, composition, and nutrients in boreal interior Alaska, Can. J. Fore. Res., 35, 2178–2187, 2005.

Neris, J., Tejedor, M., Fuentes, J., and Jimenez, C.: Infiltration, runoff and soil loss in andisols affected by forest fire (Canary Islands, Spain), Hydrol. Process., 27, 2814–2824, 2013.

Nguyen, T. T. and Marschner, P.: Addition of a fine-textured soil to compost to reduce nutrient leaching in a sandy soil, Soil Res., 51, 232–239, 2013.

Noordman, W. H. and Janssen, D. B.: Rhamnolipid stimulates uptake of hydrophobic compounds by *Pseudomonas aeruginosa*, Appl. Environ. Microbiol., 68, 4502–4508, 2002.

Novara, A., Gristina, L., Bodí, M. B., and Cerdà, A.: The impact of fire on redistribution of soil organic matter on a mediterranean hillslope under maqui vegetation, Land Degrad. Dev., 22, 530–536, 2011.

Novara, A., Gristina, L., Ruhl, J., Pasta, S., D'Angelo, G., La Mantia, T., and Pereira, P.: Grassland fire effect on soil organic carbon reservoirs in a semiarid environment, Solid Earth, 4, 381–385, 2013.

Pausas, J. and Keeley, J.: A burning story: The role of fire in the history of life, BioScience, 59, 593–601, 2009.

Pereira, P., Úbeda, X., Martin, D., Mataix-Solera, J., and Guerrero, C.: Effects of a low prescribed fire in ash water soluble elements in a Cork Oak (Quercus suber) forest located in Northeast of Iberian Peninsula, Environ. Res., 111, 237–247, 2011.

Pereira, P., Cepanko, V., Vaitkute, D., Pundyte, N., Pranskevicius, M., Zuokaite, E., Úbeda, X., Mataix-Solera, J., and Cerdà, A.: Grassland fire effects on ash properties and vegetation restoration in Lithuania (North-Eastern Europe), Flamma, 3, 3–8, 2012a.

Pereira, P., Mataix-Solera, J., Ubeda, X., Cerdà, A., Cepanko, V., Vaitkute, D., Pundyte, N., Pranskevicius, M., and Zuokaite, E.: Spring Grassland fire effects on soil organic matter, soil moisture and soil water repellency in Lithuania (North-Eastern Europe). First results. 4th International Congress, Eurosoil 2012, Soil science for the benefit of mankind and environment, 2012b.

Pereira, P., Cerdà, A., Jordán, A., Bolutiene, V., Úbeda, X., Pranskevicius, M., and Mataix-Solera, J.: Spatio-temporal vegetation recuperation after a grassland fire in Lithuania, Procedia Environ. Sci., 19, 856–895, 2013a.

Pereira, P., Cerdà, A., Úbeda, X., Mataix-Solera, J., Martin, D., Jordán, A., and Burguet, M.: Spatial models for monitoring the spatio-temporal evolution of ashes after fire – a case study of a burnt grassland in Lithuania, Solid Earth, 4, 153–165, doi:10.5194/se-4-153-2013, 2013b.

Pereira, P., Pranskevicius, M., Cepanko, V., Vaitkute, D., Pundyte, N., Úbeda, X., Mataix-Solera, J., Cerdà, A., and Martin, D. A: Short time vegetation recovery after a spring grassland fire in Lithuania. Temporal and slope position effect, Flamma, 4, 13–17, 2013c.

Post, D. F., Bryant, R. B., Batchily, A. K., Huete, A. R., Levine, S. J., Mays, M. D., and Escadafal, R.: Correlations between field and laboratory measurments of soil color, edited by: Bigham, J. M. and Ciolkosz, E. J., Soil Science Society of America, 35–49, 1983.

Raison, R. J. and McGarity, J. W.: Effects of ash, heat, and the ash-heat interaction on biological activities in two contrasting soils, Plant Soil, 55, 363–376, 1980.

Ravi, S., D'Odorico, P., Wang, L., White, C. S., Okin, G. S., Macko, S. A., and Collins, S. L.: Post-fire resource distribution in desert grasslands: A possible negative feedback on land degradation, Ecosystems, 12, 434–444, 2009a.

Ravi, S., D'Odorico, P., Zobeck, T. M., and Over, T. M.: The effect of fire-induced soil hydrophobicity on wind erosion in a semiarid grassland: Experimental observations and theoretical framework, Geomorphology, 105, 80–86, 2009b.

Rodriguez-Alleres, M., Varela, M. E., and Benito, E.: Natural severity of wáter repellency in pine forest soils from NW Spain and influence of wildfire severity on its persistence, Geoderma, 191, 125–131, 2012.

Sankey, J. B., Germino, M. J., Sankey, T. T., and Hoover, A. N.: Fire effects on spatial patterning of soil properties in sagebrush steppe, USA: a meta analysis, Int. J. Wildland Fire, 21, 545–556, 2012.

Scharenbroch, B. C., Nix, B., Jacobs, K. A., and Bowles, M. L.: Two decades of low-severity prescribed fire increases soil nutrient availability in a Midwestern, USA oak (*Quercus*) forest, Geoderma, 183–184, 80–91, 2012.

Schmidt, M. W. I., Knicker, P. G., Hatcher, P. G., and Kogel-Knaber, I.: Improvement of ^{13}C and ^{15}N CPMAS NMR spectra of bulk soils, particle size fractions and organic material treatment with 10 % hydrofluoric acid, Eur. J. Soil Sci., 48, 319–328, 1997.

Schulze, D. G., Nagel, J. L., Van Scoyoc, G. E., Henderson, T. L., Baumgardner, M. F., and Scoot, D. E.: Significance of organic matter in determining soil colors, edited by: Bigham, J. M. and Ciolkosz, E. J., Soil Science Society of America, 71–90, 1993.

Shakesby, R. A.: Post-wildfire soil erosion in the Mediterranean: Review and future research directions, Earth-Sci. Rev., 105, 71–100, 2011.

Shapiro, S. and Wilk, M.: An analysis of variance test for normality, Biometrika, 52, 591–561, 1965.

Shields, J. A., Paul, E. A., St. Arnaud, R. J., and Head, W. K.: Spectrophotometric measurement of soil colour and its relationship to moisture and organic matter, Can. J. Soil. Sci., 48, 271–280, 1968.

Spielvogel, S., Knicker, H., and Kogel-Knaber, I.: Soil organic matter composition and soil lightness, J. Plant Nutr. Soil Sci., 167, 545–555, 2004.

StatSoft Inc.: STATISTICA, Version 6.0., Tulsa, OK, 2006.

Stoof, C. R., Moore, D., Ritsema, C. J., and Dekker, L. W.: Natural and fire induced soil water repellency in a Portuguese shrubland, Soil Sci. Soc. Am. J., 75, 2283–2295, 2011.

Terefe, T., Mariscal-Sancho, I., Gomez, M. V., and Espejo, R.: Relationship between soil color and temperature in the surface horizon of Mediterranean soils: A laboratory study, Soil Sci., 170, 495–403, 2005.

Terefe, T., Mariscal-Sancho, I., Peregrina, F., and Espejo, R.: Influence of heating on various properties of six Mediterranean soils. A Laboratory study, Geoderma, 143, 273–280, 2008.

Tessler, N., Wittenberg, L., and Greenbaum, N.: Soil water repellency persistence after recurrent forest fires on Mount Carmel, Israel, Int. J. Wildland Fire, 22, 515–526, 2012.

Thwaites, R.: Color, in: Encyclopedia of Soil Science, second Ed., edited by: Lal, R., 211–214, 2002.

Torrent, J. and Barron. V.: Laboratory measurment of soil colour, edited by: Bigham, J. M. and Ciolkosz, E. J., Soil Science Society of America, 21–23, 1993.

Torrent, J., Schwertmann, U., and Schulze, D. G.: Iron oxide mineralogy of some soils of two river terrace sequences in Spain, Geoderma, 23, 191–208, 1980.

Úbeda, X., Lorca, M., Outeiro, L. R., and Bernia, S.: Effects of s prescribed fire on soil quality in Mediterranean grassland (Prades Mountains, north-east, Spain), Int. J. Wildland Fire, 14, 379–384, 2005.

Úbeda, X., Pereira, P., Outeiro, L., and Martin, D. A.: Effects of fire temperature on the physical and chemical characteristics of the ash from plots of cork oak (Quercus suber), Land Degrad. Dev., 20, 589–608, 2009.

Ulery, A. L. and Graham, R. C.: Forest fire effects on soil colour and texture, Soil Sci. Soc. Am. J., 57, 135–140, 1991.

Urbanek, E., Hallet, P., Feeney, D., and Horn, R.: Watter repellency and distribution and of hydrophilic and hydrophobic compounds in soil aggregates from different tillage systems, Geoderma, 140, 147–155, 2007.

USDA: Soil Survey Laboratory Methods Manual. Soil Survey Investigation Report No. 42. Version 4.0, USDA-NCRS, Lincoln, NE, 693 pp., 2004.

Van Bellen, S., Garneau, M., and Bergeron, Y.: Impact of climate change on forest fire severity and consequences for carbon stocks in boreal forest stands of Quebec, Canada: A Synthesis, Fire Ecol., 6, 16–44, 2010.

Vanha-Majamaa, I., Lilija, S., Ryoma, R., Kotiaho, J. S., Laaka-Lindberg, S., Lindberg, H., Puttonen, P., Tamminen, P., Toivanen, T., and Kuuluvainen, T.: Rehabilitating boreal forest structure and species composition in Finland through logging, dead wood creation and fire: The EVO experiment, Forest Ecol. Manag., 250, 77–88, 2007.

Varela, M. E., Benito, E., and Blas, E.: Impact of wildfires on surface water repellency in soils of northwest Spain, Hydrol. Process., 19, 3649–3657, 2005.

Vargas, R., Collins, S. L., Thomey, M. L., Johnson, J. E., Brown, R. F., Natvig, D. O. and Friggens, M. T.: Precipitation variability and fire influence the temporal dynamics of soil CO_2 efflux in an arid grassland, Glob. Change Biol., 18, 1401–1411, 2012.

Vergnoux, A., Di Rocco, R., Domeizel, M., Guiliano, M., Doumenq, P., and Theraulaz, F.: Effects of forest fires on water extractable organic matter and humic substances from Mediterranean soils: UV-vis and fluorescence spectroscopy approaches, Geoderma, 160, 434–443, 2011.

Vessey, J. K.: Plant growth promoting rhizobacteria as biofertilizers, Plant Soil, 255, 571–586, 2003.

Viscarra Rossell, R. A., Minasny, B., Roudier, P., and McBratney, A. B.: Colour space models for soil science, Geoderma, 133, 320–337, 2006.

Vogelmann, E. S., Reichert, J. M., Prevedello, J., Barros, C. A. P., de Quadros, F. L. F., and Mataix-Solera, J.: Soil hydro-physical changes in natural grassland of southern Brazil subjected to burning management, Soil Res., 50, 465–472, 2012.

Vogelmann, E. S., Reichert, J. M., Prevedello, J., Consensa, C. O. B., Oliveira, A. E., Awe, G. O., and Mataix-Solera, J.: Threshold water content beyond which hydrophobic soils become hydrophilic: The role of soil texture and organic matter content, Geoderma, 209–210, 177–187, 2013.

Wessel, A. T.: On using the effective contact angle and the water drop penetration time for classification of water repellency in dune soils, Earth Surf. Process. Landf., 13, 555–562, 1988.

Whitford, W. G. and Steinberger, Y.: Effects of seasonal grazing, drought, fire, and carbon enrichment on soil microarthropods in a desert grassland, J. Arid. Environ. 83, 10–14, 2012.

Woods, S. and Balfour, V.: The effects of soil texture and ash thickness on the post-fire hydrological response from ash-covered soils, J. Hydrol., 393, 274–286, 2011.

WRB: World Reference base for soil resources 2006, 2nd edition, World Soil Resources Reports, No 103, Food and Agriculture Organization of the United Nations, Rome, 2006.

Wu, G. L., Zhao, L. P., Wang, D., and Shi, Z. H.: Effects of Time-Since-Fire on Vegetation Composition and Structures in Semi-Arid Perennial Grassland on the Loess Plateau, China, CLEAN-Soil Air-Water, 42, 98–103, 2014.

Wu, H., Wiesmeier, M., Yu, Q., Steffans, M., Han, X., and Kogel-Knabner, I.: Labile organic C and N mineralization of soil aggregates size classes in semi-arid grasslands as affected by grazing management, Biol. Fertil. Soils, 48, 305–313, 2012.

Xu, W. and Wan, S.: Water- and plant mediated responses of soil respiration to topography, fire, and nitrogen fertilization in a semiarid grassland in northern China, Soil Biol. Biochem., 40, 679–689, 2008.

Zhao, H., Tong, Q., Lin, Q., Lu, X., and Wang, G.: Effect of fires on soil organic carbon pool and mineralization in Northeasthern China wetland, Geoderma, 189–190, 532–539, 2012.

Characterization of hydrochars produced by hydrothermal carbonization of rice husk

D. Kalderis[1], M. S. Kotti[1], A. Méndez[2], and G. Gascó[3]

[1]Department of Environmental and Natural Resources Engineering, Technological and Educational Institute of Crete, Chania, 73100 Crete, Greece
[2]Departamento de Ingeniería de Materiales, E.T.S.I. Minas, Universidad Politécnica de Madrid, C/Ríos Rosas no. 21, 28003 Madrid, Spain
[3]Departamento de Edafología, E.T.S.I. Agrónomos, Universidad Politécnica de Madrid, Ciudad Universitaria, 28004 Madrid, Spain

Correspondence to: D. Kalderis (dkalderis@chania.teicrete.gr)

Abstract. Biochar is the carbon-rich product obtained when biomass, such as wood, manure or leaves, is heated in a closed container with little or no available air. In more technical terms, biochar is produced by so-called thermal decomposition of organic material under limited supply of oxygen (O_2), and at relatively low temperatures ($< 700\,°C$). Hydrochar differentiates from biochar because it is produced in an aqueous environment, at lower temperatures and longer retention times. This work describes the production of hydrochar from rice husks using a simple, safe and environmentally friendly experimental set-up, previously used for degradation of various wastewaters. Hydrochars were obtained at $200\,°C$ and $300\,°C$ and at residence times ranging from 2 to 16 h. All samples were then characterized in terms of yield, surface area, pH, conductivity and elemental analysis, and two of them were selected for further testing with respect to heating values and heavy metal content. The surface area was low for all hydrochars, indicating that porous structure was not developed during treatment. The hydrochar obtained at $300\,°C$ and 6 h residence times showed a predicted higher heating value of $17.8\,MJ\,kg^{-1}$, a fixed carbon content of 46.5 % and a fixed carbon recovery of 113 %, indicating a promising behaviour as a fuel.

1 Introduction

Subcritical water is hot water (100–$374\,°C$) under enough pressure to maintain the liquid state. It is an environmentally friendly and inexpensive solvent that exhibits a wide range of properties that renders it very effective in solvating and decomposing moderately polar or non-polar substances from a wide range of environmental matrices. Subcritical water can decompose naturally occurring substances and materials, such as complex amino acids, proteins and carbohydrates (sucrose, fructose and sorbose), sodium alginate, and brown coal, to produce more valuable and useful products. Additionally, subcritical water has been proved to decompose hazardous organic substances and materials such as pentachlorophenol (PCP), fluorochemicals, dioxins, polycyclic aromatic hydrocarbons (PAHs), polychlorinated biphenyls (PCBs) and polyvinyl chloride (PVC) (Kalderis et al., 2008, and references therein).

The hydrothermal treatment of biomass at temperatures in the range of 100–$374\,°C$ gives rise to water-soluble organic substances and a carbon-rich solid product, commonly known as hydrochar (Sevilla and Fuertes, 2009). Typically, the main components of biomass resources are 40–45 wt% cellulose, 25–35 wt% hemicellulose, 15–30 wt% lignin and up to 10 wt% for other compounds (Toor et al., 2011). The treatment of biomass in subcritical water has received considerable attention over the last few years. The degradation mechanisms of lignin, cellulose and hemicellulose during hydrothermal treatment and the effect of the experimental parameters (residence time, temperature, type of biomass) have been thoroughly described elsewhere (Sevilla and Fuertes, 2009; Jamari et al., 2012; Wahyudiono et al., 2012; Wiedner et al., 2013; Gao et al., 2013; Parshetti et al., 2013; Lu et al., 2013). The products obtained during treatment are gases (about 10 % of the original biomass, mainly CO_2), bio-oil consisting primarily of sugars, acetic acid, and other organic acids and the solid product (char), which contains about 41–90 % of the mass of the original feedstock. The produced char has a higher energy density and is more hydrophobic than the original biomass (Tufiq Reza et al., 2013).

Compared to other thermochemical processes such as pyrolysis, hydrogenation or gasification, aqueous conversion using subcritical water has the significant advantage of not requiring a drying process for feedstock and therefore can be conducted at high moisture content typical of biomass feedstocks. The hydrothermal carbonization temperature is usually much lower than that of pyrolysis, gasification, and flash carbonization. The water present can be used as the reaction solvent, whereas at the same time some off-gases, such as CO_2, nitrogen oxides, and sulfur oxides, are dissolved in water, forming the corresponding acids and/or salts, making further treatment for air pollution possibly unnecessary. Finally, it is an environmentally friendly method as it requires no additives and in most cases is simple to set up and operate.

Compared to hydrochars, biochars produced through conventional pyrolysis methods have been more thoroughly applied and tested. Research on the applications of hydrochar is still in its early stages. Lately, hydrochars have been tested as soil conditioners and heavy metal immobilization means (Abel et al., 2013; Wagner and Kaupenjohann, 2014), as electrochemical supercapacitor electrode materials (Ding et al., 2013a, 2013b) and as anode materials for lithium ion batteries (Unur et al., 2013), with promising results. Ongoing work also focuses on their environmental impact and compatibility with agricultural and horticultural systems (Gajic and Koch, 2012; Busch et al., 2013; Bargmann et al., 2013).

Around 20 % of the whole rice grain weight is rice husk. In 2008 the world rice grain production was 661 million tons, and consequently 132 million tons of rice husk were also produced. While there are some established uses for rice husk, it is still considered a waste product in the rice mill industry and it is often either burned in the open or disposed of in landfills. Rice husk has been extensively studied for the production of activated carbon through conventional pyrolysis routes (Kalderis et al., 2008, and references therein). However, the studies that deal with hydrothermal carbonization of rice husk are few. The scope of this study was to use a simple, safe, effective, environmentally friendly method to produce hydrochars from rice husk and characterize the products. Two of the produced hydrochars were selected, and their behaviour as fuels was examined.

Table 1. Properties of the rice husk (used in this study) and rice husk ash.

Rice husk	
Moisture (%)	4.2
Ash content (%)	16.1
Volatile matter (%)	62
Carbon (%)	36.1
Fixed carbon (%)	17.7
Higher heating value (MJ kg^{-1})	15.1
Rice husk ash	**wt%**
SiO_2	84.7
K_2O	2.51
CaO	0.74
Al_2O_3	0.36
Na_2O	0.20
MgO	0.76
P_2O_5	0.62
SO_3	0.38
Fe_2O_3	0.28
Cl	0.18

2 Materials and methods

Rice husk (RH) was obtained from Janta Rice Mill in Gurdaspur (32.0333° N–75.40° E) in India. Rice husk was initially washed thoroughly with water to remove any impurities, dried at 110 °C for 6 h and then ground with a micro-hammer cutter mill and sieved to a 32-mesh (500 μm) particle size. The properties of the feedstock material are shown in brief in Table 1 and described in detail in Kalderis et al. (2008).

The experiments described here are under static conditions: no flow is required and no additional use of water. Additionally, monitoring of the process is not essential, since the oven can be pre-set at the required temperature and residence time. Finally, no pumping system is needed to maintain the system pressure, since pressure is automatically controlled by the steam/water equilibrium inside the reactor cell. The experimental set-up is described in detail in Kalderis et al. (2008) and Daskalaki et al. (2011). Briefly, one type of small laboratory reactor was used for hydrothermal treatment studies. The 25 mL reactors were constructed from (6 inches long, 0.64 inches i.d.) 316 stainless steel pipe with male national pipe threads (npt) and female end caps sealed with Teflon tape (Swagelok Company, USA).

A sample of the raw material was mixed with distilled water at a ratio of approximately 1/5. The mixture was then stirred and heated to become homogenized and impregnated at a temperature of 85 °C until a thick uniform paste was obtained. A sample of wet rice husk paste (~75 % moisture) was weighed before placing inside the reactor. Each reactor was loaded with 25 g of wet paste. This procedure left ~5 mL of headspace in the cell. All static (non-flowing) reaction cells must contain a sufficient headspace so that the pressure inside the cell is controlled by the steam/liquid equilibrium. A full cell must never be used since the pressures could reach several thousand bar. The reactors were placed in a (pre-heated at the required temperature) gas chromatography oven (Hewlett-Packard 5890, series II) for heating. Zero time was taken when the reactors were placed in the oven. The experiments were performed at 200 and 300 °C and residence times of 2, 4, 6, 8, 12 and 16 h (a total of 12 hydrochars). All runs were performed in triplicate. At the end of each experimental time, each reactor was removed from the oven and was allowed to cool in room temperature. The solid product (hydrochar) was recovered by filtration, washed with acetone and then with distilled water to remove all traces of acetone and air-dried for 24 h. From now on, the hydrochars produced will be referred to as H-temperature-residence time, e.g. H-200-2 for the sample obtained at 200 °C and 2 h residence time.

3 Analysis and characterization

The hydrochar yield was determined as the ratio of the produced hydrochar weight (after washing and drying) to the dry weight of rice husk subjected to hydrothermal treatment:

$$\text{Hydrochar yield } (\%) = (W_2/W_1) \times 100, \tag{1}$$

where W_1 is the dry weight of the rice husk sample prior to the treatment and W_2 is the hydrochar weight.

For measuring pH and electrical conductivity (EC) of hydrochars, suspensions of $0.01 \, \text{mol} \, \text{l}^{-1}$ CaCl$_2$ and distilled H$_2$O (1 : 5) were prepared. The mixtures were shaken for 1 h on a low-speed shaker at room temperature. After sedimentation of hydrochar material for another hour, EC and pH were determined in the supernatant (Wiedner et al., 2013). Hydrochar nitrogen adsorption analysis to determine BET surface area and pore structure was carried out at 77 K in a Micromeritics Tristar 3000. The content in C, H, N and S was analysed by an elemental microanalyser LECO CHNS-932, and the oxygen content was determined by difference. The parameters of yield, residence time and specific surface area were used to determine the optimum preparation conditions and the corresponding two hydrochar samples (one at each experimental temperature) to analyse further. As a result, the analyses described below were only applied to the selected hydrochars.

Metal content (Cu, Ni, Zn, Cd, and Pb) was determined using a Perkin Elmer 2280 atomic absorption spectrophotometer after sample extraction by digestion with 3 : 1 (v/v) concentrated HCl/HNO$_3$ following USEPA-3051a method (USEPA, 1997). The theoretical higher heating value (HHV$_p$) was calculated using an empirical correlation developed by Channiwala and Parikh (2002):

$$\text{HHV}_{\text{predicted}}(\text{MJ kg}^{-1}) = 0.3491\text{C} \tag{2}$$
$$+ 1.1783\text{H} + 0.1005\text{S} - 0.1034\text{O} - 0.0015\text{N} - 0.0211\text{A}.$$

Equation 2 is used to predict the HHV, where C, H, S, O, N, and A represent the weight percentages of carbon, hydrogen, sulfur, oxygen, nitrogen, and ash in hydrochars, respectively (Channiwala and Parikh, 2002). Kang et al. (2012) and He et al. (2013) used this formula, and the relative error between the calculated and predicted values was less than 6 %. The selected hydrochars were also subjected to derivative thermogravimetric analysis (TG-dTG) in a Labsys Setaram thermobalance under air atmosphere and 15 °C min^{-1} heating rate. Proximate analysis was performed in the same Labsys Setaram thermobalance using N$_2$ atmosphere and 30 °C min^{-1} heating rate. Moisture content was calculated as the weight loss from the initial temperature to 150 °C. The volatile fraction (VM) was determined as the weight loss from 150 °C to 600 °C under N$_2$ atmosphere. At this temperature, air was introduced in order to determine the ash content. The fixed carbon percentage content was calculated as 100 % – volatile matter percentage content – ash percentage content. Fixed carbon recovery is the percent of the fixed carbon content in the biomass that is maintained in the final processed product. It is an indication of the carbon sequestration potential and was determined as follows:

$$\text{Fixed C recovery } (\%) = \tag{3}$$
(% of fixed C in hydrochar/% of fixed C in rice husk)
$\times$ % yield.

4 Results and discussion

The effect of temperature and residence time in hydrochar yield is presented in Fig. 1. It can be seen that hydrochar

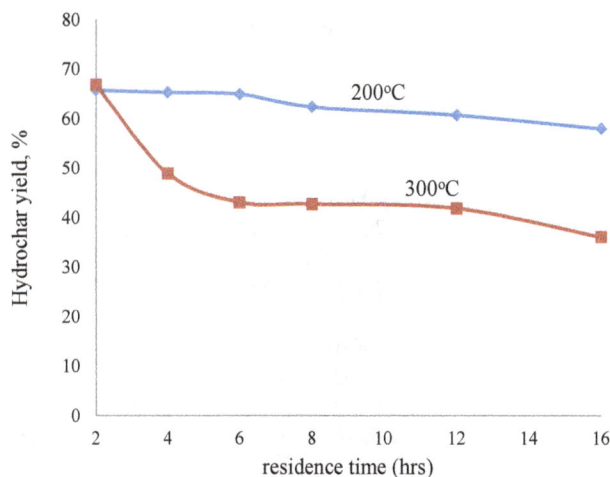

Figure 1. Hydrochar yields.

Table 2. Characterization of the hydrochar samples obtained at 200 and 300 °C.

	pH	EC (mS)	C^2 (%)	H (%)	N (%)	S (%)	H/C atomic ratio	Surface area^2 ($m^2 g^{-1}$)	Pore volume ($cm^3 g^{-1}$)
H-200-2	4.42	0.98	37.82	4.82	0.34	0.05	1.53	14.6	0.034
H-200-4	4.34	1.01	39.83	4.51	0.40	0.04	1.36	19.4	0.056
H-200-6[1]	4.19	1.02	40.81	4.31	0.44	0.04	1.27	20.7	0.064
H-200-8	4.22	1.01	42.6	4.17	0.50	0.08	1.17	21.5	0.076
H-200-12	4.33	0.99	44.08	4.06	0.53	0.01	1.11	22.6	0.092
H-200-16	4.35	0.98	43.13	3.83	0.53	0.02	1.07	29.7	0.128
H-300-2	3.41	1.18	41.87	2.97	0.62	0.01	0.85	14.5	0.065
H-300-4	3.43	1.23	42.43	3.55	0.64	0.06	0.84	24.9	0.082
H-300-6[1]	3.41	1.35	45.56	3.2	0.69	0.04	0.84	20.3	0.074
H-300-8	3.41	1.17	46.01	3.26	0.75	0.03	0.85	19.1	0.069
H-300-12	3.46	1.34	46.19	3.26	0.75	0.03	0.85	23.3	0.073
H-300-16	3.43	1.36	47.32	3.13	0.74	0.04	0.79	18.7	0.049

[1] These samples were selected for further analyses.
[2] Average values of triplicate measurements.

yields decrease as the temperature is raised from 200 to 300 °C. This decrease is closely connected with deoxygenating reactions (e.g. dehydration, decarboxylation) and volatile matter conversion, as the oxygen and hydrogen contents become lower at higher temperatures (Table 2). The hydrochar yields obtained from the hydrothermal carbonization are in the 66–58 wt% range at 200 °C and 66–36 wt% at 300 °C. At both temperatures, it can be seen that after 6 h of treatment, the yield remains somewhat constant. This indicates that any major transformations and structural rearrangements do occur in the first 6 h, after which the products became structurally stable. Gao et al. (2013) and He et al. (2013) observed the same trend during the production of hydrochars from water hyacinth and sewage sludge, respectively. The values of carbon content and surface area were used as reference points for the reproducibility of the production method. At each temperature and residence time, triplicate samples were measured in terms of carbon content and surface area, and the relative standard deviation was found to be 9 and 6 %, respectively.

The hydrothermal treatment of rice husk led to an increase in the carbon content of the solid residue from 36.1 % (rice husk, Table 1) to 43 and 47 % in the case of H-200-16 and H-300-16, respectively (Table 2). This shows that the rice husk was only partially carbonized during the process. The increasing trend at 300 °C suggests that a more complete carbonization of the product can be achieved at longer residence times. The H/C atomic ratio decreased steadily with time, at 200 °C (from 1.53 to 1.07). At 300 °C, in the first 12 h the ratio is practically the same and only after 16 h a small decline was observed. This indicates that at the higher temperature the structural rearrangements and reaction pathways occur at a faster rate and the product becomes stable in a smaller amount of time. Therefore, temperature has a more predom-

inant role than time during hydrochar production. This behaviour is consistent with the formation of a well-condensed material, especially at 300 °C (Sevilla and Fuertes, 2009).

The pH values were acidic, approximately 4.4 and 3.4 for the 200 and 300 °C hydrochars respectively. Electrical conductivity was slightly increased with temperature, from a mean of 1 mS cm^{-1} at 200 °C to a mean of 1.2 mS cm^{-1} at 300 °C, indicating high salinity for all samples.

Surface areas and pore volumes were low and very similar for all hydrochar samples. The slightly increasing trend at 200 °C can be attributed to the surface roughness because the pore volume remains practically the same (Unur et al., 2013). Based on the yield, residence time and surface area, two hydrochar samples were selected for further analyses. Since the yield remains nearly constant after the 6 h mark and surface area is practically the same for all samples at both temperatures, H-200-6 and H-300-6 were selected for further tests. It is important to remember that hydrothermal treatment is an energy-consuming process; thus reducing treatment time may have a positive economic effect when scaling-up occurs.

Table 3 shows the results obtained from the analyses of H-200-6 and H-300-6, where an important influence of temperature can be observed. Volatile matter significantly decreased from 43.06 to 15.1 wt % in H-200-6 and H-300-6, respectively. With respect to fixed carbon, it increased from 29.43 % in H-200-6 to 46.57 % in H-300-6, indicating polymerization/condensation reactions during treatment of rice husk at 300 °C. These results were similar to those obtained during pyrolysis of wastes (Méndez et al., 2013). High fixed carbon recoveries (108 and 113 % for the selected samples at 200 and 300 °C, respectively) were obtained. Considering the principle of mass conservation, this indicates that the decrease of volatile matter at the higher temperature is converted to other products, probably CO_2 and other gases

Table 3. Properties of hydrochar samples H-200-6 and H-300-6.

	H-200-6	H-300-6
Volatile organic content VM (%)	43.06	15.10
Fixed carbon content FC (%)	29.43	46.57
Fixed carbon recovery (%)	108	113
Ash content (%)	24.54	40.14
Micropore area ($m^2\,g^{-1}$)	0.8714	1.0036
HHV predicted ($MJ\,kg^{-1}$)	15.7	17.8
O (%)[1]	29.86	10.37
Cu ($mg\,kg^{-1}$)	nd[2]	nd
Ni ($mg\,kg^{-1}$)	nd	nd
Cd ($mg\,kg^{-1}$)	nd	nd
Zn ($mg\,kg^{-1}$)	0.80	1.32
Pb ($mg\,kg^{-1}$)	nd	nd

[1] calculated by difference
[2] non-detected

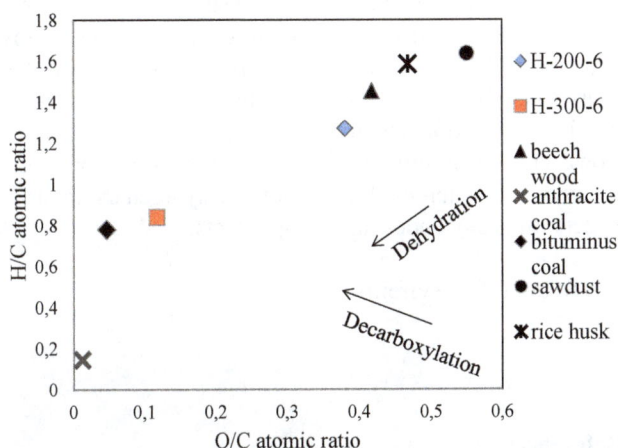

Figure 2. Van Krevelen diagram showing the position of H-200-6 and H-300-6 hydrochars among known fuels and biomass materials.

(a)

(b)

Figure 3. TG-dTG curves for the combustion profiles of (**a**) H-200-6 and (**b**) H-300-6.

(Kang et al., 2012). Furthermore, it is worth mentioning that, at the end of hydrothermal process, once the reactor has cooled down, there is only a slight overpressure inside the vessel, which suggests that a small number of gaseous products are generated during hydrothermal carbonization. The heavy metal contents after acid digestion were below detection limit, except for Zn^{2+}, which increased a 65 % with temperature as compared to biochars obtained by conventional pyrolysis (Méndez et al., 2012).

The atomic H/C and O/C ratios were calculated using the elemental composition data. Results from this analysis are presented in a Van Krevelen diagram (Fig. 2). Van Krevelen diagrams allow for delineation of reaction pathways and offer a clear insight into the chemical transformations of the carbon-rich material, which are demethanation (production of methane), dehydration (production of water) and decarboxylation (production of carbonyls including carboxylic acids). Figure 2 shows that both the H/C and O/C ratios de-

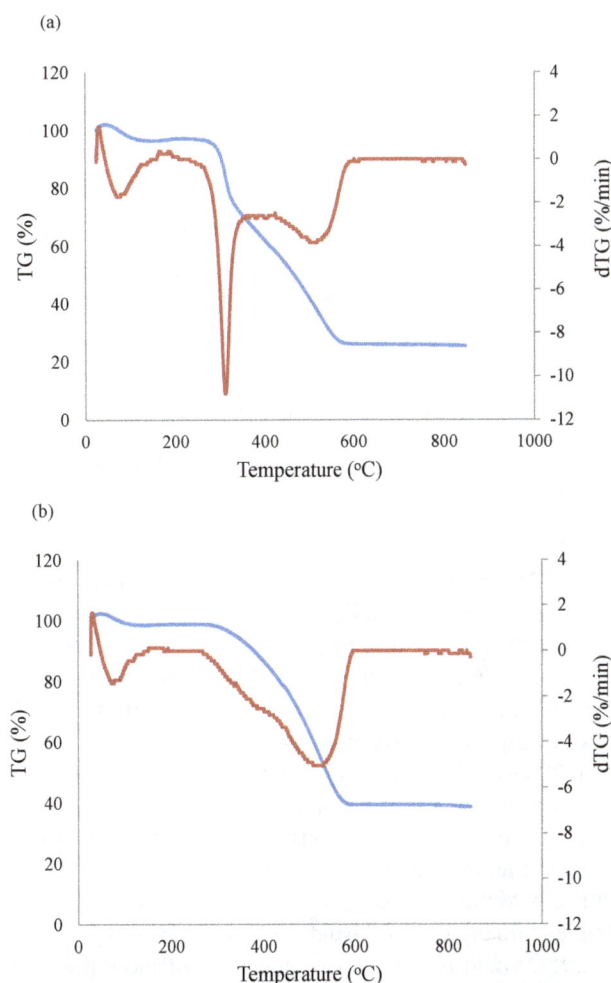

creased when the temperature was raised. At high temperature operation, the dehydration path is predominant as compared to the lower temperature operation. It is suggested that a side reaction, which is decarboxylation, occurs during the hydrothermal process because a complete dehydration reaction removes water molecules from the samples (Lu et al., 2013; Falco et al., 2011a, 2011b; Parshetti et al., 2013). Toor et al. (2011) provide a comprehensive review on the basic reaction pathways involved in the hydrothermal conversion of the main biomass components (carbohydrates, lignin, protein and lipids) to bio-products. The Van Krevelen diagram suggests an improvement in the fuel properties from the H-200-6 to the H-300-6 sample. This is confirmed by the predicted HHVs, which indicate a 11.8 % increase as the temperature is raised from 200 to 300 °C (15.7 and 17.8 $MJ\,kg^{-1}$ for the H-200-6 and H-300-6, respectively). These values are comparable to the calorific value of lignite (16.3 $MJ\,kg^{-1}$) and are in good agreement with those measured by Liu et al. (2014) for a range of hydrochars. As suggested by Danso-Boateng et

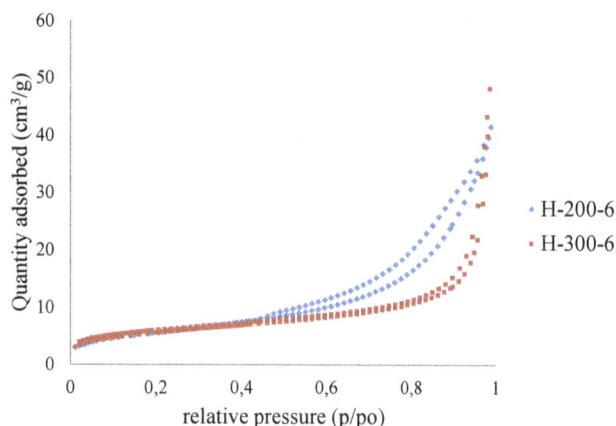

Figure 4. Nitrogen adsorption and desorption isotherms for H-200-6 and H-300-6.

al. (2013), energy densification of the hydrochars occurs as a result of decreases in solid mass caused by the dehydration and decarboxylation reactions.

The combustion profiles of H-200-6 and H-300-6 are shown in Fig. 3. At temperatures lower than 150 °C, weight loss corresponds to water release from the samples. Then, from 200 to 600 °C, weight loss was related to volatilization and combustion of organic matter. The dTG curve of H-200-6 showed three distinctive bands: the first related to humidity release at temperatures lower than 150 °C; the second band with a maximum at 310 °C was typical of cellulose combustion; and finally the third band with maximum weight loss at ~520 °C could be related to combustion of more polymerized organic matter. Comparing with the H-300-6 curve, the peak related to the presence of cellulose diminishes considerably whereas the peak at ~520 °C slightly increases.

Figure 4 shows the N_2 isotherms for H-200-6 and H-300-6. In both cases the isotherms could be classified as type II according to the IUPAC classification. Type II isotherms are typically obtained in cases of non-porous or macroporous materials, where unrestricted monolayer–multilayer adsorption can occur. Improvements in the porosity of hydrochars and the surface area are therefore necessary to enable their use as adsorbents of contaminants, hydrogen storage or electrical energy storage (supercapacitors). Such improvements have been achieved with a combination of thermal and chemical activation methods (Sevilla et al., 2011b).

5   Conclusions

Rice husk was treated in subcritical water (hydrothermal carbonization) in order to obtain hydrochars. A safe and simple to set up and operate system was used, consisting of a stainless steel reactor, caps and a source of heat. Two sets of hydrochars were obtained, corresponding to experimental temperatures of 200 and 300 °C and residence times in

the range of 2–16 h. The carbon contents of the products increased with temperature, whereas the hydrogen and oxygen contents decreased. The surface area was low for all hydrochars, indicating that porous structure was not developed during treatment. Of the two hydrochars tested further (H-200-6 and H-300-6), the latter showed improved fuel properties as indicated by the Van Krevelen diagram and the predicted higher heating value. However, the high ash content of hydrochars from rice husk should be taken into consideration when such materials are to be used as fuels, due to potential slagging or fouling of boiler tubes and corrosion of metal surfaces. Since additional steps (such as activation) are required to increase the surface area – and therefore the adsorption capacity – of the hydrochars, their production for fuel purposes may be a more suitable pathway. The fact that hydrothermal carbonization takes place in an aqueous reaction medium means that wet biomass can be used, thus eliminating any energy-consuming pre-drying steps before treatment. An additional advantage is that unlike dry pyrolysis, any gaseous emissions produced during hydrothermal carbonization are largely dissolved in the char–water slurry. For this reason, hydrothermal carbonization is more flexible and has fewer technical considerations. However, there is still need for a full characterization of the acetone- and water-soluble fractions of hydrochars, in order to determine any undesirable by-products.

Edited by: J. Paz-Ferreiro

References

Abel, S., Peters, A., Trinks, S., Schonsky, H., Facklam, M., and Wessolek, G.: Impact of biochar and hydrochar addition on water retention and water repellency of sandy soil, Geoderma, 202–203, 183–191, 2013.

Bargmann, I., Rillig, M. C., Buss, W., Kruse, A., and Kuecke, M.: Hydrochar and biochar effects on Germination of Spring Barley, J. Agron. Crop Sci., 199, 360–373, 2013.

Busch, D., Stark, A., Kammann, C. I., and Glaser, B.: Genotoxic and phytotoxic risk assessment of fresh and treated hydrochar from hydrothermal carbonization compared to biochar from pyrolysis, Ecotox. Environ. Safe., 97, 59–66, 2013.

Channiwala, S. A. and Parikh, P. P.: A unified correlation for estimating HHV of solid, liquid and gaseous fuels, Fuel, 81, 1051–1063, 2002.

Danso-Boateng, E., Holdich, R. G., Shama, G., Wheatley, A. D., Sohail, M., and Martin, S. J.: Kinetics of faecal biomass hydrothermal carbonisation for hydrochar production, Appl. Energ., 111, 351–357, 2013.

Daskalaki, V. M., Timotheatou, E. S., Katsaounis, A., and Kalderis, D.: Degradation of Reactive Red 120 using hydrogen peroxide in subcritical water, Desalination, 274, 200–205, 2011.

Ding, L., Zou, B., Li, Y., Liu, H., Wang, Z., Zhao, C., Su, Y., and Guo, Y.: The production of hydrochar-based hierarchical porous carbons for use as electrochemical supercapacitor electrode materials, Colloids Surfaces A, 423, 104–111, 2013a.

Ding, L., Zou, B., Liu, H., Li, Y., Wang, Z., Su, Y., Guo, Y., and Wang, X.: A new route for conversion of corncob to porous carbon by hydrolysis and activation, Chem. Eng. J., 225, 300–305, 2013b.

Falco, C., Baccile, N., and Titirici, M.-M.: Morphological and structural differences between glucose, cellulose and lignocellulosic biomass derived hydrothermal carbons, Green Chem., 13, 3273–3281, 2011a.

Falco, C., Caballero, F. P., Babonneau, F., Gervais, C., Laurent, G., Titirici, M. M., and Baccile, N.: Hydrothermal carbon from biomass: structural differences between hydrothermal and pyrolyzed carbons via ^{13}C solid state NMR, Langmuir, 27, 14460–14471, 2011b.

Gao, Y., Wang, X., Wang, J., Li, X., Cheng, J., Yang, H., and Chen, H.: Effect of residence time on chemical and structural properties of hydrochar obtained by hydrothermal carbonization of water hyacinth, Energy, 58, 376–383, 2013.

Gajić, A. and Koch, H.-J.: Sugar beet (*Beta vulgaris* L.) growth reduction caused by hydrochar is related to nitrogen supply, J. Environ. Qual., 41, 1067–1075, 2012.

He, C., Giannis, A., and Wang, J.-Y.: Conversion of sewage sludge to clean solid fuel using hydrothermal carbonization: hydrochar fuel characteristics and combustion behaviour, Appl. Energ., 111, 257–266, 2013.

Jamari, S. S. and Howse, J. R.: The effect of the hydrothermal carbonization process on palm oil empty fruit bunch, Biomass Bioenergy, 47, 82–90, 2012.

Kalderis, D., Hawthorne, S. B., Clifford, A. A., and Gidarakos, E.: Interaction of soil, water and TNT during degradation of TNT on contaminated soil using subcritical water, J. Hazard. Mater., 159, 329–334, 2008.

Kang, S., Li, X., Fan, J., and Chang, J.: Characterization of hydrochars produced by hydrothermal carbonization of lignin, cellulose, D-xylose, and wood meal, Indust. Engin. Chem. Res., 51, 9012–9023, 2012.

Liu, Z., Quek, A., Kent Hoekman, S., and Balasubramanian, R.: Production of solid biochar fuel from waste biomass by hydrothermal carbonization, Fuel, 103, 943–949, 2013.

Liu, Z., Quek, A., and Balasubramanian, R.: Preparation and characterization of fuel pellets from woody biomass, agro-residues and their corresponding hydrochars, Appl. Energ., 113, 1315–1322, 2014.

Lu, X., Pellechia, P. J., Flora, J. R. V., and Berge, N. D.: Influence of reaction time and temperature on product formation and characteristics associated with the hydrothermal carbonization of cellulose, Bioresource Technol., 138, 180–190, 2013.

Méndez, A., Gómez, A., Paz-Ferreiro, J., and Gascó, G.: Effects of sewage sludge biochar on plant metal availability after application to a Mediterranean soil, Chemosphere, 89, 1354–1359, 2012.

Méndez, A., Terradillos, M., and Gascó, G.: Physicochemical and agronomic properties of biochar from sewage sludge pyrolysed at different temperatures, J. Anal. Appl. Pyrol., 102, 124–130, 2013.

Parshetti, G. K., Kent Hoekman, S., and Balasubramanian, R.: Chemical, structural and combustion characteristics of carbonaceous products obtained by hydrothermal carbonization of palm empty fruit bunches, Bioresource Technol., 135, 683–689, 2013.

Sevilla, M. and Fuertes, A. B.: The production of carbon materials by hydrothermal carbonization of cellulose, Carbon, 47, 2281–2289, 2009.

Sevilla, M., Macia-Agullo, J. A., and Fuertes, A. B.: Hydrothermal carbonization of biomass as a route for the sequestration of CO_2: chemical and structural properties of the carbonized products, Biomass Bioen., 35, 3152–3159, 2011a.

Sevilla, M., Fuertes, A. B., and Mokaya, R.: High density hydrogen storage in superactivated carbons from hydrothermally carbonized renewable organic materials, Energ. Environ. Sci., 4, 1400–1410, 2011b.

Toufiq Reza, M., Lynam, J. G., Helal Uddin, M., and Coronella, C. J.: Hydrothermal carbonization: fate of inorganics, Biomass Bioen., 49, 89–94, 2013.

Toor, S. S., Rosendahl, L., and Rudolf, A.: Hydrothermal liquefaction of biomass: A review of subcritical water technologies, Energy, 36, 2328–2342, 2011.

Unur, E., Brutti, S., Panero, S., and Scrosati, B.: Nanoporous carbons from hydrothermally treated biomass as anode materials for lithium ion batteries, Micropor. Mesopor. Mat., 174, 25–33, 2013.

USEPA: Method 3051a: Microwave Assisted Acid Dissolution of Sediments, Sludges, Soils and Oils, 2° edn., US Gov. Print Office, Washington, USA, 1997.

Wagner, A. and Kaupenjohann, M.: Suitability of biochars (pyro- and hydrochars) for metal immobilization on former sewage-field soils, Eur. J. Soil Sci., 65, 139–148, 2014.

Wiedner, K., Naisse, C., Rumpel, C., Pozzi, A., Wieczorek, P., and Glaser, B.: Chemical modification of biomass residues during hydrothermal carbonization – What makes the difference, temperature or feedstock?, Org. Geochem., 54, 91–100, 2013.

17

Variations of soil profile characteristics due to varying time spans since ice retreat in the inner Nordfjord, western Norway

A. Navas[1], K. Laute[2], A. A. Beylich[2], and L. Gaspar[3]

[1]Estación Experimental de Aula Dei (EEAD-CSIC), Department of Soil and Water, Avda. Montañana 1005, 50059 Zaragoza, Spain
[2]Geological Survey of Norway (NGU), Geo-Environment Division, 7491 Trondheim, Norway
[3]School of Geography, Earth and Environmental Sciences, Plymouth University, Plymouth, Devon, PL4 8AA, UK

Correspondence to: A. Navas (anavas@eead.csic.es)

Abstract. In the Erdalen and Bødalen drainage basins located in the inner Nordfjord in western Norway the soils were formed after deglaciation. The climate in the uppermost valley areas is sub-arctic oceanic, and the lithology consists of Precambrian granitic orthogneisses on which Leptosols and Regosols are the most common soils. The Little Ice Age glacier advance affected parts of the valleys with the maximum glacier extent around AD 1750. In this study five sites on moraine and colluvium materials were selected to examine main soil properties, grain size distribution, soil organic carbon and pH to assess if soil profile characteristics and patterns of fallout radionuclides (FRNs) and environmental radionuclides (ERNs) are affected by different stages of ice retreat. The Leptosols on the moraines are shallow, poorly developed and vegetated with moss and small birches. The two selected profiles show different radionuclide activities and grain size distribution. The sampled soils on the colluviums outside the LIA glacier limit became ice-free during the Preboral. The Regosols present better-developed profiles, thicker organic horizons and are fully covered by grasses. Activity of ^{137}Cs and ^{210}Pb$_{ex}$ concentrate at the topsoil and decrease sharply with depth. The grain size distribution of these soils also reflects the difference in geomorphic processes that have affected the colluvium sites. Significantly lower mass activities of FRNs were found in soils on the moraines than on colluviums. Variations of ERN activities in the valleys were related to characteristics of soil mineralogical composition. These results indicate differences in soil development that are consistent with the age of ice retreat. In addition, the pattern distribution of ^{137}Cs and ^{210}Pb$_{ex}$ activities differs in the soils related to the LIA glacier limits in the drainage basins.

1 Introduction

Glacial retreat in the cold regions of Northern Europe is a general trend that has intensified over the last decades. In the Nordfjord region (western Norway) this trend is also observed and the magnitude of glacial retreat in the Erdalen and Bødalen valleys has reached its fastest rate over the last century in recent years (e.g. Winkler et al., 2009; Laute and Beylich, 2013). The retreat of ice from glaciated valleys (Mavlyudov et al., 2012) causes important changes in geomorphic processes of glacial erosion, but also has an impact on the hydrological resources by changing runoff and associated sediment transport as well as on the formation of soils on the newly exposed surfaces.

The landscapes of Norwegian fjords reveal the inheritance of glacial processes since the Last Glacial Maximum (LGM). The Little Ice Age (LIA) glacier advance also affected parts of the Norwegian valleys (Bickerton and Matthews, 1993; Laute and Beylich, 2012, 2013). Amongst the main glacial deposits colluviums and moraines are surface formations resulting from the evolution of slopes and the ice retreat. In the glacial valleys these formations are representative of newly exposed material conditioned by former glaciations and deglaciation. Moraine ridges formed during and after the maximum extent of the Little Ice Age (LIA) with the maximum glacier extent around AD 1750 (Bickerton and

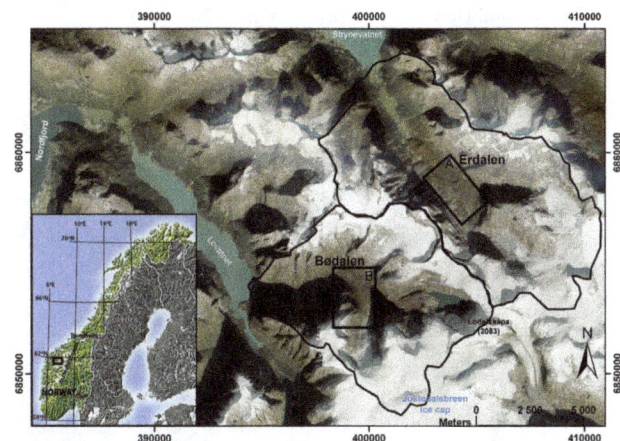

Figure 1. Location of the Nordfjord region (western Norway), aerial photograph of the Erdalen and Bødalen glacial valleys and situation of the study areas.

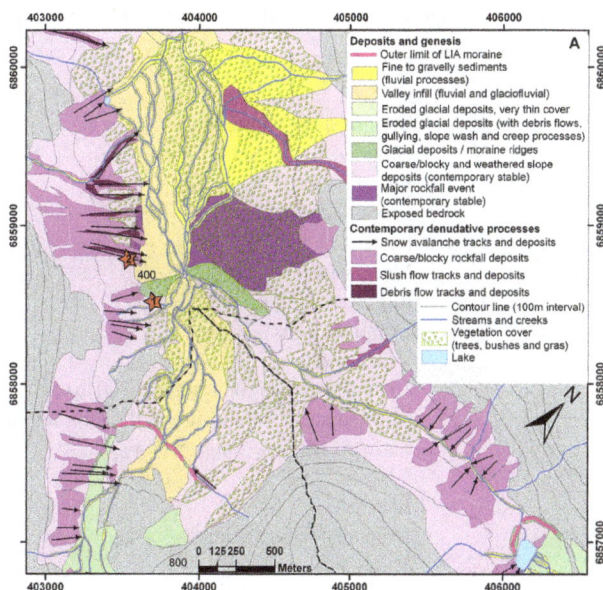

Figure 2. Geomorphological map of the Erdalen drainage basin (modified after Laute and Beylich, 2012) and location of the study profiles 1) PE1 and 2) PE2.

Matthews, 1993). The soils have been forming on the newly exposed glacial deposits, and as a result of processes related to ice retreat, different soil types have developed. Soil properties might be characteristic of materials and processes and could reflect different ages of ice retreat. In this study we aim to characterize the soils formed on colluvium and moraine material. We applied a multiproxy approach in order to examine if the soil elemental composition (Li, K, Na, Be, Mg, Ca, Sr, Cr, Cu, Mn, Fe, Al, Zn, Ni, Co, Cd, Tl, Bi, V, Ti, Pb, B, Sb, As, P, S, Mo and Se) and radionuclide tracers, environmental (ERNs: ^{238}U, ^{210}Pb, ^{226}Ra, ^{232}Th, ^{40}K, ^{210}Pb) and fallout (FRNs: ^{137}Cs and ^{210}Pb$_{ex}$), together with other properties (grain size distribution, soil organic carbon, pH) could be indicative of soil development, and if it was related to stages of ice retreat. To this purpose we selected two glacial valleys representative of the inner Nordfjord region, the valleys of Erdalen and Bødalen, and sampled five sites on two main geomorphic elements, moraine and colluvium, to examine within the soil profile main soil properties, elemental composition and fallout (FRNs) and environmental radionuclides (ERNs).

2   Material and methods

2.1   The study area

The inner Nordfjord region is located in western Norway on the western side of the Jostedalsbreen ice cap (Fig. 1). Climate is sub-arctic oceanic in the uppermost parts of tributary valleys draining into the fjord. The mean annual air temperature at 360 m a.s.l. is 5.5 °C and the mean annual areal precipitation is 1500 mm in the drainage basins of Erdalen and Bødalen (Table 1) (Beylich and Laute, 2012; Laute and Beylich, 2014).

The lithology in the valleys consists of Precambrian granitic orthogneisses with migmatic and dioritic composition (Table 1). Within the higher part in Bødalen (glacier area) there is a small area of quartz monzonite outcrops (Lutro and Tveten, 1996).

The landforms and processes in the study area are characteristic of glacial valleys. The detailed geomorphological maps (Laute and Beylich, 2012) shown in Figs. 2 and 3 present the main deposits identified in the valleys and their genesis. The main contemporary denudational surface processes, the limits of the LIA moraines and the soil profiles studied are identified in the maps.

Based on intensive field pre-investigations and existing knowledge, sites which are representative for the relevant deposits existing in the valleys and which are close to defined and dated moraine ridges were considered for this preliminary study. Furthermore, an important criterion for the selection was that the sampling sites were nearly horizontal, and no relevant delivery of new material was currently affecting these sites. A total of five sites corresponding to soils developed on moraine and colluvium deposits, the most common formations of the glacier valleys and which represent different stages of ice retreat, were selected for collecting soil profiles. Two profiles were established on moraines in the Bødalen valley (PB1 and PB2) and the other three profiles were established on colluvium materials in the Erdalen (PE1, PE2) and Bødalen (PB3) valleys. The profiles on the moraines are located inside the LIA glacier limit; the moraine material is in general characterized by diamicton. The two sites represent different stages of ice retreat, thus PB1 became ice-free starting from ca. AD 1930, but PB2 became ice-free earlier,

Table 1. Physiographic and climatic characteristics of the drainage basins of Erdalen and Bødalen (Nordfjord region, Norway).

	Erdalen	Bødalen
Geographical coordinates	61°50′ N, 07°10′ E	61°48′ N, 07°05′ E
Drainage basin area [km^2]	79.5	60.1
Elevation min [m a.s.l.]	20	52
Elevation max [m a.s.l.]	1888	2083
Relief [m]	1868	2031
Mean slope [°]	32	34
Lithology	Precambrian granitic orthogneisses	
Dominant soils	Leptosols, Regosols	
Vegetation	Birch, grey alder, grass, moos, lichens	
Annual precipitation [mm] (at 360 m a.s.l.)	1500	1500
Mean annual air temperature [°C] (at 360 m a.s.l.)	5.5	5.5
Surface area percentages[a] [%]		
Glacier	18	38
Bedrock	45	43
Slope deposits/regolith	32	16
Valley infill	5	2
Lake	< 1	1

[a] As % of the total drainage basin surface area.

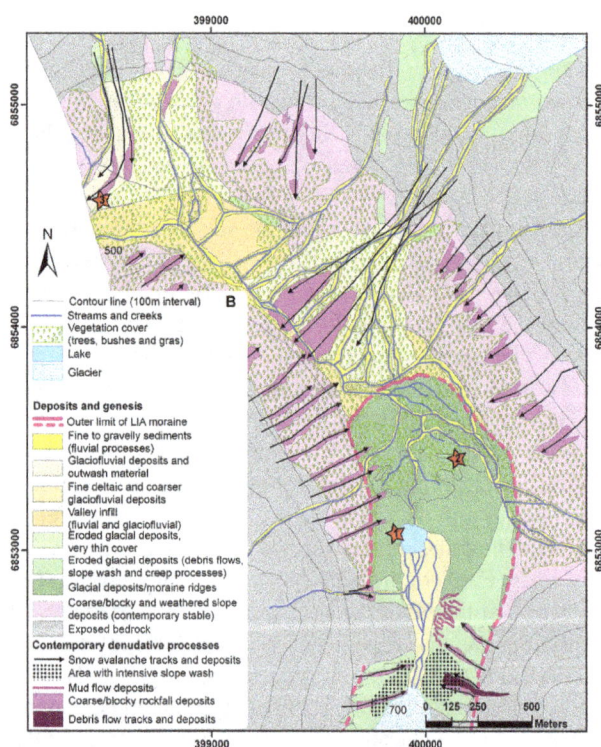

Figure 3. Geomorphological map of the Bødalen drainage basin (modified after Laute and Beylich, 2012) and location of the study profiles 1) PB1, 2) PB2 and 3) PB3.

at about AD 1800 (see Bickerton and Matthews, 1993). Both sites have a vegetation cover composed of mosses and small birches and can be considered quite stable regarding surface

soil processes, as they are on gentle slopes. The soils on the moraines are Leptosols (FAO classification); they are poorly developed with almost no horizon differentiation and with a high content of rock fragments.

The colluvium sites, located outside the LIA glacier limit, are characterized by slope processes accumulating both fine and coarser material at the slope foot derived from slope wash, avalanches, debris flows and rock falls. The soils correspond to Regosols that are deeper and better developed than soils on the moraines and those covered by grass. The sites became ice-free during the Preboral deglaciation. Considering the age of ice retreat, the oldest is PB3 in Bødalen, which became ice-free around 10 000 years ago. This profile, located on a hillslope beneath the Tindefjell glacier, is more influenced by glaciofluvial and outwash processes rather than rockfall activity and presents high avalanche activity. Of the studied sites PB3 had the strongest anthropogenic impact through animal husbandry, as sheep grazing occurred starting from approximately 1800, albeit with less intensity since 1930 (T. Lopez, personal communication, 2011). Of the profiles in Erdalen, PE4 had higher rockfall activity and became ice-free earlier than PE3, which had not been ice-free since ca. 9800 BP (see Nesje, 1984; Matthews et al., 2008).

2.2 Soil sampling and analyses

To collect the soil samples of the soil profiles, five pits of 20 × 20 cm were excavated down to a depth of 20–27 cm. Samples were extracted at depth intervals of approx. 5–6 cm by using a 5 cm diameter cylinder. Soils in the study area are shallow and according to field observations depths vary from few centimetres for Leptosols to a range of 30 to 50 cm for Regosols.

The soil samples were air-dried, ground, homogenized and quartered to pass through a 2 mm sieve. The grain size fraction > 2 mm containing stones and rock fragments was separated and weighted to estimate the percentages of coarse fractions in the soil profiles. General soil properties analysed in the fraction < 2 mm were pH, soil organic carbon (SOC %) at 310° (active carbon fraction, ACF) and 550° (stable carbon fraction, SCF) and soil texture. Analysis of the clay, silt and sand fractions was performed using a laser diffraction particle size analyser. Prior to the analysis the organic matter was eliminated with H_2O_2 (10 %) heated to 80 °C and samples were disaggregated with sodium hexametaphosphate (40 %), stirred for 2 h and dispersed with ultrasound for a few minutes. The pH (1 : 2.5, soil : water) was measured using a pH-meter.

The contents of SOC, both active and stable carbon fractions, were analysed by the dry combustion method using a LECO *RC-612* multiphase carbon analyser designed to differentiate forms of carbon by oxidation temperature (LECO, 1996) in a sub-sample of the < 2 mm fraction that had been ground to a very fine powder with a mortar and pestle. According to López-Capel et al. (2008), the decomposition of the most thermally labile components of SOC is released at approximately 300–350 °C during thermal decomposition because they are rapidly and easily burnable; the active and decomposable fraction (ACF), while decomposition of more refractory and stable carbon (SCF), occurs at higher temperatures (420–550 °C).

The analysis of the total elemental composition was carried out after total acid digestion with HF (48 %) in a microwave oven (Navas et al., 2002c). Samples were analysed for the following 28 elements: Lithium (Li), Potassium (K), Sodium (Na) (alkaline), Beryllium (Be), Magnesium (Mg), Calcium (Ca), Strontium (Sr) (light metals), Chromium (Cr), Copper (Cu), Manganese (Mn), Iron (Fe), Aluminum (Al), Zinc (Zn), Nickel (Ni), Cobalt (Co), Cadmium (Cd), Thallium (Tl), Bismuth (Bi), Vanadium (V), Titanium (Ti) and Lead (Pb) (heavy metals), Boron (B), Antimony (Sb), Arsenic (As) (metalloids), and Phosphorus (P), Sulfur (S), Molybdenum (Mo) and Selenium (Se). Analyses were performed by atomic emission spectrometry using inductively coupled plasma ICP. Concentrations, obtained after three measurements per element, are expressed in $mg\,kg^{-1}$.

The methods used in the analysis of radionuclides are described in detail in previous works (Navas et al., 2005a, b). Radionuclide activity in the soil samples was measured using a Canberra high-resolution, low-background, hyperpure germanium coaxial gamma detector coupled to an amplifier and multichannel analyser. The detector had a relative efficiency of 50 % and a resolution of 1.9 keV (shielded to reduce background), and was calibrated using standard certified samples that had the same geometry as the measured samples. Subsamples of 50 g were loaded into plastic containers. Count

times over 24 h provided an analytical precision of about ±3– 10 % at the 95 % confidence level. Activities were expressed as $Bq\,kg^{-1}$ dry soil.

Gamma emissions of Uranium-238 (^{238}U), Radium-226 (^{226}Ra), Thorium-232 (^{232}Th), Potassium-40 (^{40}K), Lead-210 (^{210}Pb), and Cesium-137 (^{137}Cs) (in $Bq\,kg^{-1}$ air-dried soil) were measured in the bulk soil samples. Considering the appropriate corrections for laboratory background, ^{238}U was determined from the 63 keV line of ^{234}Th (lower limit of detection (LLD): 2.6 $Bq\,kg^{-1}$), the activity of ^{226}Ra was determined from the 352 keV line of ^{214}Pb (LLD: 0.5 $Bq\,kg^{-1}$) (Van Cleef, 1994); ^{210}Pb activity was determined from the 47 keV photopeak (LLD: 3.2 $Bq\,kg^{-1}$), ^{40}K from the 1461 keV photopeak (LLD: 2 $Bq\,kg^{-1}$); ^{232}Th was estimated using the 911 keV photopeak of ^{228}Ac (LLD: 0.5 $Bq\,kg^{-1}$), and ^{137}Cs activity was determined from the 661.6 keV photopeak (LLD: 0.2 $Bq\,kg^{-1}$). The ^{210}Pb (half-life = 22.26 yr) is integrated by the "in situ"-produced fraction from the decay of ^{226}Ra (Appleby and Oldfield, 1992) and the upward diffusion of ^{222}Rn in the atmosphere, which is the source of $^{210}Pb_{ex}$. Spectrometric measurements were performed a month after the samples were sealed, which ensured a secular equilibrium between ^{222}Rn and ^{226}Ra. The $^{210}Pb_{ex}$ activities were estimated from the difference between the total ^{210}Pb activity and the ^{226}Ra activity.

A one-way analysis of variance (ANOVA) was applied to analyse the statistical significance of the differences in the means of the study parameters ($p < 0.05$) using the least significant difference (LSD Fisher) test. The normality of data was tested using the Wilk–Shapiro test ($p < 0.05$). Pearson's linear correlations were also established to assess the relationships between the study elements and between the radionuclides and soil properties and significance was set at $p < 0.05$.

3   Results

3.1   Characteristics of the soil profiles

The soil profiles in the Bødalen and Erdalen drainage basins were acidic, with a pH ranging from 4.45 to 5.85, and the predominant soil textures were sandy loam. The percentages of rock fragments and stones (> 2 mm) were much higher in the moraine than in the colluvium profiles. Contents in SOC were low on average (2.03 %, SD: 3.16), range between 0.03 to 14.39 % and had large variability (CV : 156 %). The contents of ACF were always higher than the SCF, especially in the colluvium profiles. There were differences between the properties of the soil profiles on the moraines and on colluviums (Table 2). The latter were significantly less stony and more acidic. Comparing with the Leptosols on the moraines that did not present horizon differentiation, the soils on the colluvium were deeper and better developed with a rich organic A horizon, and relatively higher SOC and

Table 2. Summary statistics of main properties in the soil samples on moraines and colluvium. Different letters indicate significant differences at the p level 0.05 between moraines and colluvium soils.

	Moraines $n = 9$						Colluvium $n = 12$					
	Mean		SD	Min	Max	CV	Mean		SD	Min	Max	CV
SOC %	0.48	a	0.44	0.03	1.03	91	3.18	a	3.82	0.43	14.39	120
ACF %	0.28	a	0.31	0.03	0.79	137	2.78	a	3.40	0.38	12.69	123
SCF %	0.07	a	0.05	0.01	0.12	276	0.36	a	0.31	0.03	1.10	84
Stones %	60.21	b	20.30	28.59	87.59	34	18.00	a	23.55	0.13	80.59	131
pH	5.45	b	0.33	5.02	5.85	5	4.80	a	0.24	4.45	5.16	5
2000–50 μm %	72.58	a	21.84	42.40	95.00	30	60.28	a	23.52	13.50	81.60	39
50–2 μm %	23.86	a	18.51	4.60	49.20	78	36.22	a	20.36	17.40	75.70	56
< 2 μm %	3.52	a	3.36	0.40	8.40	96	3.51	a	3.21	1.00	10.80	92

SD: Standard deviation, CV: Coefficient of variation %.

carbon fractions. Soil samples of the moraines had higher sand contents and lower silt contents than samples of the colluvium, but clay contents were similar and low.

The most abundant elements in the studied profiles were Al, Fe, Na, and K ($\overline{x}$: 47 071, 27 608, 25 205 and 23 723 mg kg^{-1}), Ca and Mg were also major components ($\overline{x}$: 10 492, 7497 mg kg^{-1}), followed by Ti, B and P ($\overline{x}$: 3942, 1948 and 1068 mg kg^{-1}), Mn, Sr and S ($\overline{x}$: 448, 279 and 266 mg kg^{-1}), and Cr, V, Zn, Ni, Tl, Pb and Bi ($\overline{x}$: 90, 54, 47, 43, 39, 28 and 23 mg kg^{-1}) whereas Li, Sb, Be and Mo had the lower contents ($\overline{x}$: 8.4, 2.4, 1.3 and 0.7 mg kg^{-1}); As and Cd were not detected in the study samples (Table 3).

The mean contents of Al were similar in colluvium and moraine profiles, although the variation range was higher in the colluvium profiles (Table 3). Of the major elements, mean Fe, Ca and Mg contents were significantly lower in the moraine profiles, whereas the opposite was observed for Na and K. In lower concentration ranges (91–6500 mg kg^{-1}), the mean contents of Sr, S, Mn, P and Ti were significantly lower in samples of the moraines than in the colluvium ones, but the opposite was found for B which was much higher in the moraine samples. In minor concentrations (ranges between 5 and 330 mg kg^{-1}), the contents of Tl, V, Zn, Ni and Cr were significantly lower in moraine samples than in colluvium soils. On the other hand, significantly higher contents of Pb and Be were found in colluvium soils. Other trace elements like Mo and Sb had significantly lower contents in moraine samples, and likewise for Li, Bi and Cu, although for the latter differences were not significant. For all samples most elements were directly correlated among them. However, Na, K, Pb, and Be were mostly inversely correlated with the rest of the elements and directly correlated between them (Table 4). When considering the moraines and colluviums separately, the Pearson's coefficients of correlation of the moraine profiles showed similar patterns (Table 5), although correlations had lower significances. For the colluvium profiles (whilst Pb and Na maintained inverse relationships with

the rest of the elements) Be and K showed different trends and few correlations were significant (Table 6).

The radioisotope mass activities ranges (Bq kg^{-1}) were 28.1–64.9 for ^{238}U, 12.6–47.7 for ^{226}Ra, 12.6–83.7 for ^{232}Th, 652–1320 for ^{40}K, b.d.l.(below detection limit)–118 for ^{210}Pb, b.d.l.–102.4 for ^{210}Pb$_{ex}$, and 1.2–346 for ^{137}Cs (Table 7). The mean contents of FRNs were much lower in the soils of the moraines, although differences were not significant. The range of variation of ^{137}Cs and ^{210}Pb$_{ex}$ in the samples of the colluviums was much larger than that of the moraines. Apart from ^{210}Pb, the ERN contents were significantly higher in the moraine samples.

The correlations established among the FRNs with the soil properties showed that in the moraine profiles ^{137}Cs was related directly and significantly with SOC, ACF and sand contents, but inversely with the pH and the silt and clay fractions (Table 8). However, in the colluvium profiles ^{137}Cs was only significantly correlated with SCF content and the correlations with the fine fractions were direct but not significant. Concerning the ^{210}Pb$_{ex}$ the only significant correlation was that with the SCF content in the colluvium profiles.

Few correlations of ERNs with soil properties were significant in the colluvium profiles (Table 8). The type of correlations were similar for ^{232}Th and ^{40}K, as both radionuclides were directly related with the fine fractions but inversely related with the soil organic carbon and the sand contents; whereas the opposite was true for ^{226}Ra and ^{238}U. However, in the moraine profiles the type of correlations was different and not significant.

3.2 Distribution of soil properties and elements in the profiles

The vertical distribution of main soil properties, radionuclide and element contents in the study soils showed very distinctive patterns in the moraine and the colluvium profiles. The SOC distribution down the profiles of moraine soils was quite homogeneous and contents were much lower than in the

Table 3. Summary statistics of the elemental composition ($mg\,kg^{-1}$) in the samples of the soil profiles of the moraines and colluviums. Different letters indicate significant differences at the p level 0.05 between moraines and colluvium soils.

| | Moraines $n = 9$ | | | | | | Colluvium $n = 12$ | | | | | |
	Mean		SD	Min	Max	CV	Mean		SD	Min	Max	CV
Be	1.5	b	0.0	1.4	1.5	2	1.1	a	0.1	0.8	1.3	12
Mo	0.5	a	0.2	b.d.l.	0.7	41	0.8	b	0.1	0.6	1.0	17
Sb	1.4	a	0.3	0.9	1.8	18	3.1	b	1.1	1.6	5.1	35
Li	7.6	a	1.0	5.2	8.5	13	9.0	a	6.0	0.8	17.4	67
Bi	21.6	a	1.6	18.6	23.8	8	24.4	a	7.2	12.1	34.2	30
Cu	17.3	a	10.5	7.1	41.2	61	22.7	a	13.2	5.6	47.3	58
Pb	31.0	b	1.0	29.5	32.6	3	26.6	a	3.6	19.3	31.8	14
Tl	24.6	a	3.4	19.2	29.7	14	49.6	b	26.0	13.0	86.4	53
V	34.9	a	4.9	25.9	40.3	14	69.1	b	21.3	32.8	99.3	31
Zn	29.8	a	2.5	23.9	33.1	8	58.9	b	34.8	9.6	110.0	59
Ni	27.0	a	2.9	22.4	31.4	11	55.6	b	37.7	14.0	121.0	68
Cr	25.1	a	4.6	17.1	30.5	19	139.0	b	99.0	30.6	323.9	71
Sr	241.9	a	16.3	212.7	261.1	30	306.2	a	94.3	172.1	465.5	31
S	125.1	a	37.2	90.8	216.2	30	371.3	b	232.9	191.5	1005.0	63
Mn	275.3	a	40.2	211.7	336.7	15	571.7	b	349.3	117.5	1088.0	61
P	839.3	a	109.3	613.6	942.8	13	1239.9	b	356.0	545.4	1791.0	29
B	3197.8	b	237.5	2970.0	3770.0	7	1010.1	a	796.4	300.7	3090.0	79
Ti	3092.2	a	419.0	2240.0	3840.0	14	4580.0	b	1070.0	2720.0	6500.0	23
Mg	2710.8	a	610.1	1771.0	3633.0	23	11086.1	b	8748.6	1206.0	25100.0	79
Ca	8241.4	a	778.3	6516.0	8982.0	9	12180.0	b	2496.8	6728.0	16140.0	21
K	28358.9	b	1423.5	26140.0	30120.0	5	20246.7	a	3394.7	12810.0	24980.0	17
Na	30607.8	b	1579.4	29000.0	33940.0	5	21153.3	a	4763.4	13520.0	27270.0	23
Fe	19300.0	a	2500.9	15490.0	22960.0	13	33838.4	b	15949.9	9471.0	56000.0	47
Al	45526.7	a	4215.0	40330.0	53590.0	9	48230.0	a	10710.9	27050.0	61550.0	22

SD: Standard deviation, CV: Coefficient of variation %, b.d.l.: below detection limit

Table 4. Pearson correlation coefficients among elements ($mg\,kg^{-1}$) in the samples of all soil profiles. Bold face numbers are significant at the 95 % confidence level, underlined numbers are significant at the 99 % confidence level.

	Mo	Sb	Li	Bi	Cu	Pb	Tl	V	Zn	Ni	Cr	Sr	S	Mn	Pb	B	Ti	Ca	Mg	K	Al	Na	Fe
Be	**-0.682**	**-0.535**	0.050	-0.014	-0.150	**0.661**	-0.360	**-0.522**	-0.310	-0.363	**-0.508**	-0.052	**-0.817**	-0.353	-0.341	**0.765**	-0.431	-0.337	-0.417	**0.914**	0.138	**0.809**	-0.351
Mo		**0.573**	0.156	0.296	0.088	-0.371	**0.466**	**0.537**	0.451	0.426	**0.524**	0.189	**0.548**	0.494	0.406	**-0.502**	0.389	**0.433**	**0.470**	**-0.565**	0.062	**-0.755**	0.458
Sb			**0.702**	**0.743**	0.499	-0.481	**0.934**	**0.935**	**0.886**	**0.876**	**0.952**	0.257	**0.885**	**0.537**	**-0.659**	**0.766**	**0.772**	**0.944**	-0.453	**0.635**	0.175	**-0.815**	**0.895**
Li				**0.953**	**0.685**	-0.097	**0.877**	**0.720**	**0.902**	**0.746**	**0.726**	0.537	-0.134	**0.847**	0.430	-0.105	**0.551**	0.490	**0.826**	0.175	**0.895**	-0.463	**0.893**
Bi					**0.659**	-0.045	**0.903**	**0.794**	**0.926**	**0.759**	**0.745**	**0.627**	-0.119	**0.887**	0.519	-0.124	**0.636**	**0.638**	**0.830**	0.110	**0.916**	-0.503	**0.913**
Cu						-0.386	**0.658**	0.548	**0.689**	**0.663**	**0.616**	0.233	0.096	**0.685**	0.248	-0.161	0.393	0.316	**0.631**	0.469	**0.641**	-0.465	**0.689**
Pb							-0.357	-0.357	-0.286	-0.418	**-0.500**	0.054	**-0.550**	-0.331	-0.317	**0.681**	-0.187	-0.310	-0.413	**0.692**	0.058	**0.642**	-0.376
Tl								**0.911**	**0.986**	**0.919**	**0.946**	**0.493**	-0.431	**0.975**	0.529	-0.431	**0.688**	**0.699**	**0.980**	-0.263	**0.794**	**-0.744**	**0.973**
V									**0.895**	**0.774**	**0.864**	**0.633**	0.226	**0.892**	**0.612**	**-0.658**	**0.909**	**0.796**	**0.892**	-0.377	**0.676**	**-0.816**	**0.910**
Zn										**0.869**	**0.896**	0.541	0.155	**0.978**	**0.552**	-0.384	**0.685**	**0.657**	**0.953**	-0.202	**0.796**	**-0.735**	**0.980**
Ni											**0.974**	0.145	0.087	**0.908**	0.223	-0.341	**0.489**	**0.540**	**0.949**	-0.305	**0.643**	**-0.641**	**0.834**
Cr												0.253	0.233	**0.921**	0.366	-0.519	**0.611**	**0.626**	**0.976**	-0.430	**0.620**	**-0.774**	**0.882**
Sr													0.006	0.428	**0.834**	-0.352	**0.733**	**0.754**	0.357	**0.641**	-0.379	**0.582**
S														0.137	0.423	**-0.545**	0.162	0.071	0.188	**-0.796**	-0.310	**-0.688**	0.219
Mn															0.458	-0.398	**0.662**	**0.621**	**0.963**	-0.239	**0.738**	**-0.742**	**0.953**
Pb																**-0.530**	**0.604**	**0.755**	0.464	**-0.614**	**0.630**
B																	**-0.677**	**-0.580**	-0.476	**0.728**	-0.024	**0.772**	-0.449
Ti																		0.749	**0.662**	-0.278	0.536	**-0.668**	**0.728**
Ca																			**0.615**	**0.624**	-0.355	**-0.556**	**0.681**
Mg																				-0.315	**0.708**	**-0.766**	**0.944**
K																					0.246	**0.665**	-0.227
Al																						-0.289	**0.782**
Na																							**-0.782**

colluvium profiles (Fig. 4). The PB1 profile had the lowest SOC (below 0.1 %), most in the active form with almost negligible amounts of the stable carbon fraction, which was also very low in PB2. The vertical distribution of SOC contents in the colluvium profiles showed decreasing trends with depth, which were more marked in PB3. The PE2 profile showed a large SOC enrichment at the 10–14 cm interval depth, most of it in active form (ACF reaches 12.7 %). The pH profiles showed little variations with depth, both in the moraine and the colluvium profiles, and values were slightly lower at the topsoil than at deeper layers.

Table 5. Pearson correlation coefficients among elements ($mg\,kg^{-1}$) in the samples of the moraine profiles. Bold face numbers are significant at the 95 % confidence level, underlined numbers are significant at the 99 % confidence level.

	Mo	Sb	Li	Bi	Cu	Pb	Tl	V	Zn	Ni	Cr	Sr	S	Mn	P	B	Ti	Ca	Mg	K	Al	Na	Fe
Be	-0.613	-0.632	-0.460	**-0.747**	0.401	-0.146	**-0.784**	**-0.697**	-0.479	-0.178	**-0.786**	<u>-0.823</u>	-0.062	**-0.696**	**-0.693**	-0.311	-0.408	**-0.641**	**-0.745**	-0.062	-0.301	0.628	**-0.679**
Mo		<u>0.890</u>	-0.210	0.293	**-0.728**	0.521	0.106	0.369	-0.141	-0.615	0.329	0.474	0.108	0.057	0.314	-0.221	0.186	0.487	0.077	0.452	-0.090	<u>-0.915</u>	0.135
Sb			-0.135	0.483	-0.447	0.317	0.135	0.541	-0.136	-0.378	0.457	0.502	0.351	0.132	0.420	-0.139	0.546	0.651	0.045	0.260	-0.143	<u>-0.822</u>	0.374
Li				0.464	0.211	-0.513	**0.802**	0.280	<u>0.947</u>	0.555	0.491	0.265	0.060	**0.798**	0.249	0.496	0.058	-0.104	**0.784**	-0.361	0.468	0.230	0.603
Bi					-0.074	0.176	**0.787**	<u>0.970</u>	0.445	0.422	<u>0.971</u>	**0.829**	0.098	<u>0.844</u>	<u>0.891</u>	0.639	**0.769**	**0.721**	**0.671**	0.094	0.515	-0.235	<u>0.952</u>
Cu						**-0.763**	-0.132	-0.138	0.176	0.611	-0.185	-0.425	0.580	-0.108	-0.298	0.035	0.288	-0.262	-0.235	**-0.711**	-0.330	0.578	0.152
Pb							-0.027	0.296	-0.398	-0.459	0.199	0.414	-0.577	0.007	0.377	0.218	-0.016	0.387	0.014	**0.786**	0.395	-0.323	-0.055
Tl								0.653	**0.797**	0.560	**0.809**	**0.767**	-0.170	<u>0.956</u>	**0.684**	0.662	0.282	0.409	<u>0.978</u>	-0.118	**0.687**	-0.070	**0.776**
V									0.260	0.280	<u>0.962</u>	<u>0.805</u>	0.113	**0.738**	<u>0.928</u>	0.547	**0.810**	**0.764**	0.533	0.256	0.412	-0.336	**0.891**
Zn										0.502	0.256	0.067	**0.757**	0.170	0.572	-0.002	-0.155	**0.769**	-0.324	0.479	0.208		0.597
Ni											0.340	0.286	0.514	0.300	0.548	0.358	0.212	0.518	**-0.631**	0.343	0.505		0.526
Cr												<u>0.831</u>	0.023	<u>0.877</u>	<u>0.934</u>	0.216	**0.673**	**0.720**	0.216	0.516	-0.305		<u>0.908</u>
Sr													-0.221	**0.710**	<u>0.881</u>	0.422	0.499	**0.860**	**0.731**	0.123	0.540	-0.455	0.651
S														-0.172	-0.173	-0.340	0.511	0.023	-0.336	-0.438	**-0.699**	-0.253	0.262
Mn															**0.750**	**0.742**	0.366	0.373	<u>0.919</u>	0.036	**0.742**	0.018	<u>0.843</u>
P																0.503	0.649	<u>0.802</u>	0.630	0.318	0.518	-0.326	**0.746**
B																	0.271	0.164	0.633	0.150	<u>0.828</u>	0.424	**0.686**
Ti																		**0.731**	0.107	-0.082	-0.021	-0.233	**0.762**
Ca																			0.333	0.069	0.185	-0.533	0.532
Mg																				-0.070	**0.735**	-0.040	0.641
K																					0.257	-0.289	-0.048
Al																						0.330	0.439
Na																							-0.086

Table 6. Pearson correlation coefficients among elements ($mg\,kg^{-1}$) in the samples of colluvium profiles. Bold face numbers are significant at the 95 % confidence level, underlined numbers are significant at the 99 % confidence level.

	Mo	Sb	Li	Bi	Cu	Pb	Tl	V	Zn	Ni	Cr	Sr	S	Mn	P	B	Ti	Ca	Mg	K	Al	Na	Fe
Be	-0.486	0.267	0.383	0.448	0.053	0.336	0.270	0.338	0.265	0.071	0.058	**0.695**	<u>-0.777</u>	0.185	0.465	0.077	0.423	<u>0.857</u>	0.129	<u>0.759</u>	**0.601**	0.401	0.247
Mo		0.157	0.168	0.238	0.532	-0.126	0.315	0.172	0.369	0.398	0.371	-0.262	0.482	0.452	-0.011	0.258	-0.160	-0.216	0.334	-0.416	-0.046	**-0.618**	0.313
Sb			<u>0.891</u>	<u>0.851</u>	**0.675**	-0.073	<u>0.953</u>	<u>0.875</u>	<u>0.899</u>	<u>0.910</u>	<u>0.946</u>	0.234	-0.308	<u>0.894</u>	0.179	-0.075	0.555	0.518	<u>0.969</u>	0.421	<u>0.827</u>	-0.562	<u>0.892</u>
Li				<u>0.971</u>	**0.802**	0.018	<u>0.959</u>	<u>0.919</u>	<u>0.964</u>	**0.772**	**0.816**	0.531	-0.288	<u>0.901</u>	0.434	0.053	**0.644**	0.580	<u>0.899</u>	0.628	<u>0.928</u>	<u>-0.600</u>	<u>0.971</u>
Bi					0.795	0.149	<u>0.950</u>	<u>0.922</u>	<u>0.967</u>	**0.759**	**0.781**	**0.586**	-0.351	<u>0.913</u>	0.454	0.190	**0.647**	**0.678**	<u>0.863</u>	**0.655**	<u>0.951</u>	-0.529	<u>0.949</u>
Cu						-0.278	**0.806**	**0.729**	**0.817**	**0.754**	**0.756**	0.234	-0.108	**0.830**	0.233	0.090	0.362	0.331	**0.762**	0.365	**0.647**	**-0.716**	**0.824**
Pb							-0.024	0.184	0.042	-0.187	-0.193	0.447	-0.287	-0.030	0.071	0.365	0.433	0.251	-0.113	0.367	0.192	0.352	-0.073
Tl								**0.905**	<u>0.985</u>	**0.901**	<u>0.927</u>	0.341	-0.250	<u>0.967</u>	0.295	0.104	0.539	0.535	<u>0.972</u>	0.462	**0.881**	-0.635	<u>0.966</u>
V									**0.916**	**0.734**	**0.776**	0.520	-0.376	**0.898**	0.289	-0.076	**0.836**	0.558	**0.873**	0.676	**0.865**	-0.580	**0.909**
Zn										**0.833**	**0.865**	0.426	-0.193	<u>0.972</u>	0.378	0.117	0.578	0.518	<u>0.939</u>	0.489	**0.873**	-0.677	<u>0.980</u>
Ni											<u>0.990</u>	-0.063	-0.254	**0.885**	-0.080	0.141	0.283	0.347	<u>0.940</u>	0.193	**0.679**	-0.542	**0.789**
Cr												-0.013	-0.200	**0.900**	-0.014	0.055	0.348	0.335	<u>0.970</u>	0.215	**0.700**	-0.612	**0.836**
Sr													-0.324	0.272	**0.804**	0.020	0.691	**0.717**	**0.819**	0.655	-0.069		0.465
S														-0.219	0.127	-0.081	-0.425	**-0.649**	-0.191	<u>-0.724</u>	-0.504	-0.468	-0.134
Mn															0.219	0.081	0.528	0.442	<u>0.952</u>	0.406	**0.790**	-0.685	<u>0.938</u>
P																-0.052	0.317	0.568	0.134	0.395	0.461	-0.283	0.458
B																	-0.297	0.157	-0.001	-0.043	0.199	0.247	0.007
Ti																		0.498	0.501	**0.793**	**0.652**	-0.308	**0.600**
Ca																			0.388	**0.722**	**0.794**	0.083	0.516
Mg																				0.337	**0.778**	-0.679	<u>0.926</u>
K																					**0.763**	0.026	0.497
Al																						-0.325	<u>0.864</u>
Na																							<u>-0.736</u>

The abundance of sand fractions was general in the study profiles, but differences in the depth distribution arose; thus, sand contents decreased with depth in PB1 that was paralleled with clay content increases in this moraine profile and characterized by diamicton, which is matrix supported. This pattern was not observed in the other moraine profile (PB2) which had very high contents of sand with homogeneous depth distribution. The colluvium profiles of Erdalen had comparable depth distributions of the grain size fractions, with large predominance of sand fractions. However, the PB3 profile of Bødalen showed a quite distinct distribution, with increasing sand contents down the profile and conversely decreasing clay contents with depth.

The depth distributions of the chemical elements were considerably more homogeneous in the soil profiles of the moraines than those of the colluviums (Fig. 5). The contents of most chemical elements in the moraine profiles almost did not vary with depth. Exceptions were some trace elements, namely Sb, Mo, Cu, and B and major elements K, Al, and Fe. In the profile PB2 Cu, V and Fe varied more than in PB1, where Al varied the most with depth. In contrast, the profiles of the colluviums exhibited larger variations in the element depth distributions and showed clearly distinctive patterns among profiles. The profile PB3 showed decreasing trends in the contents of most elements, apart from Na and B. In profile PE2 the contents of most elements decreased sharply at 10–15 cm depth, but the opposite was seen for Mo, Cu and S, whose contents increased. In profile PE1 the largest variations in the element contents appeared at the soil surface where there was high SOC (4%) in combination with low sand content. Most elements showed lower contents at the top layer and decreases were high for Sb, Li, Bi, Tl, V, Zn, Ni, Cr, Mg, Fe, but conversely, increases were recorded for B and Pb.

Table 7. Summary statistics of radionuclide contents in the samples of the soil profiles of the moraines and colluviums. Different letters indicate significant differences at the p level 0.05 between moraines and colluvium soils.

Bq kg^{-1}	Moraines $n = 9$					Colluvium $n = 12$				
	Mean		SD	Min	Max	Mean		SD	Min	Max
^{137}Cs	34.63	a	36.96	2.33	85.10	63.38	a	117.55	1.23	346.00
^{210}Pb$_{ex}$	2.63	a	5.26	b.d.l.	13.00	17.93	a	36.64	b.d.l.	102.40
^{40}K	1177.78	b	142.02	820.00	1320.00	890.42	a	124.45	652.00	1070.00
^{226}Ra	36.84	b	6.68	24.60	47.70	18.68	a	6.87	12.60	30.70
^{232}Th	56.97	b	13.91	33.70	83.70	39.27	a	13.86	12.60	61.20
^{238}U	52.22	b	7.65	41.50	64.90	38.99	a	6.46	28.10	47.10
^{210}Pb	25.82	a	12.74	3.91	45.40	29.65	a	43.13	b.d.l.	118.00

SD: Standard deviation, b.d.l.: below detection limit.

Table 8. Pearson correlation coefficients among soil properties FRN and ERNs in the samples of the soil profiles of the moraines and colluviums. Bold face numbers are significant at the 95 % confidence level, underlined numbers are significant at the 99 % confidence level.

FRNs, ERNs Bq kg^{-1}	SOC %	ACF %	SCF %	pH	2000-50 μm %	50-2 μm %	<2 μm %
Moraines $n = 9$							
^{137}Cs	**_0.902_**	**_0.939_**	0.929	**_-0.95_**	**0.822**	**-0.822**	**-0.808**
^{210}Pb$_{ex}$	0.491	0.264	0.087	-0.32	0.348	-0.342	-0.376
^{40}K	-0.185	0.216	0.275	-0.06	0.005	-0.017	0.056
^{226}Ra	0.201	0.524	0.084	-0.37	0.363	-0.368	-0.336
^{232}Th	0.298	0.459	-0.143	-0.5	0.496	-0.503	-0.454
^{238}U	0.101	0.033	-0.259	-0.17	0.24	-0.246	-0.203
^{210}Pb	**0.718**	0.516	0.057	**-0.71**	**0.731**	**-0.726**	**-0.75**
Colluvium $n = 12$							
^{137}Cs	0.178	0.211	**_0.852_**	-0.39	-0.228	0.218	0.282
^{210}Pb$_{ex}$	0.188	0.211	**0.765**	-0.46	-0.383	0.371	0.451
^{40}K	**-0.648**	**-0.671**	-0.275	0.175	-0.485	0.486	0.469
^{226}Ra	0.014	0.047	0.537	0.287	0.503	-0.502	-0.5
^{232}Th	-0.043	-0.066	-0.446	-0.2	**_-0.799_**	**_0.811_**	**_0.709_**
^{238}U	0.319	0.334	0.118	0.463	0.358	-0.346	-0.426
^{210}Pb	0.193	0.22	**0.809**	-0.34	-0.303	0.293	0.361

3.3 The vertical distribution of radionuclides

The mass activities of ^{137}Cs and ^{210}Pb$_{ex}$ down the profiles showed different patterns in the moraine and the colluvium profiles (Fig. 6). The moraine profiles did not show exponential decreases of ^{137}Cs with depth and lower contents of ^{137}Cs were found in PB1 (range: 2.3–30.2 Bq kg^{-1}) than in PB2 (20–84 Bq kg^{-1}). The ^{210}Pb$_{ex}$ that was almost negligible in the moraine profiles was only detected in the topsoil of PB1 and at 10–20 cm in PB2. Therefore, the typical decay pattern of the FRNs mass activities with depth was not found in these moraine profiles.

The colluvium profiles of ^{137}Cs showed an exponential decrease with the depth. The decay was more marked in PB3 and PE1 profiles, the ^{137}Cs mass activities decreased sharply from the topsoil that had very high values (266–346 Bq kg^{-1})

to low values in deeper layers (1.64–2.75 Bq kg^{-1}). The mass activities of ^{210}Pb$_{ex}$ were considerably lower and the radionuclide was only found in the upper layers, therefore the penetration of ^{210}Pb$_{ex}$ was much lower than that of ^{137}Cs that was detected at 25 cm in PE1 (Fig. 6).

The depth distributions of ERNs showed different patterns, thus that of ^{226}Ra and ^{232}Th were very similar in the moraine profiles, but this was not the case in the colluvium profiles which even exhibited opposite trends in profile PE1. The ^{40}K mass activities varied largely and values were higher in the profiles of Bødalen. The ^{238}U was the lesser variable, and it did not show any clear pattern in its depth distribution.

All profiles showed disequilibrium in the U-Th series. Under secular equilibrium the activity ratios of ^{238}U / ^{226}Ra will be approximately 1, and 1.1 for ^{232}Th / ^{226}Ra (Evans et al., 1997). However, values in the profiles largely exceeded 1

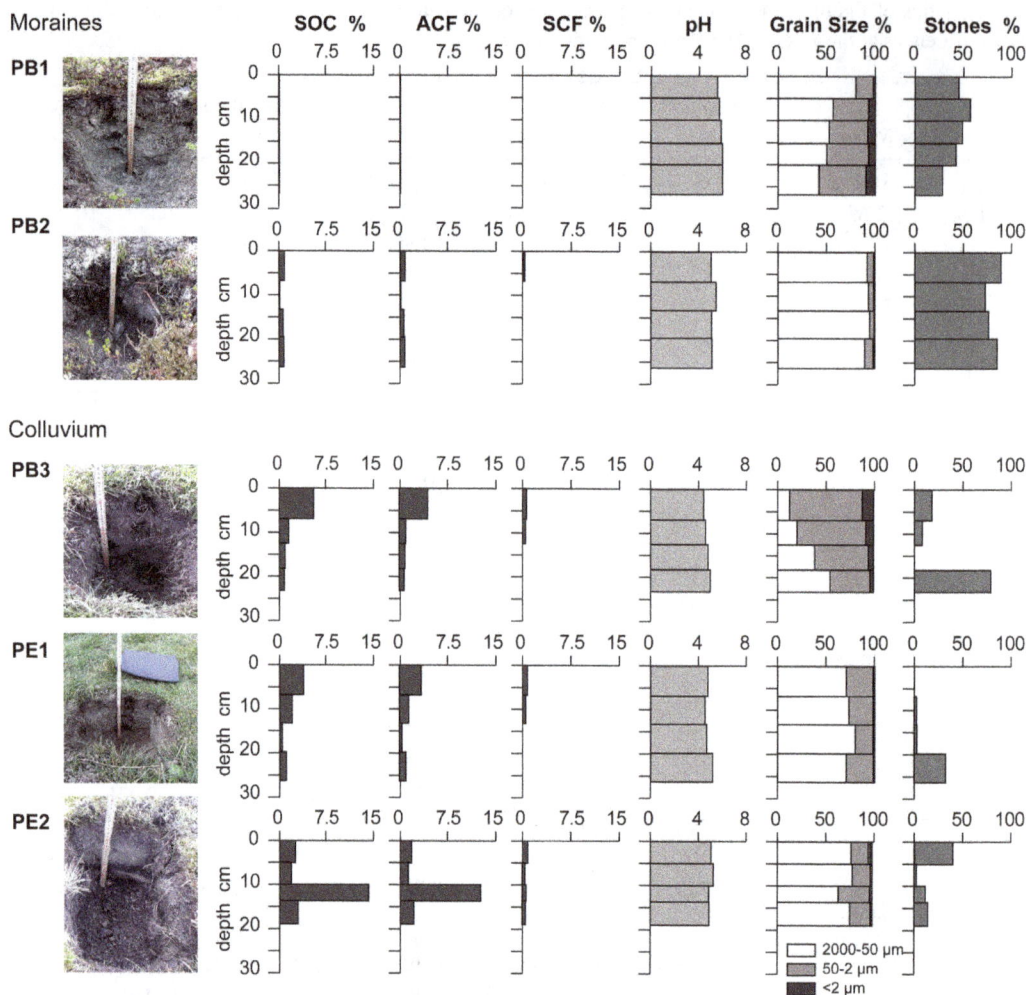

Figure 4. The study profiles on the Leptosols of the moraines and the Regosols of the colluvium and the depth distribution of main soil properties.

(range 1.27–2.98) and the colluvium profile PB3 showed the greater disequilibrium. Similarly, all soil profiles had ^{232}Th / ^{226}Ra activity ratios higher than 1.1. Deviations were much higher in profile PB3 (3.34–4.25) despite the depleted levels of ^{232}Th (Fig. 6) in topsoil layers of the colluvium profiles.

4   Discussion

On the colluvium the Regosols are better developed with a rich organic A horizon, whereas the Leptosols on the moraines are shallow and do not present horizon differentiation. The higher content of coarse fractions in the Leptosols is related to the original till parent material, but is also in accordance with the physical processes (such as rock disintegration by the action of ice) that are more important in the moraines than in the colluviums. This is likely because moraine materials have been subjected until very recently to

ice action, but also because they are more exposed to physical disintegration, as moraines have less continuous vegetation cover than colluviums. These features result in distinctive soil properties that affect the pattern distribution of stable elements and radionuclides in the profiles. In spite of the fact that studies on the vertical distribution of elements in soils of cold regions are scarce, results from previous research in Maritime Antarctica (Navas et al., 2005a, 2008) have also evidenced the variation of FRN and ERN contents in relation to processes affecting soils in different morphoedaphic environments. Moreover, similar to what it was observed in the study valleys, the variability in some radionuclides and elements was also related to the mineral composition of parent materials and to cryogenic and soil processes influencing the depth distribution of soil properties, such as that of granulometric fractions and organic matter.

The type of relationship between stable elements (either direct or inverse) suggests common or different origins, respectively, from minerals contained in the granitic

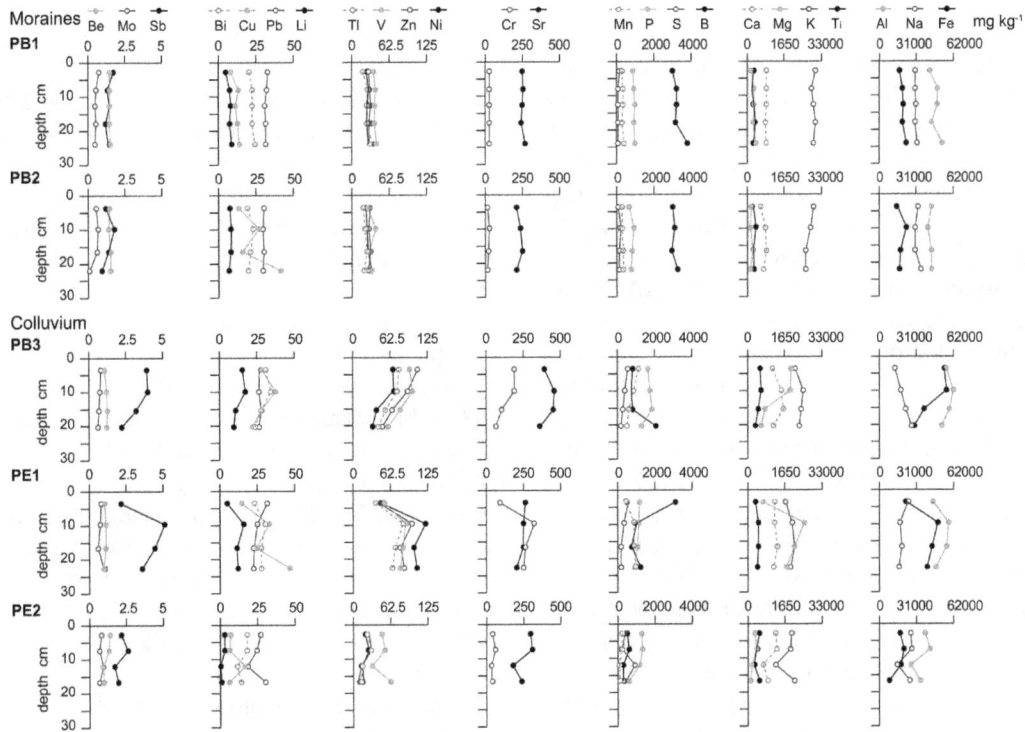

Figure 5. Vertical distribution of the chemical elements (mg kg^{-1}) in the soil profiles on the moraines and colluviums.

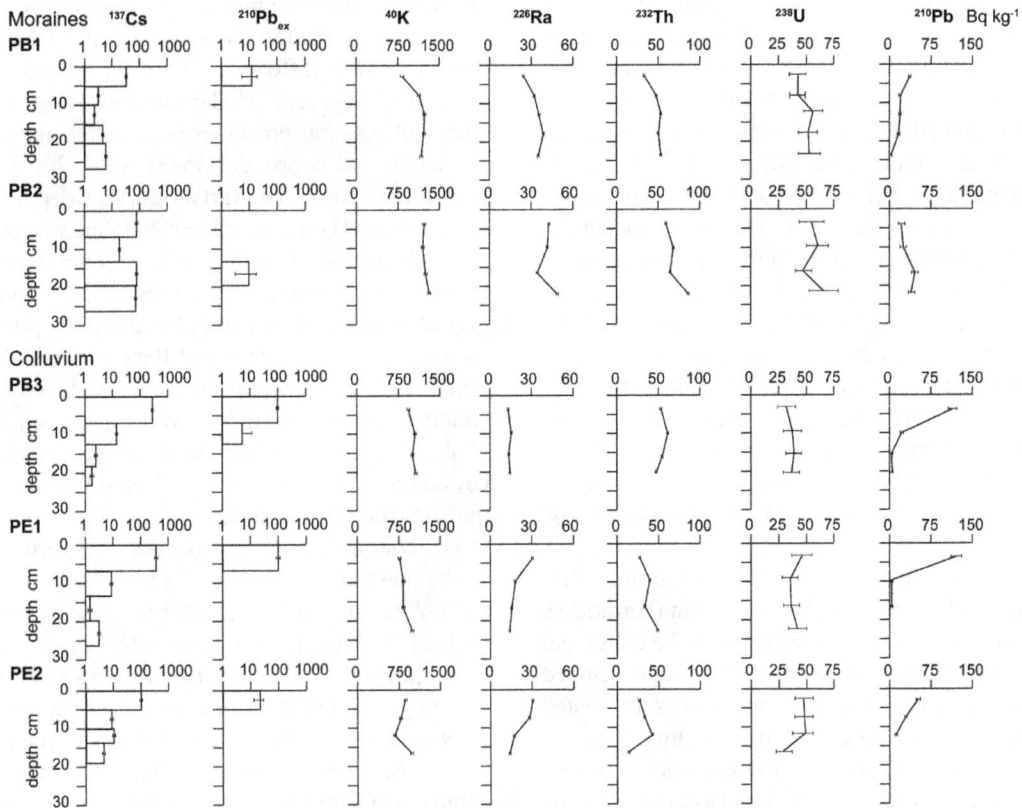

Figure 6. Depth distribution of the mass activities (Bq kg^{-1}) of FRNs and ERNs in the soil profiles on the moraines and colluviums.

orthogneisses, on which soils are developed. Furthermore, differences in the development of the study soils might have an influence on the elements and their abundance. Thus, elements present in Leptosols would reflect more closely the composition of parent materials than those in Regosols, which are more developed soils. This fact may also have an effect on element accumulation, as most elements present higher contents in Regosols than in Leptosols. High correlation coefficients between elements would indicate similar transport, accumulation and sources (Acosta et al., 2011). The inverse correlations found between Na, K, Pb, and Be with the rest of elements and the direct correlations between them suggest different mineralogical sources. These elements, which are more abundant in Leptosols on the moraines, might be mainly derived from tectosilicates, whereas the rest of the elements likely originated from other types of silicates. Evidences of the main control of substrate mineralogy and on the element transfer from rock to soil were also found on a variety of substrates and environments (Wang and Chen, 1998; Navas et al., 2002c). Moreover, sharp decreases in the contents of most elements, apart from Mo, Cu and S, are associated with the high increase of SOC content found at 10–15 cm depth in PE2, supporting the close links of most elements with the mineral contents.

The range of variation of ^{137}Cs and ^{210}Pb$_{ex}$ is linked to the high contents recorded in the organic horizons of the Regosols of the colluviums, while the highest content of ERNs, in the Leptosols, apart from ^{210}Pb, is related to mineralogical and geochemical differences in the composition of the moraine materials. Significant positive correlations of ^{137}Cs with SOC and ACF contents and inverse correlations with pH and the silt and clay fractions found in Leptosols were not observed in the Regosols on the colluviums outside the LIA. Therefore, correlations among the radionuclides with the soil properties support the different characteristics of FRNs in the Leptosols of recently deglaciated moraines affected by LIA compared with the Regosols outside the LIA.

In the colluvium profiles the type of correlations that were similar for ^{232}Th and ^{40}K suggest that both radionuclides are related with clay minerals, whereas the opposite was true for ^{226}Ra and ^{238}U which might indicate a common and different mineralogical source to that of ^{232}Th and ^{40}K. However, in the moraine profiles the type of correlations was different and not significant.

The direct and significant correlations among the ERNs in the moraine profiles denote a common origin (Fairbridge, 1972). However, the contrary was observed in the colluvium profiles, suggesting that soil processes in the more developed soils, either by accumulation of the radionuclides associated with fine minerals or due to differential mobility, have affected the distribution of ERNs. Thus ERNs are not as closely linked to their primary mineralogical origin as in the moraine profiles, where radionuclides might be internally bound in primary minerals as found by Nielsen and Murray (2008) in sandy sediments of Jutland (Denmark).

The higher values and decreasing distribution of SOC in Regosols illustrates a higher degree of soil evolution, reflecting the oldest age in terms of ice-free retreat of the colluvium sites that became deglaciated approximately 10 000 years ago. This is further confirmed by the higher percentages of the SCF compared to its content in the moraine profiles. The SOC-rich layer at the 10–14 cm interval depth in PE2 (see Fig. 4) might correspond to a buried soil that is likely to have occurred at this site due to intense and frequent rockfall activity.

The grain size distribution in the profiles helps to interpret the role of physical processes such as rock disintegration on soil development in cold environments (Navas et al., 2008). In general, these soils are characterized by the abundance of sand content, especially in the topsoil. However, the low values and increasing contents of sand with depth in PB3 (see Fig. 4) are related to the decreasing trends in the contents of most elements, apart from Na and B. This site, which is in a hillslope located beneath the Tindefjell glacier, is more influenced by glaciofluvial and outwash processes rather than rockfall activity that likely affects the particle size distribution in the profile. In addition, compared to the other sites, this site is more impacted through animal husbandry, as sheep grazing has taken place since approximately 1800, but with lower intensity since 1930.

The reason for the homogeneous vertical distribution of chemical elements in moraine profiles, rather than the larger variations observed in colluvium profiles (see Fig. 5), is the lack of horizon differentiation in the recently formed Leptosols on till material. Under the cold climate existing in the study valleys, soil processes are limited and slow, resulting in shallow and poorly developed soils with almost no horizon differentiation. The variations in the depth distribution of SOC and sand contents are related to the vertical distribution of the elements and are responsible for the contrasts observed between colluvial and moraine soil profiles. Furthermore, the geochemical variability found in the study profiles is linked with the parent materials and their mineralogical composition. In agreement with what was found by other authors in a variety of environments (e.g. Wang and Chen, 1998; Acosta et al., 2011), relationships between elements evidenced the key control of the mineralogy of the substrate on the element transfer from rock to soils.

The contrasting patterns between the vertical distribution of the mass activities of ^{137}Cs and ^{210}Pb$_{ex}$ in the moraine and colluvium profiles are likely related to the different periods of ice retreat. Even for shorter periods, as in the case of moraine profiles, differences seen between the study profiles might be related to the age of ice retreat, as the lower FRN contents in PB1 (that is the less developed soil since the site became ice-free at ca. AD 1930) might suggest via comparison with PB2, which became ice-free earlier around AD 1800. Furthermore, the absence of the typical decay pattern of the FRN mass activities with depth in moraine profiles could be due to several reasons. The till material, the lack of horizon differentiation, and the predominance of coarse fractions in the soil matrix of the moraines may have favoured the rapid infiltration of water carrying the FRNs to deeper layers. In spite that some types of clays may be more efficient in the adsorption of the radionuclides (e.g. Staunton and Roubaud, 1997) especially in the frayed edge sites (Sawhney, 1972), the fixation of the FRNs by the organic matter that may inhibit that of clays is an efficient non-specific mechanism fixing ^{137}Cs and ^{210}Pb$_{ex}$(Takenaka et al., 1998, Gaspar and Navas, 2013; Gaspar et al., 2013). Therefore, the very low content of SOC in the moraine profiles would also contribute to the low fixation of the FRNs. Another reason may be the cryogenic processes and the disturbance of soil by the ice action, which in the moraine sites can be more intense than in the colluviums. Meanwhile, the high values of ^{137}Cs mass activity in the topsoil of the exponential decay profiles in colluvium soils is likely because of the influence of the Chernobyl accident in 1986 (Gjelsvik and Steinnes, 2013).

In relation to the depth distributions of ERNs, the larger variability of ^{226}Ra and ^{232}Th in this environment can be explained by the lack of carbonates as opposed to those observed in Mediterranean carbonate-rich soils, where high carbonate contents restricted the mobility of these radionuclides (Navas et al., 2002a, b). The depleted levels of ^{40}K at the topsoil of PB1 are related to lower contents in clay and silt fractions. As it is widely known in the literature (e.g. Jasinska et al., 1982, Baeza et al., 1995, VandenBygaart and Protz, 1995), the environmental radionuclides are associated with clay minerals or they are fixed within the lattice structure. The significantly higher values of ^{40}K in the profiles of Bødalen are likely related to differences in mineralogical composition, which can be further confirmed by its sharp depletion in the rich organic layer of PE2. Baeza et al. (1995) indicate that radioactivity increases as particle size decreases. In Antarctica, profile increases in clay contents were paralleled with ERN enrichments (Navas et al., 2005a). In general, bedrock composition appears to be the main factor of variation of the ERNs in the study profiles.

Although most environmental samples have ^{232}Th$/^{226}$Ra activity ratios around 1.1 (Evans et al., 1997) and ^{238}U$/^{226}$Ra ratios are approximately 1 under secular equilibrium, all soil profiles deviated from these values, indicating disequilibrium in the uranium and thorium series. Disequilibrium can be due to the active hillslope processes such as intense outwash processes and rockfall. Besides, differential mobility of the radionuclides (e.g. Collerson et al., 1991) may have also had an influence. Harmsen and de Haan (1980) indicate that U and Th form hydrated cations (UO$_2^{2+}$, Th^{4+}) that are easily mobilized over a broad range of soil pH from less than 4 to 9.

In spite of differences with the climatic conditions existing in the southern circumpolar region, processes of soil formation similar as those operating in extreme cold regions (Bockheim and McLeod, 2006) can be expected to affect the areas recently deglaciated in the upper Erdalen and Bødalen val-

leys. Thus, in the LIA moraines (and especially in the areas that became ice-free in the past century) the role of freeze–thaw and wetting and drying cycles seems to be more important than other weathering mechanisms. In cold regions such as in Antarctica, freeze–thaw weathering is generally recognized to be the most important process, causing rock disintegration and soil formation (Serrano et al., 1996, Hall 1997; Navas et al., 2008). However, in the areas outside the LIA influence, where the colluvium profiles are located physical processes, as rockfall, glaciofluvial and outwash processes and chemical weathering are main soil-forming processes.

5 Conclusions

The higher horizon differentiation in the more evolved Regosols developed on colluvium, in comparison to the Leptosols on the moraines, determine the larger variability in the elemental composition down the Regosol profiles. Radionuclide activities in the soils differed as a function of the characteristics of geomorphic elements and the processes occurring on the different geomorphic elements and substrates that have became ice-free at different ages. The distribution of ENRs is linked to the mineral composition of the parent materials. Geomorphic and soil processes that trigger the enrichment of the fine fractions containing sheet silicates determine the abundance of ^{40}K and ^{232}Th, whereas ^{238}U and ^{226}Ra are rather associated to minerals included in coarser fractions. In this environment the transference of the radionuclides and elements down the profile might be time restricted to the periods in which water circulates down the soil profile, which can further influence the differences in soil processes found between colluvium and moraine profiles.

In cold regions, such as the one in this study, information on the period of ice retreat could be derived by the pattern distribution of FRNs, as the typical decay pattern of FRN mass activities with depth is not found in moraine profiles of the LIA compared to the typical decay patterns found in more evolved soils of the colluviums that were deglaciated around 10 000 BP. In addition, other soil properties can be combined to discern the main geomorphic processes related to the ice age retreat.

Acknowledgements. Financial support from CICYT project EROMED (CGL2011-25486/BTE) is gratefully acknowledged. Fieldwork in Erdalen und Bødalen was funded by the SedyMONT-Norway project (Norwegian Research Council NFR, grant 193358/V30 to A.A. Beylich).

Edited by: P. Pereira

References

Acosta, J. A., Martínez-Martínez, S., Faz, A., and Arocena, J.: Accumulations of major and trace elements in particle size fractions of soils on eight different parent materials, Geoderma, 161, 30–42, 2011.

Appleby, P. G. and Oldfield, F.: Application of lead-210 to sedimentation studies, in: Uranium-series disequilibrium: Applications to earth, marine, and environmental sciences, edited by: Ivanovich, J. and Harman, R. S., Clarendon Press, Oxford, United Kingdom, 731–738, 1992.

Baeza, A., del Río, M., Jiménez, A., Miró, C., and Paniagua, J.: Influence of geology and soil particle size on the surface-area/volume activity ratio for natural radionuclides, J. Radioanal. Nucl. Ch., 189, 289–299, 1995.

Beylich, A. A. and Laute, K.: Spatial variations of surface water chemistry and chemical denudation in the Erdalen drainage basin, western Norway, Geomorphology, 167–168, 77–90, 2012.

Bickerton, R. W. and Matthews, J. A.: Little Ice Age variations of outlet glaciers from the Jostedalsbreen ice-cap, southern Norway: a regional lichenometric-dating study of ice-marginal moraine sequences and their climatic significance, J. Quaternary Sci., 8, 45–66, 1993.

Bockheim, J. G. and McLeod, J. G.: Soil formation in Wright Valley, Antarctica since the late Neogene, Geoderma, 137, 109–116, 2006.

Collerson, K. D., Gregor, D. J., McNaughton, D., and Baweja, A. S.: Effect of coal dewatering and coal use on the water quality of the East Poplar River, Saskatchewan. A literature review. Inland Waters Directorate, Environment Canada, Scientific Series, No. 177, 1991.

Evans, C. V., Morton, L. S., and Harbottle, G.: Pedologic assessment of radionuclide distributions: use of a radio-pedogenic index, Soil Sci. Soc. Am. J., 61, 1440–1449, 1997.

Fairbridge, R. W.: The encyclopedia of geochemistry and environmental sciences, edited by: Van Nostrand, R., Van Nostrand Reinhold Co, New York, vol. 4A, 1215–1228, 1972.

Gaspar, L. and Navas, A.: Vertical and lateral distributions of ^{137}Cs in cultivated and uncultivated soils on Mediterranean hillslopes, Geoderma, 207–208, 131–143, 2013.

Gaspar, L., Navas, A., Machín, J., and Walling, D. E.: Using [210]Pb$_{ex}$ measurements to quantify soil redistribution along two complex toposequences in Mediterranean agroecosystems, northern Spain, Soil Till. Res., 130, 81–90, 2013.

Gjelsvik, R. and Steinnes, E.: Geographical trends in ^{137}Cs fallout from the Chernobyl accident and leaching from natural surface soil in Norway, J. Environ. Radioactiv., 126C, 99–103, 2013.

Hall, K.: Rock temperatures and implications for cold region weathering. I: New data from Viking Valley, Alexander Island, Antactica, in: Permafrost Periglac. Proc., vol. 8, 69–90, 1997.

Harmsen, K. and de Haan, F. A, M.: Occurrence and behavior of uranium and thorium in soil and water, Neth. J. Agr. Sci., 28, 40–62, 1980.

Jasinska, M., Niewiadomski, T. and Schwbenthan, J.: Correlation between soil parameters and natural radioactivity, in: Natural radiation environment, edited by: Vohra, K., Mishra, U. C., Pillai, K. C., and Sadasivan, S., John Wiley and Sons, New York, 206–211, 1982.

Laute, K. and Beylich, A. A.: Influences of the Little Ice Age glacier advance on hillslope morphometry and development in paraglacial valley systems around the Jostedalsbreen ice cap in Western Norway, Geomorphology, 167–168, 51–69, 2012.

Laute, K. and Beylich, A. A.: Holocene hillslope development in glacially formed valley systems in Nordfjord, western Norway, Geomorphology, 188, 12–30, 2013.

Laute, K. and Beylich, A. A.: Morphometric and meteorological controls on recent snow avalanche distribution and activity at hillslopes in steep mountain valleys in western Norway, Geomorphology, 218, 16–34, 2014.

Lopez-Capel, E., Krull, E. S., Bol, R., and Manning, D. A. C.: Influence of recent vegetation on labile and recalcitrant carbon soil pools in central Queensland, Australia: evidence from thermal analysis quadrupole mass spectrometry-isotope ratio mass spectrometry, Rapid Commun. Mass Sp., 22, 1751–1758, 2008.

Lutro, O. and Tveten, E.: Bedrock map ÅRDAL M 1 : 250.000. Norges geologiske undersøkelse, Trondheim, 1996.

Matthews, J. A., Shakesby, R. A., Schnabel, C., and Freeman, S.: Cosmogenic [10]Be and [26]Al ages of Holocene moraines in southern Norway I: testing the method of confirmation of the date of the Erdalen event (c. 10 ka) at its type site, Holocene, 18, 1155–1164, 2008.

Mavlyudov, B. R., Savatugin, L. M., and Solovyanova, I. Yu.: Reaction of the Glaciers of Nordenskiold Land (Spitsbergen archipelago) on climate change, Problems of Arctic and Antarctic, 1 (91), AARI, Sankt-Petersburg, 67–77, 2012 (in Russian).

Navas, A., and Machín, J.: Spatial distribution of heavy metals and arsenic in soils of Aragón (northeast Spain): controlling factors and environmental implications. Appl.Geochem., 17, 961-973, 2002c.

Navas, A., Soto, J., and Machín, J.: Edaphic and physiographic factors affecting the distribution of natural gamma-emitting radionuclides in the soils of the Arnás catchment in the Central Spanish Pyrenees, Eur. J. Soil Sci., 53, 629–638, 2002a.

Navas, A., Soto, J., and Machín, J.: [238]U, [226]Ra, ^{210}Pb, [232]Th and [40]K activities in soil profiles of the Flysch sector (Central Spanish Pyrenees), Appl. Radiat. Isotopes, 57, 579–589, 2002b.

Navas, A., Soto, J., and López-Martínez, J.: Radionuclides in soils of Byers Peninsula, South Shetland Islands, Western Antarctica, Appl. Radiat. Isotopes, 62, 809–816, 2005a.

Navas, A., Soto, J., and Machín, J.: Mobility of natural radionuclides and selected major and trace elements along a soil toposequence in the central Spanish Pyrenees, Soil Sci., 170, 743–757, 2005b.

Navas, A., López-Martinez, J., Casas, J., Machín, J., Durán, J. J., Serrano, E., Cuchí, J. A., and Mink, S.: Soil characteristics on varying lithological substrates in the South Shetland Islands, maritime Antarctica, Geoderma, 144, 123–139, 2008.

Nesje, A.: Kvartærgeologiske undersøkningar i Erdalen, Stryn, Sogn og Fjordane, M. Sc. thesis, University of Bergen, 201 pp., 1984.

Nielsen, A. H and Murray, A. S.: The effects of Holocene podzolisation on radionuclide distributions and dose rates in sandy coastal sediments, Geochronometria, 31, 53–63, 2008.

Sawhney, B. L.: Selective adsorption and fixation of cations by clay minerals: a review, Clay. Clay Miner., 20, 93–100, 1972.

Serrano, E., Martínez de Pisón, E., and López-Martínez, J.: Periglacial and nival landforms and deposits, in: Geomorphological map of Byers Peninsula, Livingston Island, edited by: López-Martínez, J., Thomson, M. R. A., and Thomson J. W. BAS Geomap Series 5-A. British Ant Surv, Cambridge, 15–19 pp. 1996.

Serrano, E. and López-Martínez, J.: Rock glaciers in the South Shetland Islands, Western Antarctica, Geomorphology, 35, 145–162, 2000.

Staunton, S. and Roubaud, M.: Adsorption of ^{137}Cs on montmorillonite and illite: effect of charge compensanting cation, ionic strength, concentration of Cs, K and fulvic acid, Clay Clay Miner., 45, 251–260, 1997.

Takenaka, C., Onda, Y., and Hamajima, Y.: Distribution of cesium 137 in Japanese forest soils: correlation with the contents of organic carbon, Sci. Total Environ., 222, 193–199, 1998.

Van Cleef, D. J.: Determination of [226]Ra in soil using [214]Pb and [214]Bi inmediately after sampling, Health Phys., 67, 288–289, 1994.

VandenBygaart, A. J. and Protz, R.: Gamma radioactivity on a chronosequence, Pinery Provincial Park, Ontario, Can. J. Soil Sci., 75, 73–84, 1995.

Wang, X. J. and Chen, J. S.: Trace element contents and correlation in surface soils in China's eastern alluvial plains, Environ. Geol., 36, 277–284, 1998.

Winkler, S., Elvehøy, H., and Nesje, A.: 2009: Glacier fluctuations of the Jostedalsbreen, western Norway, during the past 20 years: the sensitive response of maritime mountain glaciers, Holocene, 19, 395–414, 2009.

18

Crop residue decomposition in Minnesota biochar-amended plots

S. L. Weyers[1] and K. A. Spokas[2]

[1] USDA Agricultural Research Service, North Central Soil Conservation Research Lab, Morris, MN, USA
[2] USDA Agricultural Research Service, Soil and Water Management Unit, University of Minnesota, Saint Paul, MN, USA

Correspondence to: S. L. Weyers (sharon.weyers@ars.usda.gov)

Abstract. Impacts of biochar application at laboratory scales are routinely studied, but impacts of biochar application on decomposition of crop residues at field scales have not been widely addressed. The priming or hindrance of crop residue decomposition could have a cascading impact on soil processes, particularly those influencing nutrient availability. Our objectives were to evaluate biochar effects on field decomposition of crop residue, using plots that were amended with biochars made from different plant-based feedstocks and pyrolysis platforms in the fall of 2008. Litterbags containing wheat straw material were buried in July of 2011 below the soil surface in a continuous-corn cropped field in plots that had received one of seven different biochar amendments or a uncharred wood-pellet amendment 2.5 yr prior to start of this study. Litterbags were collected over the course of 14 weeks. Microbial biomass was assessed in treatment plots the previous fall. Though first-order decomposition rate constants were positively correlated to microbial biomass, neither parameter was statistically affected by biochar or wood-pellet treatments. The findings indicated only a residual of potentially positive and negative initial impacts of biochars on residue decomposition, which fit in line with established feedstock and pyrolysis influences. Overall, these findings indicate that no significant alteration in the microbial dynamics of the soil decomposer communities occurred as a consequence of the application of plant-based biochars evaluated here.

1 Introduction

Biochar is the solid product that comes from a variety of thermolytic conversion processes creating a carbon-rich material, which is intended for carbon sequestration purposes.

Biochar, when used as a soil amendment, has been hypothesized to provide nutrients for plant growth, counteract soil acidity, or induce positive effects on soil properties such as cation exchange capacity, bulk density and water-holding capacity (Atkinson et al., 2010; Sohi et al., 2010; Dai et al. 2013). Biochar additions have been theorized to improve soil biological activity (Paz-Ferreiro and Fu, 2014) and improve agricultural production in drought and water-stressed regions in combination with other water conservation practices (Blackwell et al., 2010; Kammann et al., 2011; Artiola et al., 2012; Ibrahim et al., 2013). Various studies have hypothesized, through meta-analysis, that a crop yield improvement of 10–12 % is expected when biochar addition is made to typically acidic coarse-textured soils (Biederman and Harpole, 2013; Crane-Droesch et al., 2013; Liu et al., 2013). Biochar may also improve soil structure and reduce soil losses through erosion (García-Orenes et al., 2012; Stavi et al., 2012). Regardless of all of these isolated cases of noted soil improvements, no universal correlation between yield improvement and biochar properties has been elucidated (Crane-Droesch et al., 2013), which leaves scientific-based guidance on its use indeterminate. Despite this, biochar is perceived as a beneficial soil amendment product with multiple advantages (Laird, 2008).

Biochar can have positive effects on soil biota as well (Lehmann et al., 2011). Addition of biochar might alter properties that regulate soil organic matter (SOM) decomposition – which are decomposer organism diversity and abundance, resource availability, and the physio-chemical environment, particularly soil aeration and moisture content (Swift et al., 1979; Heal et al., 1997). Microorganisms are the primary decomposers of SOM. The majority of studies evaluating biological effects of biochars observe positive stimulation of microbial abundance, which has been correlated with the improved soil conditions (Lehmann et al., 2011) and the concept of biochar being a beneficial habitat for microbes (Warnock et al., 2007). On the other hand, recent studies have not detected this microbial colonization of biochar (Quilliam et al., 2013; Jaafar et al., 2014). Laboratory studies indicate biochar addition can change resource availability and induce priming effects, which are short-term changes in the mineralization of SOM due to stimulated microbial processing (Luo et al., 2011; Zimmerman et al., 2011).

Variable effects on residue decomposition dynamics can be expected when evaluating dissimilar biochars applied to the same or similar soils. Nutrient composition, pH, volatile components, density, porosity and other characteristics of biochar are affected by the feedstock and the conditions of the thermolytic conversion process used (Spokas et al., 2012; Lee et al., 2013; Sigua et al., 2014). In particular, the soluble, leachable components also differ among biochars (Jaffé et al., 2013). Different biochars can have unique effects on composition of the microbial community (Lehmann et al., 2011). For instance, some biochars might stimulate bacteria and others fungi (Steinbeiss et al., 2009). Altered microbial community composition in this sense could have cascading effects on higher levels of the soil food web that could result in significant functionality differences in later years, such as that observed under different tillage regimes (Hendrix et al., 1986). Further, biochar may alter nutrient availability (Noguera et al., 2010). In particular for N, biochar may reduce the N limitation that results in slower C mineralization rates (Vitousek and Howarth, 1991).

A majority of studies to evaluate biochar's impact on organic matter decomposition have been conducted in the laboratory. Most of these studies use freshly made biochar, small amounts of finely ground or sieved organic material, and short time frames in laboratory incubations. For example, Novak et al. (2010) determined that a fresh pecan shell-derived biochar primed the mineralization of 0.25 mm sieved switchgrass residues in a 67-day incubation. Similarly, Awad et al. (2012) also observed an increased rate of maize residue decomposition in a laboratory study following biochar addition, with the observed rate a function of the soil texture and biochar production temperature (Awad et al., 2013). On the other hand, Bruun and EL-Zehery (2012) found an insignificant increase in laboratory C mineralization of uncharred barley straw in the presence of fresh barley straw-derived biochar (0.15 % w/w), and Zavalloni et al. (2011) also observed no significant difference in the degradation of wheat straw residues in the presence of 5 % hardwood biochar. It is already known that biochar's surface chemistry and reactivity changes with time, largely believed due to the reactivity to oxygen (Puri et al., 1958) and water (Pierce et al., 1951) at ambient conditions. These differences in surface and bulk chemistries can lead to various responses in microbial mineralization dynamics following biochar additions (Liu et al., 2013; Cely et al., 2014), particularly since the term "biochar" does not contain any information on the actual chemical composition of the material (Spokas et al., 2012).

On the other hand, only limited field-based biochar studies have been conducted. Wardle et al. (2008) evaluated mass loss of humus encapsulated with fresh wood charcoal (1 : 1) in mesh bags in field plots over 10 years. They observed that charcoal mixed with humus possessed a greater synergetic mass loss over the 10 years than expected from charcoal and soil humus alone (Wardle et al., 2008). From the laboratory studies, fresh biochar appears to prime the decomposition of soil organic matter. In the limited field experiments, biochar had a long-term impact on humus decomposition, resulting in overall greater cumulative mass loss over time. Despite these findings, the impact of biochar on the decomposition of freshly added organic matter, in particular crop residue in agricultural soils, is still unknown.

The objectives of this study were to determine (1) whether field-weathered biochar can affect the field decomposition of freshly added crop residue, (2) whether any impact on field decomposition rates can be related to biochar feedstock or pyrolysis method, and (3) whether microbial biomass was influenced by biochar applications. Based on the findings of Wardle et al. (2008), Novak et al. (2010) and others, accelerated decomposition of freshly added organic material was expected in field-weathered biochar plots. We further hypothesized that there would be differences in observed decomposition rates in field plots as a function of biochar type.

2 Materials and methods

2.1 Site description and biochar treatments

The research site is located at the University of Minnesota Research and Outreach Center in Rosemount, MN, USA (44° N, 93° W). Soil at the site is a low-slope (< 2 %) Waukegan silt loam (fine-silty over skeletal mixed, super active, mesic Typic Hapludoll) containing approximately 22 % sand, 55 % silt, and 23 % clay with a pH of 6.4 and total organic C of 26 g kg^{-1}. Seven different biochar treatments, a raw biomass (uncharred wood pellet), and a zero-amendment control treatment were applied in triplicate to 27 completely randomized plots in the fall of 2008 (Table 1). The plots measured 4.88 m on a side with a 3 m buffer zone between plots. Feedstocks for these biochars were hardwoods, pine chips, macadamia nut shells, and wheat middlings (or wheat midds, which are the by-product from milling wheat), and all were produced by thermal pyrolysis (Table 1). All biochars and the wood-pellet amendment used in the test plots were applied at a rate of 22.4 Mg ha^{-1} (as received), thus providing total C additions ranging 14.4 to 19.9 Mg C ha^{-1}. Since these biochars were produced in different pyrolysis units, they lack the overall relationship between properties and production processes (e.g., temperature and residence time) that have been correlated by previous studies when they use the

Table 1. Treatment designations by production source and biochar characteristics.

| Treatment designation | Application rate (kg ha⁻¹) | Biochar source[a] | Feed-stock | Pyrolysis method[b] | Pyrolysis temperature (°C) | Ultimate analysis[c] (% by dry mass basis) | | | | | | Proximate analysis[d] | | % Moisture (air dry) | Molar ratios | | pH[e] | Surface area (BET-N_2) (m² g⁻¹) | Particle size (cm) |
						C	N	O	H	S	Ash	Volatile matter	Fixed C		O:C	H:C			
Control WP	–	Somerset Wood Pellets (US)	100% Hardwood Pellet	Uncharred	–	23.50	0.20	70.00	5.60	0.10	0.60	76.90	17.00	5.5	2.23	2.86	4.77	<0.1	2
BC1	22400	Dynamotive BC (Canada)	Hardwood	fast	500	63.86	0.22	11.78	3.02	0.01	21.11	26.07	48.97	3.85	0.14	0.57	6.76	0.77	<0.001
BC2	22400	Chip Energy (US)	Hardwood Pellet	slow (updraft gasifier)	n/a (>500)	73.37	0.21	18.75	1.28	0.01	6.38	12.36	71.41	9.85	0.19	0.21	9.75	62.44	1.5
BC3	22400	Best Energies (US)	Mixed hard and softwoods	Slow	550	71.09	0.11	20.57	3.44	0.02	4.77	34.75	57.17	3.31	0.22	0.58	5.01	24.24	1
BC4	22400	Cowboy Charcoal (US)	Hardwood	slow	538	88.28	0.25	7.02	2.41	0.01	2.03	18.12	76.19	3.66	0.06	0.33	6.55	85.60	5
BC5	22400	ICM (US)	Wheat middlings	slow	540–600	81.83	0.52	4.75	0.32	0.05	12.53	10.06	73.79	3.62	0.04	0.05	8.86	1.10	<0.5
BC6	22400	ICM (US)	Pine chip (bark + wood)	slow	600–700	64.33	3.11	6.16	1.16	0.04	25.20	14.96	48.57	11.27	0.07	0.22	9.54	0.40	0.5
BC7	22400	Biochar Brokers (US)	Macadamia nut shell	fast	650	93.15	0.67	1.68	2.56	0.02	1.92	16.84	71.70	9.54	0.01	0.33	6.20	0.35	2.5

Notes:
[a] Names are necessary to report factually on available data; however, the USDA neither guarantees nor warrants the standard of the product, and the use of the name by USDA implies no approval of the product to the exclusion of others that may also be suitable.
[b] Abbreviations: fast indicates less than 2 s resident time; slow greater than 2 s.
[c] Ultimate analysis (ASTM D3176), where % C + % N + % S + % H + % Ash = 100 − % O, percentages expressed as percentage of dried material.
[d] Proximate analysis (ASTM D3172), where % Ash + % VM + % Fixed C + % Moisture = 100 %.
[e] pH was determined on a 1 : 5 biochar : distilled water slurry.

same pyrolysis unit (i.e., Zimmerman, 2010; Mašek et al., 2013). These non-universal trends have also been observed in the chemical composition of volatile matter across different biochars (Spokas et al., 2011). Amendments were incorporated into the soil by rotary tillage to a 15 cm depth starting in the fall of 2008. After incorporation, plots were annually planted with corn (*Zea mays*), and the residue was managed with spring rotary tillage prior to planting. Fertilization was applied uniformly and annually to all test plots, according to the control plot soil test rates, which amounted to between 100 and 125 kg N ha^{-1} (urea) being broadcasted prior to tillage and planting. This fertilization and corn planting occurred prior to residue bag placement. There have been no observed statistical differences between the yield of corn from the biochar-amended and control plots in any year over the duration of the experiment (2009–2013; unpublished data).

2.2 Litterbag preparation and processing

Freshly harvested and baled wheat (*Triticum aestivum* L.) straw was the organic material used in this study. Straw was cut into 10 cm lengths and included stem nodes but not grain or grain heads. Air dry litter weights were corrected to a 50 °C oven dry weight equivalent. Approximately 3.0 ± 0.3 g dry weight equivalent of wheat straw material was placed in 15 cm × 15 cm fiberglass mesh (∼ 1.5 mm) bags. At the beginning of July 2011 (approximately 45 days after planting), 10 bags were inserted into 15 cm deep vertical slits in the ground along a center transect in each plot. Bags were randomly retrieved after 1, 3, 5, 7, 10 and 14 weeks in the field. On week 5 and 14, three replicate bags per plot (nine per treatment) were retrieved. For all other weeks only one bag per plot (three per treatment) was retrieved. Bags were brushed free of dirt and dried at 50 °C before processing. Litter material was manually cleaned of extraneous dirt, roots and other visible contaminants. Following this final cleaning, litter was dried again at 50 °C to obtain final oven dry weights. Mass loss was calculated as initial weight minus final weight of individual litterbags. To account for differences in initial weights among litterbags, data were analyzed as a percent litter mass remaining (% LMR), where

% LMR=((initial weight−final weight)/(initial weight))×100.

2.3 Microbial biomass

Soil sampling of the surface 0–10 cm in each plot was conducted in October 2010 prior to the litterbag decomposition study. Three soil cores were homogenized from each plot and sieved to 2 mm. Microbial biomass (μg C g^{-1} soil) in all treatment plots was determined by the chloroform fumigation-incubation technique (Anderson and Domsch, 1978) on 5 g of soil, with CO_2 production measured by gas chromatography (Koerner et al., 2011). The microbial biomass carbon was calculated as the μg CO_2-C g^{-1} soil of

fumigated soil minus the μg CO_2-C g^{-1} soil from unfumigated soil divided by an efficiency factor of 0.411 (Anderson and Domsch, 1978). Some studies have observed impacts of high surface area biochars on the determination of biomass through chloroform fumigation/extraction procedures (Durenkamp et al., 2010). Though we hypothesize that this effect is minimized since the biochars used in the current study had low surface areas (< 86 m^2 g^{-1}), and that the biochar was exposed to and sorbed DOC in the soil environment, we chose the incubation technique to measure respiration instead of the direct extraction of liberated biomass from fumigation.

2.4 Statistical analysis

The rate of litter mass loss was fit to a first-order decomposition kinetics model (Aber et al., 1990), since this is the most commonly used kinetic model. The data were fit to the following decomposition equation:

$$\% \text{LMR} = 100e^{-kt},$$

where % LMR is the percent of litter mass remaining over time for each treatment, k is the unknown simple first-order decomposition constant, and t is time (Karberg et al., 2008). The decomposition constant, k, and 95 % confidence intervals were determined across the experiment, by treatments and by replicates within treatments using the non-linear platform in JMP 10.0 software (SAS Institute, 2012). Percent litter mass remaining for each sampling week, calculated decomposition rate (k), and microbial biomass were analyzed by a one-way analysis of variance (ANOVA) on treatment (Wider and Lang, 1982) with PROC GLM in SAS 9.2 software (SAS Institute, 2009), using an $\alpha = 0.05$. Differences of means were tested with Bonferoni adjustment to p values of multiple comparison tests, Tukey's honestly significant difference, and with Dunnet's test for comparison to control. The correlation between microbial biomass and k was determined using the pairwise correlation procedure in JMP 10.0 software (SAS Institute, 2012).

3 Results

Despite the short duration of this study (14 weeks), the average mass loss over all the treatments was greater than 50 % (Fig. 1). The estimated decomposition constants, k, ranges from 7.5×10^{-3} to 9.8×10^{-3} d^{-1} (Table 2). Compared to the control, decomposition rates were stimulated in the wood-pellet amendment (WP; +18 %) and the fast pyrolysis hardwood sawdust biochar (BC1; +18 %), 16 % faster in the slow pyrolysis pine chip biochar (BC6), and 11 % faster in the slow pyrolysis wood-pellet biochar (BC2). On the other hand, a decrease in the rate of decomposition was observed in the fast pyrolysis macadamia nut biochar (BC7; −10 %). However, the differences in the k or % LMR

Table 2. Decomposition rate constant, k, with standard error (SE) and 95 % lower and upper confidence limits (LCL, UCL), model fit (r^2), and microbial biomass carbon (MBC) with SE

Treatment	k ($\times 10^{-3}\,\mathrm{d}^{-1}$)	SE ($\times 10^{-3}\,\mathrm{d}^{-1}$)	95 % LCL ($\times 10^{-3}\,\mathrm{d}^{-1}$)	95 % UCL ($\times 10^{-3}\,\mathrm{d}^{-1}$)	r^2	MBC (μg g^{-1} soil)	SE (μg g^{-1} soil)
Control	8.3	0.3	7.5	9.0	0.76	142	19.2
WP	9.8	0.4	8.9	10.7	0.72	835	53.4
BC1	9.8	0.6	8.6	10.9	0.63	232	31.0
BC2	9.2	0.6	7.9	10.5	0.53	277	64.5
BC3	8.0	0.4	7.1	9.0	0.50	136	10.7
BC4	8.9	0.4	8.0	9.9	0.71	133	19.6
BC5	8.8	0.4	7.8	9.9	0.58	239	54.3
BC6	9.6	0.5	8.5	10.8	0.56	435	48.0
BC7	7.5	0.3	6.7	8.4	0.63	117	24.5
Mean						283	44.3

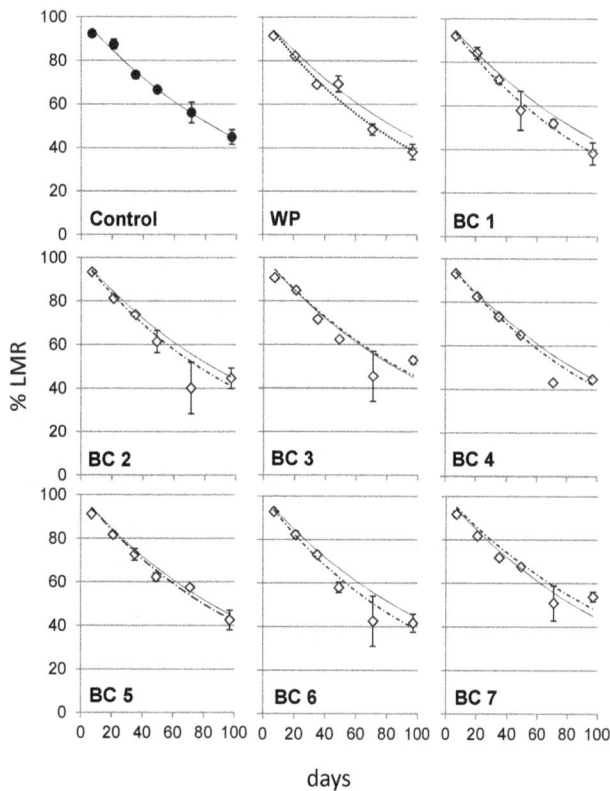

Figure 1. Average percent litter mass remaining (% LMR), over days of incubation, by treatments given in Table 1. Modeled exponential decay curves are shown for each treatment (broken lines) compared to control (solid line). Bars indicate one standard error of the mean ($n = 3$ or $n = 9$; see text).

treatments as a whole. Contrary to the hypothesis that pyrolysis conditions and feedstock are deterministic variables for biochar, the decomposition dynamics did not display distinct overall patterns related to these two variables. Microbial biomass averaged $283\,\mu$g C g^{-1} soil, with a high of $835 \pm 53\,\mu$g C g^{-1} soil for the wood-pellet amendment (WP), a mean of $142 \pm 19\,\mu$g C g^{-1} soil for the control, and a low of $117 \pm 25\,\mu$C g^{-1} for macadamia nut biochar (BC7) (Table 2). Microbial biomass was not significantly different among the treatments ($p > 0.05$). In spite of the lack of statistical significance between treatments, microbial biomass was positively correlated to k, the observed litter decomposition constant, with a pairwise correlation coefficient of 0.698 ($p < 0.001$).

4 Discussion

The decomposition rate of wheat straw observed in our control plots was similar to the rate observed by prior studies (Christensen, 1985). Wang et al. (2012) also observed similar decomposition rates in their 2-year study, with degradation rates spanning from 3.8 to 8.1 yr^{-1}. Though particulate mass can be lost from litterbags over time and other difficulties in the analysis of litterbag results are encountered (Wider and Lang, 1982), similarity of decomposition rates to prior studies and the condition of the wheat straw remaining over the course of the experiment indicated that the majority of the material was retained inside the litterbag and decomposed in situ.

The litterbag method was purposely chosen for its ability to integrate mesofaunal contributions, a component that has not been examined in biochar-amended systems, with the microbial dynamics primarily responsible for decomposition of organic material (Coleman et al., 1999). Thus, the litterbag evaluation allowed a functional determination of biochar influence on dynamics of the decomposer community as a whole. Macrofaunal activity was evident at the field

were not significant across all treatments due to high spatial variability among replicates, which exists in natural field settings. Therefore, the data contradict our initial hypothesis that there would be detectable differences in the observed degradation rates between the biochar and control plots, in particular as visible surface earthworm activity and castings. However, a macrofaunal sampling established that earthworm abundance was not significantly different at the time of litterbag placement (Weyers and Spokas, 2011). This would be in agreement with other studies illustrating short-term impacts on macrofaunal activity observed in short-term laboratory studies (i.e., months) (Domene et al., 2014; Marks et al., 2014), but these short-term effects are not persistent in the field (Domene et al., 2014). This litterbag analysis did not investigate any further impact of biochar application on mesofauna activity.

The lack of significant differences in decomposition rates among the biochar and control treatments indicated that 2.5 years after application biochar did not result in any statistically significant chronic priming effect for the decomposition of freshly added coarse wheat residues, since the observed differences could be attributed to natural spatial variability. Our results are in direct contrast to Wardle et al. (2008), who stated that charcoal maintained an influence on decomposition of soil humus for 10 years. The exact reasons for these differences could be related to the fact that the Wardle et al. (2008) study was conducted in an acidic forest soil, where the liming effect of biochar could play a more critical role than in our more neutral Midwest agricultural soil. Furthermore, upon closer inspection of their data, the mass loss rates of humus versus humus–charcoal mixtures after the first year appear similar, suggesting that the influence was not continuous but only a carryover effect from the initial impacts. It is interesting to note that the highest microbial biomass occurred in the plots with the raw biomass additions, as it is well established that adding a degradable substrate stimulates the microbial activity (Hadas et al., 2004). While on the other hand, adding biochar alone has not stimulated the soil microbial community in the longer term (> 1 year) due to the lack of microbial utilization of the biochar (Rutigliano et al., 2014).

Wardle et al. (2008) cited the absorption of organic compounds on the charcoal as the leading cause of the increased microbial activity and enhanced decomposition they observed. This hypothesis can be traced back to the early 1950s, with Turner (1955) suggesting this as a potential explanation for the increased growth of clover following biochar additions. According to Bruun et al. (2011) an incomplete conversion of feedstock into biochar, as would result from a natural fire or a fast pyrolysis platform, can leave behind decomposable labile material that can sorb to the biochar. The impact of these sorbed volatiles on ash has been reviewed recently by Nelson et al. (2012). Accessibility to this labile component might stimulate soil microbial activity, which may have led to the greater turnover of soil C and N observed with fast pyrolysis biochars in comparison to slow pyrolysis biochars made from the same feedstock (Bruun et al., 2012). These sorbed volatiles could be a potential mechanism behind the short-term impacts that have been observed following biochar additions, such as the impact on microbial communities resulting in decreased greenhouse gas production in incubations that have not been correspondingly observed from field plots (Castaldi et al., 2011; Suddick and Six, 2013).

Along the same lines, Zimmerman et al. (2010, 2011) determined a greater effect on soil processes from labile components released from freshly added low-temperature pyrolysis biochars made from grass and pinewood feedstocks as compared to slow pyrolysis hardwood biochars. Luo et al. (2011) also determined that this priming effect declined with increasing pyrolysis temperatures. Although not statistically significant, the somewhat higher decomposition of the wheat straw in the wheat middlings biochar (BC5) and pine chip biochar (BC6) treatments compared to the slow pyrolysis hardwood biochars falls in line with these evaluations.

These studies all indicated that sorbed compounds and not the actual biochar structure were responsible for the impact on microbial communities. Though the present study still indicated the absence of an effect on microbial biomass and decomposition rates, the significant correlation between the two could be a residual of an impact that might have occurred when the biochar was freshly added. Regardless, the current data indicated that any potential impact from initial application is not likely to last beyond 3 years in the field. This would be in agreement with current meta-analysis of the yield improvements of biochar in soil that cannot be directly correlated to any specific biochar property or characteristic (Crane-Droesch et al., 2013). This further emphasizes the need to understand the mechanisms and impacts before extrapolating any biochar impact to the field. The positive correlation between microbial biomass and decomposition rate was notable, particularly as it relates to the low measurements in the macadamia nut biochar treatment (BC7). Using fresh samples of this biochar, a reduction of CO_2 production rates in the laboratory was found (Spokas and Reicosky, 2009) and correlated to elevated ethylene levels (Spokas et al., 2010). Ethylene can inhibit soil microbial processes (Augustin, 1991; McCarty and Bremner 1991; Wheatley, 2002), plant growth (Deenik et al., 2010), and soil greenhouse gas production (Spokas et al., 2009). Though weathering in the field may have reduced the impact of ethylene, such that the results were not significant, the lower decomposition rates observed here could be the residual of this earlier impact.

Changes in soil physical and chemical characteristics, such as higher moisture content, reduced soil bulk density and increased nutrient availability, have been noted with fresh biochar additions (Atkinson et al., 2010; Sohi et al., 2010; Spokas et al., 2012), though these potential changes from multiple biochars in field plots are rarely compared (Brockhoff et al., 2010; Laird et al., 2010; Meyer et al., 2012). Biochars greater than 1 cm in size are likely to influence soil bulk density, which includes some of the biochars used in this study. These effects may have contributed to the increased variability in our results, thus negating our ability to detect differences.

5   Conclusions

In this study we evaluated the impact of seven different biochars and one non-biochar wood-pellet amendment on the degradation rate of wheat straw in Minnesota field plots. The results indicated that 2.5 years after application these biochars had no significant impact on the decomposition of freshly added organic residues. The variability in decomposition rates among the biochars could be correlated to disappearing impacts observed with fresh biochar (sorbed volatile components), thus providing some indication these slight differences might be limited in duration as the compounds volatilize or are mineralized. Although not statistically affirmed here, soil microbial biomass changes were the most likely drivers of the variability in the decomposition rates observed. These observations suggest that a one-time biochar application has little potential for chronic influences on degradation rates of freshly applied organic matter. Further long-term field studies using charred and uncharred feedstocks, fresh and weathered, are necessary to confirm this result.

Acknowledgements. We appreciate the help of our research assistants, Alan Wilts, USDA-ARS Morris, Martin Dusaire, USDA-ARS St. Paul, and undergraduate assistants, Natalie Barnes UMN-Morris, and the following students at UMN-Twin Cities: Eric Nooker, Edward Colosky, Tia Phan, Michael Ottman, Amanda Bidwell, Lindsay Watson, Kia Yang, Vang Yang, and Lianne Endo. We thank Hal Collins, USDA-ARS Prosser, WA, for a constructive review of the manuscript. In addition, the authors would like to acknowledge the partial funding from the Minnesota Department of Agriculture Specialty Block Grant program and the Minnesota Corn Growers Association/Minnesota Corn Research Production Council. This research is part of the USDA-ARS Biochar and Pyrolysis Initiative and USDA-ARS GRACEnet (Greenhouse Gas Reduction through Agricultural Carbon Enhancement Network) programs. The USDA is an equal opportunity provider and employer.

Edited by: A. Cerdà

References

Aber, J. D., Melillo, J. M., and McClaugherty, C. A.: Predicting long-term patterns of mass loss, nitrogen dynamics, and soil organic matter formation from initial fine litter chemistry in temperate forest ecosystems, Can. J. Bot., 68, 2201–2208, 1990.

Anderson, J. P. E. and Domsch, K. H.: Mineralization of bacteria and fungi in chloroform-fumigated soils, Soil Biol. Biochem., 10, 207–213, 1978.

Artiola, J. F., Rasmussen, C., and Freitas, R.: Effects of a biochar-amended alkaline soil on the growth of romaine lettuce and bermudagrass, Soil Sci., 177, 561–57, 2012.

Atkinson, C., Fitzgerald, J., and Hipps, N.: Potential mechanisms for achieving agricultural benefits from biochar application to temperate soils: A review, Plant Soil, 337, 1–18, 2010.

Augustin, S.: Antimicrobial properties of tannins, Phytochem., 30, 3875–3883, 1991.

Awad, Y. M., Blagodatskaya, E., Ok, Y. S., and Kuzyakov, Y. Y.: Effects of polyacrylamide, biopolymer, and biochar on decomposition of soil organic matter and plant residues as determined by 14C and enzyme activities, Eur. J. Soil Biol., 48, 1–10, 2012.

Awad, Y. M., Blagodatskaya, E., Ok, Y. S., and Kuzyakov, Y.: Effects of polyacrylamide, biopolymer and biochar on the decomposition of ^{14}C-labelled maize residues and on their stabilization in soil aggregates, Eur. J. Soil Sci., 64, 488–499, 2013.

Biederman, L. A. and Harpole, W. S.: Biochar and its effects on plant productivity and nutrient cycling: a meta-analysis, GCB Bioenergy, 5, 202–214, 2013.

Blackwell, P., Krull, E., Butler, G., Herbert, A., and Solaiman, Z.: Effect of banded biochar on dryland wheat production and fertiliser use in south-western Australia: an agronomic and economic perspective, Aust. J. Soil Res., 48, 531–54, 2010.

Brockhoff, S. R., Christians, N. E., Killorn, R. J., Horton, R., and Davis, D. D.: Physical and mineral-nutrition properties of sand-based turfgrass root zones amended with biochar, Agron. J., 102, 1627–1631, 2010.

Bruun, E. W., Hauggaard-Nielsen, H., Norazana, I., Egsgaard, H., Ambus, P., Jensen, P. A., and Dam-Johansen, K.: Influence of fast pyrolysis temperature on biochar labile fraction and short-term carbon loss in a loamy soil, Biomass Bioenerg., 35, 1182–1189, 2011.

Bruun, E. W., Ambus, P., Egsgaard, H., and Hauggaard-Nielsen, H.: Effects of slow and fast pyrolysis biochar on soil C and N turnover dynamics, Soil Biol. Biochem., 46, 73–79, 2012.

Bruun, S. and EL-Zehery, T.: Biochar effect on the mineralization of soil organic matter, Pesq. Agropec. Bras., 47, 665–671, 2012.

Castaldi, S., Riondino, M., Baronti, S., Esposito, F. R., Marzaioli, R., Rutigliano, F. A., Vaccari, F. P., and Miglietta, F.: Impact of biochar application to a Mediterranean wheat crop on soil microbial activity and greenhouse gas fluxes, Chemosphere 85, 1464–1471, 2011.

Christensen, B. T.: Wheat and barley straw decomposition under field conditions: Effect of soil type and plant cover on weight loss, nitrogen and potassium content, Soil Biol. Biochem., 17, 691–697, 1985.

Cely, P., Tarquis, A. M., Paz-Ferreiro, J., Méndez, A., and Gascó, G.: Factors driving carbon mineralization priming effect in a soil amended with different types of biochar, Solid Earth Discuss., 6, 849–868, doi:10.5194/sed-6-849-2014, 2014.

Coleman, D. C., Blair, J. M., Eliott, E. T., and Freckman, D. W.: Soil invertebrates, in: Standard soil methods for long-term ecological research, edited by: Robertson G. P., Bledsoe, C. S., Coleman, D. C., and Sollins, P., Oxford University Press, New York, 349–377, 1999.

Crane-Droesch, A., Abiven, S., Jeffery, S., and Torn, M. S.: Heterogeneous global crop yield response to biochar: a meta-regression analysis, Environ. Res. Lett., 8, 44–49, 2013.

Dai, Z., Meng, J., Muhammad, N., Liu, X., Wang, H., He, Y., Brooks, P. C., and Xu, J.: The potential feasibility for soil improvement, based on the properties of biochars pyrolyzed from different feedstocks, J. Soils Sed., 13, 989–1000, 2013.

Deenik, J. L., McClellan, T., Uehara, G., Antal, M. J., and Campbell, S.: Charcoal volatile matter content influences plant growth and soil nitrogen transformations, Soil Sci. Soc. Am. J., 74, 1259–1270, 2010.

Domene, X., Mattana, S., Hanley, K., Enders, A., and Lehmann, J.: Medium-term effects of corn biochar addition on soil biota activities and functions in a temperate soil cropped to corn, Soil Biol. Biochem., 72, 152–162, 2014.

Durenkamp, M., Luo, Y., and Brookes, P. C.: Impact of black carbon addition to soil on the determination of soil microbial biomass by fumigation extraction, Soil Biol. Biochem., 42, 2026–2029, 2010.

García-Orenes, F., Roldán, A., Mataix-Solera, J., Cerdà, A., Campoy, M., Arcenegui, V., and Caravaca, F.: Soil structural stability and erosion rates influenced by agricultural management practices in a semi-arid Mediterranean agro-ecosystem, Soil Use Manage., 28, 571–579, 2012.

Hadas, A., Kautsky, L., Goek, M., and Erman Kara, E.: Rates of decomposition of plant residues and available nitrogen in soil, related to residue composition through simulation of carbon and nitrogen turnover, Soil Biol. Biochem., 36, 255–266, 2004.

Heal, O. W., Anderson, J. M., and Swift, M. J.: Plant litter quality and decomposition: an historical overview, in: Driven by Nature: Plant Litter Quality and Decomposition, edited by: Cadisch, G. and Giller, K. E., CAB International Wallingford, UK, 3–29, 1997.

Hendrix, P. F., Parmelee, R. W., Crossley Jr., D. A., Coleman, D. C., Odum, E. P., and Groffman, P. M.: Detritus food webs in conventional and no-tillage agroecosystems, Bioscience, 36, 374–380, 1986.

Ibrahim, H. M., Al-Wabel, M. I., Usman, A. R. A., and Al-Omran, A.: Effect of Conocarpus Biochar Application on the Hydraulic Properties of a Sandy Loam Soil, Soil Sci., 178, 165–173, 2013.

Jaafar, N. M., Clode, P. L., and Abbott, L. K.: Microscopy observations of habitable space in biochar for colonization by fungal hyphae from soil, J. Integr. Agric., 13, 483–490, 2014.

Jaffé, R., Ding, Y., Niggemann, J., Vähätalo, A. V., Stubbins, A., Spencer, R. G. M., Campbell, J., and Dittmar, T.: Global charcoal mobilization from soils via dissolution and riverine transport to the oceans, Science, 340, 345–347, 2013.

Kammann, C., Linsel, S., Gößling, J., and Koyro, H.-W.: Influence of biochar on drought tolerance of *Chenopodium quinoa* Willd and on soil–plant relations, Plant Soil, 345, 195–210, 2011.

Karberg, N. J., Scott, N. A., and Giardina, C. P.: Methods for estimating litter decomposition, in: Field measurements for forest carbon monitoring, edited by: Hoover, C. M., Springer, New York, 103–111, 2008.

Koerner, B., Calvert, W., and Bailey, D. J.: Using GC-MS for the quick analysis of carbon dioxide for soil microbial biomass determination, Anal. Meth., 3, 2657–2659, 2011.

Laird, D. A.: The charcoal vision: A win–win–win scenario for simultaneously producing bioenergy, permanently sequestering carbon, while improving soil and water quality, Agron. J., 100, 178–181, 2008.

Laird, D. A., Fleming, P., Davis, D. D., Horton, R., Wang, B., and Karlen, D. L.: Impact of biochar amendments on the quality of a typical Midwestern agricultural soil, Geoderma, 158, 443–449, 2010.

Lee, Y., Park, J., Ryu, C., Gang, K. S., Yang, W., Park, Y.-K., Jung, J., and Hyun, S.: Comparison of biochar properties from biomass residues produced by slow pyrolysis at 500 °C, Bioresource Technol., 148, 196–201, 2013.

Lehmann, J., Rillig, M., Thies, J., Masiello, C. A., Hockaday, W. C., and Crowley, D.: Biochar effects on soil biota: A review, Soil Biol. Biochem., 43, 1812–1836, 2011.

Liu, X., Zhang, A., Ji, C., Joseph, S., Bian, R., Li, L., Pan, G., and Paz-Ferreiro, J.: Biochar's effect on crop productivity and the dependence on experimental conditions – a meta-analysis of literature data, Plant Soil, 373, 583–594, 2013.

Luo, Y., Durenkamp, M., De Nobili, M., Lin, Q., and Brooks, P. C.: Short term soil priming effects and the mineralization of biochar following its incorporation to soils of different pH, Soil Biol. Biochem., 43, 2304–2314, 2011.

Marks, E. A. N., Mattana, S., Alcañiz, J. M., and Domene, X.: Biochars provoke diverse soil mesofauna reproductive responses in laboratory bioassays, Eur. J. Soil Biol., 60, 104–111, 2014.

Mašek, O., Brownsort, P., Cross, A., and Sohi, S.: Influence of production conditions on the yield and environmental stability of biochar, Fuel, 103, 151–155, 2013.

McCarty, G. W. and Bremner, J. M.: Inhibition of nitrification in soil by gaseous hydrocarbons, Biol. Fert. Soil., 11, 231–233, 1991.

Meyer, S., Bright, R. M., Fischer, D., Schulz, H., and Glaser, B.: albedo impact on the suitability of biochar systems to mitigate global warming, Environ. Sci. Technol., 46, 2726–2734, 2012.

Nelson, D. C., Flematti, G. R., Ghisalberti, E. L., Dixon, K. W., and Smith, S. M.: Regulation of seed germination and seedling growth by chemical signals from burning vegetation, Plant Biol., 63, 107–130, 2012.

Noguera, D., Rondón, M., Laossi, K. R., Hoyos, V., Lavelle, P., Cruz de Carvalho, M. H., and Barot, S.: Contrasted effect of biochar and earthworms on rice growth and resource allocation in different soils, Soil Biol. Biochem., 42, 1017–1027, 2010.

Novak, J. M., Busscher, W. J., Watts, D. W., Laird, D. A., Ahmedna, M. A., and Niandou, M. A. S.: Short-term CO_2 mineralization after additions of biochar and switchgrass to a Typic Kandiudult, Geoderma, 154, 281–288, 2010.

Paz-Ferreiro, J. and Fu, S.: Biological indices for soil quality evaluation: perspectives and limitations, Land Degrad. Develop., in press, doi:10.1002/ldr.2262 2014.

Pierce, C., Smith, R. N., Wiley, J., and Cordes, H.: Adsorption of water by carbon, J. Am. Chem. Soc., 73, 4551–4557, 1951.

Puri, B., Singh, D. D., Nath, J., and Sharma, L.: Chemisorption of oxygen on activated charcoal and sorption of acids and bases, Indust. Eng. Chem., 50, 1071–1074, 1958.

Quilliam, R. S., Glanville, H. C., Wade, S. C., and Jones, D. L.: Life in the "charosphere" – Does biochar in agricultural soil provide a significant habitat for microorganisms?, Soil Biol. Biochem., 65, 287–293, 2013.

Rutigliano, F. A., Romano, M., Marzaioli, R., Baglivo, I., Baronti, S., Miglietta, F., and Castaldi, S.: Effect of biochar addition on soil microbial community in a wheat crop, Eur. J. Soil Biol., 60, 9–15, 2014.

Sohi, S. P., Krull, E., Lopez-Capel, E., and Bol, R.: A review of biochar and its use and function in soil, Adv. Agron., 105, 47–82, 2010.

SAS Institute Inc.: SAS 9.2 Users Guide, 2nd edition, SAS Institute, Inc., Cary, NC, 2009.

SAS Institute Inc.: JMP® 10 Modeling and Multivariate Methods, SAS Institute, Inc., Cary, NC, 2012.

Sigua, G. C., Novak, J. M., Watts, D. W., Cantrell, K. B., Shumaker, P. D., Szögi, A. A., and Johnson, M. G.: Carbon mineralization in two ultisols amended with different sources and particle sizes of pyrolyzed biochar, Chemosphere, 103, 313–321, 2014.

Spokas, K. A. and Reicosky, D. C.: Impacts of sixteen different biochars on soil greenhouse gas production, Ann. Environ. Sci., 3, 179–193, 2009.

Spokas, K. A., Koskinen, W. C., Baker, J. M., and Reicosky, D. C.: Impacts of woodchip biochar additions on greenhouse gas production and sorption/degradation of two herbicides in a Minnesota soil, Chemosphere, 77, 574–581, 2009.

Spokas, K. A., Baker, J. M., and Reicosky, D. C.: Ethylene: potential key for biochar amendment impacts, Plant Soil, 333, 443–452, 2010.

Spokas, K. A., Novak, J. M., Stewart, C. E., Cantrell, K. B., Uchimiya, M., Dusaire, M. G., and Ro, K. S.: Qualitative analysis of volatile organic compounds on biochar, Chemosphere, 85, 869–882, 2011.

Spokas, K. A., Cantrell, K. B., Novak, J. M., Archer, D. W., Ippolito, J. A., Collins, H. P., Boateng, A. A., Lima, I. M., Lamb, M. C., McAloon, A. J., Lentz, R. D., and Nichols, K. A.: Biochar: a synthesis of its agronomic impact beyond carbon sequestration, J. Environ. Qual., 41, 973–989, 2012.

Stavi, I., Lal, R., Jones, S., and Reeder, R. C.: Implications of cover crops for soil quality and geodiversity in a humid-temperate region in the Midwestern USA, Land Degrad. Dev., 23, 322–330, 2012.

Steinbeiss, S., Gleixner, G., and Antonietti, M.: Effect of biochar amendment on soil carbon balance and soil microbial activity, Soil Biol. Biochem., 41, 1301–1310, 2009.

Suddick, E. C. and Six, J.: An estimation of annual nitrous oxide emissions and soil quality following the amendment of high temperature walnut shell biochar and compost to a small scale vegetable crop rotation, Sci. Total Environ., 465, 298–307, 2013.

Swift, M. J., Heal, O. W., and Anderson, J. M.: Decomposition in terrestrial ecosystems, Blackwell Scientific Publications, Oxford, 1979.

Turner, E. R.: The effect of certain adsorbents on the nodulation of clover plants, Ann. Botany, 19, 149–60, 1955.

Vitousek, P. M. and Howarth, R. W.: Nitrogen limitation on land and in the sea: How can it occur?, Biogeochemistry, 13, 87–115, 1991.

Wang, X., Sun, B., Mao, J., Sui, Y., and Cao, X.: Structural convergence of maize and wheat straw during two-year decomposition under different climate conditions, Environ. Sci. Technol., 46, 7159–7165, 2012.

Wardle, D. A., Nilsson, M. C., and Zackrisson, O.: Fire-derived charcoal causes loss of forest humus, Science, 320, 629–629, 2008.

Warnock, D. D., Lehmann, J., Kuyper, T. W., and Rillig, M. C.: Mycorrhizal responses to biochar in soil – concepts and mechanisms, Plant Soil, 300, 9–20, 2007.

Wheatley, R. E.: The consequences of volatile organic compound mediated bacterial and fungal interactions, Anton. Leeuw., 81, 357–364, 2002.

Weyers, S. L. and Spokas, K. A.: Impact of biochar on earthworm populations: A review, Appl. Environ. Soil Sci., 2011, 541–592, doi:10.1155/2011/541592, 2011.

Wider, R. K. and Lang, G. E.: A critique of the analytical methods used in examining decomposition data obtained from litter bags, Ecology, 63, 1636–1642, 1982.

Zavalloni, C., Alberti, G., Biasiol, S., Vedove, G. D., Fornasier, F., Liu, J., and Peressotti, A.: Microbial mineralization of biochar and wheat straw mixture in soil: A short-term study, Appl. Soil Ecol., 50, 45–51, 2011.

Zimmerman, A. R.: Abiotic and microbial oxidation of laboratory-produced black carbon (biochar), Environ. Sci. Technol., 44, 1295–1301, 2010.

Zimmerman, A. R., Gao, B., and Ahn, M. Y.: Positive and negative carbon mineralization priming effects among a variety of biochar-amended soils, Soil Biol. Biochem., 43, 1169–1179, 2011.

Factors driving the carbon mineralization priming effect in a sandy loam soil amended with different types of biochar

P. Cely[1], A. M. Tarquis[2,3], J. Paz-Ferreiro[1], A. Méndez[4], and G. Gascó[1]

[1]Departamento de Edafología. E.T.S.I. Agrónomos. Universidad Politécnica de Madrid, Ciudad Universitaria, 28004 Madrid, Spain
[2]CEIGRAM, Universidad Politécnica de Madrid, Ciudad Universitaria, 28004 Madrid, Spain
[3]Departamento de Matemática aplicada a la Ingeniería Agronómica. Universidad Politécnica de Madrid, Ciudad Universitaria, 28040 Madrid, Spain
[4]Departamento de Ingeniería de Materiales. E.T.S.I. Minas. Universidad Politécnica de Madrid, C/Ríos Rosas no. 21, 28003 Madrid, Spain

Correspondence to: G. Gascó (gabriel.gasco@upm.es)

Abstract. The effect of biochar on the soil carbon mineralization priming effect depends on the characteristics of the raw materials, production method and pyrolysis conditions. The goal of the present study is to evaluate the impact of three different types of biochar on physicochemical properties and CO_2 emissions of a sandy loam soil. For this purpose, soil was amended with three different biochars (BI, BII and BIII) at a rate of 8 wt % and soil CO_2 emissions were measured for 45 days. BI is produced from a mixed wood sieving from wood chip production, BII from a mixture of paper sludge and wheat husks and BIII from sewage sludge. Cumulative CO_2 emissions of biochars, soil and amended soil were well fit to a simple first-order kinetic model with correlation coefficients (r^2) greater than 0.97. Results show a negative priming effect in the soil after addition of BI and a positive priming effect in the case of soil amended with BII and BIII. These results can be related to different biochar properties such as carbon content, carbon aromaticity, volatile matter, fixed carbon, easily oxidized organic carbon or metal and phenolic substance content in addition to surface biochar properties. Three biochars increased the values of soil field capacity and wilting point, while effects over pH and cation exchange capacity were not observed.

1 Introduction

Biochar is a carbonaceous material obtained from biomass pyrolysis or gasification processes. Biochar production emits carbon dioxide and other greenhouse gases, but combined with proper waste disposal or biofuel production it offers a practical way of mitigating global warming (Barrow, 2012). For many years now, it has been researched as a significant means of improving soil productivity, carbon storage, and the filtration of soil percolating water (Lehmann and Joseph, 2009). In fact, land degradation is a worldwide phenomenon that affects soil quality, water resources, human societies and economic development (Zhao et al., 2013; Omutu et al., 2014). Biochar as a source of organic matter (Paz Ferreiro and Fu, 2014) can improve the quality of soils in crop and rangeland (Yan-Gui et al., 2013) and then the development of societies (Srinivasarao et al., 2014) and reduce the impact of climate change (Barbera et al., 2013).

Nowadays, biochar production is attracting more attention because it is a safer method of organic waste management. Many types of biomass can be transformed into biochar, including wood wastes, crop residues, switch grass, wastewater sludge or deinking sludges (Méndez et al., 2012; Paz-Ferreiro et al., 2014; Sohi et al., 2010). If enough farmers, larger agricultural enterprises, biofuel producers, and waste treatment plants established biochar production methods, it could reduce CO_2 emissions related to agriculture while improving soil productivity.

Biochar is a highly recalcitrant organic material, with a long-term stability in soil, which is on the scale of millennia or longer (Kuzyakov et al., 2014). The response that soil exhibits to biochar addition has global consequences for carbon cycling. Depending on the interaction between soil and biochar, the ecosystem could become a sink or source of carbon.

The term priming effect refers to increases or decreases in the mineralization of native soil organic matter due to the addition of substrates and has been observed in many studies, both in the field and under laboratory conditions (Paz-Ferreiro et al., 2012; Zavalloni et al., 2009; Zimmerman et al., 2011). While it is generally regarded that biochar addition results in a reduction in soil carbon emissions from the soil, the fact is that the results are biochar and soil specific. Indeed, previous works have shown that there is no clear trend in CO_2 emissions after biochar application. For example, Zimmerman et al. (2011) found that carbon mineralization was generally less than expected (negative priming) for soils combined with biochars produced at high temperatures (525 and 650 °C) and from hard woods, whereas carbon mineralization was greater than expected (positive priming) for soils combined with biochars produced at low temperatures (250 and 400 °C) and from grasses, particularly during the early incubation stage and in soils of lower organic carbon content. Luo et al. (2011) used biochar from plant residues and found during the first 13 days of incubation experiment that biochar obtained at 350 °C causes a large positive priming effect, while biochar prepared at higher temperatures (700 °C) caused a relatively small positive priming effect. These authors hypothesized that the priming effect was probably caused by labile organic matter remaining in the biochar after pyrolysis, which in turn activated the soil microorganism. Jones et al. (2011) hypothesized that the increment in soil respiration is due to a different mechanism than changes in soil physical properties (bulk density, porosity, moisture); biological breakdown of organic carbon released from the biochar; abiotic release of inorganic carbon contained in the biochar and a stimulation of decomposition of soil organic matter. Zavalloni et al. (2011) have shown that the amount of soil carbon respired was similar between the control and soil treated with biochar from coppiced woodland pyrolysis in a short-term incubation experiment. Also, Wardle et al. (2008) reported a priming effect from a boreal soil after biochar addition, although the results of this experiment have been disputed by others (Lehmann and Sohi, 2008). If a strong positive priming effect occurs after biochar addition to the soil, then the beneficial effects attained by biochar addition to the soil becomes mitigated. Furthermore, although the use of biochar measuring soil respiration has been evaluated (Méndez et al., 2012; Zimmerman et al., 2011), fewer studies have studied the role of biochar addition of native soil organic matter (Zimmermann et al., 2011; Cross and Sohi, 2011; Gascó

et al., 2012). For example, Gascó et al. (2012) observed using thermal methods that there is a degradation of more complex structures after application of a sewage sludge biochar to a Haplic Cambisol. The final chemical composition and physical properties of biochar, and thus its potential for having a positive or negative priming effect, depends on the characteristics of the raw materials, production method and pyrolysis conditions. Different studies has been performed in order to study the influence of feedstock, production method and pyrolisis temperature on biochar properties and uses (Calvelo Pereira et al., 2011; Méndez et al., 2012; Zimmermann et al., 2011; Paz-Ferreiro et al., 2014).

In the present work, three different biochars were used in order to study their influence on soil properties and CO_2 emissions. Biochars were obtained from pyrolysis of different types of biomass: mixed wood sievings from wood chip production, paper sludge and wheat husks and sewage sludge at temperatures between 500 and 620 °C using slow pyrolysis processes.

2   Materials and methods

2.1   Soil selection and characterization

The selected soil was taken from the northeast of Toledo (Spain) and the soil was air-dried, crushed and sieved through a 2 mm mesh prior to analysis. The initial pH and electrical conductivity (EC) were determined with a soil : water ratio of 1 : 2.5 $(g\,m\,L^{-1})$ using a Crison micro-pH 2000 (Thomas et al., 1996) and a Crison 222 conductivimeter (Rhoades, 1996) respectively. CEC was determined by $NH_4OAc/HOAc$ at pH 7.0 (Sumner and Miller, 1996). Total organic matter (TOM) was determined using the dry combustion method at 540 °C (Nelson and Sommers, 1996). Soil metal content was determined using a Perkin Elmer 2280 atomic absorption spectrophotometer after sample extraction by digestion with concentrated HCl/HNO_3 following method 3051a (USEPA, 1997). Soil texture was determined following the methodology of Bouyoucos (1962). These analyses were performed in triplicate.

2.2   Biochar characterization

Three different biochar samples were selected and used for the present work: biochar I (BI) was produced by Swiss Biochar (Lausanne, Switzerland) from mixed wood sievings from wood chip production at 620 °C; biochar II (BII) was produced by Sonnenerde (Austria) from a mixture of paper sludge and wheat husks at 500 °C; and biochar III (BIII) was produced by Pyreg (Germany) from sewage sludge at 600 °C. The pyrolisis duration was 20 min in all cases. All biochar samples were produced using Pyreg500-III pyrolysis (Germany) units that can work until 650 °C in a continuous process.

The pH, EC, CEC and metal content in biochars were performed as in Sect. 2.1. Proximate analysis was determined by thermogravimetry using Labsys Setaram equipment. The sample was heated to a temperature of 600 °C under N_2 atmosphere and 30 °C min^{-1} heating rate. Humidity was calculated as the weight loss from the initial temperature to 150 °C. The volatile matter (VM) was determined as the weight loss from 150 °C to 600 °C under N_2 atmosphere. At this temperature, air atmosphere was introduced and fixed carbon (FC) was calculated as the weight produced when the final sample was burned. The ashes were determined as the final weight of the samples. The content in C, H, N and S was analyzed by an elemental microanalyzer LECO CHNS-932 and the oxygen content was determined by difference. Biochar nitrogen adsorption analysis to determine BET surface area was carried out at 77 K in a Micromeritics Tristar 3000. Also, biochar CO_2 adsorption analysis to determine both CO_2 micropore surface area and monolayer capacity was performed at 273 K in a ASAP 2020 V3.01.

Finally, biochar phenolic substances were determined using a Folin–Ciocalteu reagent (Martín-Lara et al., 2009).

2.3 Treatments and soil respiration

The selected soil (S) was amended with the three biochar samples at 8 wt % (S+BI, S+BII, S+BIII) and mixtures were incubated at constant temperature (28 ± 2 °C) and humidity (60 % FC) for 45 days. Additionally, it was studied if the application of the different amendments had an additive or synergistic effect in the soil (priming effect); in this way each biochar (BI, BII, BIII) was incubated individually in the experimental conditions.

Each sample (100 g) was introduced into a 1 L airtight jar and the CO_2 produced during incubation was collected in 50 mL of a 0.3 N NaOH solution, which was then titrated using 0.3 N HCl after the $BaCl_2$ precipitation of the carbonates. All treatments were performed in triplicate.

Organic carbon oxidized with dichromate from initial and final biochars was determined by the Walkley–Black method (Nelson and Sommers, 1996).

After incubation time, the next soil properties were determined: pH, EC, CEC, field capacity (FC), wilting point (WP) and available water (AW). pH, EC and CEC were determined as in Sect. 2.1. Field capacity (FC) and wilting point (WP) were determined as the soil moisture content at 33 kPa (FC) and 1500 kPa (WP) (Richards, 1954). Available water (AW) was calculated as the difference between FC and WP. All analyses were performed in triplicate.

In addition, thermal analysis (TG, dTG and DTA) of soil was performed in a thermogravimetric equipment Labsys Setaram. About 50 mg of each sample were heated at 15 °C min^{-1} until 850 °C in the air atmosphere using a flow rate of 40 mL min^{-1}.

2.4 Mineralization model

The cumulative mineralization data were fitted to a first-order kinetic model, which is widely used to model soil respiration data (Méndez et al., 2013). The kinetic model used to calculate the evolved CO_2–C soil is described as follows:

$$Y = Ct^m \tag{1}$$

where Y is the cumulative CO_2–C (mg CO_2–C 100 g^{-1} soil), t is the cumulative time of incubation (d), and C and m are the mineralization constants, with $C \cdot m$ representing the initial mineralization rate. The convexity shape of Y in this model is defined mainly by m, with $m \leq 1$ and $C \geq 0$. This equation was fitted to describe the C mineralization in S, the biochars (BI, BII and BIII) and the amended soils (S+BI, S+BII and S+BIII). The mineralization rate parameters of Eq. (1) were estimated by a non-linear model method, minimizing RMSA.

To quantify the priming effect of the three raw materials, the model was fitted to the experimental data (Experiment) and to the respiration data with the addition of 92 g of soil with 8 g of biochars (Addition). Also, C_{10} was calculated as the evolved CO_2–C after 10 days according to the model.

3 Results and discussion

Table 1 shows the main properties of the soil and three biochars. The soil texture was sandy loam, it had a slightly alkaline pH, the EC value indicated that the soil has no risk of salinization and soil organic matter content was 6.30 %.

With respect to biochars, BI and BII showed basic pH, whereas BIII had a pH value near 7. Proximate analysis of three biochar samples showed differences in their composition. The ash content of biochars followed the next sequence BIII > BII > BI depending on the feedstock, i.e., BI is prepared from woodchip, BII from paper sludge and wheat husk and BIII from sewage sludge presenting a higher mineral content. Indeed, BIII had the highest EC and metal content. Biochar metal content did not exceed the limit values for concentrations of metals in soil set up by the European Union (European Community, 1986), with BIII presenting the highest content, which can be explained according to its origin. All biochars presented a similar CEC, which can be related to the comparable temperature of preparation. The volatile matter content of BI and BIII was similar and lower than that of BII. Fixed carbon of BI was significantly higher than that of BII and BIII. Combining VM and FC, the ratio FC/(FC + VM) could be indicative of the carbon stability. According to this, BI was a very recalcitrant carbon material, whereas BIII showed the lowest ratio. The molar H/C ratio was used as an indicator of the degree of aromatization. This ratio shows the sequence BI < BII < BIII. The O/C ratio was indicative of the degree of carbonization following the same trend as the H/C ratio, BI < BII < BIII. According to

Table 1. Main properties of the soil (S) and biochars.

	S	BI	BII	BIII
pH (1 : 2.5)	7.66 ± 0.10	10.19 ± 0.12	9.40 ± 0.19	7.66 ± 0.13
EC (1 : 2.5 (dS m^{-1}, 25 °C)	70 ± 10	1776 ± 44	2330 ± 50	3700 ± 157
TOM (%)	6.30 ± 0.15	87.71 ± 0.71	59.90 ± 0.89	25.15 ± 0.40
CEC (cmol$_{(c)}$ kg^{-1})	15.87 ± 0.25	23.77 ± 0.36	20.97 ± 0.24	24.19 ± 0.30
Cd (mg kg^{-1})	–	0.43 ± 0.05	0.72 ± 0.08	4.98 ± 0.01
Cr (mg kg^{-1})	–	21 ± 2	32 ± 4	76 ± 8
Cu (mg kg^{-1})	–	61 ± 9	37 ± 8	406 ± 25
Ni (mg kg^{-1})	–	18 ± 1	30 ± 1	78 ± 10
Pb (mg kg^{-1})	–	4 ± 1	24 ± 3	141 ± 10
Zn (mg kg^{-1})	–	47 ± 5	134 ± 9	1350 ± 49
Phenolic substances (mg gallic acid g^{-1})		0.93 ± 0.05	1.01 ± 0.07	0.49 ± 0.04
Sand (%)	77.78	–	–	–
Silt (%)	17.78	–	–	–
Clay (%)	4.44	–	–	–
Soil textural class	Sandy loam	–	–	–
FC (%)		113 ± 1	122 ± 1	36 ± 1
WP (%)		52 ± 1	63 ± 1	31 ± 1
AW (%)		61 ± 1	59 ± 1	5 ± 1
BET Surface Area (m^2 g^{-1})	–	332.138	92.6115	59.1572
Micropore area (m^2 g^{-1})	–	305.9972	66.9119	30.9545
Adsorption average pore width (Å)	–	21.2622	32.9697	77.1478
CO$_2$ micropore surface area (m^2 g^{-1})		414.206	229.399	86.329
CO$_2$ monolayer capacity (cm^3 g^{-1})		90.672	50.217	18.898
Proximate analysis				
VM (%)[a]	–	14.88	22.43	13.68
FC (%)[b]	–	77.25	42.72	12.77
Ash (%)	–	7.87	34.85	73.55
FC/(FC + VM)	–	0.84	0.66	0.48
Elemental analysis				
C (%)		82.00	50.75	18.45
H (%)		1.49	1.73	1.19
N (%)		0.33	1.36	2.10
O (%)		5.76	12.08	7.69
H / C atomic ratio		0.018	0.034	0.064
O / C atomic ratio		0.070	0.238	0.417

[a] VM: Volatile matter.
[b] FC: Fixed carbon.

previous studies on biochars (Kuhbusch and Crutzen, 1995; Hammes et al., 2006) the H / C ratio of ≤ 0.3 (like BI) indicates a highly condensed aromatic ring system, whereas the H / C ratio of ≥ 0.7 (like BIII) represents a non-condensed structure.

Table 2 shows the changes in pH, EC and CEC after the 45 days of the incubation experiment. Instead, biochar pHs were different (Table 1); pH did not change after biochar application, though BI and BII presented pH 2 units higher than soil. Conversely, other studies have shown pH increments after biochar application. For example, Méndez et al. (2012) observed a pH increment on a Haplic Cambisol after the ad-

Table 2. pH, electrical conductivity (EC), cation exchange capacity of treated soils after the incubation experiment.

	pH	EC (μS cm^{-1})	CEC (cmol$_{(c)}$ kg^{-1})
S	7.45ab	496a	15.71a
S+BI	7.68b	535a	16.28a
S+BII	7.47ab	624b	16.08a
S+BIII	7.29a	764c	17.07a

Values in column followed by the same letter are not significantly different ($P = 0.05$) using the Duncan test.
The number of replicates was 3 for each determination.

dition of sewage sludge-derived biochar, Kloss et al. (2014) described a slight increment of soil pH (0.3 units) in an acid soil after application of woodchip-derived biochar, and Jien and Wang (2013) observed a significant increase in Ultisol pH from 3.9 to 5.1 after addition of biochar made from the waste wood of white lead trees, so both biochar and soil composition influence the pH changes. Biochar addition slightly increased soil EC (Table 1), but the risk of salinization was negligible at the applied dose (USDA, 1999). The increase in soil EC is very common in soils treated with biochar prepared from sludge, which is the case for BII and BIII, as reported in other studies (Hossain et al., 2010 or Méndez et al., 2012). With respect to CEC, biochars did not increase soil CEC, a result according to previous works (Méndez et al., 2012) and which can be related to the low CEC of biochar with respect to soil OM (Lehmann, 2007).

Biochars increased the values of soil FC and WP following the sequence S < S+BIII < S+BI ≈ S+BII for both properties. Also, there was an increment in the AW when the soil was treated by BI and BII. This improvement in water retention is in accordance with the results previously obtained by Méndez et al. (2012), which found the same trend in a soil with a similar sand content treated with biochar prepared for sewage sludge at 600 °C. The higher increment of FC, WP and AW in S+BI and S+BII treatments could be related to the higher values of FC and WP of these biochars according to their high surface area and porosity (Table 1).

In the last years, thermal analysis has been proposed as an interesting technique in the characterization of organic matter stabilization processes. Additionally, it has been applied to soil characterization to assess proportions of labile and recalcitrant organic matter (Plante et al., 2009) and to study the evolution of organic matter in amended soils (Barriga et al., 2010; Gascó et al., 2012). Thermal analysis has the advantage of providing information about the chemical characteristics of soil organic matter without any extraction step as all samples were analyzed. Figure 1 shows dTG (Fig. 1a) and DTA (Fig. 1b) of S, S+BI, S+BII and S+BIII samples after the incubation period. Different peaks were observed in Fig. 1; at temperatures lower than 150 °C, water releases were observed, then at temperatures from 200 to 650 °C, oxidation of organic matter takes place. Initially, weight loss corresponds to less humified matter (from 200 to 400 °C), whereas the peaks observed at temperatures higher than 400 °C correspond to more humified organic matter. At temperatures higher than 550 °C weight loss could be attributed to refractory carbon from biochars and clay decomposition (Gascó et al., 2012).

From the DTA curve, the first endothermic peak could be observed at temperatures lower than 150 °C due to moisture release from soil samples. Then, two small exothermic peaks could be observed between 200 and 650 °C due to combustion reactions of soil organic matter. It is established that the first peak was associated with combustion of less humified organic matter, whilst the second one was related to the

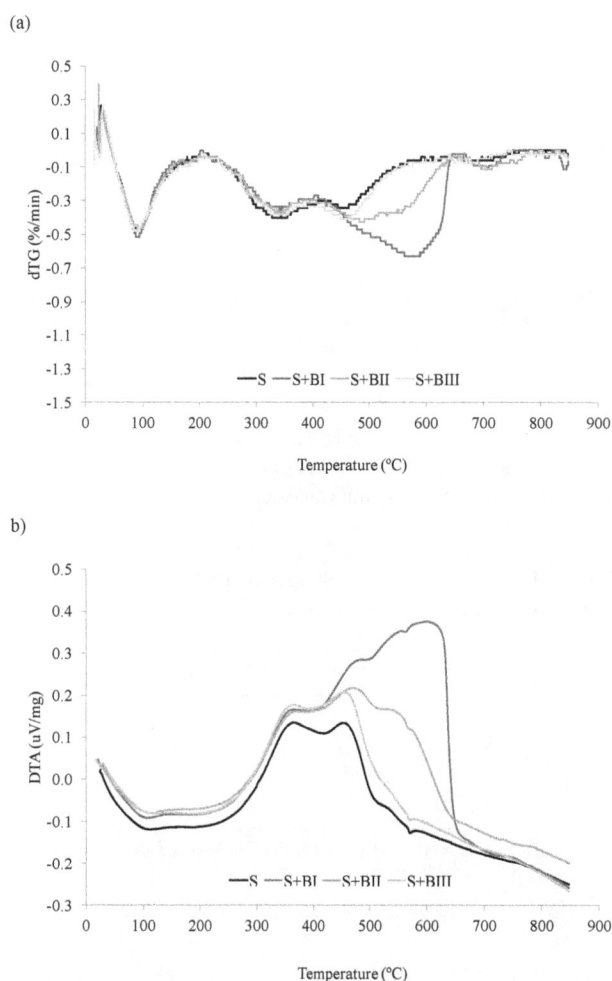

Figure 1. dTG (1.a) and DTA curves (1.b) of soil and soil amended with biochar after the incubation period.

more humified matter. Four samples are shown at 573 °C, the characteristic small endothermic peak due to the quartz α–β inversion. Comparison of four samples in Fig. 1a and b shows the influence of different biochars in soil organic matter composition. Biochar addition increases the amount of more humified or thermally stable organic matter following the sequence S+BI > S+BII > S+BIII. It was interesting to note that S+BIII shows a thermal behavior similar to that of unamended soil (S), indicating a similar organic matter composition to original soil.

With respect to biochar CO_2 emissions, these were higher in BI while significant differences between BII and BIII were not found. This fact can be attributed to the elevated TOM of BI (87.71%) with respect to BII (59.90%) and BIII (25.15%). In order to explain the similar CO_2 emissions of BII and BIII other factors need to be accounted for (Jones et al., 2011). Calvelo Pereira et al. (2011) found that dichromate oxidation reflects the degree of biochar carbonization and could therefore be used to estimate the labile fraction of

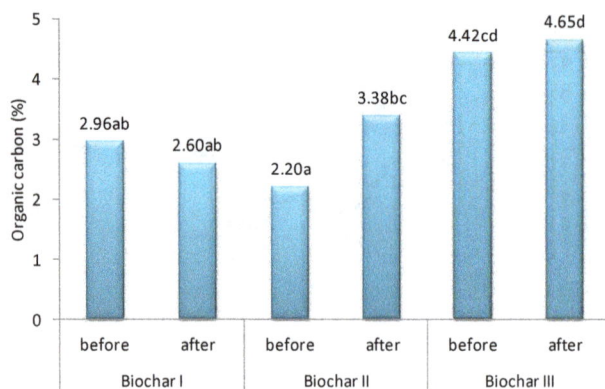

Figure 2. Evolution of organic carbon oxidized with dichromate. Values in columns followed by the same letter are not significantly different ($P = 0.05$) using the Duncan test.

Table 3. Field capacity (FC), wilting point (WP) and available water (AW) after the incubation experiment.

	FC(%)	WP(%)	AW(%)
S	13.54a	11.04a	2.49a
S+BI	20.41c	13.79c	6.61b
S+BII	20.24c	13.91c	6.33b
S+BIII	16.31b	12.72b	3.60a

Values in column followed by the same letter are not significantly different ($P = 0.05$) using the Duncan test.
The number of replicates was 3 for each determination.

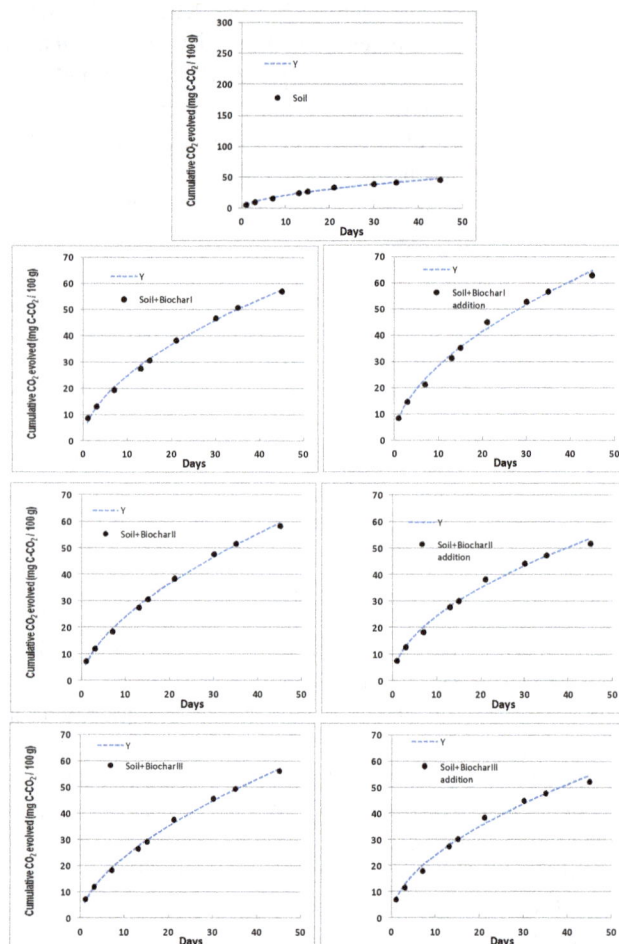

Figure 3. Exponential model of measured C mineralized (as CO_2) and that calculated by the addition of soil and BI, BII and BIII effects.

carbon in biochar. Figure 2 shows that BIII, with the highest ash content and the lowest C content and consequently, expected lowest CO_2 emissions, has the highest content of labile, so the H/C and O/C ratios have shown that BIII has non-condensed organic structures. After incubation, the labile carbon of BI decreases, whereas that of BII and BIII slightly increases, indicating that some of the more stable organic structures were transformed into labile carbon. This result was in accordance with that obtained previously by Gascó et al. (2012) using thermal analysis and biochar form sewage sludge. However, for BI the labile carbon slightly decreases after incubation.

Results show that biochar addition increased CO_2 soil emissions by approximately 25 %, but there were no differences between the different treatments (Fig. 3). On the other hand, Zavalloni et al. (2011) found that respiration rates in soil with coppiced woodland-derived biochar were not significantly different from control soil. This matter can be attributed to a combination of different factors, not only to one. Méndez at al. (2013) found that higher CO_2 emissions can be related to a higher content of VM and lower values of ratio FC/(FC + VM) from biochars. Also, the CO_2 evolved can be related to the labile carbon content of biochars (Fig. 2). On the other hand, different authors (Méndez et al., 2013;

Thies and Rillig, 2009) observed that the reduction in CO_2 emissions can be attributed to chemisorptions of the respired CO_2 on the biochar surface. Indeed, BI had a CO_2 micropore surface area and CO_2 monolayer capacity more than 44 % higher than BI and BII, so their labile carbon content was lower. Also, H/C, O/C and FC/(FC + VM) ratios indicate that instead of their high carbon content it was a more stable carbon material. Finally, the electrical conductivity, metal and phenolic substances of biochar can have a negative effect on soil microbial activity, reducing the respired CO_2. Table 4 summarizes the qualitative influence of different factors on CO_2 emissions and it shows an orientation about the influence of different biochar properties on the increment of soil CO_2 emissions after biochar application. pH limits have been fixed following the classes of soil pH of USDA (1998) and the guidelines to biochar production according to Schmidt et al. (2012). It must be pointed out that pH of 6.6 to 7.3 is favorable for microbial activities that contribute to the availability of nitrogen, sulfur, and phosphorus in soils (USDA,

Table 4. Influence of different biochar properties on the increment of soil CO_2 emissions after biochar application.

Value	pH	Electrical conductivity	Organic carbon	Metal content	Phenolic substances	Volatile matter	Fixed carbon	BET surface area
High[b]	− [a]	−	+	−	−	+	+	−
Normal	+	+	+	+	+	+	+	−
Low	−	+	−	+	+	−	−	−

[a] +: positive effect; −: negative effect

[b]: **pH** (USDA, 1998; Schmidt et al., 2012): High: > 10, Normal: 6–10, Low: < 6. **Electrical conductivity** (Richards, 1958): High: > 4 dS m^{-1}, Normal: 4–2 dS m^{-1}, Low < 2 dS m^{-1}. **Metal content** (European Community, 1986): High: Cd > 40 mg Kg^{-1}, Cu > 1750 mg Kg^{-1}, Ni > 400 mg Kg^{-1}, Pb > 1200 mg Kg^{-1}, Zn > 4000 mg Kg^{-1}, Hg > 25 mg Kg^{-1}; Normal: Cd 20–40 mg Kg^{-1}, Cu > 1000–1750 mg Kg^{-1}, Ni > 300–400 mg Kg^{-1}, Pb > 750–1200 mg Kg^{-1}, Zn > 2500–4000 mg Kg^{-1}, Hg > 16–25 mg Kg^{-1}; Low: Cd < 20 mg Kg^{-1}, Cu < 1000 mg Kg^{-1}, Ni < 300 mg Kg^{-1}, Pb < 750 mg Kg^{-1}, Zn < 2500 mg Kg^{-1}, Hg < 16 mg Kg^{-1}. **Organic carbon** (International Biochar Initiative, 2011): High: > 50 %, Normal: 30–60 %, Low < 10 %. **Phenolic substances** (Kuiters and Sarink, 1986): High: > 10 μg g^{-1}, Normal: 10–1 μg g^{-1}, Low: < 1 μg g^{-1}. **Volatile matter:** High: > 20 %, Normal: 20–10 %, Low: < 10 %. **Fixed carbon:** High: > 40, Normal: 40–20, Low: < 20. **BET surface area** (Schmidt et al., 2012): High: > 750 m^2 g^{-1}, Normal: 750–150 m^2 g^{-1}, Low: < 150 m^2 g^{-1}.

Table 5. CO_2−C evolved (mg CO_2 100 g^{-1} dry weight) during incubation experiment and parameters estimated according to a simple first-order kinetic model to describe the C mineralization in soil (S), biochars (BI, BII, BIII) and amended soils (S+ BI, S+ BII, S+ BIII). Root mean square deviation (RMSD), correlation coefficient (r^2) and coefficient of determination (R^2) of the fitted model are shown.

Substrate		CO_2 evolved (mg C−CO_2/100 g)	m	C	RMSD	r^2	C_{10}[b] (mg C−CO_2/100 g)
S		45.8	0.5524	5.81	1.23	0.996	20.72
BI		261.2	0.5513	32.15	10.94	0.989	114.41
BII		120.1	0.4092	25.51	6.69	0.975	65.46
BIII		125.6	0.5046	19.34	6.26	0.985	61.79
S+ BI	Experiment	57.1	0.5606	6.83	0.94	0.998	24.83
	Addition[a]	63.0	0.5521	7.91	1.34	0.997	28.22
S+ BII	Experiment	58.3	0.5987	6.07	0.86	0.999	24.10
	Addition	51.7	0.5262	7.22	1.22	0.997	24.25
S+ BIII	Experiment	56.1	0.5872	6.08	0.82	0.999	23.50
	Addition	52.2	0.5434	6.87	1.40	0.996	23.99

[a] The addition of the experimental data has been made taking into account a dose of 8 %.

[b] C_{10} is the evolved CO_2−C after 10 days according the model.

1998), and a pH value exceeding 10 can have negative effects on soil pH, but it must noted that only the application of larger amounts of biochar will lead to changes in a soil's pH value (Schmidt et al., 2012). With respect to electrical conductivity, limits have been fixed according to the limits fixed by Richards (1954) where the high value (4 dS m^{-1}, 25 °C) is the limit between normal and saline soils. The organic carbon limits have been fixed according to the International Biochar Initiative (2012) and the recommendations of Schmidt et al. (2012), who described that the organic carbon content of pyrolysed chars fluctuates between 10 % and 95 % of the dry mass, depending on the feedstock and process temperature used. With respect to volatile matter (VM) and fixed carbon (FC), values over 20 % and 40 % of VM and FC can be considered high according to biochar prepared from different feedstocks as sewage sludge (Gascó et al., 2012; Méndez et al., 2012), rice husk (Kalderis et al., 2014), euca-

lyptus wood or poultry litter (Paz-Ferreiro, 2012; Lu et al., 2014). Finally, BET surface area values should be preferably higher than 150 m^2 g^{-1} (Schmidt et al., 2012), values over 750 m^2 g^{-1} being very high and of the same order as montmorillonite. It must stand out that the negative effects are usually due to a combination of different factors and cannot be attributed to a unique factor. Table 5 and Fig. 3 show the parameters estimated according to simple first-order kinetic model to describe the C mineralization in soil (S), biochars (BI, BII, BIII) and amended soils (S+BI, S+BII, S+BIII). The kinetics of CO_2 evolved from biochars was well fit to the proposed model, presenting r^2 values higher to 0.97. With respect to the amended soils, the fit presented a root mean square deviation (RSMD) lower than 2 and r^2 values higher than 0.99. In fact, this model of a simple first-order kinetic model has been successfully used to estimate CO_2 emissions

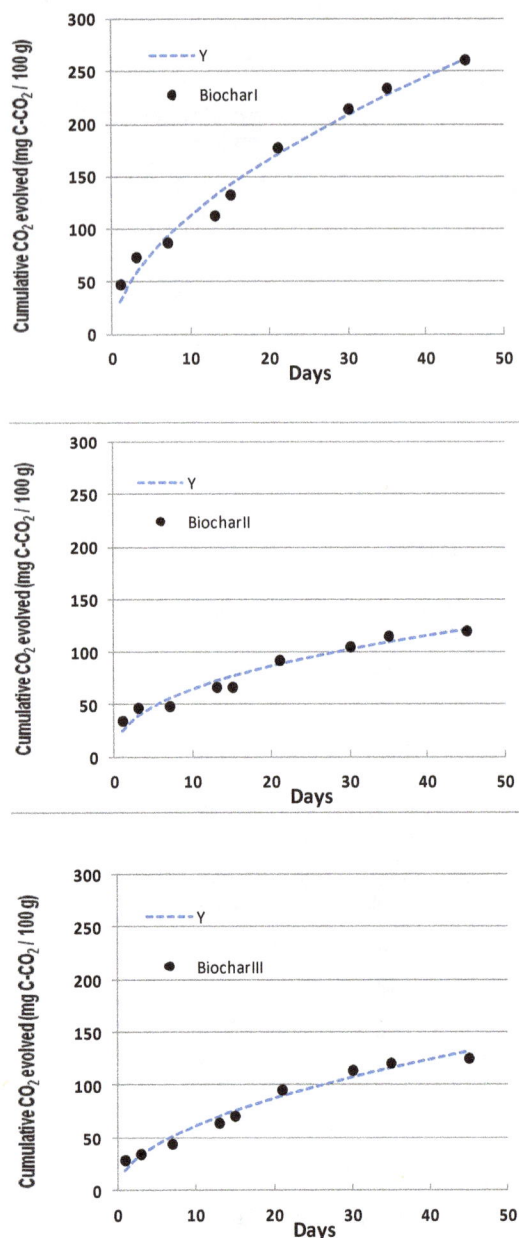

Figure 4. Exponential model of measured C mineralized (as CO_2) in BI, BII and BIII biochars.

from biochar and biochar amended soil in a short-term incubation experiment (Méndez et al., 2013).

Also, results show that the application of BI had a negative priming effect if data of the experiment (57.1 mg C–CO_2/100 g) and addition (63.0 mg C–CO_2/100 g) are compared (Table 4), according with the similar values of model parameters (m and C). This result was in accordance with that obtained by Zimmerman et al. (2011), who found that biochars produced at high temperatures and from hard woods like BI show negative priming. With respect to the application of BII and BIII to soil, results showed a positive priming

effect. It is interesting to note that both biochars increase their labile carbon content during individual incubation (Fig. 2), whereas for BI, their content slightly decreases. The initial organic matter mineralization was very similar in all cases (C parameter ranged from 6.07 to 7.91) according to Méndez et al. (2012), who found an increment of CO_2 emissions after application at the same rate after application of biochar prepared from sludge to a similar sandy soil or results obtained by Smith (2010). Nevertheless, Paz-Ferreiro et al. (2012) found a negative priming effect after sewage sludge biochar application (prepared at 650 °C) to an Umbrisol. Indeed, Zimmerman et al. (2011) concluded that discrepancies in C mineralization of biochar-treated soils are likely due to the type of both soil and biochar, the duration of the experiment and the dose of used biochar.

Finally, the C_{10} parameter, i.e. evolved CO_2–C after 10 days according to the model, is related to the labile fraction of biochar to be released by microbial activity. Results show that experimental data were very similar, and the difference between experiment and addition (Table 4) in the case of S+BI could suggest a toxic effect of biochar.

4 Conclusions

The effect of biochar on the soil carbon mineralization priming effect depends on the characteristics of the raw materials, production method and pyrolysis conditions. Indeed, results show a negative priming effect in the soil after addition of BI (prepared at 620 °C from a mixed wood sieving from wood chip production) and a positive priming effect in the case of soil amended with BII (prepared at 500 °C from a mixture of paper sludge and wheat husks) and BIII (prepared at 600 °C from sewage sludge). These facts can be related to different biochar properties such as carbon content, carbon aromaticity, volatile matter, fixed carbon, easily oxidized organic carbon, metal and phenolic substance content and surface biochar properties. In addition, experimental results show that cumulative CO_2 emissions were well fit to a simple first-order kinetic model for the different biochar and amended soils. Also, biochars additionally improved water soil retention. Finally, further research is required to determine the importance of the different biochar properties involved in soil CO_2 emissions.

Acknowledgements. We are very grateful to the Delinat Institute for Ecology and Climate Framing (Switzerland) for providing the biochar samples.

Edited by: A. Cerdà

References

Barbera, V., Poma, I., Gristina, L., Novara, A., and Egli, M.: Long-term cropping systems and tillage management effects on soil organic carbon stock and steady state level of C sequestration rates in a semiarid environment, Land Degrad. Dev., 23, 82–91, 2013.

Barriga, S., Méndez, A., Cámara, J., Guerrero, F., and Gascó, G.: Agricultural valorisation of de-inking paper sludge as organic amendment in different soils: thermal study, J. Therm. Anal. Calorim., 99, 981–986, 2010.

Barrow, C.: Biochar: potential for countering land degradation and for improving agriculture, Appl. Geogr., 34, 21–28, 2012.

Bouyoucos, G. J.: Hydrometer method improved for makins paricle size analyses of soil, Agron. J., 54, 464–465, 1962.

Calvelo Pereira, R., Kaal, J., Camps Arbestain, M., Pardo Lorenzo, R., Aitkenhead, W., Hedley, M., Macías, F., Hindmarsh, J., and Maciá-Agulló, J. A.: Contribution to characterisation of Biochar to estimate the labile fraction of carbon, Org. Geochem., 42, 1331–1342, 2011.

Cross, A. and Sohi, S. P.: The priming potential of biochar products in relation to labile carbon contents and soil organic matter status, Soil Biol. Biochem., 43, 2127–2134, 2011.

European Community: Council Directive of 12 June 1986 on the protection of the environment, and in particular of the soil, when sewage sludge is used in agriculture, Directive 86/276/EEC, Official Journal of the European Communities, L 181, 6–12, 1986.

Gascó, G., Paz-Ferreiro, J., and Méndez, A.: Thermal analysis of soil amended with sewage sludge and biochar from sewage sludge pyrolysis, J. Therm. Anal. Calorim., 108, 769–775, 2012.

Hossain, M. K., Strezov, V., Chan, K. Y., Ziokowski, A., and Nelson, P. F.: Influence of pyrolysis temperature on production and nutrient properties of wastewater sludge biochar, J. Environ. Manage., 92, 223–228, 2011.

International Biochar Initiative: Standardized product definition and product testing guidelines for biochar that is used in soil, International Biochar Initiative, Westerville, United States of America, 2012.

Jien, S. H. and Wang, C. S.: Effects of biochar on soil properties and erosion potential in a highly weathered soil, Catena 110, 225–233, 2012.

Jones, D. L., Murphy, D. V., Khalid, M., Ahmad, W., Edwards-Jones, G., and DeLuca, T. H.: Short-term biochar-induced increase in soil CO_2 release is both biotically and abiotically mediated, Soil Biol. Biochem., 43, 1723–1731, 2011.

Kalderis, D., Kotti, M. S., Méndez, A., and Gascó, G.: Characterization of hydrochars produced by hydrothermal carbonization of rice husk, Solid Earth Discuss., 6, 657–677, doi:10.5194/sed-6-657-2014, 2014.

Kloss, S., Zehetner, F., Oburger, E, Buecker, J., Kitzler, B., Wenzel, W., Wimmer, B., and Soja, G.: Trace element concentrations in leachates and mustard plant tissue (Sinapis alba L.) after biochar application to temperate soil, Sci. Total Environ. 481, 498–508, 2014.

Kuiters, A. T. and Sarink, H. M.: Leaching of phenolic compounds from leaf and needle litter of several deciduous and coniferous trees, Soil Biol. Biochem., 18, 475–480, 1986.

Kuzyakov, Y., Bogomolova, I., and Glaser, B.: Biochar stability in soil: decomposition during eight years and transformation as as-sessed by compound-specific ^{14}C analysis, Soil Biol. Biochem., 70, 229–236, 2014.

Lehmann, J.: A handful of carbon, Nature, 447, 143–144, 2007.

Lehmann, J. and Sohi, S. P.: Comment on "Fire-derived charcoal causes loss of forest humus", Science, 321, 1295, doi:10.1126/science.1160005, 2008.

Lehmann, J. and Joseph, S. (Eds.): Biochar for Environmental Management: Science and Technology, Earthscan, London, 2009.

Lu, H, Li, Z., Fu, S., Méndez, A., Gascó, G., and Paz-Ferreiro, J.: Can biochar and phytoextractors be jointly used for cadmium remediation? Plos One, 9, e95218, 2014.

Martín-Lara, M. A., Hernáinz, F., Calero, M., Blázquez, G., and Tenorio, G.: Surface chemistry evaluation of some solid wastes from olive-oil industry used for lead removal from aqueous solutions, Biochem. Eng. J., 44, 151–159, 2009.

Méndez, A., Gómez, A., Paz-Ferreiro, J., and Gascó, G.: Effects of biochar from sewage sludge pyrolysis on Mediterranean agricultural soils, Chemosphere, 89, 1354–1359, 2012.

Méndez, A., Tarquis, A., Saa-Requejo, A., Guerrero, F., and Gascó, G.: Influence of pyrolysis temperature on composted sewage sludge biochar priming effect in a loamy soil, Chemosphere, 93, 668–676, 2013.

Nelson, D. and Sommers, L.: Total carbon, organic carbon and organic matter, in: Methods of Soil Analysis, Part 3, Chemical Methods, SSSA, Madison, USA, 1996.

Omuto, C. T., Balint, Z., and Alim, M. S.: A framework for national assessment of land degradation in the drylands: A case study of Somalia, Land Degrad. Dev., 25, 105–119 2014.

Paz-Ferreiro, J., Gascó, G., Gutierrez, B., and Méndez, A.: Soil activities and the geometric mean of enzyme activities after application of sewage sludge and sewage sludge biochar to soil, Biol. Fert. Soils, 48, 512–517, 2012.

Paz-Ferreiro, J., Fu, S., Méndez, A., and Gascó, G.: Interactive effects of biochar and the earthworm Pontoscolex corethrurus on plant productivity and soil enzyme activities, J. Soil. Sediment., 14, 483–494, 2014.

Paz-Ferreiro, J. and Fu, S.: Biological indices for soil quality evaluation: perspectives and limitations, Land Degrad. Dev., doi:10.1002/ldr.2262, 2014.

Plante, A. F., Fernández, J. M., and Leifeld, J.: Application of thermal analysis techniques in soil science, Geoderma, 153, 1–10, 2009.

Rhoades, J. D.: Salinity, Electrical conductivity and total dissolved solids, in: Methods of Soil Analysis, Part 3, Chemical Methods, SSSA, Madison, USA, 1996.

Richards, L. A.: Diagnosis and Improvement of Saline and Alkali Soils, Handbook No. 60, USDA, Washington, USA, 1954.

Schmidt, H. P., Abiven, S., Kammann, C., Glaser, B., Bucheli, T., and Leifeld, J.: Guidelines for biochar production, Delinat Institute und Biochar Science Network, Arbaz, Switzerland, 2012.

Smith, J. L., Collins, H. P., and Bailey, V. L.: The effect of young biochar on soil respiration, Soil Biol. Biochem., 42, 2345–2347, 2010.

Sohi, S. P., Krull, E., Lopez-Capel, E., and Bol, R.: A review of biochar and its use and function in soil, Adv. Agron., 105, 47–82, 2010.

Srinivasarao, C. H., Venkateswarlu, B., Lal, R., Singh, A. K., Kundu, S., Vittal, K. P. R., Patel, J. J., and Patel, M. M.:Long-term manuring and fertilizer effects on depletion of soil organic carbon stocks under pearl millet-cluster bean-castor rotation in Western India, Land Degrad. Dev., 25, 173–183, 2014.

Sumner, M. and Miller, W.: Cation exchange capacity and exchange coefficients, in: Methods of Soil Analysis, Part 3, Chemical Methods, SSSA, Madison, USA, 1996.

Thies, J. E. and Rillig, M. C.: Characteristics of biochar: biological properties, in: Biochar for Environmental Management: Science and Technology, Earthscan, London, 2009.

Thomas, G. W.: Soil pH and soil acidity, in: Methods of Soil Analysis, Part 3, Chemical Methods, SSSA, Madison, USA, 1996.

USDA: Soil QualityTest Kit Guide, USDA, Washington, USA, 1999.

USEPA: Method 3051a: microwave assisted acid dissolution of sediments, sludges, soils, and oils, USDA, Washington, USA, 1997.

Wardle, D. A., Nilsson, M. C., and Zackrisson, O.: Fire-derived charcoal causes loss of forest humus, Science, 320, 5876, doi:10.1126/science.1154960, 2008.

Yan-Gui, S., Xin-Rong, L., Ying-Wu, C., Zhi-Shan, Z., and Yan, L.: Carbon fixaton of cyanobacterial-algal crusts after desert fixation and its implication to soil organic matter accumulation in Desert, Land Degrad. Dev., 24, 342–349, 2013.

Zavalloni, C., Alberti, G., Biasiol, S., Delle Vedove, G., Fornasier, F., Liu, J., and Peressotti, A.: Microbial mineralization of biochar and wheat straw mixture in soil: a short-term study, Appl. Soil Ecol., 50, 45–51, 2011.

Zhao, G., Mu, X., Wen, Z., Wang, F., and Gao, P.: Soil erosion, conservation, and Eco-environment changes in the Loess Plateau of China, Land Degrad. Dev., 24, 499–510, 2013.

Zimmerman, A. R., Gao, B., and Ahn, M.: Positive and negative carbon mineralization priming effects among a variety of biochar amended soils, Soil Biol. Biochem., 43, 1169–1179, 2011.

Permissions

List of Contributors

S. H. R. Sadeghi
Department of Watershed Management Engineering, Faculty of Natural Resources, Tarbiat Modares University, P.O. Box 46417-76489, Noor, Iran

L. Gholami
Department of Watershed Management Engineering, Faculty of Natural Resources, Tarbiat Modares University, P.O. Box 46417-76489, Noor, Iran

E. Sharifi
Department of Watershed Management Engineering, Faculty of Natural Resources, Tarbiat Modares University, P.O. Box 46417-76489, Noor, Iran

Khaledi Darvishan
Department of Watershed Management Engineering, Faculty of Natural Resources, Tarbiat Modares University, P.O. Box 46417-76489, Noor, Iran

M. Homaee
Department of Soil Science, Faculty of Agriculture, Tarbiat Modares University, Tehran, Iran
now at: Department of Rangeland and Watershed Management, Faculty of Natural Resources, Sari Agricultural Sciences and Natural Resources University, Sari, Iran

K. Erkan
Department of Civil Engineering, Marmara University, 34722, Göztepe, Istanbul, Turkey

K. Zhang
State Key Laboratory of Urban and Regional Ecology, Research Center for Eco-Environmental Sciences, Chinese Academy of Sciences, Beijing 100085, China

H. Zheng
State Key Laboratory of Urban and Regional Ecology, Research Center for Eco-Environmental Sciences, Chinese Academy of Sciences, Beijing 100085, China

F. L. Chen
State Key Laboratory of Urban and Regional Ecology, Research Center for Eco-Environmental Sciences, Chinese Academy of Sciences, Beijing 100085, China

Z. Y. Ouyang
State Key Laboratory of Urban and Regional Ecology, Research Center for Eco-Environmental Sciences, Chinese Academy of Sciences, Beijing 100085, China

Y. Wang
State Key Laboratory of Urban and Regional Ecology, Research Center for Eco-Environmental Sciences, Chinese Academy of Sciences, Beijing 100085, China

Y. F. Wu
Guangxi Dongmen Forest Farm, Fusui 532108, Guangxi, China

J. Lan
Guangxi Dongmen Forest Farm, Fusui 532108, Guangxi, China

M. Fu
Guangxi Dongmen Forest Farm, Fusui 532108, Guangxi, China

X. W. Xiang
Guangxi Dongmen Forest Farm, Fusui 532108, Guangxi, China

L. Parras-Alcántara
Department of Agricultural Chemistry and Soil Science, Faculty of Science, Agrifood Campus of International Excellence –ceiA3, University of Córdoba, 14071 Córdoba, Spain

B. Lozano-García
Department of Agricultural Chemistry and Soil Science, Faculty of Science, Agrifood Campus of International Excellence –ceiA3, University of Córdoba, 14071 Córdoba, Spain

A. Galán-Espejo
Department of Agricultural Chemistry and Soil Science, Faculty of Science, Agrifood Campus of International Excellence –ceiA3, University of Córdoba, 14071 Córdoba, Spain

M. Kaleeem Abbasi
Department of Soil and Environmental Sciences, University of Poonch, Rawalakot Azad Jammu and Kashmir, Pakistan

M. Mahmood Tahir
Department of Soil and Environmental Sciences, University of Poonch, Rawalakot Azad Jammu and Kashmir, Pakistan

N. Sabir
Department of Soil and Environmental Sciences, University of Poonch, Rawalakot Azad Jammu and Kashmir, Pakistan

M. Khurshid
Department of Soil and Environmental Sciences, University of Poonch, Rawalakot Azad Jammu and Kashmir, Pakistan

B. K. Rajashekhar Rao
Department of Agriculture, Papua New Guinea University of Technology, Lae 411, Papua New Guinea

J. Hedo
Department of Plant Production and Agricultural Technology, School of Advanced Agricultural Engineering, Castilla La Mancha University, Campus Universitario s/n, CP 02071, Albacete, Spain

M. E. Lucas-Borja
Department of Agroforestry Technology and Science and Genetics, School of Advanced Agricultural Engineering, Castilla La Mancha University, Campus Universitario s/n, CP 02071, Albacete, Spain

C. Wic
Department of Agroforestry Technology and Science and Genetics, School of Advanced Agricultural Engineering, Castilla La Mancha University, Campus Universitario s/n, CP 02071, Albacete, Spain

M. Andrés-Abellán
Department of Agroforestry Technology and Science and Genetics, School of Advanced Agricultural Engineering, Castilla La Mancha University, Campus Universitario s/n, CP 02071, Albacete, Spain

J. de Las Heras
Department of Plant Production and Agricultural Technology, School of Advanced Agricultural Engineering, Castilla La Mancha University, Campus Universitario s/n, CP 02071, Albacete, Spain

F. Peng
Key laboratory of desert and desertification, Cold and Arid Regions Environmental and Engineering Research Institute (CAREERI), Chinese Academy of Sciences (CAS), Lanzhou, China

Y. Quangang
Key laboratory of desert and desertification, Cold and Arid Regions Environmental and Engineering Research Institute (CAREERI), Chinese Academy of Sciences (CAS), Lanzhou, China

X. Xue
Key laboratory of desert and desertification, Cold and Arid Regions Environmental and Engineering Research Institute (CAREERI), Chinese Academy of Sciences (CAS), Lanzhou, China

J. Guo
Key laboratory of desert and desertification, Cold and Arid Regions Environmental and Engineering Research Institute (CAREERI), Chinese Academy of Sciences (CAS), Lanzhou, China

T. Wang
Key laboratory of desert and desertification, Cold and Arid Regions Environmental and Engineering Research Institute (CAREERI), Chinese Academy of Sciences (CAS), Lanzhou, China

S. Arjmand Sajjadi
Department of Soil Science, Agriculture Faculty, Shahid Bahonar University of Kerman, Kerman, Iran

M. Mahmoodabadi
Department of Soil Science, Agriculture Faculty, Shahid Bahonar University of Kerman, Kerman, Iran

M. Lago-Vila
Department of Plant Biology and Soil Science, University of Vigo, 36310 Vigo, Spain

D. Arenas-Lago
Department of Plant Biology and Soil Science, University of Vigo, 36310 Vigo, Spain

A. Rodríguez-Seijo
Department of Plant Biology and Soil Science, University of Vigo, 36310 Vigo, Spain

M. L. Andrade Couce
Department of Plant Biology and Soil Science, University of Vigo, 36310 Vigo, Spain

F. A. Vega
Department of Plant Biology and Soil Science, University of Vigo, 36310 Vigo, Spain

N. Seco-Reigosa
Department of Soil Sciences and Agricultural Chemistry, Higher Polytechnic School, University Santiago de Compostela, 27002 Lugo, Spain

L. Cutillas-Barreiro
Department of Plant Biology and Soil Sciences, Faculty of Sciences, University Vigo, 32004 Ourense, Spain

J. C. Nóvoa-Muñoz
Department of Plant Biology and Soil Sciences, Faculty of Sciences, University Vigo, 32004 Ourense, Spain

M. Arias-Estévez
Department of Plant Biology and Soil Sciences, Faculty of Sciences, University Vigo, 32004 Ourense, Spain

E. Álvarez-Rodríguez
Department of Soil Sciences and Agricultural Chemistry, Higher Polytechnic School, University Santiago de Compostela, 27002 Lugo, Spain

M. J. Fernández-Sanjurjo
Department of Soil Sciences and Agricultural Chemistry, Higher Polytechnic School, University Santiago de Compostela, 27002 Lugo, Spain

A.Núñez-Delgado
Department of Soil Sciences and Agricultural Chemistry, Higher Polytechnic School, University Santiago de Compostela, 27002 Lugo, Spain

R. M. S. P. Vieira
Instituto Nacional de Pesquisas Espaciais, São José dos Campos, Brazil

J. Tomasella
Instituto Nacional de Pesquisas Espaciais, São José dos Campos, Brazil
Centro Nacional de Monitoramento e Alertas de Desastres Naturais, Cachoeira Paulista, Brazil

R. C. S. Alvalá
Centro Nacional de Monitoramento e Alertas de Desastres Naturais, Cachoeira Paulista, Brazil

M. F. Sestini
Instituto Nacional de Pesquisas Espaciais, São José dos Campos, Brazil

A. G. Affonso
Instituto Nacional de Pesquisas Espaciais, São José dos Campos, Brazil

D. A. Rodriguez
Instituto Nacional de Pesquisas Espaciais, São José dos Campos, Brazil

A. A. Barbosa
Instituto Nacional de Pesquisas Espaciais, São José dos Campos, Brazil

A. P. M. A. Cunha
Centro Nacional de Monitoramento e Alertas de Desastres Naturais, Cachoeira Paulista, Brazil

G. F. Valles
Instituto Nacional de Pesquisas Espaciais, São José dos Campos, Brazil

E. Crepani
Instituto Nacional de Pesquisas Espaciais, São José dos Campos, Brazil

S. B. P. de Oliveira
Fundação Cearense de Meteorologia e Recursos Hídricos, Fortaleza, Brazil

M. S. B. de Souza
Fundação Cearense de Meteorologia e Recursos Hídricos, Fortaleza, Brazil

P. M. Calil
Secretaria de Agricultura Agropecuária e Abastecimento de Goiás, Goiânia, Brazil

M. A. de Carvalho
Centro Nacional de Monitoramento e Alertas de Desastres Naturais, Cachoeira Paulista, Brazil

D. M. Valeriano
Instituto Nacional de Pesquisas Espaciais, São José dos Campos, Brazil

F. C. B. Campello
Secretaria de Extrativismo e Desenvolvimento Rural Sustentável, Brasília, Brazil

M. O. Santana
Secretaria de Extrativismo e Desenvolvimento Rural Sustentável, Brasília, Brazil

J. Paz-Ferreiro
Key Laboratory of Vegetation Restoration and Management of Degraded Ecosystems, South China Botanical Garden, Chinese Academy of Sciences, Guangzhou 510650, China

H. Lu
Key Laboratory of Vegetation Restoration and Management of Degraded Ecosystems, South China Botanical Garden, Chinese Academy of Sciences, Guangzhou 510650, China
University of Chinese Academy of Sciences, Beijing 100049, China

S. Fu
Key Laboratory of Vegetation Restoration and Management of Degraded Ecosystems, South China Botanical Garden, Chinese Academy of Sciences, Guangzhou 510650, China

A.Méndez
Departamento de Edafología, ETSI Agrónomos, Universidad Politécnica de Madrid, Avenida Complutense 3, Madrid 28050, Spain

G. Gascó
Departamento de Edafología, ETSI Agrónomos, Universidad Politécnica de Madrid, Avenida Complutense 3, Madrid 28050, Spain

J. León
Dept. of Geography and Land Management, University of Zaragoza, Spain

M. Seeger
Soil Physics and Land Management, Wageningen University, the Netherlands Physical Geography, Trier University, Germany

D. Badía
Dept. of Agricultural Science and Environment, University of Zaragoza, Spain

P. Peters
Soil Physics and Land Management, Wageningen University, the Netherlands

M. T. Echeverría
Dept. of Geography and Land Management, University of Zaragoza, Spain

P. Pereira
Environmental Management Center, Mykolas Romeris University, Ateities g. 20, 08303 Vilnius, Lithuania

X. Úbeda
GRAM (Mediterranean Environmental Research Group), Department of Physical Geography and Regional Geographic Analysis, University of Barcelona, Montalegre, 6, 08001 Barcelona, Spain

J. Mataix-Solera
Environmental Soil Science Group, Department of Agrochemistry and Environment, Miguel Hernández University, Avda. dela Universidad s/n, Elche, Alicante, Spain

M. Oliva
Institute of Geography and Territorial Planning, University of Lisbon Alameda da Universidade, 1600-214, Lisbon, Portugal

A. Novara
Dipartimento di Scienze agrarie e forestali, University of Palermo, 90128 Palermo, Italy

D. Kalderis
Department of Environmental and Natural Resources Engineering, Technological and Educational Institute of Crete, Chania, 73100 Crete, Greece

M. S. Kotti
Department of Environmental and Natural Resources Engineering, Technological and Educational Institute of Crete, Chania, 73100 Crete, Greece

A. Méndez
Departamento de Ingeniería de Materiales, E.T.S.I. Minas, Universidad Politécnica de Madrid, C/Ríos Rosas no. 21, 28003 Madrid, Spain

G. Gascó
Departamento de Edafología, E.T.S.I. Agrónomos, Universidad Politécnica de Madrid, Ciudad Universitaria, 28004 Madrid, Spain

A. Navas
Estación Experimental de Aula Dei (EEAD-CSIC), Department of Soil and Water, Avda. Montañana 1005, 50059 Zaragoza, Spain

K. Laute
Geological Survey of Norway (NGU), Geo-Environment Division, 7491 Trondheim, Norway

A. A. Beylich
Geological Survey of Norway (NGU), Geo-Environment Division, 7491 Trondheim, Norway

L. Gaspar
School of Geography, Earth and Environmental Sciences, Plymouth University, Plymouth, Devon, PL4 8AA, UK

S. L. Weyers
USDA Agricultural Research Service, North Central Soil Conservation Research Lab, Morris, MN, USA

K. A. Spokas
USDA Agricultural Research Service, Soil and Water Management Unit, University of Minnesota, Saint Paul, MN, USA

P. Cely
Departamento de Edafología. E.T.S.I. Agrónomos. Universidad Politécnica de Madrid, Ciudad Universitaria, 28004 Madrid, Spain

A.M. Tarquis
CEIGRAM, Universidad Politécnica de Madrid, Ciudad Universitaria, 28004 Madrid, Spain
Departamento de Matemática aplicada a la Ingeniería Agronómica. Universidad Politécnica de Madrid, Ciudad Universitaria, 28040 Madrid, Spain

J. Paz-Ferreiro
Departamento de Edafología. E.T.S.I. Agrónomos. Universidad Politécnica de Madrid, Ciudad Universitaria, 28004 Madrid, Spain

A.Méndez
Departamento de Ingeniería de Materiales. E.T.S.I. Minas. Universidad Politécnica de Madrid, C/Ríos Rosas no. 21,28003 Madrid, Spain

G. Gascó
Departamento de Edafología. E.T.S.I. Agrónomos. Universidad Politécnica de Madrid, Ciudad Universitaria, 28004 Madrid, Spain